THE CHEMICAL ANALYSIS OF FOODS

THE CHEMICAL ANALYSIS OF FOODS

by

H. E. COX, Ph.D., D.Sc.

and

DAVID PEARSON, M.Sc.

1962

CHEMICAL PUBLISHING CO., INC.

212 Fifth Avenue New York, N. Y.

The Chemical Analysis of Foods

ISBN: 978-0-8206-0141-0

Chemical Publishing Company:
www.chemical-publishing.com
www.chemicalpublishing.net

First American Edition: Chemical Publishing New York 1962

Printed in the United States of America

PREFACE TO THE FIRST AMERICAN EDITION

The sudden death of the eminent Public Analyst, Dr. H. E. Cox, in 1951 came as a shock to all who were connected with food analysis. Long before this, "Cox" meant to members of the profession a standard work to which to refer for advice on analytical methods and interpretation. Also his tables of analytical data based on long experience have been widely used and quoted throughout the English speaking countries ever since the first edition was published. Although it was considered advisable to make various changes to bring this new edition up-to-date, an attempt has been made to retain Dr. Cox's style, which has been described as "readable and to the point."

The first American edition corresponds to the fifth British edition with the modification that instead of a summary of food legislation in England and Wales, the Requirements for food of the United States Food, Drug, and Cosmetic Act are given in Chapter 1. As the body of the text was retained in its original form the reference to British food legislation could not be eliminated from special sections. The users of this book should refer to Chapter 1 for American requirements whenever necessary. Three new basic chapters have been inserted at the beginning of this edition.

Many of the general methods such as those which are available for the determination of moisture, protein, fat, etc., have been brought together in Chapter 2. Similarly, the methods which may be employed for the detection and determination of trace elements, preservatives, colouring matters, etc., have been incorporated in Chapter 3, instead of being included in various parts of the book with specific foods. The titles of Chapters 4–14 are very similar to Chapters 1–11 in the previous edition, but much of the material has been re-written and many foods have been added to those previously covered, *e.g.*, mincemeat, numerous herbs and spices, ice-cream, table jellies. Also entirely new sections on butter confectionery, iodised salt, saccharin tablets and soups have been included in the "Miscellaneous" Chapter 15.

With the tremendous advances being made in the subject, it is now more difficult than ever to decide what is not to be included in a fairly small volume of this type. Also, it is well known that food analysts differ considerably in their choice of methods and any book on the subject therefore is liable to reflect to a greater or lesser extent the personal choice of the author. In this volume the aim has been to bring the material more in line with current practice in the average food laboratory, which is equipped for dealing with routine work. At the same time, a large number of references are quoted, which should enable the reader to seek out further information as to alternative methods which are available. In this connection, it may be noted that most of the recently introduced methods are included in the reviews which appear periodically in the U.S. journal "Analytical Chemistry."

As previously, the matter contained in most sections is largely based on the work carried out by Public Analysts. With many materials, a suggestion is made as to the type of tests which are called for in routine work, and an indication is also given of other methods which may be performed when a more searching examination is necessary. One difference from the previous edition, however, is that in many instances mention is made of other methods which are likely to be used in the food industry for control purposes. Similarly, although the standardised methods of the Analytical Methods Committee of the Society for Analytical Chemistry (formerly the Society of Public Analysts) are given major prominence, those published by the British Standards Institution (which appear to be increasingly used in industry) are also included where appropriate. Such additions should, it is hoped, make the book more useful to a wider section of readers.

It is obviously impossible to thank all of one's past and present colleagues, and others one has met, who have helped in different ways with the preparation of a book of this type. The author feels, however, that this opportunity should not be allowed to pass without according special thanks to Mr. R. G. Minor, F.R.I.C., F.P.S. (Deputy Public Analyst and Lecturer-in-Charge of the Food and Drugs Section at Chelsea College of Science and Technology), whose experience, encouragement and good humour at all times have been invaluable. Special thanks are also due to another " Chelsea " colleague, Mr. E. C. Apling, B.Sc., A.R.I.C. (Chief Chemist, D. W. Kent-Jones and A. J. Amos), for first-rate assistance, particularly during proof reading, and to Mr. E. C. Tubb, B.A., B.Sc. (Chief Chemist, Tate and Lyle Ltd.), for supplying much useful information relating to the composition and properties of cane and beet sugar. Also the author would like to thank the numerous official Associations, Societies and other organisations who readily gave permission to use published information, among whom are the following:

American Association of Cereal Chemists, Minnesota, U.S.A.
Association of Official Agricultural Chemists, Washington, U.S.A.
Association of Public Analysts, London.
British Pharmacopoeia Commission (General Medical Council), London.
Canada Department of Agriculture, Summerland, B.C.
C.S.I.R.O. (Division of Food Preservation), Homebush, N.S.W.
Her Majesty's Stationery Office, London.
National Bureau of Standards, U.S.A.
Society for Analytical Chemistry, London.
Society of Chemical Industry, London.

Finally, the new author would like to repeat a statement made by Dr. Cox in the preface to the first edition, that he " will be grateful to anyone directing his attention to errors, which are sure to have crept in, so that they may be eliminated."

D. PEARSON.

National College of Food Technology,
St. George's Avenue,
Weybridge, Surrey.

CONTENTS

LIST OF ABBREVIATIONS

A.M.C.	—	Analytical Methods Committee (of the Society of Public Analysts, or the Society for Analytical Chemistry).
A.O.A.C.	—	Official Methods of the Association of Official Agricultural Chemists, U.S.A.
J.A.O.A.C.	—	Journal of the A.O.A.C. (see above).
A.P.A.	—	Association of Public Analysts, London.
B.P.	—	British Pharmacopoeia.
B.P.C.	—	British Pharmaceutical Codex.
B.S.	—	British Standard (of the British Standards Institution).
B.S.I.	—	British Standards Institution, London.
F.C.	—	Full Cream.
f.f.	—	Fat-free material.
F.M.F.	—	Food Manufacturers' Federation, London
F.S.C.	—	Food Standards Committee.
G.M.S.	—	Glyceryl monostearate.
max.	—	maximum.
min.	—	minimum.
M.O.F.	—	Ministry of Food.
M.O.H.	—	Ministry of Health.
M.R.C.	—	Medical Research Council.
M.S.N.F.	—	Milk-solids-not-fat.
N.F.M.S.	—	Non-fatty milk solids.
N.F.S.	—	Non-fatty solids.
p.p.m.	—	parts per million.
P.S.	—	Proof Spirit.
R.I.	—	Refractive index.
S.A.C.	—	Society for Analytical Chemistry, London.
S.I.	—	Statutory Instrument.
S.N.F.	—	Solids-not-fat.
S.P.A.	—	Society of Public Analysts, London.
S.R. & O.	—	Statutory Rule and Order.
T.M.S.	—	Total milk solids.
U.S.F.D.A.	—	Food and Drug Administration of the United States.

Chapter 1

REQUIREMENTS FOR FOOD OF THE UNITED STATES FOOD, DRUG, AND COSMETIC ACT

Section 201 (f) of the Federal Food, Drug, and Cosmetic Act defines food as follows:

The term "food" means (1) articles used for food or drink for man or other animals, (2) chewing gum, and (3) articles used for components of any such article.

PRINCIPAL REQUIREMENTS OF FOOD LAW

Below is a synopsis of the principal requirements of the act relating to foods, in nonlegal language. The symbols in parentheses are the pertinent sections of the statute itself.

Health Safeguards

A food is illegal if it contains a natural or added deleterious substance which may render it injurious to health or unsafe (402 (a) (1) and (2)).

The act authorizes placing a limit on the amount in a food of an added deleterious substance that cannot be avoided in the manufacture of the food (406; 402 (a) (2)).

Raw agricultural products containing residues of pesticides not authorized by, or in excess of, a tolerance established by regulation are illegal (408).

A food is illegal if it is prepared, packed, or held under insanitary conditions whereby it may be rendered injurious to health (402 (a) (4)).

Food containers must be free from any poisonous or deleterious substance which may cause the contents to be injurious to health (402 (a) (6)).

Coal-tar colors contained in food must come from batches certified by the Food and Drug Administration as being harmless (402 (c); 406).

Confectionery, including candy, must not contain any alcohol except alcohol not in excess of 0.5 percent derived solely from the use of flavoring extracts, and must not contain a nonnutritive substance (talc, for example) except harmless color, flavor, or not over 0.4 percent resinous glaze (402 (d)).

Sanitary Safeguards

A food is illegal if it is filthy, putrid, or decomposed (402 (a) (3)).

A food is illegal if it is prepared, packed, or held under insanitary conditions whereby it may have become contaminated with filth (402 (a) (4)).

A food is illegal if it is the product of a diseased animal or one that has died otherwise than by slaughter (402 (a) (5)).

Prohibited Deceptions

Food labels or labeling (circulars, etc.) must not be false or misleading in any particular (403 (a)). In determining whether labeling is misleading, consideration is given not only to the representations made or suggested, but also to the extent to which there is a failure of the labeling to reveal material facts. Half truths are not enough (201 (n)).

Damage or inferiority in a food must not be concealed in any manner (402 (b) (3)). Example: A diluted fruit juice artificially colored to resemble the undiluted juice.

No substance may be added to a food to increase its bulk or weight or make it appear of greater value than it is (402 (b) (4)). Example: In the artificially colored diluted fruit juice, the added water increases its bulk or weight and the added color makes it appear to be fruit juice which is better and of greater value.

A food must not be sold under the name of another food (403 (b)). Example: Canned bonito fish represented by label or otherwise as tuna fish.

A substance which is recognized as being a valuable constituent of a food must not be omitted or abstracted in whole or in part, nor may any substance be substituted for the food in whole or in part (402 (b) (1) and (2)). Example: An article labeled or represented as "milk" or "whole milk" from which part of the valuable constituent butterfat has been skimmed, since skimmed milk is substituted for milk or whole milk.

Food containers must not be so made, formed, or filled as to be misleading (403 (d)). Example: A closed package filled to less than its capacity.

Required Label Statements

Required label information must not only be conspicuously displayed, but it must be in terms that the ordinary consumer is likely to read and understand under ordinary conditions of purchase and use (403 (f)). To insure this, the required statements should appear in prominent positions on the principal display panels of labels.

If the label of a food bears representations in a foreign language, the label must bear all of the required statements in the foreign language, as well as in English. (Note—The Tariff Act of 1930 requires all imported articles to be marked with the English name of the country of origin.)

The following statements must appear on the label in the English language:

1. If the food is packaged, the name and address of either the manufacturer, or the packer, or the distributor, and an accurate statement of the amount of food in the package (403 (e)).

   a. If the food is not manufactured by the person or company whose name appears on the label, the name must be qualified by "Manufactured for," "Distributed by," or similar expression.

   b. The statement of quantity of the contents must express the quantity of food in the package, exclusive of wrappers. The statement shall be expressed in the terms of weight, measure, numerical count, or a combination of numerical count and weight or measure, which are generally used by consumers to express quantity of such food and which give accurate information as to the quantity thereof. But if no such general consumer usage exists, the statement shall be in terms of liquid measure if the food is liquid; or in terms of weight if the food is solid, semisolid, viscous, or a mixture of solid and liquid; or it may be in terms of dry measure if the food is a fresh fruit, fresh vegetable, or other dry commodity. A statement of weight must be in terms of the avoirdupois pound and ounce. A statement of liquid measure must be in terms of the United States gallon of 231 cubic inches and subdivisions thereof (quarts, pints, and fluid ounces). The statement must be in terms of the largest unit of measure, such as "1 pound" not "16 ounces," and "1 pint" not "16 fluid ounces."

2. The common or usual name of the food and the common or usual name of each ingredient, except foods for which identity standards have been promulgated, which must make label declarations of ingredients in accordance with the prescriptions of the standards (403 (h) and (i); 401). The word "ingredients" does not refer to the chemical composition, but means the individual food components of a mixed food. The ingredients should be listed in the order of their predominance in the food.

3. The labeling of foods intended for special dietary uses must bear certain prescribed additional information concerning their vitamin, mineral, and other dietary properties which is necessary to inform purchasers fully as to their value for such uses. Regulations that have been issued under this section prescribe the specific additional mandatory label information (403(j)).

4. Foods must bear labeling stating the presence of any artificial flavoring, artificial coloring (except butter, cheese, or ice cream), or chemical preservative (403 (k)).

5. Imitations must be labeled as such (403 (c)).

Standards for Foods

The act authorizes promulgation of definitions and standards of identity, standards of quality, and standards of fill of container for foods (401). A food which is represented as or purports to be a food for which a standard of identity has been promulgated must comply with the specifications of the standard in every respect (403 (g)). If a food for which a standard of quality or fill of container has been promulgated falls below such standard, it must bear label statements specified in the standard showing it to be substandard in quality or in fill of container (403 (h)). Standards for foods are discussed more fully elsewhere in this publication.

SANITATION AND FILTH

The Food, Drug, and Cosmetic Act is often referred to by the public as the "Pure Food and Drug Act," thus emphasizing one of its basic purposes—the protection of the consuming public from articles that may be deleterious, that are unclean or decomposed, or have been exposed to insanitary conditions that may contaminate the article with filth or may render it injurious to health.

In determining whether or not conditions surrounding preparation, packing, and storage of foods are insanitary, it is necessary to evaluate the possibilities of contamination both with foreign matter that may be harmless but disgusting and with elements which involve danger to health. In its sanitary provisions, the Food, Drug, and Cosmetic Act goes further than to prohibit commerce in products that are carriers of causative agents of disease. It prohibits the distribution and sale of foods which may contain repulsive or offensive matter classed as filth regardless of whether such objectionable substances can be detected by laboratory procedures or are very likely to be present because of the conditions under which the goods were prepared and handled.

However, there is nothing highly technical in the concept of sanitation applied in the enforcement of the act. Ability to conform to standards of production that will result in a clean product does require a careful study of sources and routes of contamination. It is expected, of course, that establishments in which foods are prepared will be properly constructed, lighted, and ventilated, and that employees will be properly attired. Conditions which give rise to denial of entry to importations into this country because a product is filthy or has been produced under insanitary conditions go far deeper than that. Such contaminating elements as rat and mouse hairs and excreta, whole insects, insect parts and excreta, maggots, larvae, and parasitic worms, pollution from the excrement of man and animals, as well as other extraneous material which, because of their repulsiveness would not knowingly be eaten or used, constitute filth. The presence of such filth renders foods adulterated, whether or not harm to health can be shown. Insanitary conditions which permit the contamination by rodents, such as rats and mice, infestation by insects, and pollution from human and animal wastes, are those with which the act is primarily concerned.

The maintenance of sanitary conditions requires extermination and exclusion of rodents, inspection and sorting of raw materials to eliminate the insect-infested and decomposed portions, fumigation, quick handling and proper storage to prevent insect development or contamination, the use of clean equipment, control of possible sources of sewage pollution, and supervision of the conduct of those who prepare foods so that acts of misconduct may not defile the products they handle. Fumigation of commodities already infested with insects will not result in a legal product since dead insects or evidence of past insect activity are objectionable. Fumigation should be employed where necessary, to *prevent* infestation.

Many food materials imported into this country are intended for further manufacture into finished foods. This fact in no way relieves such imported raw materials from these requirements of cleanliness and freedom from deleterious impurities. It must be remembered also that articles of foods that are free from contamination at the time they leave foreign shores sometimes become contaminated en route and must be detained. This emphasizes the importance of insisting upon proper storage conditions in vessels, railroad cars, or other conveyances. Many detentions have been made of shipments of foods that may have become contaminated, insect-infested, moldy, or otherwise illegal during the journey from the country of origin. While the foreign exporter may be wholly blameless, the law requires action against illegal merchandise no matter where it may have become illegal. Foreign exporters should pack their products for shipment so as to protect them against contamination en route, and should urge transportation agencies to protect the merchandise by maintaining sanitary conditions and segregating food from other cargo which might contaminate it. For example, vessels transporting green coffee in burlap bags have also carried ore concentrates and poisonous insecticides. Improper cargo handling or disasters at sea have resulted in large lots of coffee becoming seriously contaminated, with detentions required.

TOLERANCES FOR FILTH

Innumerable inquiries are received by the Food and Drug Administration as to permitted variations from complete cleanliness or soundness in foods. The act does not authorize "tolerances" for filth or decomposition in foods. It states that a food is adulterated if it consists *in whole or in part* of a filthy, putrid, or decomposed substance. It is obvious that tolerances cannot be recognized, from the additional fact that the mere production of a food under insanitary conditions which may contaminate it with unclean foreign matter renders such food adulterated under the law.

This does not mean that a food is necessarily condemned because of the presence of foreign matter in amounts below the irreducible minimum after all precautions humanly possible have been taken to prevent contamination. In some instances the Food and Drug Administration has informally advised importers of the basis upon which actions are taken against foods which may have been subjected to attack by insect pests or subjected to deterioration due to climatic conditions. As commercial practices improve or an insect infestation is brought under control, the basis of action may be lowered. This should not be a hardship on those who prepare the foods in foreign countries, because whatever may be the basis upon which actions are taken at the United States ports on filth and decomposition, it represents only what the careful producer or manufacturer can meet.

FOOD STANDARDS

Section 401 of the Food, Drug, and Cosmetic Act provides:

> Whenever in the judgment of the Secretary such action will promote honesty and fair dealing in the interest of consumers, he shall promulgate regulations fixing and establishing for any food, under its common or usual name so far as practicable, a reasonable definition and standard of identity, a reasonable standard of quality, and/or reasonable standards of fill of container: *Provided,* That no definition and standard of identity and no standard of quality shall be established for fresh or dried fruits, fresh or dried vegetables, or butter, except that definitions and standards of identity may be established for avocadoes, cantaloupes, citrus fruits, and melons.

It should be borne in mind that these are standards of identity (what the particular food is), of quality (whether above or below standard), and fill of container (how full the package must be). These standards contemplate in every case that the food is properly prepared from clean, sound materials. They do not relate to factors such as deleterious impurities, filth, and decomposition.

The standards of quality established under the Food, Drug, and Cosmetic Act must not be confused with "standards for grades" that are established from time to time by the United States Department of Agriculture. Under the Food, Drug, and Cosmetic Act one minimum standard of quality is established for each food standardized. A food that falls below the minimum standard established must bear a special substandard label reading, for example, "Below Standard in Quality. Good Food—Not High Grade" in a prescribed size and style of type.

The United States Department of Agriculture grades which have been established are usually designated "Grade A" or "Fancy," "Grade B" or "Choice," or "Grade C" or "Standard." These grade designations are not required by the Food, Drug, and Cosmetic Act to be stated on the labels, but if they are stated, the product must comply with the specifications for the declared grade.

The various foods for which standards of identity, quality, or fill of container have been promulgated under the Food, Drug, and Cosmetic Act are listed in the Appendix.

The standards govern labeling and composition and should be consulted for detailed specifications. All standards are published in the Federal Register of the United States. The promulgation of food standards is a continuous task and standards for additional foods will be prescribed in the future. Except in emergencies, 90 days must elapse between the date of promulgation and the date on which standards go into effect.

In addition to the standards authorized to be promulgated under the act, the Act of Congress of March 4, 1923, provides a definition for butter which is quoted under Dairy Products.

SPICES, SPICE SEEDS, AND HERBS

This group represents food materials that are subject to the depredations of various animal and insect pests from which they must be protected. They may also become moldy or otherwise decomposed unless properly prepared and stored. In any discussion of methods of preparing and storing foods intended for human consumption, emphasis must be placed on the principle of "clean" food, not "cleaned" food. This is particularly true with spices. One of the most serious consequences of failure to protect spices is contamination with excreta from rats, mice, birds, chickens, or other animals. Imported spices have been found to contain excreta not only of these animals, but of bats, goats, and camels. Insects, larvae, weevils, moths, mites, beetles, flies, etc., and their excreta constitute another problem.

Emphasis should be placed on harvesting, storing, handling, packing, and shipping under conditions which will *prevent* contamination; that means keeping rats, mice, goats, and other animals away from the food product. Besides shortening the time of exposure in the fields, or in drying or curing spaces which are often in the open, all the necessary preventive measures should be taken to keep animals from walking over or having access to the food material. Separate enclosures may be necessary; places under trees or other perches where birds might roost should be avoided. Barns, sheds, cribs, warehouses, and other storage spaces should be made rodentproof. Experience in the United States has indicated that only by such measures as metal flashing around baseboards, use of concrete floors and foundations, screening of openings and windows, elimination of false ceilings or spaces between inner and outer walls, and of rat and mice harborages of all sorts inside and outside of buildings, can rats and mice be effectively kept away from these products. Where it is not feasible to rodentproof the entire building, for example, a barn on the farm, it might be possible to rodentproof a section or room where the foodstuff is being held.

The same basic principle of prevention of contamination applies in the case of insects. Gauze netting spread over foods drying in the open may be necessary to keep insects away. Screening of windows and doors will help to keep insects out of storage or packing places. Careful cleaning and fumigation of premises and equipment before a new crop is put into a storage space may save it from contamination by insects that are left over and have been multiplying unmolested since the old crop was removed. The use of infested secondhand bags is another common source of trouble.

While insecticides and fumigants have their function (for example, in preparing a storage space for the reception of spices or other foods, and in preventing the development of infestation in a lot), a product which is already infested is not made acceptable for food by fumigation because the insects are no less objectionable just because they may have been killed by the fumigation.

Many insecticides and fumigants are poisonous or deleterious substances, and if they contaminate food, the food becomes adulterated and subject to action under the requirements of the act. While most of the fumigants are volatile, they may, nevertheless, result in contamination of the food with deleterious residues.

In some cases spices may be used for drug purposes and they then become subject to the drug provisions of the act discussed in Part II. Those spices or spice oils which are listed in the United States Pharmacopeia or the National Formulary are subject to the standards set forth in these compendia when used for drug purposes. Spice products should be sold under their correct names; for example, in the United States the spice known to consumers as sage is *Salvia officinalis* L. Other related sages should be labeled to show they are not this variety. Spice products, when offered for importation, should be reasonably free from foreign matter, such as pebbles, dirt, stems, wood chaff, or other extraneous substance peculiar to each type of product.

No standards of identity for food spice products have been promulgated under the Food, Drug, and Cosmetic Act. Solely as an advisory guide as to IDENTITY of food spice products, the Food and Drug Administration uses the following definitions:

SPICES. Aromatic vegetable substances used for the seasoning of food. They are true to name, and from them no portion of any volatile oil or other flavoring principle has been removed.

ALLSPICE, PIMENTO. The dried, nearly ripe fruit of *Pimenta officinalis* Lindl.

ANISE, ANISEED. The dried fruit of *Pimpinella anisum* L.

BAY LEAVES. The dried leaves of *Laurus nobilis* L.

CAPERS. The flower buds of *Capparis spinosa* L.

CARAWAY, CARAWAY SEED. The dried fruit of *Carum carvi* L.

CARDAMON. The dried, nearly ripe fruit of *Elettaria cardamomum* Maton.

CARDAMON SEED. The dried seed of cardamom.

CINNAMON. The dried bark of cultivated varieties of *Cinnamomum zeylanicum* Nees or of *C. cassia* (L.) Blume, from which the outer layers may or may not have been removed.

CEYLON CINNAMON. The dried inner bark of cultivated varieties of *Cinnamomum zeylanicum* Nees.

SAIGON CINNAMON, CASSIA. The dried bark of cultivated varieties of *Cinnamomum cassia* (L.) Blume.

CLOVES. The dried flower buds of *Caryophyllus aromaticus* L.

CORIANDER SEED. The dried fruit of *Coriandrum sativum* L.

CUMIN SEED. The dried fruit of *Cuminum cyminum* L.

GINGER. The washed and dried, or decoriated and dried, rhizome of *Zingiber officinale* Roscoe.

MACE. The dried arillus of *Myristica fragrans* Houtt.

MACASSAR MACE, PAPUA MACE. The dried arillus of *Myristica argentea* Warb.

MARJORAM, LEAF MARJORAM. The dried leaves, with or without a small proportion of the flowering tops, of *Marjorana hortensis* Moench.

NUTMEG. The dried seed of *Myristica fragrans* Houtt, deprived of its testa, with or without a thin coating of lime (CaO).

MACASSAR NUTMEG, PAPUA NUTMEG, MALE NUTMEG, LONG NUTMEG. The dried seed of *Myristica argentea* Warb, deprived of its testa.

PAPRIKA. The dried, ripe fruit of *Capsicum annuum* L.

BLACK PEPPER. The dried, immature berry of *Piper nigrum* L.

WHITE PEPPER. The dried mature berry of *Piper nigrum* L. from which the outer coating or the outer and inner coatings have been removed.

SAFFRON. The dried stigma of *Crocus sativus* L.

SAGE. The dried leaf of *Salvia officinalis* L.

TARRAGON. The dried leaves and flowering tops of *Artemisia dracunculus* L.

THYME. The dried leaves and flowering tops of *Thymus vulgaris* L.

CANNED FRUITS AND VEGETABLES

Canned Fruits

Standards of identity, quality, and fill of container have been promulgated under section 401 of the act for a number of canned fruits. The specific standards should be consulted by anyone intending to ship canned fruits to the United States. (Definitions and Standards for Foods, part 27).

The standards of quality are minimum standards. If a canned fruit falls below the minimum standard of quality prescribed for it, the label must bear a substandard statement in the manner and form prescribed in the standard. Labels on canned fruits which meet the minimum quality standards, and on canned fruits for which no standards of quality have been promulgated, need not make reference to quality, but if they do, the product must correspond to the usual understanding of the labeled grade. Particular care must be taken not to use the terms "Fancy" or "Grade A" on products that do not meet the established understanding of these terms in the United States.

Fill-of-container standards have been promulgated for only a few specified canned fruits, but in packing any other canned fruit the container must be well filled with fruit with only enough packing medium added to fill the interstices, otherwise the container may be deceptive and the product prohibited by the act. In judging the finished product, due allowance is made for natural shrinkage in processing.

Canned fruits for which identity and quality standards have not yet been promulgated are sometimes imported. Fruit used for canning should be mature and should be sound, that is, free from insect infestation, moldiness, or other form of decomposition. In some growing areas blueberries are infested with larvae; such fruit must be discarded when selecting fruit for canning or other food use.

Pitted and stuffed olives found to contain pits and parts of pits were the basis of a notice issued to dealers, importers, and shippers of pitted and stuffed olives. The notice read in part:

> While a complete elimination of pit fragments may not be possible in all cases, certainly consumers do not expect to encounter them in so-called pitted olives, to the extent to which they now occur. Shipments of stuffed or so-called pitted olives showing an appreciable percentage of olives containing pits or fragments of pits must be regarded as in violation of the Food and Drugs Act.
>
> Packers and dealers in these commodities are hereby requested to give immediate attention to the correction of this condition in order to obviate that action which otherwise must be taken by the Administration under both the import and domestic provisions of the Act. Opinions have been expressed by dealers to the effect that pitted olives should not contain at the most in excess of 1 percent of olives showing pits or fragments of pits. While this is not announced as a tentative tolerance, immediate attention should be given to eliminating pits and fragments of pits to the extent that it is reasonably feasible. Importers should take up this matter promptly with their contacts abroad, with a view to ensuring that shipments actually are "pitted" rather than "partially pitted." This condition has been found to be more prevalent in the so-called Queen olives than in the smaller, or Manzanilla olives, possibly due to the fact that the pit of the larger olive appears to adhere more closely to the flesh of the olive. In not a few of the lots of the larger olives examined, the percentage of olives containing fragments has been found to be surprisingly high.

Canned Vegetables

Canned vegetable products must be prepared from sound, wholesome raw materials free from decomposition. It has been necessary on occasions to deny entry of canned pimientos and other vegetable products because they had been prepared from sour, fermented raw materials. The definitions and standards of identity which have been promulgated under the act for a wide variety of canned vegetables provide that "the food is sealed in a container and so processed by heat as to prevent spoilage." The importance of adequate heat processing of canned vegetables, particularly the nonacid types, is emphasized by the danger that spoilage may be caused by *Clostridium botulinum*, the organism responsible for a highly fatal type of food poisoning. It has sometimes been necessary to detain or deny entry to shipments of canned vegetables, canned mushrooms, canned pimientos, and other canned food products because of spoilage resulting from underprocessing or occasionally from the use of defective containers.

Cans of food which have become swells or otherwise abnormal should be destroyed.

Standards of quality have been promulgated for several vegetables. The detailed requirements of the standards, should be consulted before making import shipments. For each canned vegetable there is one minimum standard of quality established. If the product falls below this minimum quality standard when judged by the criteria laid down in the standard, its label must show the required substandard statement in the manner and form prescribed.

If a fill-of-container standard has not been promulgated for a canned vegetable, the container must nevertheless be well filled with the vegetable, with only enough packing medium to fill the interstices. In the case of canned tomatoes, no excess water is needed or permitted by the fill-of-container standard.

Canned Tomato Paste, Puree, Juice, and Catsup

Shippers of tomato products (Tomato juice, catsup, puree, and paste) should note the standards of identity for these items in Definitions and Standards for Foods, part 53. Particular attention is called to the salt-free tomato solids requirements for puree and paste, and to the fact that neither artificial color nor preservatives are permitted in any of these products. Tomato juice is unconcentrated; tomato puree must contain not less than 8.37 and tomato paste not less than 25 percent salt-free tomato solids.

These tomato products are occasionally contaminated with rot because of failure to remove decayed tomatoes from the raw material entering the cannery. Flies and worms are also filth contaminants of tomato products. The preparation of a clean tomato product requires proper washing, sorting, and trimming of the tomatoes and frequent cleaning of the cannery equipment, such as tables, utensils, vats, and pipelines.

In judging whether tomato products have been properly prepared to eliminate rot and decay, the Food and Drug Administration uses the Howard mold-count test, and refuses admission to import shipments and takes action against domestic shipments if mold filaments are present in more than 40 percent of the microscopic fields in the case of puree, paste, catsup, or sauce or more than 20 percent in the case of tomato juice. Methods of testing tomato products are given in the Methods of Analysis of the Association of Official Agricultural Chemists.

DRIED FRUITS AND VEGETABLES

Much, if not all, that has been said about prevention of contamination of spices is applicable to other perishable foods such as dried fruits and vegetables which are subject to attack by insects, other animals, or to deterioration resulting in moldiness or other forms of decomposition.

Dried Figs

Dried figs, both domestic and foreign, are subject to insect infestation during their growth and during storage where improper storage conditions exist. They are also at times subject to conditions which may result in development of mold which must be guarded against. The domestic industry and to a large extent foreign producers have made great progress in recent years in elimination of conditions responsible for contamination or deterioration of figs. Some importations, however, are encountered which bear evidence of failure of producers to eliminate objectionable conditions and these must necessarily be refused admission. Figs are classed as objectionable if upon examination they show internal or external insect excreta or internal or external evidence of larvae, dead or alive. Figs are classed as objectionable also if they show internal or external evidence of mold or of a webby condition or of sourness or fermentation.

In order to insure a suitable product for importation, the attacks of the fig moth or similar pests should be guarded against as much as possible, as well as those conditions tending to produce mold growth or fermentation. If the figs are found to be so affected to some extent upon harvesting, they should be thoroughly sorted to remove objectionable figs before packing or shipping is undertaken. To insure against infestation or deterioration between the time of harvest and exportation to this country, the movement from the farms or orchards to the packing houses and thence to vessels should be prompt. Storage of figs before shipment should be for as short a time as possible and should be under sanitary conditions so as to prevent infestation.

Dried Dates

Dried dates constitute another important article of import commerce as well as domestic commerce, and again both domestic and foreign producers are confronted with problems involving infestation and deterioration. As in the case of figs, importations of dates are examined for evidence of insect infestation or moldiness and decomposition, or other objectionable conditions.

Dried Orange Peel

Importations of dried orange peel are occasionally encountered which must be denied entry because of extensive insect infestation and moldiness. Any orange peels which are to be dried or otherwise prepared for subsequent use in any food product must be scrupulously guarded against contamination.

Dried Mushrooms

Inclusion in wild mushrooms of poisonous species is a matter of concern and steps should be taken to insure that only edible species are offered for entry into the United States. The most common bar to entry of mushrooms, especially the wild dried product, is infestation with flies and maggots, since mushrooms sometimes are attacked by maggots during the course of their growth. If infestation is heavy in certain areas and removal of the insects is difficult, the elimination of mushrooms from these areas for importation into this country may be necessary. Infestation may possibly occur in some seasons and not in others. Experience in the examination of importations of dried mushrooms has shown a wide divergence in the condition of different importations from the standpoint of infestation, and it appears that infestation either may not be universal throughout mushroom-growing areas, or it may be sporadic.

It has been necessary also to deny entry to mushrooms for such other causes as evidence of storage under insanitary conditions resulting in contamination by rats or mice or storage insects, and even in some cases by mites in large numbers. After having satisfied himself that he has eliminated field-infested mushrooms from his stock ready for drying, the mushroom producer should then concern himself with the problem of carrying out his drying and storage under conditions which will keep his product clean and free from infestation or from contamination by mice, rats, or other animals.

FRESH FRUITS

Apples and other fruits bearing excessive residues from insecticide sprays or dusts are classed as adulterated under the Federal law. See paragraph entitled "Pesticidal Residues on Raw Agricultural Products."

Pineapples showing or likely to show the internal condition known as "brown heart" or "black heart," should not be offered for entry into the United States. All of the factors causing this condition have not been established but it can be stated that time, and storage and shipping temperatures are important factors.

Blueberries and huckleberries sometimes contain small larvae which render them unfit for food. Fruit from infested areas should be avoided. Fresh blueberries should be held and transported under conditions which will prevent deterioration and decomposition evidenced by mold or other indications of unfitness.

PESTICIDAL RESIDUES ON RAW AGRICULTURAL PRODUCTS

Raw agricultural products include unprocessed fruits and vegetables, grains, meat, and poultry. Products of this kind containing residues of poisonous or deleterious pesticides are in violation of the Federal Food, Drug, and Cosmetic Act unless a tolerance has been established for the particular pesticide on the specific food, and the residue does not exceed the established tolerance. In some instances, a zero tolerance may be established for a pesticide, which means that no residue is permitted on the food as it is marketed.

Tolerances for pesticidal residues on many raw agricultural commodities have been established under the law. Additional tolerances will be established as the facts warrant such action. Firms considering offering for entry into the United States raw agricultural commodities which may contain pesticidal residues should write to the Food and Drug Administration, U. S. Department of Health, Education, and Welfare, Washington 25, D. C., for current information concerning the status of tolerances for residues on raw agricultural products.

FRUIT JAMS (PRESERVES), JELLIES, FRUIT BUTTERS, MARMALADES

Standards of identity have been promulgated for jams, jellies, and fruit butters, copies of which will be sent by the Food and Drug Administration upon request. These standards should be consulted before manufacturing these foods for importation. The standards for jams and jellies require that there be in the batch not less than 45 parts by weight of fruit (or fruit juice in the case of jelly) to each 55 parts by weight of sugar or other optional sweetening ingredient. Only sufficient pectin may be added to jams and jellies to compensate for deficiency, if any, of the natural pectin content of the particular fruit. The standards also require a certain degree of concentration. In the case of jellies the finished product must be concentrated to not less than 65 percent soluble solids. In the case of jams (preserves) the finished jam should have not less than 65 percent soluble solids if made from certain specified fruits. With other designated fruits the batch must be concentrated to 68 percent soluble solids. Importations of a product complying with the standard for pineapple preserve except that it was somewhat less concentrated have been refused admission on the ground they purported by their appearance to be the standardized product and failed to comply with the standard.

Fruit butters have been defined in the standards of identity established for them as the smooth, semi-solid foods, each of which is made from a mixture composed of not less than five parts by weight of fruit ingredient to each two parts of sweetening ingredient.

Foreign manufacturers should consult the standards themselves before preparing these products for shipment to the United States.

Jams, jellies, and similar fruit products should, of course, be prepared only from sound fruit. Decayed or otherwise decomposed fruits such as berries and objectionable portions of larger fruits should be sorted out and discarded as should insect-contaminated fruit.

CONFECTIONERY, HONEY, SUGARS, AND SIRUPS

Confectionery

The utmost in cleanliness and sanitation should govern the manufacture of confectionery because both the raw materials and finished products tend to attract rodents and insects. It is not possible to manufacture a satisfactory product if rats or mice are not excluded from the establishment. Rodents will, if permitted, run over and contaminate equipment, stored materials, and finished products. Stored materials should be inspected frequently and carefully to see that they do not become infested with insects or contaminated with rodents. Equipment should be washed thoroughly and kept free from accumulation of materials that would attract vermin.

Nonnutritive substances, with the exception of certain ingredients specifically named in the act, and alcohol in excess of 0.5 percent, are specifically prohibited in confectionery. The presence of nonpermitted coal-tar colors and misbranding by the incorrect listing of ingredients also constitute grounds for detention of importations. Fruit-type confectionery should not be misrepresented with respect to fruit content. Artificial flavor, when present, should be declared on the label and the names of the fruits imitated should be preceded by the word "Imitation."

Under the Food and Drugs Act of 1906, talc was specifically named as one of the substances prohibited in confectionery. Under the present law, talc is included in the broad prohibition against all nonnutritive substances. The former practice of dusting candy with talc to prevent sticking has long since been abandoned in the United States in favor of starch or some other nutritive material that serves such purpose. Occasional importations of candy bearing this prohibited coating of talc have been the basis of detention.

Manufacturers who use artificial colors should assure themselves, in the case of coal-tar color, that they are from a batch certified as suitable for use in foods by the U. S. Food and Drug Administration. Information on coal-tar color certification is given in a separate discussion in this publication.

Honey

Generally the mixing of honey and sugar sirup of a density similar to honey results in deception and is regarded as adulteration. Shipments of honey have frequently been denied entry because they were found to contain various forms of filth and trash. Honey should be gathered and prepared under clean conditions and packed for shipment in clean and tightly closed containers.

Maple Sugar and Maple Sirup

Maple sugar and maple sirup containing significant amounts of lead are adulterated. The sources of lead contamination should be ascertained and conditions corrected. Metal buckets, spiles, or other equipment containing lead have been found responsible for lead contamination in many cases.

CACAO PRODUCTS

Standards of identity have been promulgated under the Federal Food, Drug, and Cosmetic Act for the following 12 cacao products:

Cacao nibs, cocoa nibs, cracked cocoa
Chocolate liquor, chocolate, baking chocolate, bitter chocolate, cooking chocolate, chocolate coating, bitter chocolate coating
Breakfast cocoa, high-fat cocoa
Cocoa, medium-fat cocoa
Low-fat cocoa
Sweet chocolate, sweet chocolate coating
Milk chocolate, sweet milk chocolate, milk chocolate coating, sweet milk chocolate coating
Skim milk chocolate, sweet skim milk chocolate, skim milk chocolate coating, sweet skim milk chocolate coating
Buttermilk chocolate, buttermilk chocolate coating
Mixed dairy product chocolates, mixed dairy product chocolate coatings
Sweet chocolate and vegetable fat (other than cacao fat) coating
Sweet cocoa and vegetable fat (other than cacao fat) coating

Since the standards prescribe the composition and labeling which must be conformed to, they should be consulted before manufacturing any of these products for importation into the United States.

Excess shell is frequently the cause of detention of various standardized cacao products.

Cocoa beans offered for import sometimes show damage which bars them from entry. Beans with mold permeating the bean or showing clouds of spores when cracked are objectionable, as are wormy, insect-infested beans which may show live or dead insects, webbing, or insect excreta. Moldy and insect-infested beans should be removed from the lot before shipment. Care should be taken to store beans before and during shipment so that mold and insect contamination are prevented. Some shipments have been encountered which contained goat and other animal excreta which is evidence of storage or handling under insanitary conditions.

BEVERAGES AND BEVERAGE MATERIALS

Alcoholic

Beer, wines, liquors, and other alcoholic beverages are subject to the Federal Food, Drug, and Cosmetic Act but since they are also subject to the more specific Federal Alcohol Administration Act which is enforced by the Bureau of Internal Revenue of the U. S. Treasury Department, questions of labeling and of normal composition should be taken up with the Bureau of Internal Revenue rather than with the Food and Drug Administration. Questions dealing with the presence of deleterious substances in alcoholic beverages (such as excess fusel oil and excess aldehydes in whisky, methyl alcohol in brandy, glass splinters from bottles, toxic ingredients such as arsenic, lead, or fluorine, resulting from the spraying of the fruit used in wine manufacture, and residues of toxic clarifying substances) are dealt with by the Food and Drug Administration, as are questions relating to sanitation and filth. The importation of absinth and any other liquors or liqueurs that contain an excess of *Artemisia absinthium*, is prohibited.

Large amounts of brandy, wine, and in some cases whisky and champagne have been detained when offered for entry because of the presence of small, sharp splinters of glass. In some cases the cause was attributed to the use of defective bottles that "blistered" on the inner surface. In other cases chips from the neck of the bottle apparently fell into the bottle before filling.

Nonalcoholic

Nonalcoholic beverages and the bases of flavoring sirups used to prepare nonalcoholic beverages should not be artificially colored if the use of artificial color makes them appear to contain more fruit juice than is actually present. Usually, the mixing together of an insignificant amount of fruit juice, artificial color, sugar, water, and an acidulant will constitute adulteration under the act, regardless of the labeling employed. The labels of such products should not show pictures of fruit, and claims of fruit content should not be made in the labeling, unless the amount of fruit juice that the consumer would expect from the labeling and appearance of the product is actually present. Products sold as fruit juices may be sweetened if the label states the presence of the added sugar or other wholesome, nutritive sweetening agent (saccharin is not a nutritive sweetening agent), but they should not contain added water.

Fruit juices should be manufactured only from clean, sound fruit. Equipment used for manufacture of fruit juices should be cleaned thoroughly, otherwise a mold growth may form which will contaminate the product. Proper preparation of fruit for juice manufacture involves thorough washing, sorting to remove wormy or rotten or otherwise spoiled fruit, and trimming. Flies should not come in contact with the fruit or equipment.

Bottled Mineral Waters

The listing on the labels of mineral waters of minerals which are present in insignificant amounts is misleading. Such a listing of ingredients may convey an impression of therapeutic benefit not possessed by the water. All bottled waters offered for import should be—

1. Obtained from sources free from pollution.
2. Bottled or otherwise prepared for the market under sanitary conditions, with special reference to the condition of the bottles or other containers and of their stoppers.
3. Free from micro-organisms of the coliform group.
4. Of good sanitary quality when judged by the results of a bacteriological examination to determine the numbers of bacteria growing on gelatin at 20° C. and on agar at 37° C.
5. Of good sanitary quality when judged by the results of a sanitary chemical analysis.

Some of the types of misbranding found on mineral water labels are—

1. Objectionable therapeutic claims.
2. False and misleading statements other than therapeutic claims.
3. Label statement of analysis inaccurate or otherwise misleading.
4. Undeclared addition of salts, carbon dioxide, etc.
5. Labeling of artificially prepared waters as "Mineral water."

Tea (*Thea sinensis* L.)

Tea is not only subject to the Federal Food, Drug, and Cosmetic Act but also to the Tea Importation Act. Under the latter law, tea offered for entry must meet the standards of purity, quality, and fitness for consumption prescribed under the act.

Beverages brewed from the leaves of other plants cannot properly be labeled as tea but should be labeled for what they are. For example, maté or matté, or the beverage prepared from it, should not be labeled unqualifiedly as tea or South American tea, since such names might mislead consumers in the United States who may not be familiar with the fact that it is an entirely different plant and beverage from the beverage tea of commerce in the United States. Like tea or coffee, maté may be mildly stimulating because of its caffein content, but it should not be represented as capable of producing health, energy, endurance, or other physiological effects.

Coffee

Green coffee imported into the United States should be held at all times under sanitary conditions to prevent contamination by storage insects, rats, and mice. At times much coffee has been detained because of the presence of insect-bored or insect-infested berries. Occasionally shipments are encountered in which the coffee is contaminated with rat, mice, and bird droppings, which is clear evidence of storage under insanitary conditions rendering such coffee adulterated.

The following conditions class coffee as objectionable and subject to refusal of entry: Coffee berries which are black, part black, moldy, water-damaged, insect-eaten or infested, brown ("sour"), "Quakers" (immature berries), shriveled, "sailors," or contain foreign material, such as pods, sticks, stones, trash, and "sweepings" of spilled coffee in ships' holds and on docks. Contamination by ores and other poisonous materials in the cargo also has resulted in detentions.

SEAFOODS

Fish and shellfish are imported in the fresh, frozen, canned, salted, dried, or smoked condition. The raw materials are by nature extremely perishable and must be handled rapidly under adequate refrigeration if decomposition is to be avoided. The conditions of handling are also likely to result in contamination of the raw materials unless particular safeguards are imposed. Decomposition problems are not peculiar to any one type or class of fish products but may occur in any or all types. In some instances examination of imported seafood products has disclosed evidence of the packing of decomposed raw materials. In occasional cases decomposition is encountered which is the result of active spoilage caused by insufficient processing or preservation.

Sometimes preservatives have been used to prevent or retard spoilage in fish. Any preservative used must be harmless and must be declared on the label.

Labels must bear the common or usual name of the fish. The word "fish" on the label is not sufficient as a designation of a product; the kind of fish must be stated plainly. The selection of the common or usual name of fish products is not always a simple matter because of the variety of common names applied locally to some fish erroneously. In deciding what constitutes the common name, full consideration must be given to the biological classification of the fish. Generally there is sufficient recognition of fish under names that are common or usual within the industry and among consumers to make it possible to conform with this requirement without confusion or conflict.

Canned Fish

Failure to declare the presence of added salt or the kinds of oil used as the packing medium in canned fish has resulted in the detention of fish products. If permitted artificial colors or chemical preservatives are used, their presence must be conspicuously declared in the labeling. Artificial coloring is not permitted if it conceals damage or inferiority or if it makes the product appear better or of greater value than it is.

The packing of canned fish and fish products with excessive amounts of packing medium has resulted in many detentions. The can, in canned food products, serves not only as a container but also as an index of the quantity of food therein. Where the fish are in a packing medium such as anchovies in oil, the container should be as full as possible of fish with the minimum amount of oil. The fact that the oil may be equal in value or even more expensive than the fish does not affect this principle. The purchaser of the fish should not be called upon to buy large quantities of oil which could be purchased separately if desired. The addition of excess packing medium is deemed adulteration regardless of the character of the packing medium. Canned lobster paste and similar products have been encountered which were deceptively packaged because of excessive headspace, i. e., excessive space between the lid of the can and the surface of the food in the can.

Anchovies

Products represented as anchovies should consist of fish of the family *Engraulidae*. Other small fish such as small herring and herringlike fishes which may superficially resemble anchovies, cannot properly be labeled as anchovies. The product should be prepared from sound raw material and the salting or curing should be conducted in such a manner that spoilage does not occur.

Sardines

The term "sardines" is permitted in the labeling of the canned products prepared from small-size clupeoid fishes. The sea herring (*Clupea harengus*) the European pilchards (*Sardina pilchardus* or *Clupea pilchardus*), and the brisling or sprat (*Clupea sprattus*) are commonly packed in oil in small-size cans and labeled as sardines. The terms "brisling sardines" and "sild sardines" are permissible in the labeling of canned small brisling and herring, respectively. Large-size herring cannot be labeled sardines. These canned products must be free from all forms of decomposition such as "feedy," "belly-blown" fish and must be adequately processed to prevent active spoilage by micro-organisms. Fish are called "feedy fish" when their stomachs are filled with feed at the time the fish are taken from the water. Such fish deteriorate rapidly until the viscera and thin belly wall disintegrate producing a characteristic ragged appearance called "belly-blown." There have been instances where excessive amounts of lead imparted by the use of lead-containing equipment or derived from lead solder in the can closure have required action against canned sardines.

Tuna

A definition and standard of identity and standards of fill of container for canned tuna have been adopted under the Federal Food, Drug, and Cosmetic Act. The fish that may be canned under the name tuna are as follows:

Thunnus thynnus	Bluefin tuna.
Thunnus maccoyii	Southern bluefin tuna.
Thunnus orientalis	Oriental tuna.
Thunnus germo	Albacore.
Parathunnus mebachi	Big-eyed tuna.
Neothunnus macropterus	Yellowfin tuna.
Neothunnus raus	Northern bluefin tuna.
Katsuwonus pelamis	Skipjack.
Euthynnus alletteratus	Little tunny.
Euythnuss lineatus	Little tunny.
Euthynnus yaito	Kawakawa.

The standard provides for various styles of pack, including solid pack, chunk or chunk style, flakes, and grated tuna. Provision is also made for various types of packing media and certain specified seasonings and flavorings. Those interested in shipping tuna to the United States should consult the detailed specifications of the standards.

Bonita and Yellowtail

The meat canned from the fish *Sarda chilensis*, commonly known as the bonito or bonita may not be labeled as tuna since it is not a true tuna but must be labeled as bonito or bonita. The meat of *Seriola dorsalis*, commonly known as "yellowtail" must be labeled as yellowtail and may not be designated as tuna.

Fresh and Frozen Fish Fillets

These products are highly perishable and require extraordinary care if decomposition is to be avoided. The manufacturer must exercise extreme care in the selection of raw materials to remove any unfit, decomposed material and then to maintain the product in a sound, wholesome condition, to avoid the denial of entry to such importations.

An important problem has arisen in connection with the importation of certain species, such as whitefish, tullibees, lake herring, chubs, and ciscoes, which are subject to infestation with a repulsive form of parasite. This condition is in the form of cysts containing the parasite and puslike material in the flesh of the fish. The presence of such cysts renders the fish filthy and as such they may not be imported into the United States.

Other Seafood Products

Because raw shellfish, such as oysters, clams, and mussels may transmit intestinal diseases such as typhoid fever, it is extremely important that they be grown in unpolluted waters and produced, handled, and distributed under sanitary conditions. Shellfish must conform to the general requirements of the Food, Drug, and Cosmetic Act, but, in the importation of shellfish into the United States, consideration must be given also to the requirements of the States to which the shellfish are destined, if they are to be accepted under the respective State laws.

Caviar and Fish Roe

The name "caviar" unqualified may be applied only to the eggs of the sturgeon prepared by a special process. Fish roe prepared from the eggs of other varieties of fish, prepared by the special process for caviar, must be labeled to show the name of the fish from which they are prepared, for example, "whitefish caviar." If the product contains an artificial color, it must be from a certified batch if it is a coal-tar color, and its presence must be stated on the label conspicuously. No artificial color should be used which makes the product appear to be better or of greater value than it is. If a chemical preservative is employed, it must be one the harmlessness of which has been established, and it must be declared on the label.

Clams

Some difficulty has been encountered with importations of shellfish, particularly clams, containing a toxic substance *Gonyaulax catenella*, derived from a plankton organism upon which the shellfish feed. Such toxic shellfish may cause illness or even death. The sources of supply of shellfish intended for shipment to the United States should be periodically tested for the presence of gonyaulax.

Raw Oysters

There are standards of identity under the Food, Drug, and Cosmetic Act for raw oysters. Persons interested in shipping raw oysters for importation into the United States should consult these standards. They set limits to the exposure of the raw oysters to fresh water and require that the oysters be thoroughly drained before packing into containers for shipment. The standards also give specifications as to size, uniformity within a size, and the designations to be used for the sizes.

Canned Oysters

The standard of identity for canned oysters specifies that this food consists of oysters which have been removed from their shells and washed, and which may be steamed while in the shells or steamed or blanched or both after removal from their shells. The oysters together with the packing medium of water, the watery liquid draining from the oysters before or during processing, or mixtures of such liquid with water, with or without salt for seasoning, are sealed in containers and so processed by heat as to prevent spoilage. The standard prescribes labeling for whole oysters, pieces of oysters, and cut oysters. They may be labeled "Oysters" or the type of oyster may be specified by appropriate designation, such as "Pacific Oysters," "Cove Oysters," or "Olympia Oysters," as the case may be.

The standard of fill of container for canned oysters is a fill such that the drained weight of oysters taken from each container is not less than 59 percent of the water capacity of the container.

Shrimp

Shrimp should be iced from the time they are caught until they are processed, to prevent decomposition.

Standards of fill of container have been promulgated for canned wet pack shrimp and canned dry pack shrimp:

The standard of fill of nontransparent containers for canned wet pack shrimp is a fill such that the cut-out weight of shrimp taken from each can is not less than 64 percent of the water capacity of the container, and, for canned dry pack shrimp (except that packed in the nontransparent cylindrical container which is $2\frac{11}{16}$ inches in diameter and 4 inches in height), is a fill such that the cut-out weight of shrimp taken from each can is not less than 60 percent of the water capacity of the container. The standard of fill for canned dry pack shrimp packed in the nontransparent cylindrical container which is $2\frac{11}{16}$ inches in diameter and 4 inches in height is a cut-out weight of not less than 6½ avoirdupois ounces of shrimp for each container.

Rock Lobster, Spiny Lobster, Sea Crayfish

The sea crayfish, *Palinurus vulgaris*, is frequently imported into the United States in the form of frozen tails, frozen cooked meat, or canned meat. By long usage, the terms "Rock Lobster" and "Spiny Lobster" have been established as common or usual names for these products. No objection has been offered to either of these terms, providing the modifying words "rock" or "spiny" are used in direct connection with the word "lobster" in type of equal size and prominence.

In examination of imports, decomposition has sometimes been detected in all three forms of the product. In the canned product, decomposition resulted from the packing of decomposed raw material and also from active bacterial spoilage. In the frozen cooked product, detentions have been necessary also because of the presence of micro-organisms indicative of pollution with human or animal filth, as well as of decomposition.

MEAT AND POULTRY

Meat and Meat Food Products

The flesh of sheep, swine, cattle, and goats, when offered for importation, is subject to the provisions of the Federal Imported Meat Act, enforced by the Meat Inspection Division of the Bureau of Animal Industry of the U. S. Department of Agriculture. The imported product, after admission into the United States, becomes a domestic article, subject not only to the Federal Meat Inspection Act enforced by the same unit of the Department of Agriculture, but also to the Federal Food, Drug, and Cosmetic Act, to the extent to which the provisions of the Meat Inspection Act do not apply.

Poultry and Poultry Products

Poultry and poultry products offered for importation are subject to the Poultry Products Inspection Act which is enforced by the U. S. Department of Agriculture, to which inquiries concerning the importation of such products should be addressed. The term "poultry" means any live or slaughtered domesticated bird, and the term "poultry product" means any poultry which has been slaughtered for human food, from which the blood, feathers, feet, head, and viscera have been removed in accordance with rules and regulations promulgated by the Secretary of Agriculture, or any edible part of poultry, or, unless exempted by the Secretary, any human food product consisting of any edible part of poultry, separately or in combination with other ingredients.

After importation into the United States, poultry and poultry products are also subject to the Federal Food, Drug, and Cosmetic Act to the extent to which the provisions of the Poultry Products Inspection Act do not apply.

Gelatin

Gelatin is not held to be a meat food product amenable to the Federal Imported Meat Act. The provisions of the Food, Drug, and Cosmetic Act require it to be prepared from clean, sound, wholesome, raw materials, handled under sanitary conditions, containing no glue or other gelatin-like material with a low gelatinous power or having a disagreeable odor or taste. It should be so prepared and marketed as to be free from spoilage, filth, or putridity. It should be free from such preservatives, metals, and salts of metals as may render its use injurious to health.

DAIRY PRODUCTS

Milk, cream, and dairy products made therefrom offer ideal conditions for the growth and survival of micro-organisms. It is essential that raw milk used in the preparation of manufactured dairy products be obtained from animals free from diseases, and that the raw materials and products be handled under sanitary conditions at all times to prevent contamination with disease-producing and spoilage organisms. While pasteurization of the milk for manufacture of dairy products offers a considerable safeguard against disease transmission, contamination by workers suffering from disease, by rodents, or by unclean equipment after pasteurization is a real health hazard. In a like manner, spoilage organisms, such as undesirable bacteria, yeasts, and molds capable of causing decomposition of the raw materials or of the finished products, may be contributed by insanitary handling. Products prepared from decomposed raw materials, undergoing active spoilage, or those contaminated with disease-producing organisms are not suitable for importation.

Cream and Milk

Cream and milk used for the manufacture of cheese and other dairy products should be clean and free from decomposition. Butter is examined for evidence of the use of dirty cream or milk, for the presence of filth, such as flies, ants, beetles and their parts, and for mold which indicates the use of decomposed cream. The factors which control decomposition in cream are time, temperature, and sanitation, and it is important to keep the cream before butter making for as short a period as possible, in a cool place, and to use clean utensils in handling and producing the cream.

IMPORT MILK ACT. In addition to being subject to the requirements of the Food, Drug, and Cosmetic Act, milk and cream offered for import into the continental United States are subject to another Federal Act, the Import Milk Act, enforced by the Food and Drug Administration, U. S. Department of Health, Education, and Welfare. Under this act, these particular dairy products may be imported only under permit after certain sanitary and other prerequisites have been fulfilled.

Butter

Butter is defined in the standard provided by Act of Congress of March 4, 1923, as follows:

> "Butter" shall be understood to mean the food product usually known as butter, and which is made exclusively from milk or cream or both, with or without common salt, and with or without additional coloring matter, and containing not less than 80 percentum of milk fat, all tolerances having been allowed for.

Chemical preservatives and artificial flavor are not permitted.

Cheese

Cheese may be contaminated with insect and rodent filth during the handling of the milk, during the manufacture of the cheese, or during storage. This must be guarded against.

Standards of identity have been promulgated for most natural cheeses except whey cheeses. Standards have also been adopted for practically all process cheeses, cheese foods, and cheese spreads. Class standards (based on the consistency of the finished cheese) include practically all of the natural cheeses for which there are not individual definitions. Where a standard has been adopted for a particular variety of cheese, all cheeses belonging to the variety must comply with the standard and be labeled with the name prescribed in the standard. In the case of the class standards the cheeses of a class must comply with the standard for the class. A particular cheese in the class may be labeled with its common name followed by labeling to show the class to which it belongs. Most of the standards prescribe limits for moisture and for fat.

A few natural cheeses are required to be made from pasteurized milk. Most, however, may be made from either raw milk or pasteurized milk. When made from raw milk they are required to be aged for 60 days or longer. The 60 days aging is for the purpose of increasing the safety of the cheese. Requirements for longer aging are made in the cases of those cheeses which require long aging to develop the characteristics of the variety.

The standards for process cheeses set limits for moisture and for fat and specify the names to be used. Cheese foods, generally speaking, are process cheeses to which have been added other dairy ingredients. Cheese spreads are similar to cheese foods except they have a higher moisture content. Shippers of cheese and foods made of cheese, should consult these standards before making shipments to the United States.

Evaporated Milk, Sweetened Condensed Milk

Under the established standard of identity for evaporated milk, this product must contain not less than 7.9 percent milk fat and not less than 25.9 percent total milk solids. Sweetened condensed milk, for which a standard has also been established, must contain not less than 28 percent total milk solids, and not less than 8.5 percent milk fat.

NUTS AND NUT PRODUCTS

Nuts offered for entry, such as cashews, pignolas, pistachios, filberts, almonds, and brazil nuts, have been found to be adulterated because of worminess, moldiness, and rancidity. The shipper should also be on guard against the inclusion of dirty nut meats. Empty or worthless unshelled nuts should be removed by careful hand sorting or by machinery.

Care should be taken to eliminate infested, dirty, moldy, or rancid nuts from the shipment. Conditions which may cause nuts to be refused admission are described below.

Insect Infestation.—Nuts are insect-infested if they contain live or dead insects, whether larvae, pupae, or adults, or if they show definite evidence of insect feeding or cutting, or if insect excreta pellets are present. In English walnuts, insect damage, both field and storage, may be encountered. In the former, the damage is usually caused by the codling moth and is characterized by black-edged, darkened cavities in the kernels.

Dirt.—Nut meats may become dirty because of lack of cleanliness in cracking, sorting, and packaging.

Mold.—Nut meats occasionally are moldy in the shell and bear fruiting mold or mold hyphae covering the entire piece or portions thereof.

Rancidity.—Nuts in this class have an abnormal flavor characterized by rancidity or putridity. Rancid nuts are frequently soft and have a yellow, dark, or oily appearance.

Extraneous Material.—Stems, shells, stones, excreta, should not be present.

While no "tolerances" for the above-mentioned defects are provided for by the Food, Drug, and Cosmetic Act, it has been well demonstrated that nuts in the shell should not contain over 10 percent by count of all such defects, and shelled nuts (nut meats) not over 5 percent. Deliberate mixing of good and bad lots to result in defects just under these figures is a basis for refusal of entry even though the percentage of defects in the mixed lots is under the above figures.

Shipments of sweet almonds must not contain more than 5 percent of bitter almonds. Almond paste is the product obtained by cooking a mixture of blanched and ground sweet and bitter almonds with water, sugar, and/or dextrose. It should contain not more than 14 percent moisture nor more than 40 percent total sugars expressed as invert sugar. Kernel paste is similarly made from the kernels of one or more of the following: Apricots, peaches, or plums (prunes). The moisture and sugar content should be the same as for almond paste, and the name and the kind of kernel, for example, apricot kernel, should be used. In the light of the impossibility of removing from kernel paste the last minute traces of naturally occurring hydrocyanic acid that may be present, entry has not been denied to products containing less than 25 parts HCN per million parts of the paste, if they are not otherwise illegal.

Nuts and nut meats must be prepared and stored under sanitary conditions to prevent contamination by insects, rodents, or other animals. Nuts imported for pressing of food oil must be just as clean and sound as nuts intended to be eaten as such or to be used in manufactured foods.

EDIBLE OILS

Olive oil is the edible oil expressed from the sound, mature fruit of the olive tree. Refined or extracted oil is not entitled to the unqualified name "olive oil." Other vegetable oils should be labeled by their common or usual name, such as cottonseed, sunflower, peanut, and sesame. Mixtures of edible oils should be labeled to show all the oils present and the names should be listed on the labels in the descending order of predominance in the product. Pictures, designs, or statements on the labeling must not be misleading as to the kind or amount of oils present or as to their origin.

Cod liver oil is a drug as well as a food since it is recognized in the United States Pharmacopeia. Its value as a food whether intended for human or animal use depends for the major part on its vitamin D content. Articles offered for entry as cod liver oil must comply with the identity standard prescribed by the Pharmacopeia and conform to the other specifications set forth in that official compendium. These include a minimum vitamin A potency of 850 U. S. P. units and a minimum vitamin D potency of 85 U. S. P. units per gram. Nondestearinated cod liver oil likewise is recognized in the United States Pharmacopeia and except for sterin content and other minor deviations must meet the specifications for cod liver oil. Oils represented as containing more than the minimum vitamin content prescribed by the Pharmacopeia must contain the represented vitamin potencies. Cod liver oils complying with the identity and vitamin standards but differing in color or acidity from the Pharmacopeial specifications may be imported provided they are labeled as cod liver oil not U. S. P. and provided they also show on the label in what respects they differ from the official product.

Oils offered for entry as cod liver oil which fail to meet the minimum A and D vitamin potencies for U. S. P. Cod Liver Oil must be refused admission.

MAYONNAISE, FRENCH DRESSING, AND SALAD DRESSING

Standards of identity have been adopted for mayonnaise, salad dressing, and french dressing. These specify that mayonnaise must have not less than 65 percent vegetable oil, that salad dressing must have not less than 30 percent vegetable oil, and french dressing must have not less than 35 percent vegetable oil. In mayonnaise an egg-yolk-containing ingredient is the sole emulsifying ingredient recognized. In salad dressing and french dressing certain designated emulsifying ingredients, in addition to egg yolk, are permitted. The standard for salad dressing requires that it contain not less egg yolk than the equivalent of 4 percent of liquid egg yolk. The acidifying ingredients for all three of the dressings are vinegar or lemon juice, and these may be diluted with water to a limited extent. Citric acid is permitted to the extent of not more than one-fourth of the acetic acid in the vinegar. When citric acid is used the standards require that it be declared on the label. A starch paste is a required ingredient of salad dressing. It is not permitted in mayonnaise or french dressing. Mineral oil is not a permitted ingredient in any of these foods.

FOODS FOR SPECIAL DIETARY USES

Section 403 (j) of the Food, Drug, and Cosmetic Act classes a food as misbranded "If it purports to be or is represented for special dietary uses, unless its label bears such information concerning its vitamin, mineral, and other dietary properties as the Secretary determines to be, and by regulations prescribes as, necessary in order fully to inform purchasers as to its value for such uses."

Regulations referred to in Section 403 (j) were promulgated in 1941. At the same time a general regulation under the act (Regulation 2.11) was issued interpreting the term "special dietary uses" as used in the statute. This general regulation is as follows:

(a) The term "special dietary uses," as applied to food for man, means particular (as distinguished from general) uses of food, as follows: (1) Uses for supplying particular dietary needs which exist by reason of a physical, physiological, pathological or other condition, including but not limited to the conditions of disease, convalescence, pregnancy, lactation, allergic hypersensitivity to food, underweight, and overweight; (2) Uses for supplying particular dietary needs which exist by reason of age, including but not limited to the ages of infancy and childhood; (3) Uses for supplementing or fortifying the ordinary or usual diet with any vitamin, mineral, or other dietary property. Any such particular use of a food is a special dietary use, regardless of whether such food also purports to be or is represented for general use. (b) No provision of any regulation under section 403 (j) of the act shall be construed as exempting any food from any other provision of the act or regulations thereunder, including sections 403 (a) and (g) and, when applicable, the provisions of Chapter V.

The regulations under section 403 (j) prescribe label statements of the dietary properties upon which the special dietary use of the food is based. They provide that labels of foods represented as for special dietary use by reason of their vitamin or mineral content shall state the proportion of the minimum daily requirement for each such vitamin or mineral supplied by the food when consumed in a specified quantity in 1 day. The regulations specify the minimum daily nutritional requirements of certain vitamins and minerals. The regulations also prescribe label statements relating to infant food, label statements relating to food used in control of body weight or in dietary management with respect to disease, label statements relating to nonnutritive constituents, label statements relating to hypoallergenic food, and to foods intended for use in low-sodium (salt restricted) diets.

Importers and foreign shippers should consult the detailed regulations under section 403 (j) before importing foods represented by labeling or otherwise as special dietary foods.

When special dietary foods are labeled with health claims, they are subject to the drug provisions of the act (see Chapter II).

Chapter 2

GENERAL METHODS

Moisture — Ash — Nitrogen and Crude Protein — Fat — Crude Fibre — Calcium, Phosphate — "Filth" Test

MOISTURE

SEVERAL methods are available for the determination of moisture in foods:—

Indirect Distillation (or Drying) Methods

These involve the measurement of the weight lost due to the evaporation of water. Although such methods are frequently used as they usually give accurate results when considered on a comparative basis, it should be borne in mind that the figure obtained may not be a true measure of the water content of the sample. For example, volatile oil, if present, is also lost at drying temperatures, such as 100° C. On the other hand, with some foods (*e.g.*, cereals) only a proportion of the water present (*viz.* the "free" water) is lost at the drying temperature. The remainder (often referred to as the "bound" water as it is difficult to remove) appears to be associated with the proteins present. The proportion of free water lost increases as the temperature is raised, so it is especially important only to compare the results obtained using the same conditions of drying. Further if decomposition is likely to occur, as with foods which contain an appreciable proportion of sugars, it is advisable to use a lower drying temperature, such as 70° C. and to apply vaccum. Baking powder has to be dried at room temperature in a vacuum desiccator for a considerable period as heat causes appreciable losses of carbon dioxide.

As the loss in weight tends to vary according to the conditions of drying, it is important to have an accurate control of temperature within the oven used. Also as the temperature tends to be different on different shelves within ordinary " static " ovens, more consistent results can be obtained if the drying oven used is fitted with an internal fan.

Direct Distillation Methods

These involve the reflux distillation of the food with an immiscible solvent having a higher boiling point and a lower specific gravity than water, *e.g.*, toluene, heptane, xylene (Tate and Warren, *Analyst*, 1936, **61**, 367). The refluxed water falls below the solvent in a graduated tube, in which it can be measured by volume. A long wire or policeman should be pushed into the condenser tube near the end of the distillation in order to coax any adhering water into the graduated tube. Although low results are not uncommon with the distillation method, it has the advantages that (a) it needs little attention once the apparatus has been set up and (b) any volatile oils which distil over mix with the solvent and are not measured.

Rapid Physical Methods

The proportion of water in foods affects certain electrical properties and this has led to the introduction of various instruments which measure the resistance, frequency and di-electric properties. Such instruments are most useful for making rapid checks of the moisture content of materials in warehouses, factories, mills, etc., provided that an accuracy not greater than about $\pm$ 1 per cent. is required.

Specially calibrated hydrometers are frequently used in industry for assessing the concentration of total dissolved solids (and indirectly the water content) of brines and syrups. The sugar concentration of syrups, preserves, etc., can similarly be assessed by determining the refractive index.

Chemical Methods

A sensitive titration method for determining water was produced by Karl Fischer (*Angew. Chem.*, 1935, **48**, 394). This method, which is now the subject of a British Standard (B.S. 2511 : 1954), is based on the non-stoichiometric reaction of water with iodine and sulphur dioxide in pyridine-methanol solution. The titration can be carried out either visually or electrometrically. The Karl Fischer method is probably most suitable for materials in which only traces of moisture are present.

Some rapid methods for determining moisture are based on the reaction of water with calcium carbide to produce acetylene (Serger,

Chem. Ztg., 1954, **76**, 681). The moisture content can be calculated either from the loss in weight or from the increase in pressure produced. The carbide method appears to be most suitable for powdered foods which more readily come into intimate contact with the powdered carbide, *e.g.*, cereals.

For further information relating to methods for determining moisture, the reader should consult the following references:—

Geary, *Determination of Moisture in Solids*, Ref. M. 24, B.S.I.R.A., Chislehurst, Eng. (1956).

Johnson, *Chem. and Ind.*, 1949, p. 511.

Joslyn, *Methods in Food Analysis Applied to Plant Products*, 1950, pp. 47–85, Acad. Press N.Y.

Kent-Jones and Amos, *Modern Cereal Chemistry*, 1957, 5th Edn. Northern Publ. Co., Liverpool.

Pearson, *Lab. Practice*, 1958, **7**, 584.

ASH AND MINERAL MATTER

Total ash. The ash of a foodstuff is the inorganic residue remaining after the organic matter has been burnt away. The ash obtained is not necessarily of exactly the same composition as the mineral matter present in the food itself as there may be losses due to volatilisation or some interaction between constituents. The ash figure can be regarded as a general measure of quality (*e.g.*, a maximum ash is prescribed for edible gelatine) and often is a useful criterion in identifying the food. When a high ash figure suggests the presence of an adulterant, it is often advisable to determine the acid insoluble ash also.

The general method for determining the total ash involves weighing out **5** g. of the sample into a silica (or platinum) dish (about 7 cm. diam.) which has previously been ignited and cooled before weighing. Then the dish and contents are ignited, first gently and then at 550° C. (loss of chloride due to volatilisation tends to occur above 600° C.). With some foods, *e.g.*, cereals, such as barley and oats, it is difficult to burn away all the organic matter, but the ashing can sometimes be completed if the particles are broken up with a stout platinum wire. Alternatively the ash can be treated with water and the liquid filtered through an ashless filter paper, which is then thoroughly ignited in the dish. The filtrate is then evaporated in the dish, which is heated again for a short time. Care must be taken when transferring dishes containing fluffy ashes (*e.g.*, of gelatine), which tend to blow away easily. Such ashes should be covered with a Petri dish after placing in the desiccator prior to weighing.

Methods for determining the ash based on electrical conductivity measurements have also been proposed (*see* Zerban and Sattler, ***Ind. Eng. Chem. Anal. Ed.***, 1930, **2**, 32). Such methods have been applied to solutions of sugars, in which there is a relationship between the electrical conductivity and the ash.

Water soluble ash. The ash is boiled with 25 ml. water and the liquid filtered through an ashless filter paper and thoroughly washed with hot water. The filter paper is then ignited in the original dish, cooled and the *water insoluble ash* weighed:—

Water soluble ash (per cent.) = Total ash (per cent.) — Water insoluble ash (per cent).

If the filtrate is cooled and titrated with N/10 sulphuric acid using methyl orange the *alkalinity of the soluble ash* can be obtained. The alkalinity is usually calculated as K_2O, K_2CO_3 or Na_2CO_3 and is often useful when considered in conjunction with the ash figure for confirming other results in relation to the original composition. In some instances a low alkalinity of the ash indicates adulteration, *e.g.*, spent tea.

Acid insoluble ash is determined by boiling the ash with 25 ml. dilute hydrochloric acid (10 per cent. w/w) for five minutes and filtering through an ashless paper. After washing thoroughly with hot water, the paper is returned to the dish and ignited, cooled and weighed.

The acid insoluble ash is a measure of the sandy matter present, *e.g.*, in herbs and spices for which maxima are prescribed in the B.P., B.P.C. and in U.S. regulations. The Fertilisers and Feeding Stuffs Regulations require the result to be returned as " Sand and other silicious matter."

Sulphated ash. This involves moistening the ash with concentrated sulphuric acid and igniting gently to constant weight. The sulphated ash enables a " constant " figure to be obtained for ashes in which varying proportions of carbon dioxide are lost according to the ignition temperature employed, *e.g.*, when calcium is present.

For further information relating to the determination of ash and related constituents the reader should consult the following references:—

Joslyn, " Methods in Food Analysis Applied to Plant Products," 1950, p. 87–113, Acad. Press, N.Y.
Pearson, *Lab. Practice*, 1958, **7**, 654.

NITROGEN AND CRUDE PROTEIN

The Kjeldahl method for determining total nitrogen involves converting the nitrogen present to ammonium sulphate in sulphuric acid solution. After making alkaline with sodium hydroxide, the ammonia is distilled into either excess of boric acid or standard acid and is estimated by titration. For most routine purposes the "protein" in the sample is then calculated by multiplying the total nitrogen by an empirical factor. It should be borne in mind however that such a figure for "protein" often includes some non-protein nitrogenous substances and further that nitrogen in certain forms (*e.g.*, nitrates, nitrites and nitroso compounds) is not estimated in the Kjeldahl process unless the procedure is modified (see note (4) below).

The **macro Kjeldahl method** is described below:—

Weigh out (1) a portion of the sample (containing 0·03–0·04 g.N) and transfer to a Kjeldahl digestion flask. Add 8 g. of catalyst mixture (96 per cent. anhydrous sodium sulphate, 3·5 per cent. copper sulphate and 0·5 per cent. selenium

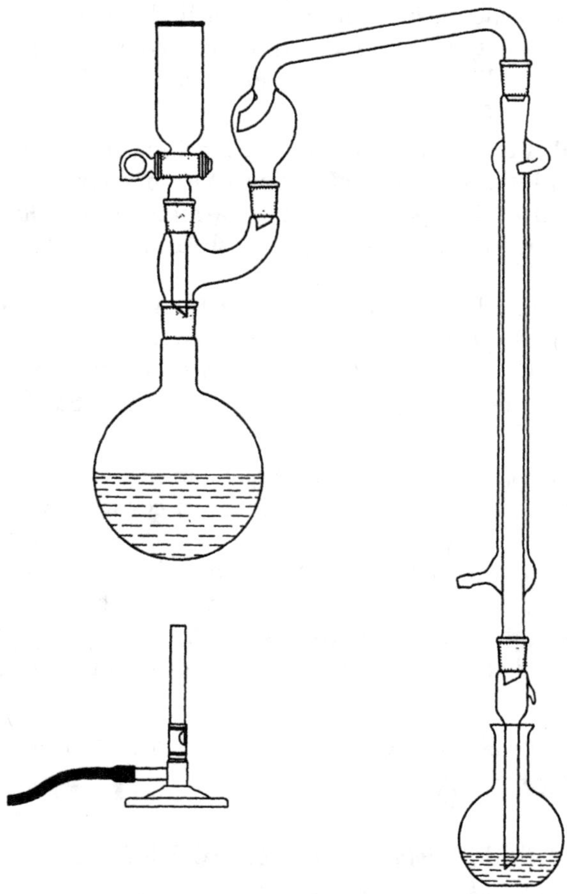

Fig. 1. Macro Kjeldahl Distillation Apparatus (by courtesy of Quickfit and Quartz, Ltd.).

dioxide) and 20 ml. of concentrated sulphuric acid (2). Heat the flask gently in an inclined position. When the initial frothing has ceased, fit a loose pear stopper in the top of the flask and then heat more strongly, so that the liquid boils at a moderate rate. Shake the flask from time to time and continue the heating for one hour after the liquid has become clear. Cool somewhat and wash the digest into the distilling flask with 400 ml. of " ammonia-free " water. To the receiving flask add 50 ml. of boric acid solution (2 per cent.) and screened methyl red indicator (0·016 per cent. methyl red and 0·083 per cent. bromocresol green in alcohol). Connect up the distillation apparatus (Fig. 1) with the delivery tube dipping below the boric acid solution. Make the diluted digest alkaline with 50 per cent. sodium hydroxide solution (at least 75 ml. should be added). Close the tap and distil the ammonia (3) into the boric acid solution. After about 300 ml. has distilled over, open the tap and wash down the condenser and delivery tube into the receiver. Titrate the distillate with N/10 sulphuric acid. The blank should not exceed 0·5 ml. N/10. Calculate the percentage of nitrogen (4) in the sample (1 ml. N/10 $\equiv$ 0·0014 g. N). The crude protein figure can be calculated using an appropriate factor: meat N $\times$ 6·25 (also the general factor); milk N $\times$ 6·38; wheat flour N $\times$ 5·7.

NOTES

(1) Semi-solid materials, *e.g.*, meat products, can be conveniently weighed out on folded filter paper and both the sample and paper dropped into the digestion flask.

(2) More acid may be required with liquids.

(3) Steam distillation is preferred by some workers.

(4) **Nitrates and nitrites,** if present, would not be included in the " total nitrogen " figure as determined by the process described. They will be included, however, if the process is modified in the following way:—

To the sample in the digestion flask, add 25 ml. of ice-cold concentrated sulphuric acid and 1 g. of salicylic acid or 1 g. of phenol. Shake the cooled flask at intervals during ten minutes and then add 8 g. of sodium sulphate (anhydrous) and 5 g. of sodium thiosulphate or 2 g. of zinc dust. Then digest and distil as in the conventional method. In this modification the nitrate nitrogen is converted to nitric acid, which nitrates the salicylic acid or phenol. The nitro compounds produced are then reduced by the thiosulphate (or zinc) to ammonia, which distils over after adding excess alkali (cf. S.I. 1960 No. 1165).

The Kjeldahl procedure can also be applied on the semi-micro or micro scale. Suitable **semi-micro distillation apparatus** have been devised by Parnas and Wagner (*Biochem. Z.*, 1921, **125**, 253; B.S. 1428, Part B1, 1953), Hoskins (*Analyst*, 1944, **69**, 271), Scandrett (*ibid*, 1953, **78**, 734) and Markham (*Biochem. J.*, 1942, **36**, 790; B.S. 1428: Part B2: 1953).

If the sample is homogeneous (*e.g.*, fine powders, liquids), it is possible to carry out the procedure on 0·15–0·2 g. of sample (more if a liquid). The digestion is carried out in a 35 ml. flask using 0·8 g. of catalyst mixture (see macro method above) and 2 ml. of concentrated sulphuric acid. Then this digest is diluted and transferred to the steamed out apparatus using a total of 20 ml. water, made alkaline with 15 ml. of sodium hydroxide solution (40 per cent.) and steam distilled into 10 ml. of boric acid solution (2 per cent.) for at least five minutes.

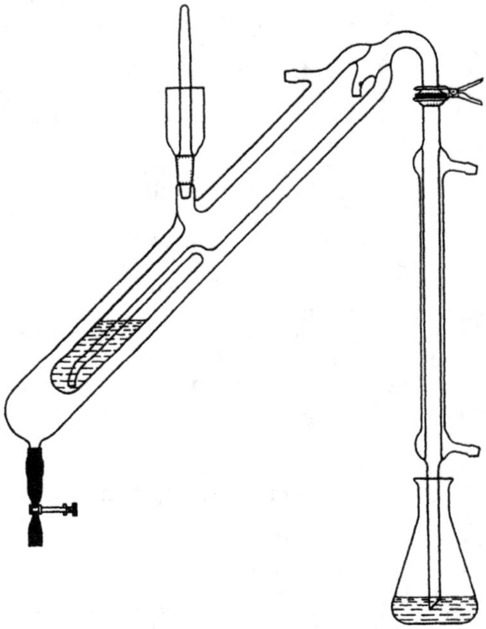

Fig. 2. Markham semi-micro Kjeldahl Distillation Apparatus (by courtesy of Quickfit and Quartz, Ltd.).

If the sample is not homogeneous, e.g., sausage, it is preferable to dilute a macro digest (on say 1·5–2 g. sample) to 100 ml. and use a 10 ml. aliquot for the semi-micro distillation. In either case the final distillate is titrated with N/50 acid (HCl). The " aliquot " method has the advantage that a considerable volume of solution is available for duplicates.

The crude protein can also be determined using the **Conway micro-diffusion technique** (*Microdiffusion Analysis and Volumetric Error*, Conway, 1957, Crosby Lockwood, Lond.). For this 2 g. of sample is digested with 4 g. of the catalyst mixture and 10 ml. of concentrated sulphuric acid in a small Kjeldahl flask (50–100 ml.). The digest is diluted to 100 ml. and 1 ml. of the solution pipetted into the outer annular space of the Conway unit (Fig. 3). In the inner chamber is placed 2 ml. of boric acid solution (2 per cent.) and screened methyl red indicator. One drop of phenolphthalein is added to the outer ring and the dish is partially closed by the greased plate. Then 1 ml. of saturated potassium hydroxide solution is added and the dish closed by sliding the plate on fully. After rotating the dish gently, and confirming that the outer ring is alkaline, the ammonia

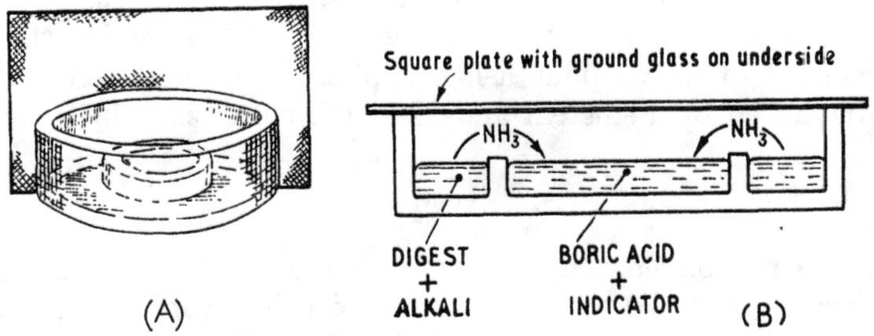

Fig. 3. Conway Microdiffusion Technique. (A) Conway Unit (dish with ground-glass joint; plate—ground-glass on one side). (B) Method of diffusion of ammonia liberated from digest into boric acid.

is allowed to diffuse over into the boric acid overnight. Then the central chamber is titrated with N/20 sulphuric acid using a 2 ml. burette. It is relatively easy to carry out the method in triplicate on further aliquots of the same digest solution. A blank should be put on at the same time.

With certain foods it is possible to estimate the protein content rapidly by means of the **formol titration.** Although amino acids are too weak to be titrated directly with alkali, if formalin is added the $-NH_2$ group is replaced by the methylene-imino group $-N{=}CH_2$ and the carboxyl group is then available for titration. The reaction is often written as:—

$$R.COOH.NH_2 + H.CHO \rightarrow H_2O + R.COOH.N{=}CH_2.$$

For foodstuffs, the formalin is added to the neutralised material and the acidity produced is titrated with standard alkali. If neutral formalin is used, the titration is proportional to the amount of protein present. It should be borne in mind that the empirical factor which is then used for calculating the percentage of total protein in the sample is dependent on the type and ratio of the protein groupings which are present (*e.g.*, the ratio of casein to albumin in milk) and the actual analytical procedure employed (*e.g.*, the type and normality of the alkali used for the titration and the degree of dilution with water).

The formol titration represents one of the most rapid methods for determining protein in certain liquids, *e.g.*, milk, or possibly the aqueous extracts of other foods, for the estimation of casein in milk, the milk solids in ice-cream and for assessing the amount of fruit juice in soft drinks. For further information relating to the formol titration the reader should refer to pages 218, 300, 344 and 412.

Rapid colorimetric methods for proteins in milk have been

described by Goiffon (*Lait*, 1948, **28**, 449) and Udy (*Nature*, 1956, **178**, 314). Also a simple turbidimetric method for determining the protein content of wheat flour using sulphosalicylic acid solution has been described by Feinstein and Hart (*Cereal Chem.*, 1959, **36** (2), 191).

FAT

Fat can be conveniently determined in foods by extracting the dried ground material with petroleum spirit in a continuous extraction apparatus of the Soxhlet or Bailey-Walker type. In order to ensure that all the fat is extracted, it is advisable to remove the material from the apparatus, regrind it with sand and return it to the extractor for a further extraction. Complete extraction of the fat is difficult if the proportion of sugar present is high.

With liquid materials which can be brought into solution in water relatively easily it may be possible to extract the fat by shaking with the solvent in a separator. At least three extractions should be carried out and the final bulked extracts should be filtered into the tared flask. The formation of emulsions, which are common with this type of extraction can sometimes be broken by the addition of alcohol.

Another useful method for extracting fat is to mix intimately the material with *anhydrous* sodium sulphate and then add ether. The ethereal solution of the fat should then be filtered before removing the solvent. This method, whereby the water is taken up by the anhydrous salt, is perhaps more suitable for the extraction of fat for the determination of oil values than for the determination of the amount of fat in the sample itself.

The presence of protein sometimes prevents the complete extraction of fat by direct methods, so it is often necessary to treat the material first with acid or alkali. In the acid method (**Werner-Schmid process**) the material is heated with hydrochloric acid to dissolve the protein and the fat separates as a layer on the top of the acid liquid (see p.342). The acid concentration during the extraction should be approximately 6 N, *e.g.*, 10 g. of milk is treated with 10 ml. of concentrated acid, or 1–2 g. of a solid food is mixed with 8–9 ml. of water and 10 ml. of the acid. The protein dissolves in the acid and the separated fat can be extracted by shaking at least three times with ether or ether and petroleum spirit. Treatment of the original material with ammonia prior to the addition of acid encourages the proteins to dissolve more readily. The acid extraction process is less suitable than the alkali method if the material contains a high proportion of sugar.

For the extraction using alkali (**Rose-Gottlieb and Mojonnier processes**), the material is treated with ammonium hydroxide and alcohol and the fat is extracted with ether-petroleum spirit. The alcohol precipitates the protein which dissolves in the ammonia and the fat can then be extracted with ether (p. 341). Petroleum spirit is then added as it reduces the solubility of non-fatty materials such as lactose which tend to be included in the extract if di-ethyl ether is used alone. The alkali extraction method gives very accurate results provided a prescribed technique is adhered to (*e.g.*, *Analyst*,

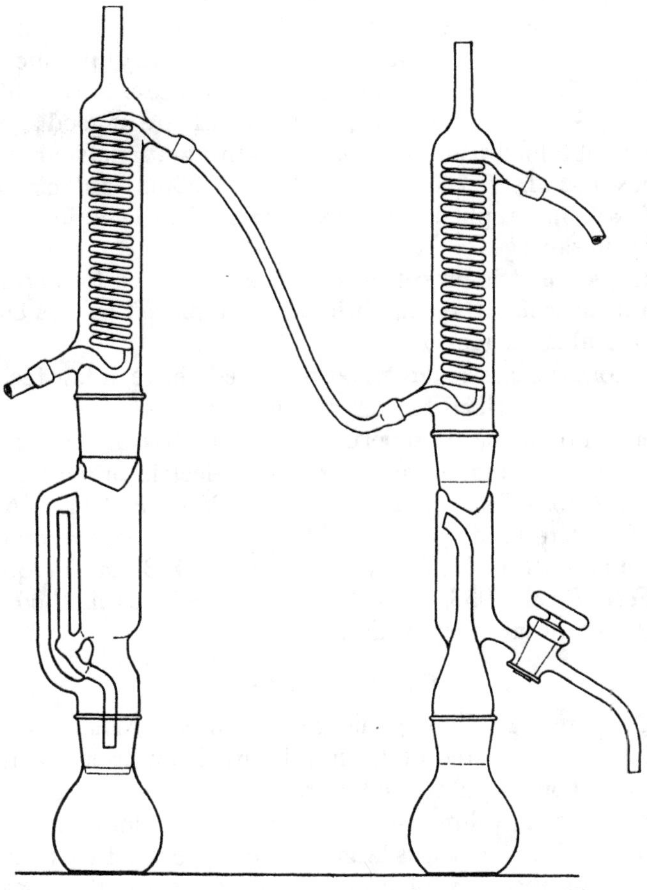

Fig. 4. A simple method of removing solvent in gravimetric determinations of fat. On the left is shown an ordinary Soxhlet extractor with ground-glass joints (B34). For removing the solvent, the extractor is replaced by the apparatus on the right which has similar joints and an internal tube which permits recovered solvent to collect prior to removal through the tap.

1927, **52**, 402; B.S. 1741: 1951; B.S. 1742 : 1951) and it is suitable for materials which contain sugar, *e.g.*, sweetened condensed milk.

For very accurate work, Muers and House (*Analyst*, 1949, **74**, 85) have stressed the advisability of using peroxide-free ether in the acid and alkali extraction methods.

A simple method for removing the solvent, which can be applied to any of the gravimetric methods mentioned above, is shown in Fig. 4.

Volumetric methods, which involve treating the material with reagents such as sulphuric acid and centrifuging out the fat in specially calibrated tubes, are very convenient for routine work. In Britain, the **Gerber method** (p. 341) is commonly used for the routine determination of fat in milk and dairy products (B.S. 696 : 1955; Parts 1 & 2). Provided the process is carefully performed, the Gerber method gives results which are in good agreement with those obtained using the more lengthy gravimetric procedures (see Crocker *et al.*, *J. Dairy Res.*, 1955, **22**, 336). Talbot (*Analyst*, 1949, **74**, 462) has applied the Gerber process for the rapid determination of fat in sausages.

Further volumetric processes have been proposed for determining fat in milk in which the sulphuric acid is replaced by less corrosive neutral and alkaline reagents

Apart from the classical methods included above, a few techniques have also been described for determining fat which are based on the measurement of physical properties, *e.g.*, refractive index (*see* Leach's " Food Inspection and Analysis ") and dielectric properties (*see J. Amer. Oil Chem. Soc.*, 1952, **29**, 258; 1956, **33**, 314). A micro method for determining fat in skim milk has been proposed by Heinemann and Rohr (*J. Dairy Sci.*, 1950, 33, 703) and Berquist and Wells (*Food Tech.*, 1956, **10**, 48) have adapted this method for the determination of fat in egg white.

CRUDE FIBRE

" Crude fibre " is the organic residue which remains after the material has been treated under standardised conditions with petroleum spirit, boiling dilute sulphuric acid, boiling dilute sodium hydroxide solution, dilute hydrochloric acid, alcohol and ether.

The " crude fibre " consists largely of cellulose together with a little lignin. As the recovery of cellulose using the specified procedure seldom exceeds four-fifths of that actually present, the " crude fibre " does not represent a measure of a specific group of substances. Also as the figure obtained tends to vary with the conditions employed (see Bidwell and Bopst, *J. Off. Agric. Chem.*, 1921, **5**, 58), it is important to adopt a standardised procedure in order to obtain consistent results. The method usually employed for foods is the procedure laid down in The Fertilisers and Feeding Stuffs Regulations but a modified method for National Flour has been described by the S.P.A.—see p. 160. The method laid down for feeding stuffs in the 1960 regulations is given below:—

PROCEDURE

Weigh to the nearest mg. about 2·7 to 3·0 g. of the sample (1), transfer to an extraction apparatus and extract with petroleum spirit (2). Alternatively, extract with petroleum spirit by stirring, settling and decanting three times. Air dry the extracted sample and transfer to a dry 1000 ml. conical flask. Add 200 ml. of 0·255 N sulphuric acid measured at ordinary temperature and brought to boiling point (3), the first 30 or 40 ml. being used to disperse the sample, and heat to boiling point within one minute. An appropriate amount of anti-foaming agent (4) may be added if necessary. Boil gently for exactly 30 minutes, maintaining a constant volume (5) and rotating the flask every few minutes in order to mix the contents and remove particles from the sides. Meantime prepare a Buchner funnel (6) fitted with a perforated plate by adjusting a piece of cut cotton cloth (7) or filter paper to cover the holes in the plate so as to serve as a support for a circular piece of suitable filter paper (8). Pour boiling water into the funnel, allow to remain until the funnel is hot and then drain by applying suction. Care should be taken to ensure that the filter paper used is of such quality that it does not release any paper fibre during this and subsequent washings. At the end of the 30 minutes boiling period, allow the acid mixture to stand for one minute and then pour immediately into a shallow layer of hot water under gentle suction in the prepared funnel. Adjust the suction so that the filtration of the bulk of the 200 ml. is completed within ten minutes. Repeat the determination if this time is exceeded. Wash the insoluble matter with boiling water until the washings are free from acid; then wash back into the original flask by means of a wash bottle containing 200 ml. of 0·313 N sodium hydroxide solution measured at ordinary temperature and brought to boiling point. Boil for 30 minutes with the same precautions as those used in the earlier boiling and treatment. Allow to stand for one minute and then filter immediately through a suitable filter paper. Transfer the whole of the insoluble material to the filter paper by means of boiling water, wash first with boiling water, then with 1 per cent. hydrochloric acid, and finally with boiling water until free from acid. Then wash twice with alcohol and three times with ether. Transfer the insoluble matter to a dried weighed ashless filter paper (9) and dry at 100° C. to a constant weight. Incinerate the paper and contents to an ash at a dull red heat (9). Subtract the weight of the ash from the increase of weight on the paper due to the insoluble material, and report the difference as fibre.

NOTES

(1) The fineness of the particles has an important bearing on the result. The regulations require the sample to pass through a sieve having apertures about 1 mm. square. Hall *et al* (*Analyst*, 1949, **74**, 445) use a B.S.S. No. 16 sieve.
(2) Treatment with petroleum spirit can be omitted if the amount of fat present is small, *e.g.*, with flour.
(3) cf. S.P.A. method for National Flour (p. 160).
(4) Frothing can also be reduced by using a "cold finger" (*see* Fill and Stock, *Analyst*, 1944, **69**, 121).
(5) The flask should be fitted with a Liebig reflux condenser.

(6) Special Buchner funnels are obtainable which are divided into two (or three) parts, so that the residue can be washed off the paper more easily with alkali (*see* Hartley, *Analyst*, 1952, **77**, 53).
(7) The A.O.A.C. prescribes the use of linen having 45 threads per inch.
(8) For the method in general, Whatman No. 54 or 531 filter papers are usually adequate.
(9) Instead of preparing a weighed filter paper, the S.P.A. method for National Flour (p. 160) suggests the transference of the residue to a dish (previously ignited and weighed) with boiling water. After evaporating off the water on the water bath, the dish is dried in the 100° C. oven, weighed, ignited and re-weighed. Alternatively a Gooch crucible may be employed.

The A.O.A.C. method for determining crude fibre is basically the same as the British method for feeding stuffs. As the official procedures are somewhat lengthy, various workers have suggested more rapid methods of determination, *e.g.*, Whitehouse *et al.*, *J. Off. Agric. Chem.*, 1945, **28**, 147; Entwhistle and Hunter, *ibid.*, 1949, **32**, 651; Weinstock and Benham, *Cereal Chem.*, 1951, **28**, 490. Such methods appear to give consistent results when considered on a comparative basis.

In general, the outer protective coatings of many foods contain considerably more fibre than the softer, more edible inner tissues. The fibre figure can be employed therefore for assessing the proportion of shell which is present in foods such as cocoa (p. **237**) and pepper (p. 253).

Similarly the fibre figure is of value in calculating the amount of the shells of nuts, the stones of fruits and sawdust which may be present in foods, as all these adulterants have a high fibre content. It has also been suggested that the fibre figure may be of value in assessing the maturity of legumes (fibre increases with maturation) and the grade of flour (the outer portions of wheat containing more fibre than the inner tissues). With feeding stuffs, the digestibility varies inversely with the fibre content.

CALCIUM

Determination of calcium by precipitation as oxalate

Calcium can be determined by precipitation as the oxalate. The precipitate is dissolved in acid and the solution titrated with permanganate:

Ash a suitable quantity of the sample (*e.g.*, 10 g.). Warm the ash with 2 ml. of 5 N hydrochloric acid, dilute with a little water and filter, if necessary, into a 150 ml. beaker. Wash through with water so that the total volume of filtrate is about 45 ml. Heat the liquid in the beaker nearly to boiling and stir in a slight excess of hot saturated ammonium oxalate solution. To the hot solution (about 80° C.) add dilute ammonia until it is neutral or faintly alkaline to methyl red and allow the liquid to stand (preferably overnight). Filter through a Whatman No. 540 filter paper and wash the precipitate with small quantities of cold water until the washings are free of oxalate.* Then pierce a hole in the filter paper and wash the calcium oxalate precipitate into a conical flask with 40 ml. of hot water followed by sufficient dilute sulphuric acid to produce an acidity of 2 N. Titrate with N/50 potassium permanganate maintaining the liquid at 60° C. throughout the titration (use a thermometer as stirring rod). Carry out a blank at the same time (*cf.* B.S. 2472 : 1954).

1 ml. N/50 potassium permanganate ≡ 0·000561 g. CaO. Alternatively the calcium oxalate precipitate can be ignited, sulphated and weighed (CaSO4 × 0·4119 = CaO).

Phosphates may interfere, but with many foodstuffs a satisfactory separation may be achieved by making the solution faintly acid with acetic acid prior to the addition of ammonium oxalate.

Determination of calcium by titration with versenate solution

Following the work of Schwarzenbach *et al.* (*Helv. Chim. Acta.*, 1946, **29**, 811) many workers have described methods in which a solution of the disodium salt of ethylenediamine tetra-acetic acid (EDTA) is employed for the direct titration of calcium (and certain other metals). Chelate compounds are formed with the EDTA salt and the metal is sequestered completely. By using murexide (ammonium purpurate) as indicator calcium in solution can be titrated directly at pH 12. Other divalent ions interfere with the titration, but removal of them is normally unnecessary with foods as the amounts present are usually extremely small. Removal of phosphate is, however, often necessary (see Moss, *J. Sci. Food Agric.*, 1961, **12**, 30).

Standard calcium solution. Dissolve 1·2486 g. calcium carbonate in the minimum volume of 5 N hydrochloric acid and dilute to 500 ml. with water (1 ml. ≡ 1 mg. Ca).

Versenate solution. Dissolve 4·0 g. crystalline disodium ethylenediamine tetra-acetate in 800 ml. water. Add 0·86 g. sodium hydroxide, dissolve and dilute to one litre. Standardise against the standard calcium solution.

METHOD

Ash a suitable quantity of the sample with the assistance of nitric acid, water and perchloric acid. Dissolve the ash in dilute nitric acid (slight excess), transfer to a 50 ml. graduated flask and make up to the mark with water. Pipette a volume of the solution containing 3–12 mg. of calcium into a white basin, dilute to 50 ml., neutralise with N/1 sodium hydroxide and add 5 ml. in excess to produce a pH of 12. Add 0·2 g. good quality murexide and titrate with the versenate solution, stirring with a glass rod. The colour change at the end-point is from red to violet red (*cf.* Banks, *Analyst*, 1952, **77**, 484; Cheng and Bray, *Soil Sci.*, 1951, **72**, 449; van Schouwenburg *Analyt. Chem.*, 1960, **32**, 709; Stephens, *J. Pharm. Pharmacol.*, 1953, **5**, 709; Moss, *loc. cit.*).

If the solution is titrated at pH 10 after making alkaline with ammonia (an ammonia-ammonium chloride buffer is preferable) and

* It is often convenient to test for the absence of chloride (provided hydrochloric acid has been used for preparing the original solution).

Eriochrome Black T is used as indicator any magnesium present is titrated in addition to the calcium (*cf.* Banks, *loc. cit.*; Stephens, *loc. cit.*; B.S. 2690 : 1956). By carrying out both titrations the amount of magnesium present can be obtained by difference.

Other methods for determining calcium

Calcium can also be determined colorimetrically (see *Analyst,* 1953 **78**, 106 and 528; *Analyt. Chem.*, 1948, **20**, 76), polarographically (*Nature, Lond.*, 1953, **172**, 257) and by flame photometry.

PHOSPHATE

Determination of phosphate by the phosphomolybdate volumetric method

When a solution of an orthophosphate is treated with excess molybdate in the presence of nitric acid a yellow precipitate of ammonium phosphomolybdate is formed. In the volumetric method for the determination of phosphorus, the precipitate (after washing) is dissolved in standard alkali and the excess titrated with standard acid:

Acidify a neutral solution of an orthophosphate containing 0·02–0·04 g. P_2O_5(*) by adding 5 ml. of nitric acid (conc.) per 100 ml. (the total volume should not exceed 180 ml.). Heat to 68° C., add 60 ml. of molybdic acid solution (see below) and maintain at 68° C. for 20 minutes. Cool and filter the precipitate on a Gooch crucible. Wash the precipitate with dilute nitric acid (1 per cent.) and a little water. Transfer the crucible containing the precipitate and asbestos to a 250 ml. beaker, add water and phenolphthalein and run in from a burette a slight excess of N/2 sodium hydroxide solution* so that the precipitate is completely dissolved. Then titrate the excess alkali with N/2 acid. If the precipitate is dissolved *in the cold* one atom of phosphorus is equivalent to 23 molecules of NaOH:

$$(NH_4)_3[PO_4.12MoO_3] + 23NaOH \rightarrow 11Na_2MoO_4 + (NH_4)_2MoO_4 + Na(NH_4)HPO_4 + 11H_2O$$

$$1 \text{ ml. N/2} \equiv 0{\cdot}001544 \text{ g. } P_2O_5$$

If the solution is boiled, the ammonia is eliminated from the phosphomolybdate and therefore three more molecules (of NaOH) are required to complete the reaction:

$$(NH_4)_3[PO_4.12MoO_3] + 26NaOH \rightarrow 12Na_2MoO_4 + Na_2HPO_4 + 3NH_3 + 14H_2O$$

$$1 \text{ ml. N/2} \equiv 0{\cdot}001365.\ P_2O_5$$

The *molybdic acid solution* is prepared as follows:

To a litre graduated flask add 125 g. molybdic acid and 100 ml. water. Add, whilst shaking 300 ml. of 8 per cent. ammonia solution (1 vol. 0·88 + 3 vols. water) to dissolve the molybdic acid. Add 400 g. ammonium nitrate, make up to a litre and add the whole solution to a litre of nitric acid (s.g. 1·19). Maintain at about 35° C. for a day and filter.

* Smaller quantities can also be estimated using N/10 solutions, but it is then more difficult to dissolve the precipitate.

The Fertilisers and Feeding Stuffs Regulations (1955) prescribed that the phosphomolybdate precipitate should be dissolved in cold dilute ammonia (2 per cent.), the solution brought to the boiling point, the beaker removed from the burner and excess of magnesia mixture slowly added. After allowing to stand the crystalline precipitate is filtered off, washed with dilute ammonia (2 per cent.), dried, ignited and weighed as magnesium pyrophosphate ($P_2O_5 = Mg_2P_2O_7 \times 0{\cdot}638$).

The *magnesia mixture* is prepared by dissolving **110** g. crystallised magnesium chloride and 140 g. ammonium chloride in 1300 ml. of water and then mixing with 700 ml. of ammonia (8 per cent.) and allowing to stand for at least three days before filtering.

Wilson (*Analyst*, 1951, **76**, 65) and Donald *et al.* (*J. Sci. Food Agric.*, 1956, **7**, 677) have critically reviewed the various methods for determining phosphate and the former paper describes a volumetric method in which the phosphate is precipitated as quinoline phosphomolybdate. This is now the prescribed method in The Fertilisers and Feeding Stuffs Regulations (1960).

Molybdenum-blue colorimetric methods for the determination of phosphate

Numerous colorimetric methods employing reagents containing molybdate have been described for the determination of phosphate. Many of such methods involve the addition of ammonium molybdate to a solution of the phosphate and the phosphomolybdate produced is partially reduced to a blue compound which probably has the formula $(MoO_2.4MoO_3)_2.H_3PO_4$. Numerous reducing agents have been suggested, *e.g.*, stannous chloride, phenylhydrazine, hydroquinone and p-methylaminophenol. The latter reagent is used in the following method, which is a modification of that proposed by Tschopp and Tschopp (*Helv. Chim. Acta.*, 1932, **15**, 793). The phosphate should be present as the orthophosphate and the solution should be neutralised before the reagents are added:—

Reagent No. 1. Add 75 ml. sulphuric acid (conc.) slowly to 50 ml. water. Cool thoroughly and slowly add 50 ml. ammonium molybdate solution (10 per cent. w/v).

Reagent No. 2. Dissolve 20 g. sodium metabisulphite, 0·5 g. sodium sulphite ($Na_2SO_3.7H_2O$) and 0·1 g. p-methylaminophenol sulphate (" metol ") in freshly boiled and cooled water and dilute to 100 ml.

Method

Weigh out a suitable quantity of the sample into a platinum basin, make alkaline with sodium carbonate solution, evaporate to dryness and ash just below red heat. Dissolve the ash by warming with 1 ml. of hydrochloric acid (conc.) and wash into a boiling tube with about 20 ml. of water. Neutralise to litmus, dilute with water to produce a total volume of 25 ml., add 5 ml. of Reagent No. 1 and immerse the tube in boiling water for 15 minutes. Then add 5 ml. of Reagent No. 2 and continue the heating for a further 15 minutes. Cool, dilute to 50 ml. and compare the colour produced with standards (over the range 0–0·05 mg. P_2O_5) prepared in a similar manner. Prepare these standards from a suitable dilution of a stock solution containing 1·917 g. potassium dihydrogen phosphate per litre (1 ml. ≡ 1 mg. P_2O_5).

The colours can be conveniently compared in 50 ml. Nessler glasses for which a B.D.H. Nessleriser disc is also available. Absorptiometric measurements should be made at about 660 mμ. A blank should be performed.

Determination of phosphate by the vanado-molybdate colorimetric method

Hanson (*J. Sci. Food Agric.*, 1950, **1**, 172) has proposed a simple colorimetric method for the determination of phosphate which is based on Misson's reaction. The acid solution containing orthophosphate is treated with an acid reagent containing molybdic acid and vanadic acid and a stable orange yellow coloured complex of vanadimolybdiphosphoric acid (H_3PO_4, VO_3, $11MoO_3$, nH_2O) is formed. The maximum absorption is at 330 mμ but Donald *et al.* (*J. Sci. Food Agric.*, 1956, **7**, 685) have reported that satisfactory results can be obtained in the region 420–480 mμ, provided the light is monochromatic. The colour development is not markedly affected by the presence of hydrochloric, sulphuric, acetic or citric acid or by fluorides provided they are not present in relatively large quantities. A similar method to that described below is prescribed in The Fertilisers and Feeding Stuffs Regulations, 1960.

Special Reagents

Vanadate—molybdate composite reagents. Dissolve 20 g. ammonium molybdate in 400 ml. warm water (50° C.) and cool. Dissolve 1·0 g. ammonium vanadate in 300 ml. boiling distilled water, cool and add 140 ml. nitric acid (conc.) gradually with stirring. Then add the molybdate solution gradually to the acid vanadate solution with stirring and dilute to a litre with water.

Standard phosphate solution. Prepare a stock solution containing 3·834 g. potassium dihydrogen phosphate (KH_2PO_4) per litre. Dilute 25 ml. to 250 ml. (1 ml. ≡ 0·2 mg. P_2O_5).

Preparation of Standard Graph

To a series of 100 ml. graduated flasks add 0, 2·5, 5, 10, 20, 30, 40 and 50 ml. of the standard phosphate solution (= 0–10 mg. P_2O_5) and dilute each to 50–60 ml. with water. Add a few drops of ammonia solution (0·88) and make just acid with nitric acid (1 : 2). Add 25 ml. of the vanadate-molybdate reagent, dilute to the mark and mix. Allow to stand for 10 minutes and measure the optical density in a 1 cm. (or 0·25 cm.) cell at 470 mμ (Ilford blue filter No. 602).

METHOD

Transfer a suitable volume of solution of the sample (containing 0·5–10 mg. P_2O_5) to a 100 ml. graduated flask. If the determination is carried out on the ash, boil the ash with 10 ml. of 5 N hydrochloric acid and wash the solution into the 100 ml. flask with water, filtering if necessary. Neutralise by the dropwise addition of 0·88 ammonia (the volume of solution at this stage should be 50–60 ml.) and then proceed as for the standard graph, *i.e.*, make just acid with dilute nitric acid, add 25 ml. of vanadate-molybdate reagent, dilute to the mark and measure the optical density after allowing to stand for 10 minutes (cf. Donald *et al, J. Sci. Food, Agric.*, 1956, **7**, 690).

Other colorimetric methods for determining phosphate have been described by Denigès (*Compt. rend.*, 1920, **171**, 802), Briggs (*J. Biol. Chem.*, 1922, **53**, 13; 1924, **59**, 255), Holman and Pollard (*J. Soc. Chem. Ind., Lond.*, 1937, **56**, 339T) and Zinzadze (*Ind. Eng. Chem., Anal. Ed.*, 1935, **7**, 227).

GENERAL REFERENCE

Craven and Schwehr ("The Determination of Phosphorus," Fertiliser Socy. and S.A.C. Symposium on Fertiliser Analysis, 1960).

THE MICRO-ANALYTICAL TEST FOR PURITY OF FOODSTUFFS ("FILTH TEST")

The F.D.A. of the U.S. have paid considerable attention to the presence of extraneous matter in foods. The "Filth Test" is mainly concerned with the detection and counting of rodent hairs and fragments of insects. Rodent hairs are invariably present in rodent fæces and it is inferred that if hairs are found in flour, say, the original grain was contaminated with rodent pellets which were later ground up with the flour on the rolls.

With foods which are soluble in water, *e.g.*, sugar products and table jelly, the method merely involves dissolving up the material, filtering and examining any matter remaining in the paper. With cereals, however, the starch and protein should preferably be digested and extraneous matter can be separated by shaking with petrol. The petrol preferentially wets the hairs, etc., which float up and congregate in the interfacial layers.

Kent-Jones *et al.* (*Analyst*, 1948, **73**, 128) have described the following method for the determination of "filth", which is particularly suitable for cereal foods (*cf.* A.O.A.C.):—

Add 100 ml. of *boiling* N/2 hydrochloric acid rapidly to 25 g. of the sample contained in a 400 ml. beaker (N.B. Foods which are relatively rich in fat should be de-fatted first). Stir thoroughly into a paste and heat on a low flame. Then boil for 10 minutes, adding boiling water if necessary if the paste seems unduly thick. Cool, add 100 ml. water and stir in 9·5 ml. of 5 N sodium hydroxide solution. Add 2·5 ml. of saturated trisodium phosphate solution, adjust to pH 7·0 using aqueous phenol red as external indicator. Add 1 g. of pancreatin (suspended in water), dilute to 350 ml., mix and incubate at 37° C. overnight. Transfer the digest to a 600 ml. tapless separator fitted with a wide rubber tube and screw clip. Add 25 ml. of petrol and shake gently. Allow to separate for 30 minutes and then run most of the lower layer back into the original beaker (leave enough to form a small layer under the interface). Add 400 ml. of water to the liquids remaining in the separator. Swirl round and allow to separate. Run off most of the lower layer. Wash the petrol layer again with more water. Withdraw almost all the water and then filter the liquids remaining in the separator on a Buchner funnel through a ruled filter paper (which should extend up the sides of the funnel), resting on a thin layer of kieselguhr so that the debris does not congregate over the holes. Return the original digest to the separator and shake it with more petrol, wash with water, etc., as before. Finally wash out the separator with alcohol and then water containing a little Teepol, collecting any extraneous matter on the filter paper throughout. Examine the debris on the paper, microscopically or by other means (see original paper) counting the fragments in each ruled square in turn. Report the result as the number of rodent hairs, insect fragments, etc., per 100 g. of samples. Rodent hairs are characterised by the striated medulla, which give them a striped appearance under the microscope. In the U.S.A. exception would appear to be taken to imported foods with a rodent hair count in excess of 5 per pound. Vegetable hairs should not be counted.

Particles of bran tend to collect in the interfacial layer and difficulties therefore often arise with the coarser cereals such as oatmeal and wheatmeal.

Bradshaw and Coppock (*Analyst*, 1949, **74**, 300) have described a modified " filth test " for the examination of dried fruits.

For further information the reader should refer to The U.S. Dept. of Health, Education & Welfare (F.D.A.) Food and Drug Tech. Bull. No. 1 entitled " Microscopic-Analytical Methods in Food and Drug Control," and also the A.O.A.C. (1960, p. 606).

Chapter 3

GENERAL METHODS

Trace Elements—Preservatives—Antioxidants—Colouring Matters

TRACE ELEMENTS

THE term " trace element " refers to the inorganic elements (mostly metals) which may be present in foods in amounts usually well below 50 p.p.m. and have some toxicological or nutritional significance. Calvery (*Food Res.*, 1942, **7**, 313) broadly classified trace elements according to their effect on life and placed them into three classes (1) *The essential nutritive elements, e.g.*, Co, Cu, Fe, I, Mn, Zn, (2) *The non-nutritive non-toxic elements, e.g.*, Al, B, Cr, Ni, Sn, which are not known to have produced harmful effects when present in quantities not exceeding 100 p.p.m. and (3) *The non-nutritive toxic elements, e.g.*, As, Sb, Cd, F, Pb, Hg, Se, which are known to have deleterious effects even when the diet contains less than 100 p.p.m. A complicating factor, however, is that elements such as copper and zinc, although essential for life processes when present in traces, have an emetic action when ingested in higher amounts. Cumulative poisoning due to ingestion of food containing elements such as lead or arsenic over a long period are extremely rare. In addition to the action on the body, however, a few elements tend to have a detrimental effect on the quality or nutritive value of the food, *e.g.*, copper causes off-flavours in milk and dairy products and tends to destroy vitamin C in fruit products. The food analyst is mainly concerned with those elements for which limits have been officially recommended or prescribed, viz. the metals arsenic, copper, lead, tin and zinc and the non-metal fluorine. The whole field has been reviewed by Monier-Williams (*Trace Elements in Food*, Chapman & Hall, Lond., 1950).

Sources of trace elements. The presence of the more undesirable trace elements in foods can usually be attributed to one of the following causes:—

(i) Natural occurrence, *e.g.*, in shellfish due to ingestion of estuary water contaminated by industrial effluents, deposition in the liver of animals.

(ii) Sprays and dusts used as insecticides during cultivation, *e.g.*, lead arsenate.

(iii) Use of impure chemicals for the manufacture of " chemical " raw materials, *e.g.*, the use of arsenic-contaminated sulphuric acid for the manufacture of glucose, tartaric acid, phosphates and food colours and the manufacture of acid phosphates from fluorine-contaminated rock phosphate.

(iv) Accidental contamination due to confusion of materials of similar appearance, *e.g.*, the use of white arsenic instead of starch for dusting confectionery.

(v) Foods (especially acid foods and those containing salt or alcohol) may dissolve metals from the equipment, *e.g.*, from tinplate, solders, galvanised iron and cheap enamels and glazes.

With improved processing techniques coupled with increased laboratory control, troubles from trace elements derived from foods are now extremely rare.

Statutory and recommended limits. The Metallic Contamination Sub-Committee was appointed by the Food Standards Committee in 1948 to consider the contamination of food by trace elements. Since that date various official reports have been published by the committee which contain the Committee's *recommended limits.* In addition *statutory limits* for certain trace metals have been *prescribed* in the Food Standards Orders relating to edible gelatine, tomato sauce and curry powder and limits for fluorine are *prescribed* for baking ingredients containing acid phosphates in The Fluorine in Food Order. Only in the case of the cumulative poisons arsenic and lead do the Committee recommend incorporation of their limits into further statutory instruments. All the limits quoted in the subsequent section are expressed as p.p.m. of the food on a w/w basis.

The Preparation of Foods for the Estimation of Trace Elements

" Direct " method. With some soluble food materials (*e.g.*, salt, sodium bicarbonate, cream of tartar, citric acid) it is possible to carry out the estimation of certain trace metals on the solution obtained by dissolving the material in water or cold acid. This is the common procedure in the British Pharmacopoeia preparatory to the determination of arsenic and lead in many materials.

Dry ashing. Dry ashing at 450–600° C. (according to the particular element) is a convenient method for the destruction of organic matter. Precautions must be taken, however, against low results which may be due to (a) volatilisation of the element, (b) combination or absorption of the element with ash constituents or the vessel and (c) incomplete extraction of the ash. Such difficulties can usually be avoided by using an accurately controlled muffle furnace, by adding an ash-aid (*e.g.*, nitrates or oxides of magnesium or calcium) to the food before ashing and by using an " approved " extraction procedure (*cf.* Middleton and Stuckey, *Analyst*, 1953, **78**, 532).

In dry ashing silica dishes are preferable. After weighing out, heat the dish gently in the fume cupboard until smoke ceases to come off in quantity and then transfer to the muffle furnace. With liquids most of the water must first be removed on a hot plate or water-bath.

Wet oxidation (conventional method). Many workers prefer to destroy the organic matter by the wet oxidation method as it is in general more reliable. The necessity for constant vigilance and the possibility of high (and sometimes uncertain) blanks render the method less suitable for routine work, however. The following procedure is suitable for most determinations:—

Into a Kjeldahl digestion flask place a suitable quantity of the sample (usually 5 or 10 g.), 20 ml. concentrated nitric acid and up to 20 ml. water (depending on the water content of the sample). Boil so that the volume is reduced to about 20 ml. (about ten minutes), cool and add 10 ml. of concentrated sulphuric acid. Boil again and add further small quantities of nitric acid immediately the liquid begins to blacken (delay may cause loss of the element, especially of arsenic). When the addition of nitric acid is no longer necessary (*i.e.*, the liquid no longer blackens), continue the heating until white fumes are well in evidence. Cool and add 10 ml. of saturated oxalate solution and boil again until copious white fumes are again produced. The ammonium oxalate treatment assists in removing yellow colorations due to nitro compounds, fats, etc., so that the final solution is colourless. The blank should be prepared by " boiling out " the same volume of nitric acid, together with sulphuric acid as used for the sample.

An alternative procedure is the addition of both the nitric and sulphuric acids at the beginning. The use of sulphuric acid alone prolongs the oxidation unduly, but the time can be shortened considerably if a substance which raises the boiling point (*e.g.*, potassium sulphate) is added (*see* Williams, *Analyst*, 1941, **66**, 228).

Wet oxidation (Middleton and Stuckey's "cold" method). Middleton and Stuckey (*Analyst,* 1954, **79**, 138) have described a modified wet oxidation procedure which is especially suitable for animal and fatty materials. The main advantages over the conventional method are that by oxidising the material at $\leqslant$ 350° C. the nitric acid is used more economically, there is little likelihood of loss of the element due to volatilisation and the process does not require such constant vigilance:—

Into a litre beaker place 5–10 g. of sample, up to 20 ml. of water (depending on the water content of the sample) and 5–10 ml. of concentrated nitric acid containing 5 per cent. of sulphuric acid. Evaporate the mixture on an electrical hotplate (or an iron sheet heated over a gas ring) maintained at 310–350° C. The temperature can be checked by placing substances of selected melting points on the plate, *e.g.*, sodium nitrate 307° C., potassium nitrate 334° C. Leave the uncovered beaker on the hotplate until there is no further visible change. Allow the beaker to cool, moisten the residue with concentrated nitric acid, cover with a clock-glass and evaporate the mixture again on the hotplate. Continue the heating for fifteen minutes after the residue has gone dry. Repeat this nitric acid treatment until the residue is whitish with dark patches. Then continue the treatment using fuming nitric acid until a white residue is produced. Usually the residue can be brought into solution by heating with dilute sulphuric acid. Many alternative procedures are suggested by the authors including (a) the omission of the sulphuric acid in the first nitric acid treatment, (b) the earlier use of fuming nitric acid if the weaker acid seems ineffective and (c) the digestion with sodium hydroxide before acidifying in order to bring the final residue into solution. For further details reference should be made to the original paper.

Garratt and Elvidge (*Analyst,* 1954, **79**, 146) have discussed the possibility of destroying organic matter using oxygen gas in a bomb colorimeter. "Methods for the Destruction of Organic Matter" have been the subject of a report from the A.M.C.'s Metallic Impurities in Organic Matter Sub-Committee (*Analyst,* 1960, **85**, 643).

General Rules for Trace Element Determinations

1. It is usually quite sufficient to weigh out the material for trace element determinations on a good rough balance (see *Lab. Practice,* 1959, **8**, 94–95).

2. Use a muffle furnace with accurate temperature control for dry ashing.

3. Use clean glassware of good quality. Separators, Nessler glasses, etc. should be rinsed out with distilled water before use. Dust contains traces of lead and tap water would introduce elements such as iron.

4. Use the purest possible reagents, *e.g.*, AsT and PbT, and carry out blanks frequently. If wet oxidation is employed, the blank should include exactly the same volumes of the oxidising agents as are used for the sample (*e.g.*, nitric acid).

Trace Elements—General Reference Books

"The B.D.H. Book of Organic Reagents" (B.D.H. Ltd. Poole).
"Organic Reagents for Metals" (Hopkin & Williams, Ltd., Chadwell Heath).
Monier-Williams' "Trace Elements in Food" (Chapman & Hall, London).

Sandell's "Colorimetric Determination of Traces of Metals" (Interscience, N.Y.).
Snell's "Colorimetric Methods of Analysis," Vol. IIA (Van Nostrand).

ARSENIC

The Arsenic in Food Regulations (*S.I.* 1959 *No.* 831; 1960 *No.* 2261)

3.—(1) Save as hereinafter provided, no food shall contain arsenic† in proportions exceeding *one part per million* (*estimated by weight*) of such food:

Provided that any specified food may contain arsenic in proportions not exceeding those specified in relation thereto in the Schedule to these regulations.

(2) Paragraph (1) of this regulation shall not apply—

(a) in the case of any food, other than fish, edible seaweed, a specified food or a product containing fish or edible seaweed, where arsenic in proportions exceeding one part per million (estimated by weight) is present by reason of the inclusion in such food as an added ingredient or ingredients of a substantial quantity* of a specified food or foods and such excess of arsenic does not exceed the maximum quantity permitted under the proviso to paragraph (1) of this regulation having regard to the quantity and description of the added specified food or foods;

(b) in the case of fish, edible seaweed or any product containing fish or edible seaweed, where arsenic in proportions exceeding one part per million (estimated by weight) is naturally present in that fish or edible seaweed or, as the case may be, in that product by reason of its fish or edible seaweed content;

(c) to hops or hops concentrates intended for use for commercial brewing.

(3) No person shall sell, consign or deliver, or import into England or Wales, any food which does not comply with this regulation.

THE SCHEDULE

Description of food	Proportions of arsenic† (expressed in terms of parts per million estimated by weight)
1. Non-alcoholic beverages, ready to drink, not otherwise specified	0·1
2. Alcoholic beverages not otherwise specified	0·2
3. Black beer and mixtures of black beer and rum	0·5
4. Soft drinks intended for consumption after dilution and undiluted fruit juices (including concentrates intended for consumption after dilution to an extent not greater than ten parts to one, calculated by volume, for use by a manufacturer of soft drinks)	0·5
5. Ice-cream (including frozen confections and any other similar commodity)	0·5
6. Onions (dehydrated)	2·0
7. Hops (dried) other than those intended for use for commercial brewing	2·0
8. Liquorice (dried extract)	2·0
9. Gelatine (edible)	2·0

* "substantial quantity" in relation to an ingredient of an article of food means a quantity weighing not less than 25 per centum of the total weight of such article.

† As elemental arsenic (As).

Description of food	Proportions of arsenic† (expressed in terms of parts per million estimated by weight)
10. Pectin (liquid)	2·0
11. Yeast and yeast products	2·0 (calculated on the dry matter)
12. Chemicals not otherwise specified but not including synthetic colourings	2·0
13. Chicory (dried and roasted)	4·0
14. Herbs (dried)	5·0
15. Finings and clearing agents	5·0
16. Hops concentrates other than those intended for use for commercial brewing	5·0
17. Pectin (solid)	5·0
18. Spices (including mustard)	5·0
19. Colourings other than synthetic colourings . .	5·0 (calculated on the dry colouring matter)
20. Chemicals for which an arsenic limit is specified in the British Pharmacopoeia or the British Pharmaceutical Codex but not including synthetic colourings	2·0 or the limit specified in the edition of the British Pharmacopoeia or the British Pharmaceutical Codex current for the time being, whichever is the higher
21. Brewers' yeast intended for use by manufacturers in the manufacture of yeast products . . .	5·0 (calculated on the dry matter)

† As elemental arsenic (As).

Formerly much of the arsenic contamination of food resulted from the use of impure sulphuric acid in the manufacture of food ingredients such as sugars (especially glucose), citric, tartaric and phosphoric acids and their salts, yeast, etc. The severe outbreaks of poisoning which occurred in N.W. England in 1900 were attributed to beer which had been manufactured from arsenic contaminated glucose. About that time also excessive amounts of arsenic in beer were attributed to the coke fumes produced from the fuel used for drying the malt. Nowadays arsenic is most likely to be found in food derived from plants due to the use of insecticidal sprays and dusts such as lead arsenate. Also fish (particularly crustacea) which have lived in estuary water contaminated with industrial effluents are liable to contain arsenic in considerable quantity. Fortunately such arsenic is present in organic combination and as it does not appear to be absorbed by humans does not give rise to poisoning. Analytically, the organically combined arsenic will only react with nascent hydrogen (zinc + acid) to form arsine if the material has been previously submitted to wet oxidation. Conversely any arsine produced by the direct action of zinc and acid on the food can be attributed to inorganic arsenic which is toxic. For further information on possible sources of arsenic in foods the reader should consult the Food Standards Committee Report on Arsenic, **1955** (H.M.S.O.).

Destruction of Organic Matter

Destruction of the organic matter preparatory to the determination of arsenic is best carried out by wet oxidation using nitric and sulphuric acids. To avoid loss of the element by volatilisation during the digestion the material should be heated with nitric acid alone to dissolve as much as possible before the addition of sulphuric acid. For the same reason it is especially important to add the additional amounts of nitric acid *immediately* charring commences. Additional treatment with ammonium oxalate to remove any residual nitrogen compounds is also advisable as their presence may retard the subsequent reduction. Alternatively the Kjeldahl-type digestion of Williams (*Analyst*, 1941, **66**, 228) is suitable provided the amount of chloride present is low. Dry ashing using an ash-aid has also been proposed. Allcroft and Green (*Biochem. J.*, 1935, **29**, 824) have pointed out, however, that it is necessary to use an abundant excess of ash-aid (magnesium nitrate). With certain chemicals which dissolve in water or acid and liquid foods (*e.g.*, beer, vinegar), the food or its solution can be transferred directly to the Gutzeit bottle (see also B.P. 1958) and this is convenient in routine work.

Determination of Arsenic

Reinsch's test in which the material is boiled with hydrochloric acid and copper foil represents a rapid method for the detection of heavy traces of arsenic. The sensitivity of the test has been increased by Griffiths (*Analyst*, 1941, **66**, 491), who compared the stain produced with those obtained from known amounts of arsenic on exactly similar pieces of copper foil.

In the *Marsh-Berzelius method*, any arsenic present reacts with nascent hydrogen to form arsine, which is decomposed by heat. The burner is placed so that the arsenic is deposited as a stain in the narrow part of the tube and its intensity is compared with standard stains similarly produced. For a discussion of the technique the reader should refer to Monier-Williams' " Trace Elements in Food " p. 175 (Chapman & Hall, Lond.).

For the *Gutzeit method* (described below), the arsine forms a yellow to brown stain on a mercuric chloride (or bromide) paper.

Antimony if present interferes in each of the three methods above. The official methods therefore require the arsenic to be separated by distillation as the chloride prior to the Gutzeit stage.

The *molybdenum-blue colorimetric method* can also be employed for estimating arsenic in foods. This sensitive reaction enables very small quantities of the element to be determined, but the method has the disadvantage that the blank tends to be appreciable.

Methods for the Determination of Arsenic employing the Gutzeit technique

Apparatus

The Gutzeit method is the official process in the British Pharmacopoeia (B.P. 1958, p. 850) for the determination of arsenic. The apparatus consists of a wide-mouthed bottle (capacity 120 ml.) fitted with a rubber bung through which passes a glass tube 200 mm. in length and of diameter 8 mm. externally and 6·5 mm. internally. The lower end is drawn out to 1 mm. diameter and a hole about 2 mm. diameter is blown in the side of the tube, near the constriction. The glass tube is lightly packed with cotton wool, previously moistened with 10 per cent. lead acetate solution and dried, so that the cotton wool is not nearer than 25 mm. from the top of the tube. The upper end of the tube is then pushed into the narrow end of a rubber bung (25 × 25 mm. with hole 6·5 mm. diameter) so that the larger end is flush with the end of the tube. A piece of mercuric chloride paper is placed flat on the bung and another similar bung placed over it and secured by clips so the two bungs meet to form a continuous tube. The papers are prepared by soaking smooth filter paper (65–120 g. per sq. m.) in saturated mercuric chloride solution and then by drying at 60° C. in the dark after pressing out the superfluous solution. The papers should be stored in the dark.

General Procedure

For the test, add the prepared solution (diluted to 50 ml.) to the bottle together with 10 ml. of stannated hydrochloric acid (see below), 10 g. of granulated zinc (AsT) and 2 ml. of amyl alcohol to reduce frothing and spraying. The tube must be attached immediately. Allow the reaction to proceed for forty minutes (preferably at about 40° C.) and compare the stain produced with those obtained using known amounts of arsenic ranging from 0·001 to 0·01 mg. (as As). The Gutzeit technique has been reviewed by Crossley (*J. Soc. Chem. Ind., Lond.* 1936, **55**, 272 T; see also *Analyst*, 1960, **85**, 629).

Special Reagents for Gutzeit Methods

Standard arsenic solution. Prepare a stock solution by dissolving 0·132 g. of arsenious oxide in 2 ml. of N/1 sodium hydroxide solution, dilute to about 50 ml. with water, neutralise with 2 ml. of N/1 hydrochloric acid and dilute to 100 ml. Prepare a dilute arsenic solution containing 5 p.p.m. (as As) by diluting 5 ml. of the stock solution to one litre with water.

All other reagents should be arsenic-free, *i.e.*, of AsT grade, if available.

Stannous chloride solution. Dissolve 33 g. of stannous chloride in 10 ml. of concentrated hydrochloric acid and dilute to 100 ml. with water. Add 100 ml. of concentrated hydrochloric acid, boil down to 100 ml. and filter.

Stannated hydrochloric acid. Mix 1 ml. of the stannous chloride solution with 100 ml. of concentrated hydrochloric acid.

Bromine solution. Dissolve 30 g. of bromine and 30 g. of potassium bromide in water and make up to 100 ml.

Brominated hydrochloric acid. Mix 1 ml. of the bromine solution with 100 ml. concentrated hydrochloric acid.

Chloride-hydrazine-bromide reagent. Mix together 100 g. of sodium chloride, 10 g. of hydrazine sulphate and 0·4 g. of potassium bromide.

Many articles of food or chemicals can be treated directly in the Gutzeit bottle by adding to a suitable quantity of the material 50 ml. of water and 10 ml. of brominated hydrochloric acid and warming to dissolve. A few drops of stannous chloride solution are then added to discharge the bromine and reduce the arsenic to the arsenious condition. The 10 g. of zinc is then added and the test proceeded with as usual (*cf.* the B.P. methods for determining arsenic in calcium, carbonate, citric and tartaric acids, lactose, sodium bicarbonate and sodium chloride). It is essential to remove or oxidise all sulphur compounds as sulphides and sulphites vitiate the test by producing yellow stains on the mercuric chloride paper.

As most foods contain organic matter it is usually necessary to carry out a wet oxidation before applying the Gutzeit procedure. It should be borne in mind, however, that if the method is applied directly to an aliquot of the wet oxidised solution, antimony is liable to interfere as it also forms a stain on the mercuric chloride paper. For this reason the S.P.A. (*Analyst*, 1930, **55**, 102) and B.S.I. (B.S. 757 : 1959) procedures separate the arsenic by distillation as the chloride (the chloride of trivalent arsenic being more volatile than those of other elements):

Determination of Arsenic by the S.P.A. and B.S. Method employing the Gutzeit procedure

Wet oxidise a suitable quantity (*e.g.*, 5 g.) of the material in the special 300 ml. Kjeldahl flask (fitted with a B 24 joint) by heating first with 10 ml. of concentrated nitric acid. When the initial vigorous reaction has ceased, cool, add 10 ml. of concentrated sulphuric acid and continue the heating with further additions of nitric acid *immediately* the liquid begins to darken. Heat until white fumes appear and the liquid is colourless. If a pale yellow colour persists heat the liquid further with a few drops of hydrogen peroxide solution (30 per cent. by weight). Cool, add 15 ml. of saturated ammonium oxalate solution and heat again until white fumes appear. Cool, add 7 ml. of water, 5 g. of the chloride-hydrazine-bromide mixture (avoiding contamination of the neck of the flask) and 10 ml. of concentrated hydrochloric acid. Fit the condenser (Fig. 5) on to the flask immediately and distil into an externally cooled mixture of *either* 2 ml. of concentrated nitric acid and 10 ml. of water *or* 10 ml. of saturated bromine water (if bromine water is used, precautions should be taken to ensure that an excess of bromine is present throughout the distillation). The exit end of the condenser must dip below the surface of the liquid. Continue distilling until the liquid in the Kjeldahl flask is reduced to about half or until five minutes after the condenser is full of steam. Evaporate the distillate to dryness in a 100 ml. beaker on the water-bath and then evaporate twice more after the addition of 5 ml. of water (in each case) to

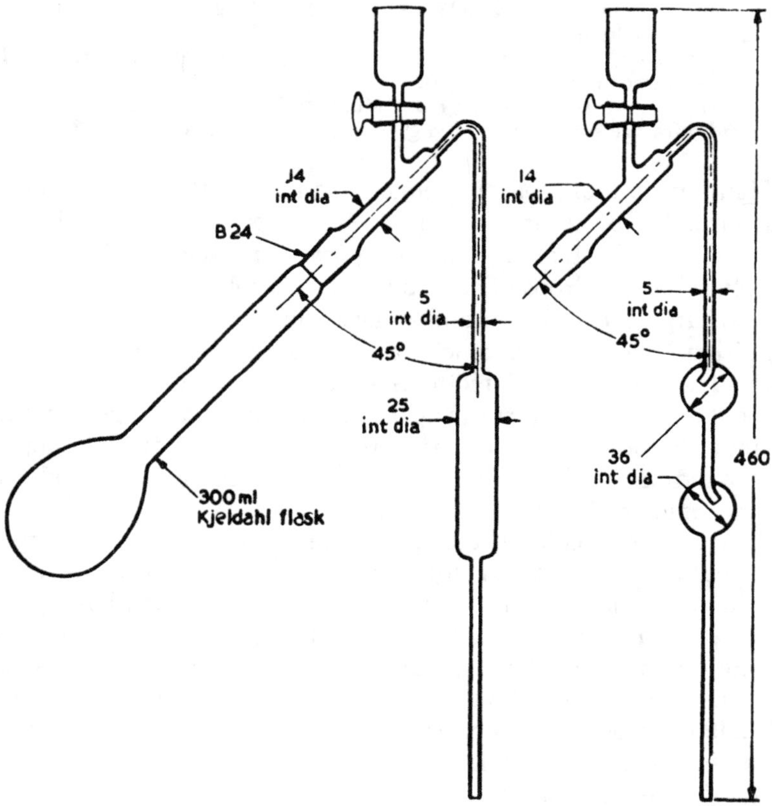

Fig. 5. Distillation Apparatus for the Determination of Arsenic (B.S.757 : 1959*), showing alternative forms of condenser (dimensions in millimetres).

* The British Standard 757 : 1959 " Sampling and Testing Gelatines " can be purchased from the B.S.I., 2 Park Street, London, W.1.

remove nitric acid. Dissolve the residue in 3 ml. of concentrated sulphuric acid and dilute to 50 ml. with water. Add 0·1 g. of sodium sulphite and boil down to 25 ml. to remove the sulphur dioxide. Add 8 ml. of stannated hydrochloric acid and make up to 60 ml. with water. Transfer to the Gutzeit apparatus, add 10 g. of granulated zinc and 2 ml. of amyl alcohol. Connect up the Gutzeit tube fitted at the top with a mercuric chloride paper and allow the reaction to proceed at about 40° C. for forty minutes. Compare the stain obtained with standard stains produced by solutions of known concentration of arsenic (0·001–0·010 mg. as As), which have been submitted to the same process as the sample.

Determination of Arsenic by means of the Molybdenum-Blue Reaction

Eastoe and Eastoe (*J. Sci. Food Agric.*, 1953, **4**, 312) replace the Gutzeit stage in the S.P.A. and B.S. procedure by a modification of

Rodden's molybdenum-blue method in which the colour produced is measured by means of the spectrophotometer (or absorptiometer). The arsenic is separated by distillation of the trichloride into dilute nitric acid. The acid solution is then evaporated to dryness and the residue dissolved in an acid solution of ammonium molybdate containing hydrazine sulphate. The blue colour is then formed by heating at 100° C. for ten minutes. In order to keep the blank as low as possible it is necessary to treat glassware with a solution for "removing" arsenic (boiling hydrochloric acid containing sodium chloride and hydrazine sulphate).

Special Reagents

Standard arsenic solution—see Gutzeit methods above.

Chloride-hydrazine-bromide reagent—see Gutzeit methods above.

Ammonium molybdate solution (1 per cent. *in* 5 *N sulphuric acid*). Add 140 ml. of concentrated sulphuric acid to 600 ml. of water and dissolve 10 g. of ammonium molybdate $(NH_4)_6Mo_7O_{24},4H_2O$, in the cooled acid solution. Dilute to 1 litre.

Molybdate-hydrazine reagent. Dilute 10 ml. of the ammonium molybdate solution to 90 ml. with water, add 1 ml. of 0·15 per cent. hydrazine sulphate solution and dilute to 100 ml. This solution should be freshly prepared.

Solution for removing arsenic from glass. Prepare a 0·1 per cent. solution of hydrazine sulphate in concentrated hydrochloric acid and saturate it with sodium chloride.

Method

Wet oxidise a suitable quantity (*e.g.*, 5 g.) of the material in a Kjeldahl flask (previously treated with the solution for removing arsenic), commencing with 10 ml. of nitric acid and then add 10 ml. of sulphuric acid. Continue adding nitric acid in the usual way until charring ceases. Then treat the digest with two portions of 15 ml. of saturated ammonium oxalate solution, heating after each addition until sulphuric acid fumes are again evolved. Cool, transfer to the distillation apparatus as used for the Gutzeit method (Fig. 5) using 7 ml. of water. Cool, add 5 g. of chloride-hydrazine-bromide mixture and 10 ml. of hydrochloric acid (s.g. 1·18). Immediately distil into a boiling tube (cooled in water) containing 10 ml. of dilute nitric acid (1 : 5) with the end of the air condenser dipping below the surface of the acid. Collect 7 ml. of distillate, transfer to a 50 ml. beaker (treated with the solution for removing arsenic), add 8 ml. of concentrated nitric acid and evaporate to dryness on a hotplate. Remove the nitric acid by heating at 130° C. for thirty minutes. Dissolve the residue in the molybdate-hydrazine reagent, transfer to a 25 ml. graduated tube and make up to the 25 ml. mark with the reagent. Mix and heat the tube in boiling water for ten minutes. Cool and measure the optical density of the blue solution in a 4 cm. cell in the absorptiometer (Ilford red filter No. 608 $\equiv$ 660 mμ) or with a spectrophotometer (at 835 mμ). A blank should be performed at the same time.

Prepare the standard curve using measured volumes of standard arsenic solution (0, 0·5, 1·0, 2·0, 3·0, 4·0 and 5·0 ml. of the dilute arsenic solution containing 5 p.p.m. as As, *i.e.*, 0 — 25 μg.). Evaporate each quantity down with nitric acid, make up to 25 ml. with the reagent; heat in boiling water for ten minutes, etc. (as for the sample). If the determination is made visually in Nessler glasses the standard blue solutions (equivalent to 0 — 20 μg.) should be freshly prepared on the day of the test (*cf. Analyst*, 1960, **85**, 629).

In addition to the reference method described, Eastoe and Eastoe (*J. Sci. Food Agric.*, **1953**, **4**, **310**) have proposed a further method employing the molybdenum-blue reaction for the determination of arsenic in gelatine in which the nitric-sulphuric wet oxidation is replaced by a rapid digestion with hydrochloric acid. The arsenic is separated, after reduction, as the hydride by the action of nascent hydrogen. The arsine is absorbed in iodine solution, ammonium molybdate and 1-amino-2-naphthol-4-sulphonic acid are added and the mixture heated at 100° C. The resulting blue colour is then measured as in the reference method (see also B.S. 757 : 1959).

COPPER

Statutory Limits

	p.p.m. (by weight)
Edible gelatine	30
Tomato ketchup, catsup, sauce or relish	20

Food Standards Committee Recommended Limits (Circular FSH 13/58)

RECOMMENDED LIMITS FOR COPPER (1)

	(*Parts per million by weight*)
General limits	
Beverages ready-to-drink (except those specified below)	2
Other foods (except those specified below)	20
Special Limits*	
(i) *Beverages*	
Wines, alcoholic liqueurs, cordials and wine or spirit cocktails, beer, cider, non-alcoholic beverages prepared from cider, and concentrated soft drinks (2)	7
(ii) *Other Foods*	
Chicory—dried or roasted	30
Cocoa powder, cocoa nib, cocoa mass and cocoa liquor	70 (on the fat-free substance)
Coffee beans	30
Colourings	30 (on the dry colouring matter)
Flavourings	30
Pectin—liquid	30
Pectin—solid	300
Tea	150
Tomato puree, paste, powder, juice and tomato juice beverages	100 (on the dried) tomato solids)
Yeast and yeast products	120 (on the dry matter)

* It is also recommended that amounts of copper in excess of the general limit should be " permitted " in shellfish, offals, etc. (see F.S.C. Report, 1956).

NOTES:

(1) other than the statutory limits for copper in edible gelatine and tomato ketchup prescribed in the Food Standards (Edible Gelatin) Order, 1951 and the Food Standards (Tomato Ketchup) (Amendment) Regulations, 1956.

(2) not including concentrates used in the manufacture of soft drinks to which the general limit of 20 p.p.m. would apply.

Copper, although emetic in large doses, is an element which is essential for growth. In plants it is necessary for respiration and in vertebrate animals traces of copper are essential for hæmoglobin formation in the blood. When present in certain foods, however, it tends to act as an oxidation catalyst and as little as 2 p.p.m. causes a tallowy flavour to develop in milk and butter and impairs the keeping qualities. Further, the presence of copper promotes the destruction of vitamin C in fruits, vegetables, etc. Copper is relatively resistant to corrosion even by acid foods, and its uptake from processing plant is not common (*cf. Food Manuf.*, 1956, **31**, 521). One possible source of contamination of plant foods, however, is the use of copper fungicides during cultivation.

Destruction of Organic Matter

If the material is wet oxidised prior to the determination of copper it is advisable to remove chloride by a preliminary heating with nitric acid before the sulphuric acid is added. High (*Analyst*, 1947, **72**, 60) has shown that dry ashing is also suitable provided nitric acid as well as hydrochloric acid are used for extracting the ash. If difficulty is experienced in destroying all the carbon matter, the first acid extraction should be passed through a filter paper. Then the paper should be returned to the dish, re-ignited, re-extracted and the two extracts combined.

Determination of Copper

Numerous colorimetric reagents have been employed for routine determinations of copper in aqueous solution, *e.g.*, potassium ferrocyanide, dithio-oxamide, sodium diethyldithiocarbamate. Iron is usually the principal interfering metal and should be filtered off as hydroxide after rendering the solution ammoniacal. If ferrocyanide is employed the reaction must be carried out in exactly neutral solution.

Methods involving the extraction of various copper carbamate complexes by organic solvents are now frequently used and two of these are described below. A review of copper determinations has been presented by Polhill (*Food Manuf.*, 1956, **31**, 519).

Determination of Copper using Sodium Diethyldithiocarbamate

Copper reacts with sodium diethyldithiocarbamate *in alkaline solution* producing a yellow to brown coloration according to the amount of the metal present. If the colour is to be measured in aqueous solution, citrate is added in order to prevent interference due to iron and precipitation of phosphate. Gum arabic is useful for stabilising the colour (see *Analyst,* **1947, 72,** 60). The yellow copper carbamate complex can be extracted by various immiscible organic solvents and Haddock and Evers (*Analyst,* 1932, **57,** 495), who employed carbon tetrachloride, found that in the presence of citrate the iron is retained in the aqueous solution. The method has been further modified by Cheng and Bray (*Anal. Chem.,* 1953, **25,** 655) who eliminate the interference due to 1 mg. quantities of iron and most other interfering metals by the addition of EDTA in addition to citrate. Versenate complexes are formed with the bi- and trivalent metals present at pH 7–10. Copper is the only metal of importance however, which can be sequestered from its versenate complex by carbamate to form a more stable copper carbamate complex. Bismuth carbamate is also extracted under such conditions, but such interference can be ignored as bismuth is only present in negligible amounts in foods. The following method includes a few modifications of the Cheng and Bray procedure:—

METHOD

Ash a suitable quantity of the material (usually 5–10 g.) at about 600° C. in a silica dish. Extract the ash by heating with 10 ml. of a mixture containing two volumes of concentrated hydrochloric acid, one volume of concentrated nitric acid and three volumes of water. Wash the extract into a separating funnel with water so that the volume is 50 ml. Cool, add 10 ml. of versenate-citrate mixture (containing 20 per cent. ammonium citrate and 5 per cent. of the disodium salt of EDTA) and two drops of cresol red. Make alkaline (pink) with 0·88 ammonia, cool and add 1 ml. of 1 per cent. of sodium diethyldithiocarbamate solution. Mix. Add 10 ml. of carbon tetrachloride from a burette to the separator, stopper and shake vigorously for two minutes. Filter the lower layer into a test tube through a plug of cotton wool pushed into the stem of the separator. Re-extract with 5 ml. of carbon tetrachloride and add the lower layer to the first extract. Mix the combined filtrates and measure the optical density (yellow) at approximately 440 mμ (Ilford No. 601 Violet filter) with an absorptiometer using a 1 cm. cell.

Alternatively the colour can be measured visually on the Lovibond Tintometer (see *Analyst,* 1932, **57,** 495). Carry out a blank on the reagents at the same time. For the standard curve, prepare a stock solution containing 0·3928 g. $CuSO_4$, $5H_2O$ per litre and dilute one hundred times before use (1 ml. = 1 μg. Cu). Then use volumes equivalent to 0 — 50 μg. of copper diluting in each case to 50 ml. before adding the versenate-citrate solution, etc. Each density reading is equivalent to the number of micrograms of copper per **15** ml. of solvent.

Determination of Copper using Salts of Dibenzyldithiocarbamic acid

Diethyldithiocarbamic acid is decomposed in acid solution. However, by using salts of dibenzyldithiocarbamic acid it is possible to extract the yellow copper complex from acid solution with carbon tetrachloride. Extraction at such a low pH possesses the additional advantage that interference from other metals is negligible. By using the minimum amounts of pure reagents, Abbott and Polhill (*Analyst*, 1954, **79**, 547) were able to reduce the blank to 0·3 μg. of copper and determine the metal in oils and fats over the range 0·02–2 p.p.m. The following method employs a solution of zinc dibenzyldithiocarbamate, which is obtainable commercially:—

METHOD

For the determination of copper in oils and fats, heat 20 g. of sample in a 200 ml. silica digestion flask until fuming commences and continue the heating until only about 2 ml. remains. Cool the flask and commence the wet oxidation with 3·5 ml. of pure sulphuric acid and 3 ml. of pure nitric acid. Continue the heating with further small quantities of nitric acid until the liquid is colourless. Then cool, add 10 ml. of water and heat until copious white fumes are evolved.

For most other foods begin the wet oxidation immediately using nitric acid and preferably not more than 4 ml. of pure sulphuric acid.

Transfer the acid solution to a separating funnel, dilute to 50 ml. with water and remove any nitrous fumes by adding 1 ml. of 5 per cent. sodium sulphite solution. Add exactly 10 ml. of the 0·05 per cent. solution of zinc dibenzyldithiocarbamate in carbon tetrachloride and shake vigorously for at least two minutes. Filter the lower layer into a test tube through a plug of cotton wool pushed into the stem of the separator and measure the optical density, preferably on a spectrophotometer at 435 mμ in a 1 cm. cell. If an absorptiometer is used the violet filter is suitable (Ilford No. 601). Carry out a blank on the reagents at the same time. For the standard curve, prepare a stock solution by dissolving 0·157 g. of $CuSO_4, 5H_2O$ in water containing 5 ml. of 5 per cent. v/v pure sulphuric acid and dilute to 200 ml. (1 ml. = 200 μg. Cu.). Then use volumes equivalent to 0–50 μg. of copper (diluting in each case to 50 ml. with 5 per cent. v/v pure sulphuric acid) and extract with 10 ml. of reagent.

In addition to the salts used by Abbott and Polhill, another acid extraction method for the copper complex using a chloroformic solution of the diethyl ammonium salt of diethyldithiocarbamic acid has been described by Wyatt (*Analyst*, 1953, **78**, 656).

LEAD

Statutory Limits. *cf*. The Lead in Food Regulations, 1961 (S.I. 1961 No. 1931)

	p.p.m. (by weight)
Edible gelatine	7 (*cf*. below)
Curry powder	20

Food Standards Committee Recommended Limits (1954)

General Limits

	p.p.m. (by weight)
Ready-to-drink beverages	0·2
Other foods	2

Foods for which special limits are recommended:—

Article of Food	*Limit Recommended in Parts per Million by Weight*
(i) *Beverages*	
Concentrated soft drinks (but not including concentrates used in the manufacture of soft drinks). Brandy, gin and geneva, rum and whisky.	0·5 p.p.m.
Wines, liqueurs, alcoholic cordials and cocktails and alcoholic liquors not otherwise specified; Beer; Cider; Fruit and vegetable juices (including tomato juice, but not including lime juice and lemon juice)	1 p.p.m.
(ii) *Other Foods*	
Anhydrous dextrose and dextrose monohydrate; Edible oils and fats; Refined white sugar (sulphated ash content not exceeding 0·03 per cent.)	0·5 p.p.m.
Ice cream, iced lollies and similar frozen confections	1 p.p.m.
Canned fish; Canned meats; Dried or dehydrated vegetables (other than onions); Tomato puree paste or powder containing 25 per cent. or more total solids; Edible gelatin; Meat extracts and hydrolysed protein	5 p.p.m.
Chemicals which may be used in foods and for which a lead limit is specified in the B.P. or B.P.C.	The limits specified in the edition of the B.P. or B.P.C. current from time to time.
All types of sugar, sugar syrups, invert sugar and direct consumption coloured sugars with a sulphated ash content exceeding 1·0 per cent.; Raw sugars except those sold for direct consumption or used for manufacturing purposes other than the manufacture of refined sugar; Edible molasses; Caramel; Liquid and solid glucose and starch conversion products with a sulphated ash content exceeding 1·0 per cent.	5 p.p.m.
Cocoa powder	5 p.p.m. on the dry fat-free substance.

Yeast and yeast products	5 p.p.m. on the dry matter
Tea	
Dehydrated onions	
Dried herbs and spices	
Flavourings	
Alginic acid, alginates, agar, carrageen and similar products derived from seaweed	10 p.p.m.
Liquid pectin	
Chemicals not otherwise specified, used as ingredients or in the preparation or processing of foods	
Food colourings (other than caramel)	20 p.p.m. on the dry *colouring* matter.
Solid pectin	50 p.p.m.

The report also states that shellfish, where lead is naturally present in varying amounts, should be the subject of a general reservation.

The cumulative poison lead is probably the most serious metallic contaminant of food. Monier-Williams has estimated that the daily intake of lead by a normal healthy individual is about 0·4 mg. including about 0·22 mg. derived from food, 0·10 mg. from water and 0·08 mg. from inhaled dust. Small amounts of lead occur in many foods naturally, but the principal source of contamination is the use of the metal and its alloys and compounds for processing materials (*e.g.*, solders, glazes, enamels, wrapping materials, piping) and for insecticides (*e.g.*, lead arsenate). Alcoholic drinks and soft water are especially liable to pick up lead from piping which conveys them. Due to action taken by the authorities and manufacturers, lead is seldom found in food nowadays in dangerous amounts (*cf.* Williams, *Roy. Soc. Hlth. J.*, 1958, **78**, 732).

Destruction of Organic Matter

Wet oxidation of the material is suitable for the determination of lead. The difficulties which were formerly experienced due to appreciable amounts of lead in the acids used have been largely overcome by the availability of " lead-free " reagents from the suppliers of chemicals. Dry ashing with or without an ash aid at a temperature not exceeding 500° C. is, in general, also satisfactory. Experiments with lead 212 as radioactive tracer have indicated, however, that (a) recoveries of the metal from dry ashing are influenced by the ash-aid and the temperature used and (b) the method may be more suitable for some foods than others (*Analyst*, 1959, **84**, 127).

Determination of Lead

Many of the methods which have been proposed for separating and determining lead in foods make use of diphenylthiocarbazone

(" dithizone "). Lead in alkaline solution can be extracted as the dithizonate complex which is soluble in organic solvents. By adding citrate and cyanide to the aqueous solution other metals are prevented from forming complexes and are not extracted. Further, the lead can be recovered from the complex in the organic layer by extraction with very dilute nitric acid.

The lead can be estimated using dithizone by various methods, viz. mono-colour methods (see S.A.C. method below), mixed colour methods (J.A.O.A.C., 1934, **17**, 108: 1936, **19**, 130), Irving's reversion technique (*Analyst*, 1953, **78**, 571) and the extractive titration procedure (*e.g.*, Lockwood, *ibid.*, 1954, **79**, 143).

By adding sodium sulphide to the lead solution in alkaline solution a brown colloidal solution of lead sulphide is formed, the intensity of which can be compared visually against standards. The sulphide method, however, lacks the sensitivity and accuracy of the absorptiometric methods employing dithizone.

The methods for determining lead in foods have been reviewed by Polhill (*Food Manuf.*, 1956, **31**, 182).

Determination of Lead by the S.A.C. (A.M.C.) reference method for amounts of lead up to 40 micrograms

The Analytical Methods Committee (of the Society for Analytical Chemistry) has approved the following method for the determination of lead which had been the subject of a report from its Metallic Impurities in Organic Matter Sub-Committee (*Analyst*, 1959, **84**, 127). After interfering substances are removed, the lead is extracted with dithizone at pH 9 to 9·5 and the red colour is measured absorptiometrically at 5200 A. Two methods are described for separating the lead.

Method A, in which the lead is extracted from an alkaline citrate and hexametaphosphate solution is usually suitable. If, however, the sample contains appreciable amounts of calcium, magnesium or phosphate the more tedious Method B is recommended. In Method B a preliminary extraction of the lead from acid solution with diethyl-ammonium diethyldithiocarbamate in chloroform replaces the hexametaphosphate treatment.

Reagents

As the method is suitable for determining minute traces of lead, it is most important to ensure that all the reagents used are free from the metal. The following " lead-free " reagents can be purchased: hydrochloric, sulphuric, nitric and citric acids, ammonium hydroxide and potassium cyanide. If special reagents are not available, ordinary quality chemicals must be rendered free from lead. Hydrochloric, sulphuric and nitric acids and ammonium hydroxide must be distilled from all-borosilicate glass. With the latter reagent the ammonia gas is trapped in ice-cold water and the solution so obtained titrated with normal acid. With most of the other reagents the material is made alkaline with ammonia (to pH 9 approx.) and shaken successively with dithizone solution (in chloroform) to remove the lead. When the chloroform layer is green (indicating that all the lead has been removed), the aqueous reagent is shaken with pure chloroform to remove the excess of dithizone. pH adjustment may also be necessary. For further details see Appendix 11 to the A.M.C. Report (*Analyst*, 1959, **84**, 133–4).

The special reagents required for the method are:—

Dilute nitric acid. Dilute 1 vol. concentrated nitric acid to 100 vols. with water.

Chloroform. Shake 250 ml. of chloroform firstly with 25 ml. of alkaline cyanide solution (containing 1 ml. of 10 per cent. potassium cyanide solution and 20 drops of 5 M ammonium hydroxide) and then with water. Filter the chloroform.

Dithizone, working solution. Shake 6 ml. of a 0·1 per cent. w/v dithizone solution in chloroform with 9 ml. water and 1 ml. of 5 M ammonium hydroxide solution. Reject the lower layer and clear the aqueous layer by filtering or centrifuging. This solution must be prepared fresh on the day of use.

Ammoniacal sulphite-cyanide solution. Mix together the following *exact* amounts: 340 ml. of 0·88 ammonia, 75 ml. of 2 per cent. w/v sodium sulphite (Na_2SO_3) solution, 30 ml. of 10 per cent. w/v potassium cyanide solution and 605 ml. of water.

Thymol blue indicator. Warm 0·1 g. thymol blue with 4·3 ml. of N/20 sodium hydroxide and 5 ml. of ethanol (90 per cent.) and dilute with 20 per cent. ethanol to produce 250 ml. of solution.

Sodium hexametaphosphate solution (10%).

Standard Curve

Dissolve 1·60 g. of lead nitrate in water, add 10 ml. of concentrated nitric acid and dilute to 1 litre. Then dilute 1 vol. to 100 vols. with water (solution b: 1 ml. = 10 μg. of lead). To a series of separators add, from a 5 ml. burette, 0, 0·5, 1·0, 1·5, 2·0, 3·0, 4·0 ml. of lead solution b and dilute each to 10 ml. with dilute nitric acid. Then treat each solution in the following manner:—

Add 30 ml. of ammoniacal sulphite-cyanide solution, *exactly* 10 ml. of chloroform (burette) and 0·5 ml. of dithizone working solution. Shake well for one minute. Filter the separated lower layer through a plug of cotton wool (pushed into the dry stem of the separator) into a 1 cm. cell, rejecting the first runnings. Measure the optical density of each chloroform solution against pure chloroform (1 cm. cell) with an absorptiometer or spectrophotometer at 520 mμ (Ilford No. 604 green gelatine filter). Construct the graph relating optical density to micrograms of lead.

DETERMINATION

A blank involving the whole procedure (omitting only the sample) must be performed.

Destruction of Organic Matter

Take an amount of sample containing not more than 40 μg. of lead.

Wet oxidation is suitable for most materials. For this not more than 5 ml. of concentrated sulphuric acid should be used. If much calcium is present, sulphuric acid should not be used and should be replaced by perchloric acid. After the completion of the "wet-ashing," cool, add 5 ml. of water and transfer to a 100 ml. conical flask, rinsing twice with 1 ml. portions of water. Then add 10 ml. of 5 M hydrochloric acid to the Kjeldahl flask, boil and wash into the conical flask, rinsing twice with 1 ml. portions of water. If the liquid in the Kjeldahl flask contains insoluble matter it should be filtered through a 7 cm. filter paper and the residue boiled with acid, passed through the filter, etc.

Dry ashing at a temperature not exceeding 500° C. is usually convenient for foods which are relatively chloride-free. Unless the food is likely to give a bulky ash, a lead-free ash aid should be used. For Method A, use 2 ml. of a solution containing 15 per cent. of ammonium nitrate and 15 per cent. of potassium sulphate. For Method B, use 10 ml. of 10 per cent. magnesium nitrate solution. Boil the resultant ash with 5 ml. of water and 10 ml. of 5 M hydrochloric acid and transfer to a 100 ml. conical flask. If insoluble matter is present the liquid should be filtered, the residue boiled with acid, etc., as for the wet-ash method above.

Separation and Estimation of Lead

A. For samples which contain only small amounts of calcium, magnesium and phosphate

To the prepared cooled solution add 5–10 ml. of 25 per cent. ammonium citrate solution and 10 ml. of fresh 10 per cent sodium hexametaphosphate solution. Add sufficient ammonium hydroxide solution (0·88) to produce a blue-green colour to thymol blue indicator (pH 9·0–9·5). Add 1 ml. of 10 per cent. potassium cyanide solution (and if much iron is present, 1 ml. of 20 per cent. hydroxylamine hydrochloride solution), cool and transfer to a separator containing 10 ml. of chloroform, rinsing with water to produce a total of about 50 ml. in the aqueous layer. Add 0·5 ml. of dithizone working solution and shake for one minute. If the lower layer is red, add more dithizone working solution until the colour (after shaking) is purple, blue or green. Add the lower layer to a second separator, washing through with 2 ml. of chloroform. Then continue extracting the lead from the first separator by shaking successively with 3 ml. of chloroform and 0·2 ml. of dithizone working solution for half minute. Add each chloroform extract to the main extract. The final extract should be green (indicating that all the lead has been extracted). Then add 10 ml. of the dilute nitric acid to the combined chloroform extracts in the second separator and shake for one minute. Reject the chloroform layer and to the acid extract add 30 ml. of ammoniacal sulphite-cyanide solution, *exactly* 10 ml. of chloroform and 0·5 ml. of dithizone working solution. Shake for one minute and measure the optical density of the filtered lower layer against pure chloroform in a 1 cm. cell at 520 mμ (see method above for Standard Curve).

B. Suitable for samples which contain relatively high amounts of calcium, magnesium and phosphate

To the prepared solution add methyl red solution and make just alkaline with ammonium hydroxide solution (0·88). Make just acid with 5 M hydrochloric acid and add a further 10 ml. Warm to 60° C., add 2 ml. of 20 per cent. sodium iodide solution and reduce any liberated iodine with 2 ml. of fresh 1·25 per cent. sodium metabisulphite solution. Cool, transfer to a separator and adjust the volume to 50–75 ml. in order to produce an acid concentration of approximately N with respect to hydrochloric acid. Add exactly 10 ml. of fresh carbamate reagent (1 per cent diethylammonium diethyldithiocarbamate in redistilled chloroform) and shake for half minute. Transfer the lower layer to a 100 ml. conical flask. Wash the aqueous layer twice with a little chloroform (without mixing) and then extract again with a further 10 ml. of carbamate reagent, adding the washings and extract to the main extract in the flask. To the combined extracts add 2 ml. of dilute sulphuric acid (1 + 1) and evaporate off the chloroform. Add 0·5 ml. of perchloric acid and heat until the strongly fuming solution is clear and colourless. Cool, add 10 ml. of water and 5 ml. of 5 M hydrochloric acid, boil for one minute, cool and add 2 ml. of 25 per cent. ammonium citrate solution.

Then continue as in A above, beginning at the second sentence " Add sufficient ammonium hydroxide solution (0·88) to produce a blue-green colour to thymol blue . . .".

N.B. Modifications are also necessary if bismuth is present (see A.M.C. Report).

Determination of Lead by Hamence's Method

Hamence (*Analyst*, 1932, **57**, 622) has described a simple method which is suitable as a routine method for determining lead in many foods. The principal interfering metal, iron, is removed as the thiocyanate in a mixture of ether and amyl alcohol and the lead is estimated by the sulphide coloration produced. Addition of cyanide forms complexes with other metals and thereby inhibits the formation of copper sulphide. The modified method is as follows:—

Ash a suitable quantity of the food (*e.g.*, 10 g.*) at a temperature not exceeding 500° C. Boil the ash gently with 3 ml. of concentrated nitric acid (if silica is present, *e.g.*, with spices, evaporate to dryness on the water bath with the acid and then heat with a further 3 ml. of acid). Add 17 ml. of water, warm and filter into a separating funnel. Wash the dish and filter with 10 ml. of water. Cool, add 20 ml. of a mixture of equal parts of ether and amyl alcohol (E.A.A.) and 5 ml. of saturated ammonium thiocyanate solution and shake vigorously. Run the lower aqueous layer into a second separator. Shake the E.A.A. layer (coloured red due to the iron thiocyanate) in the first separator with 10 ml. water and add the separated lower layer to the main aqueous liquid in the second separator. Reject the E.A.A. layer in the first separator. Shake the combined aqueous liquids successively with further 10 ml. quantities of E.A.A. until the upper layer is colourless. Run the lower layer into a 100 ml. cylinder and add 5 ml. of ammonium acetate solution (20 per cent.), 5 ml. of ammonium citrate solution (a 10 per cent. solution prepared by neutralising pure citric acid with 0·88 ammonia), 2 ml. of potassium cyanide solution (10 per cent.) and make the liquid alkaline with 10 ml. (or more) of ammonium hydroxide solution (containing 10 per cent. NH_3). If the liquid is cloudy due to phosphate precipitation, add more pure ammonium citrate solution (concentrated) to produce a clear solution. Make up to 100 ml. with water, mix and divide the liquid equally into two 50 ml. Nessler glasses. Add 2 drops of sodium sulphide solution (10 per cent.) to one glass and compare the colour against a B.D.H. Nessleriser disc, using the other glass as the " blank " (left hand compartment). The reading on the disc represents the amount of lead present in 5 g. of sample if 10 g. was taken ($10\,\gamma. = 10\,\mu g.$ Pb). The method can be readily adapted for comparison of the colour against a series of standard glasses. In that case the solution is not divided and the addition of dilute caramel solution may be necessary for equalising the colours of the solutions prior to the addition of sulphide.

* With liquids take a much larger quantity (say 100 g.) and evaporate to dryness before ashing.

REFERENCES TO OTHER METHODS FOR DETERMINING LEAD

Sulphide Coloration Methods, Monier-Williams, " Lead in Food," P.H. and Med. Subj., No. 88, H.M.S.O. (1938); see also *Analyst*, 1932, **57**, 440; 1935, **60**, 541–, 817–.

Dithizone Methods, Wichmann, *et al.*, *J. Ass. Off. Agric. Chem.*, 1934, **17**, 108; 1936, **19**, 130; see also *Analyst*, 1953, **78**, 571; 1954, **79**, 143.

Polarographic Methods, *Analyst*, 1947, **72**, 423; 1954, **79**, 731.

TIN

Limits

The Food Standards Committee have recommended a limit for canned foods of 250 *p.p.m. of tin* (*Analyst,* 1953, **78**, 187). This limit has now replaced the previously widely accepted limit of 2 grains per lb. (285 p.p.m.), which was suggested in The Report to the Local Government Board in 1908. Modern canning techniques including the use of selected lacquers are such that almost all canned food contains less than 100 p.p.m. Tin will only dissolve rapidly in the food if oxygen is present, so that provided the can is air-tight, the major attack takes place during the early period after canning, but tends to slow down when the residual oxygen is used up. Tin plate is most likely to be attacked by acid foods, especially fruits such as rhubarb and tomato and a few isolated products, *e.g.*, asparagus or foods containing sulphur dioxide or salt. Examination of the internal surface of the can is often a useful guide to the extent of the corrosion. " Sulphur-stains " which are caused by foods containing appreciable quantities of sulphur (*e.g.*, meat, fish, swedes) vary in colour from brown to purple but do not usually infer that much tin is present in the food. The pattern known as " feathering " should not be ignored, however, and cans showing dull grey patches are usually associated with the presence of tin in the food and the determination of the metal is advisable. There is no real evidence, however, that food poisoning has been caused by canned foods even if amounts of tin in excess of the recommended limit are present.

Destruction of Organic Matter

Wet oxidation using nitric acid and sulphuric acids of the organic matter is frequently employed prior to the determination of tin. The main objection which has been put forward to the use of nitric acid is that it promotes the formation of insoluble stannate, but low results are not usual in practice. Sulphuric acid may be used alone but this method of oxidation is somewhat lengthy. Dry ashing is also employed but it encourages the formation of insoluble tin oxide. The metal can be brought into solution, however, if the ash is fused with alkali and then dissolved in acid (see the dithiol method below).

Determination of Tin using the Volumetric Method

The volumetric estimation of tin in the stannous form by titration with dilute standard iodine solution is convenient for routine purposes. Interfering metals are seldom present in canned foods in any appreciable quantities so that separation of the tin as sulphide is usually unnecessary. It is important, however, to keep the reduced solution in an inert atmosphere as stannous solutions oxidise very readily:—

METHOD

To a suitable quantity of the sample (usually 10–25 g.) add 15–35 ml. of concentrated nitric acid followed by 10–15 ml. of concentrated sulphuric acid. Heat in order to oxidise the organic matter, adding small additional quantities of nitric acid immediately the liquid turns black. After cooling carefully add 10 ml. of water, 1 g. of potassium chlorate and 15 ml. of concentrated hydrochloric acid and heat the mixture until copious white fumes are evolved. Cool the oxidised solution and wash into a 350–500 ml. conical flask with successive small quantities of water. Add 0·5 g. of aluminium (either foil cut into small pieces or powder) and 25 ml. of concentrated hydrochloric acid and close the flask with a Contat-Gockel valve (see Jackson, *Analyst*, 1953, **78**, 443) filled with 10 per cent. potassium bicarbonate solution. Boil the contents of the flask until all the aluminium has dissolved and allow to cool, adding more bicarbonate solution to the valve so that it is always at least half filled with liquid. Then remove the valve, add a few pieces of marble and starch solution to the colourless liquid and titrate the reduced tin as rapidly as possible with N/100 iodine solution. Carry out a blank determination using all the reagents used for the sample (the blank titration is usually less than 0·5 ml.). Calculate the amount of tin in the sample in p.p.m. (as Sn).

1 ml. N/100 Iodine solution $\doteqdot$ 0·0005935 g. Sn.

The importance of carrying out the reduction and titration in the absence of air has been emphasised by Evans (*Analyst*, 1927, **52**, 570, 590), Willey (*ibid.*, 1924, **49**, 234) and Okell and Lumsden (*ibid.*, 1935, **60**, 803).

Determination of Tin using Dithiol

Colorimetric methods employing the reaction which takes place when stannous tin reacts with dithiol to give a red coloration were described by Clark (*Analyst*, 1936, **61**, 242; 1937, **62**, 661) and de Giacomi (*Analyst*, 1940, **65**, 216). These methods employed wet oxidation, colloids such as agar to keep the red precipitate formed in suspension and the colour was measured by reflected light in the Lovibond tintometer. Dickinson and Holt (*Analyst*, 1954, **79**, 104) have described a modified procedure which employs dry ashing, fusion of the ash so that the tin is reduced to metal, dissolution of the metal in acid and the determination of the colour produced with dithiol using transmitted light in an aborptiometer or a tintometer. The method gives accurate results even for small quantities of tin provided the directions are followed exactly:—

SPECIAL REAGENTS

Fusion mixture—30 g. sodium carbonate + 10 g. potassium cyanide.

Diethylammonium diethyldithiocarbamate—Mix together a stock solution containing 3·0 ml. of diethylamine in 7·0 ml. of chloroform with another containing 1·0 ml. of carbon disulphide in 9·0 ml. of chloroform. Store the solution in a dark bottle in a refrigerator. Dilute one volume of stock solution with 20 volumes of chloroform for the test as required.

Teepol solution—10 per cent. (v/v).

Dithiol—Dissolve 0·1 g. of dithiol (1-methyl-3 : 4-dimercaptobenzene) in 2·5 ml. of 5 N sodium hydroxide, add 0·5 ml. of thioglycollic acid and dilute with water to 50 ml. The reagent must be freshly pepared.

Method

Weigh out a suitable quantity (preferably containing 0·1–1·0 mg. of tin) of the sample into a small porcelain crucible and ash at 600° C. Scrape the white ash into the bottom of the crucible, add 1 g. (± 0·1) of fusion mixture and heat in a bunsen flame for about one minute so that the mass fuses. Cool, place the crucible in a small beaker and cover with a watch glass. Add 10 ml. of water to the beaker and run in 10 ml. of 6 N hydrochloric acid into the crucible from a pipette inserted through the lip of the beaker. Boil the liquid in the beaker for thirty minutes, cool, and filter into a separating funnel. The total volume after washing through with water should be about 35 ml. Add 5 ml. of the dilute carbamate solution, shake, allow the layers to separate and run off the chloroform layer. Continue shaking with further 1 ml. portions of the reagent until all the copper is removed (*i.e.*, when the chloroform layer is almost colourless) and then shake with pure chloroform. Finally make the aqueous layer up to 50 ml. in a graduated flask. Pipette 5 ml. of the solution into a test-tube, add 0·5 ml. of 6 N hydrochloric acid, 0·5 ml. of Teepol solution, 2·0 ml. of dithiol reagent and make up the volume to 10 ml. with water. Place in boiling water for two minutes cool and measure the absorption of the red solution using a 1 cm. cell and an Ilford No. 604 green filter (520 mμ). Alternatively measure the colour in a tintometer. Construct the standard curve from a fresh solution prepared by dissolving pure tin (metal) in hydrochloric acid. The amount of tin present in the aliquot of the *test solution* used should not exceed 150 μg.

Further Methods for the Determination of Tin

If large amounts of tin are present the *gravimetric method* is suitable. Hydrogen sulphide is passed through the wet oxidised solution and the Group II sulphides are filtered off, dissolved in sodium hydroxide solution containing a little ammonium sulphide and the tin sulphide reprecipitated with acetic acid. Then the sulphide is filtered off and ignited to the oxide ($SnO_2 \times 0{\cdot}788 = Sn$).

Polarographic methods for determining tin have been proposed by Godar and Alexander (*Analyst*, 1947, **72**, 111) and Markland and Shenton (*ibid.*, 1957, **82**, 83).

ZINC

Statutory Limit

	p.p.m.*
Edible gelatine	100

Food Standards Committee General Recommended Limits (*Analyst*, 1954, **79**, 244).

	p.p.m.*
Beverages ready-to-drink	5
Other foods	50

*w/w

The F.S. Committee did not suggest that any amendment should be made to the statutory limit for gelatine and considered that no objection should be taken to the sale of certain animal and vegetable products, *e.g.*, herrings, shellfish, crustacea, cereal offals and animal offals, even if the amount exceeds the general limit of 50 p.p.m.

High amounts of zinc are most likely to be found in foods containing acid, alcohol and salt, which are especially corrosive to zinc. Outbreaks of violent vomiting have occurred from eating stewed fruit containing several hundred p.p.m. of zinc after preparation in galvanised iron containers, but there are no records of chronic poisoning.

Destruction of Organic Matter

Although wet oxidation is preferable prior to the determination of zinc, dry ashing at 550° C. does not appear to result in loss of the metal by volatilisation and is adequate for routine work.

Determination of Zinc by the Nephelometric Method

The nephelometric method using ferrocyanide is useful as a sorting test. The ash is boiled with hydrochloric acid, then rendered ammoniacal and filtered into a 50 ml. Nessler glass. After re-acidifying the filtrate and diluting to 45 ml., 5 ml. of 2 per cent. potassium ferrocyanide is added and the turbidity compared with standard glasses containing known amounts of zinc and similar amounts of ferrocyanide, acid, etc. The conditions must be carefully standardised so that each glass is approximately N/2 to hydrochloric acid and contains about 3 g. of ammonium chloride. Further the opalescence reaches a maximum in ten minutes so the ferrocyanide must be added to the standard and test glasses at the same time. The method has been discussed by Fairhall and Richardson (*J. Amer. Chem. Soc.*, 1930, **52**, 938) and Boggs and Alben (*Ind. Eng. Chem. Anal. Ed.*, 1936, **8**, 97). For routine purposes the standard stock solution can be readily prepared from zinc sulphate (0·44 g. $ZnSO_4$, $7H_2O$ per litre: 1 ml. ≡ 0·0001 g. Zn).

Determination of Zinc by Extractive Titration using Dithizone

The following method is described in B.S. 757: 1959 for the determination of zinc in gelatine by extractive titration using dithizone (*cf.* Strafford *et al.*, *Analyst*, 1945, **70**, 238):—

Special Reagents

Dithizone extraction solution. Extract 40 ml. of a stock dithizone solution (0·1 per cent. in chloroform) with 50 ml. of 0·08 N ammonia and reject the lower layer. Acidify the ammoniacal extract with dilute hydrochloric acid and extract twice with 50 ml. portions of chloroform. Reject the aqueous layer and dilute the extract with chloroform so that 1 ml. ≡ 10 μg Zn (approx.).

Standard Zinc solution. Dilute 10 ml. of stock solution (0·44 per cent. $ZnSO_4,7H_2O$) to 1000 ml. (1 ml. ≡ 10 μg Zn).

Bromothymol blue. Warm 0·1 g. bromothymol blue with 3·2 ml. 0·05 N sodium hydroxide and 5 ml. ethyl alcohol (95 per cent.). When dissolved, dilute with 50 ml. water and 200 ml. of the ethyl alcohol.

METHOD

Destroy the organic matter in a weighed quantity of the sample by dry ashing or wet oxidation. Dilute an aliquot of the prepared solution to 20 ml. in a small pear-shaped separator, add 5 ml. of diethylammonium diethyldithiocarbamate solution (16 per cent. in chloroform) and shake for forty seconds. Separate and wash with 0·5 ml. chloroform without shaking. Repeat with 2 ml. portions of carbamate solution (and chloroform washes) until no more of the yellow complex of copper is formed and finally wash with 2 ml. chloroform. Transfer the aqueous layer to a 100 ml. beaker, add 5 ml. 3 N ammonium acetate solution and nearly neutralise with 0·88 ammonia. Add 0·2 ml. bromothymol blue and more ammonia dropwise until neutral (greenish blue) when cold. Transfer to a small pear-shaped separator and wash by shaking twice with 2 ml. of chloroform. Reject the chloroform. Add 1 ml. of dithizone solution from a 10 ml. burette and shake hard for thirty seconds. If red or pink, run off the lower layer and repeat adding more of the dithizone solution until no pink colour is formed and the chloroform layer is violet or blue (when the dithizone solution is added a few drops at a time near the end-point more pure chloroform should be added also). When titrating blanks the initial addition of dithizone should be less than 1 ml. Lead is probably measured as well as the zinc, but in most instances the amount present is negligible (*cf.* S.P.A., *Analyst*, 1948, **73**, 304).

STANDARDISATION OF DITHIZONE EXTRACTION SOLUTION

Titrate the solution blank as above and when the violet/blue stage is reached wash out all the dithizone with chloroform. Add 1 ml. of standard zinc solution and titrate as above with the dithizone.

Determination of Zinc by Sylvester and Hughes' Method

Sylvester and Hughes (*Analyst*, 1936, **61**, 734) have described a method for determining zinc which is both precise and suitable for routine purposes. The ash is dissolved in hydrochloric acid and the solution buffered to pH 4·5 with ammonium acetate. Various metals (Zn, Cu, Hg, Ag, Bi and Cd) are extracted from the solution with dithizone and the zinc (together with Bi and Cd) is removed from the organic layer by shaking with dilute hydrochloric acid. The residue obtained after evaporation of the extract is then oxidised and the zinc titrated by the micro method proposed by Lang (*Z. Anal. Chem.*, 1929, **79**, 161):—

Ash a suitable quantity of sample (say 10 g., but preferably not containing more than 0·3 mg. of zinc) in a silica dish at $\leqslant$ 550° C. Then set up four separating funnels in a row. Boil the ash with 5 ml. of 5 N hydrochloric acid, add 10 ml. of water and boil again. Transfer to the first separator using 20 ml. of water and add 10 ml. of 5 N ammonium acetate solution (386 g. per litre). Shake the aqueous extract *vigorously* with 5 ml. of fresh dithizone solution (0·15 g. per 100 ml. of chloroform) and run the lower layer into the second separator to which has previously been added 6 ml. of 5 N ammonium acetate, 3 ml. of 5 N hydrochloric acid and 10 ml. of water. Shake this second separator and run off the lower layer into the third separator to which has been previously added 20 ml. of water. Shake this third separator and run off the chloroform layer into the fourth separator. Then add another 5 ml. of dithizone solution to the first separator and repeat the extractions and separations as before. Repeat the process with more dithizone solution until the colour of the reagent in the first separator remains unchanged (*i.e.*, remains green—a reddish purple colour indicates that metals are still being extracted). Then extract the zinc from the combined chloroform extracts in the fourth separator by shaking with 10 ml. of N/2 hydrochloric acid and run off the lower layer into another separator. Re-extract the dithizone solution with a further 10 ml. of the N/2 acid and then reject the lower layer.

Add this second aqueous acid extract to the first extract, washing in with 10 ml. of water. Then shake the combined acid extracts with 5 ml. of chloroform. Reject the chloroform, pour the acid extract into a 100 ml. beaker, and wash the separator out with 10 ml. of water. Evaporate the liquid in the beaker to dryness by gentle boiling. Then oxidise the residue on a hot plate with a few drops of perchloric acid and hydrogen peroxide (100 vol.), heating down to dryness with successive portions of both until a white residue remains. Wash down the sides of the beaker with water and evaporate to dryness again. Cool and add 0·1 ml. of glacial acetic acid, a trace of ammonium hydrogen fluoride, 2 ml. of 5 per cent. potassium iodide solution and two drops of 1 per cent. starch solution. If a blue colour appears add N/500 sodium thiosulphate solution until the colour is discharged. Add 0·5 ml. of fresh 1 per cent. potassium ferricyanide solution and titrate with N/500 sodium thiosulphate solution, stirring after each addition.

0·51 ml. of N/500 sodium thiosulphate $\equiv$ 0·10 mg. Zn = 10 p.p.m. for a 10 mg. sample

The authors also describe an alternative method of titration using standard potassium ferrocyanide solution which is suitable for amounts of zinc from 0·2–1·0 mg. This titration is effected in acetic acid solution and employs diphenylbenzidine and potassium ferricyanide as internal indicators.

FLUORINE

Limits. Statutory limits relating to the amount of fluorine in acidic phosphates and foods containing such substances have been prescribed in The Fluorine in Food Regulations, 1959 (S.I. 1959 No. 2106; see also Circular FSH 15/59 and the Food Standards Committee Report, 1957):—

3. (1) No food specified in the first column of the following table shall contain any fluorine in a proportion by weight exceeding that specified in relation thereto in the second column of that table:—

Food	Maximum Fluorine Content (expressed in terms of parts per million)
1. Self-raising flour or any similar mixture (not included in item 2 below) containing a farinaceous substance and an acidic phosphate . . .	3
2. Baking powder, including golden raising powder.	15
3. Any article or substance (not included in any of the foregoing items) used as an ingredient in the preparation of food and consisting wholly or partly of an acidic phosphate	30 (of the acidic phosphates present)

In 1946, when the Inter-Departmental Committee on Food Standards published a report (*Analyst*, 1946, **71**, 382) on the question of prescribing limits for fluorine a small proportion of the acid calcium phosphate (A.C.P.) used for food purposes was manufactured from fluorine-contaminated rock phosphate.

The manufacture of A.C.P. for food purposes from rock phosphates has now been discontinued. The 1957 Report of the F.S. Committee refers to the presence of fluorine in the lime used but it seems unlikely that A.C.P. would now be manufactured with a fluorine content in excess of 27 p.p.m. The fluorine content of the other principal acid phosphate used for baking, *viz.*, acid sodium pyrophosphate is usually below 5 p.p.m.

Consideration of the amounts present in water supplies in different areas shows that at certain intake levels fluorine tends to prevent dental caries, and a certain minimum is desirable. Excessive amounts in water cause mottling of teeth, but there is no real evidence that this trouble has been encouraged by fluorine contaminated acid phosphates.

Other foods which have been shown to contain relatively high amounts of fluorine are tea (up to 100 p.p.m.) and shellfish (5–15 p.p.m.).

Determination of Fluorine

An accurate method for determining small amounts of fluorine has been described in a report of the Analytical Methods Committee of the S.P.A. (*Analyst*, 1944, **69**, 243). This involves ashing with lime, addition of perchloric acid and steam distillation of the fluorine as hydrofluosilicic acid. The determination of the element in the distillate is based on the fact that thorium nitrate and alizarin sulphonic acid form a pink lake, the colour of which is quantitatively discharged by fluorine:—

DISTILLATION APPARATUS

A Claissen flask (150–250 ml.) fitted with a 2-holed rubber stopper is used. Through the latter passes a thermometer (reading up to 150° C. at least) and a steam inlet tube. The side neck is stoppered and the side tube is connected to an efficient condenser. Steam is generated from water made alkaline with sodium hydroxide.

METHOD

Mix a suitable quantity of the sample (*e.g.*, 10 g.) with 1 g. of fluorine-free lime* and 50 ml. (or less) of water. Evaporate on a water bath, char and ignite at 600° C.

(a) Introduce into the flask some fragments of glass, 0·2 g. silver sulphate (to precipitate all the chloride), 7 ml. of water and 15 ml. of perchloric acid (60 per cent.). Assemble the apparatus and heat the flask until the temperature reaches 120–125° C. Then connect the steam supply and maintain the distillation at 137–140° C. Distil 150 ml. in thirty minutes, steaming out the condenser towards the end of the distillation. Discard the distillate.

(b) Distil a further 150 ml. as before and titrate the apparatus blank (should not exceed 1·5 μg.) by the method given below.

(c) Transfer the acid in the Claissen flask to a clean beaker and rinse the flask out with water, rejecting the rinsings. Transfer the dry ash to the flask, washing in with 5 ml. of water containing a few drops of the acid. Add the rest of the acid carefully, rinse down with 2 ml. water and distil 150 ml. as before.

(d) **Titration.** Titrate 50 ml. of distillate with N/20 sodium hydroxide in a Nessler glass using methyl orange as indicator until the colour matches that of a similar tube containing water and the same amount of methyl orange.

Transfer the remaining 100 ml. of distillate to a Nessler glass and add sufficient N/20 hydrochloric acid to make the total acidity equal to 5·0 ml. N/20 acid. Prepare a control cylinder containing 5·0 ml. N/20 acid and water and add to both the test and control cylinders 2 ml. alizarin S solution (0·01 per cent.). Then from a burette add thorium nitrate solution (0·025 per cent.) to the test cylinder, until a slight pink persists as compared with the yellow of the control cylinder. Add the same volume of thorium nitrate solution to the control cylinder, which should become more pink than the test solution. Then add slowly to the control cylinder, from a burette, standard sodium fluoride solution (0·0221 g./litre; 1 ml.$\equiv$ 10 μg. F.) until the tints of test and control solutions exactly match. From this titration, calculate the amount of fluorine in 150 ml. of distillate, subtract the apparatus blank and calculate the amount of fluorine in the sample as p.p.m.

With materials which contain a large proportion of phosphate (*e.g.*, A.C.P., A.S.P.), phosphoric acid is liable to be carried over into the distillate and interfere with the titration due to combination with

* The preparation of fluorine-free lime is described in detail in the S.P.A. Report. Each new batch should be examined for fluorine content by the method described for the sample.

the thorium nitrate. Therefore, for samples where the amount of phosphate in the distilling flask would exceed 0·5 g. (as P_2O_5), the S.P.A. (*Analyst*, 1945, **70**, 442) recommend in an addendum to the first report that two successive 150 ml. fractions of distillate should be collected. If the second fraction shows an appreciable phosphate content, the first 150 ml. is evaporated to dryness in the presence of a further gram of purified lime and redistilled to give 150 ml. of distillate. Then this purified distillate is titrated as described above to obtain the amount of fluorine in the sample. This modified procedure is also advisable for materials which are rich in sulphate.

A special apparatus for distilling fluorine which is fitted with a thermostatically controlled heater has been described by Estill and Mosier (*Anal. Chem.*, 1955, **27**, 1669).

Modified procedures for determining fluorine have also been described by Fryd (*Food Manuf.*, 1956, **31**, 236) and by Milton *et al.* (*Analyst*, 1947, **72**, 43).

PRESERVATIVES

Introduction

This section includes most of the general methods which are employed for the detection and determination of preservatives in foods. In subsequent chapters these methods are referred to together with any special pre-treatment or other modification which may seem necessary in the case of particular foods. In a few instances, however, special methods which are only applied to one group of commodities are included for convenience with the foodstuff.

The Public Health (Preservative, etc., in Food) Regulations 1925–58 permit the following preservatives to be present in certain named foods in amounts not exceeding prescribed maxima: sulphur dioxide, benzoic acid, nitrite, diphenyl and orthophenylphenol (see p. 451). During the preparation of this volume the Food Standards Committee published a Report on Preservatives in Food (1959) which they had received from the Preservatives Sub-Committee which recommends various modifications of the existing regulations. Their proposals include (a) that the list of foods which are permitted to contain preservative should be extended and (b) that additional preservatives should be permitted to be used: viz. esters of p-hydroxybenzoic acid, propionic acid, sorbic acid and the antibiotic nisin.

Apart from the matter given relating to analytical methods, much of what follows has been drawn from the F.S.C. Report. Space considerations do not permit the inclusion of the important work which has been carried out relating to the possible toxic effects of the various preservatives and for such information the reader should refer to the official Report and also to the references mentioned below.

So far as the analytical work in "official" laboratories is concerned a preliminary qualitative examination is usually carried out first before applying a quantitative procedure. The same may apply also in industry for materials received from suppliers. Most of the work carried out in industry, however, involves the estimation of known added preservatives.

Apart from the F.S.C. Report the following papers give much useful information relating to the action and other technical aspects of preservatives: Ingram *et al.* (*Chem. & Ind.*, 1956, No. 42, p. 1154; *ibid.*, 1959, No. 18, p. 552) and Barnes (*ibid.*, 1959, No. 18, p. 557).

SULPHUR DIOXIDE

Sulphurous acid may be used as a preservative in the form of sulphurous acid solution, or as the sulphites of sodium, potassium or calcium, but for the purpose of the regulations the amount present is calculated as sulphur dioxide (as SO_2), in which form (as the gas) it is also used. Sulphurous acid inhibits the growth of moulds, yeasts and aerobic bacteria and also prevents the darkening of fruits and vegetables due to enzymes. It assists in conserving vitamin C, but appears to inactivate vitamin B_1.

Sampling

As sulphur dioxide tends to be lost during storage, the sample should be kept in an air tight jar and the determinaton should be commenced as soon as possible after receiving the sample.

Qualitative Test for Sulphur Dioxide (Parkes, *Analyst*, 1926, **51**, 620)

Fit a conical flask with a small bubbler in the form of a small thistle funnel bent twice (of a similar shape to the fermentation locks used in wine making). Place a few drops of iodine-barium chloride reagent* in the bubbler. Add to the flask 5 g. of sample, 0·1 g. of copper acetate, a piece of marble and 10 ml. of concentrated hydrochloric acid and fit on the bubbler. Allow the acid to act on the marble for ten minutes and then heat to boiling. In the presence of sulphur dioxide the iodine is decolorised and a precipitate of barium sulphate settles in the tube. The formation of a turbidity is inconclusive as it may be due to other substances, *e.g.*, volatile oils.

Determination of Sulphur Dioxide

Most of the methods which have been proposed for the determination of sulphur dioxide in foods, involve (a) distillation of the sulphur

* The reagent is prepared as follows: Dissolve 3 g. of iodine in a solution containing 3 g. of potassium iodide, add 2 g. barium chloride dissolved in water and make up to 100 ml.

dioxide in the presence of acid and in an inert atmosphere, (b) absorption in an oxidising agent (*e.g.*, iodine, bromine, hydrogen peroxide), which converts the sulphurous acid to sulphuric acid, and (c) a volumetric or gravimetric estimation of the preservative.

(1) *Methods for determining Sulphur Dioxide using Iodine*

Distillation Method. A method for determining sulphur dioxide which " should be sufficiently rapid for works' control purposes " was published in the Report of the Preservatives Determination Committee of the Chemists of the Manufacturing Confectioners' Alliance, and the F.M.F. (*Analyst*, 1928, **53**, 118). The method is reliable and applicable to most foodstuffs provided that sulphides are not present.

About 200 ml. of water are placed in a round-bottomed flask, as shown in Fig. 6, and well boiled. Then 25 or 50 g. of the material are added through the funnel and followed by 20 ml. of hydrochloric acid. The burner should then be turned up and boiling effected as quickly as possible. The distillate passes down an efficient condenser and is collected, by aid of the adapter illustrated, in water containing a drop or two of 0·05 N iodine and a few drops of starch solution. The iodine solution is then run in from a burette as the

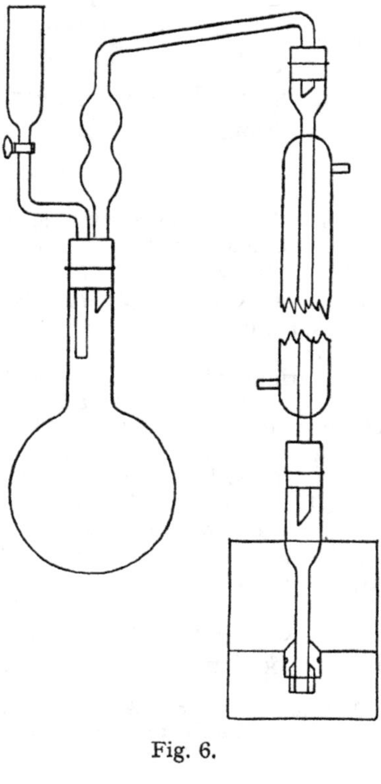

Fig. 6.

distillation proceeds, so that the blue colour just remains. Generally all the sulphur dioxide will be evolved within five minutes, but the end point is taken when the blue colour, due to 0·1 ml. of 0·05 N iodine, persists for a minute. Each 1 ml. of 0·05 N iodine = 0·0016 g. SO_2. In order to confirm that the distillate does contain sulphur dioxide, the solution remaining after titration may be acidified with hydrochloric acid, boiled to a lower bulk, and the sulphuric acid precipitated by addition of barium chloride and weighed as $BaSO_4$. In this case a " blank " on the reagents must be made and the barium sulphate therefrom deducted.

Direct Titration Method. If a liquid food is digested in the cold with alkali and then acidified, the sulphur dioxide present can be titrated directly with iodine solution. A process of this type which is suitable for wines is given on page 265 (*cf.* Jensen, *Analyst*, 1928, **53**, 133).

It should be borne in mind that sulphites combine with aldehydes, ketones and simple sugars, so that, with foods such as fruit juices, some of the sulphur dioxide is " bound " and appears to be ineffective as a preservative. It would appear that the dissociated " free " (unbound) sulphite is ineffective also. The active anti-microbial agent is the remainder—the unbound undissociated sulphite, the proportion of which increases with increasing acidity. The proportion of " free " sulphur dioxide present in fruit juices can be estimated by the method due to Downer (*J. Soc. Chem. Ind., Lond.*, 1943, **62**, 124):—

Titrate the juice with N/50 iodine to obtain the total iodine-reducing substances. Then to an equal volume of juice add 20 per cent. (by vol.) of acetone, set aside for ten minutes and titrate this also with N/50 iodine. The excess acetone combines with almost the whole of the free sulphur dioxide so that the second titration represents the iodine reducing substances other than sulphite. The difference between the two titrations is a measure of the free sulphur dioxide (*cf.* Mapson, *Biochem. J.*, 1942, **36**, 196; Braverman, " Citrus Products," p. 256, Intersci.).

(2) *Methods for determining Sulphur Dioxide using Hydrogen Peroxide*

Monier-Williams (*Analyst*, 1927, **52**, 343, 415) described a method for the determination of sulphur dioxide, in which the sulphur dioxide reacts with hydrogen peroxide in U-tubes to form sulphuric acid, which is then titrated with dilute standard alkali. This is preferable to the more rapid method using iodine (p. 68) when volatile sulphur compounds are present as in mustard oil, etc.

The apparatus is shown in Fig. 7. The flask D and tube E each contains 10 ml. of *neutralised* hydrogen peroxide solution, and F contains 5 ml. of hydrogen peroxide and barium chloride acidified with hydrochloric acid. About 500 ml. of water are put into the flask A with 20 ml. of hydrochloric acid. The solution is boiled for a short time and a current of pure carbon dioxide from a cylinder passed through the whole apparatus to remove the air. Then about 100 g. of the sample are introduced into the flask A, and the mixture is boiled

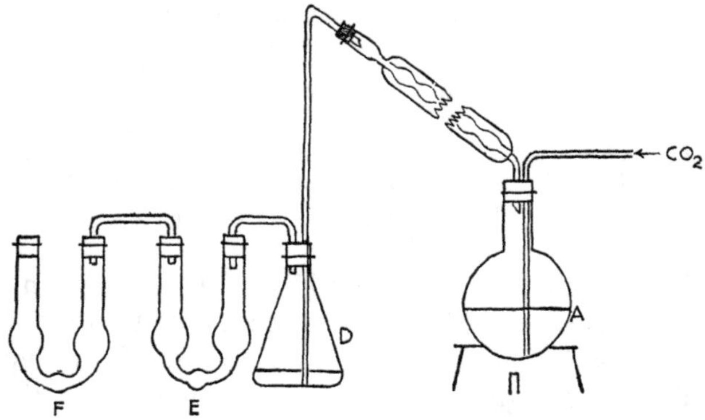

Fig. 7. Monier-Williams' distillation apparatus for the determination of sulphur dioxide.

for one hour in a slow current of carbon dioxide. At the end of this time the water supplied to the condenser is turned off so that it may become hot and any sulphur dioxide retained in it may be driven over into the receiver D, which is conveniently kept cold by standing it in a basin of water. The flask D and tube E are removed and their contents titrated with N/20 sodium hydroxide using bromphenol blue as indicator. The tube F serves to check that the absorption of the sulphur dioxide in the first vessels has been complete.

1 ml. N/20 NaOH = 0·0016 g. SO_2.

A similar method for the determination of sulphur dioxide is described in the B.P. 1958 (p. 890).

Shipton (*Food Pres. Quart.* (Australia), 1954, **14**, 54) has made a critical study of the A.O.A.C. method (which is essentially the same as that of Monier-Williams), which suggests that it is possible to shorten the procedure considerably. Shipton's main modifications (Fig. 8) are the use of (a) a vertical apparatus with ground glass joints, (b) a regulated electric heater, (c) a regulated flow of inert gas, (d) the abolition of the necessity of using previously boiled water, and (e) the omission of the heating of the condensers at the end of the refluxing. The present writer has obtained good recoveries with this method in spite of the saving in time:—

Into the conical flask and U-tube put 15 ml. and 5 ml. (respectively) of hydrogen peroxide solution (3 per cent.), adding water if necessary to ensure that bubbles will pass through each. Add to the reaction flask the sample (*e.g.*, 50 g.), 350 ml. water and 20 ml. of concentrated hydrochloric acid. Connect the flask to the apparatus and adjust the rate of flow of nitrogen so that 6–12 bubbles pass through the U-tube trap per minute. Switch on the electric heater to full heat, but turn down the heat when the liquid boils to give a slow boil. Continue boiling with the same rate of bubbling through the U-tubes for thirty minutes. Then disconnect the conical flask and U-tube. Wash the liquid in the U-tube into the flask, add three drops of bromphenol blue (0·4 per cent.) and titrate with N/20 sodium hydroxide. Titrate the free acid (if any) in a further 20 ml. of the hydrogen peroxide solution and subtract from the sample titration (1 ml. N/20 NaOH ≡ 1·6 mg. SO_2). The result can be confirmed gravimetrically if necessary by precipitating the sulphate formed with barium chloride as in other procedures.

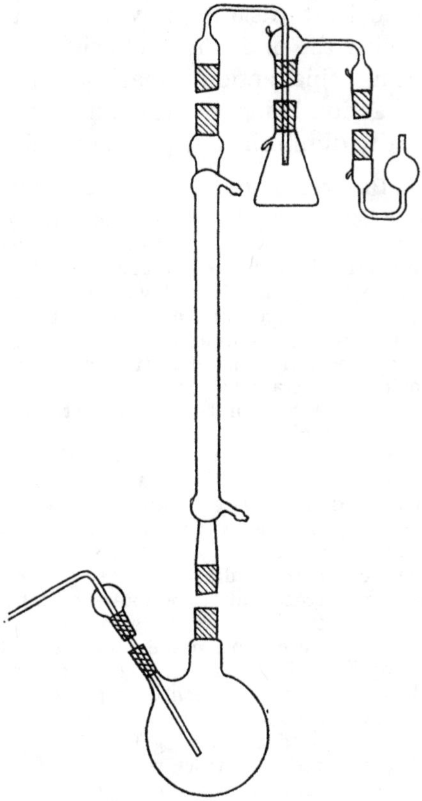

Fig. 8. Shipton's modification of Monier-Williams distillation apparatus for the determination of sulphur dioxide.

(3) *Other Methods*

Colorimetric methods for determining sulphur dioxide have been proposed by Joslyn (*J. Agric. Food Chem.*, 1955, **3**, 686) and by Brenner, Owades and Fazio (*Anal. Abstr.*, 1956, **3**, No. 2299).

BENZOIC ACID

Benzoic acid is usually used as a preservative in the form of the sodium or potassium salt, but for the purpose of the regulations the amount present is calculated as the acid itself. Weight for weight benzoic acid is only about half as effective as sulphurous acid and this difference is reflected in the maximum amounts permitted to be used, *e.g.*, either 350 p.p.m. SO_2 or 600 p.p.m. benzoic acid in squashes. Benzoic acid retards the growth of yeasts and moulds, the effective agent being the undissociated acid.

The extent of antimicrobial action is markedly affected by the pH of the medium—the action being greatest under conditions of high acidity and is markedly diminished at pH values above 5.

QUALITATIVE TEST FOR BENZOIC ACID (AND SALICYLIC ACID)

For qualitative purposes it is usually possible to shake out enough of the benzoic or salicylic acid present with ether/light petroleum mixture after acidifying with dilute sulphuric acid. Emulsions are often formed but this can be usually overcome by adopting the following procedure:—

Separate the " watery " ether layer, add more solvent and wash with water. Shake the ether extract with dilute ammonia. Transfer the layer containing the ammonium salt of any organic acid present to another separator, acidify with dilute sulphuric acid and extract with fresh ether/light petroleum mixture. Pour approximately equal volumes of the ether extract into two porcelain dishes and evaporate off the solvent. Examine the residue in one dish by dissolving in 1 ml. of water and adding ferric chloride solution. A violet coloration forms in the presence of *salicylic acid*. The amount of benzoic acid normally present in foods is usually insufficient to give a buff precipitate with ferric chloride and it is necessary to use a more sensitive test for its detection, *e.g.*, one of the modifications of Mohler's test:—

To the residue in the second basin add five drops of concentrated sulphuric acid and 0·1 g. of potassium nitrate and heat on a water bath stirring occasionally, for at least thirty minutes. This converts the acid to *m*-dinitrobenzoic acid. Dilute with 1 ml. of water and make alkaline with 0·88 ammonia. Transfer to a test-tube and boil in order to break up any ammonium nitrite. Cool and carefully add a few drops of freshly prepared ammonium sulphide solution down the side of the tube so that two distinct layers are formed (the sulphide solution can be prepared by passing H_2S into 0·88 ammonia). In the presence of benzoic acid a reddish brown ring forms at the junction of the liquids due to the formation of ammonium *m*-diaminobenzoate. The reaction is also given by salicylic acid (tested for previously), cinnamic acid (which may be present as an oxidation product of cinnamic aldehyde from cinnamon oil) and phenolphthalein (which is seldom present). If, however, after the reddish brown ring is formed the layers are mixed and then boiled the aminobenzoate is decomposed to produce a greenish yellow liquid. On the other hand the amido compounds of the interfering acids are stable and no greenish yellow colour is formed.

Parkinson (*Analyst*, 1952, **77**, 438) has described paper chromatographic methods for the detection of benzoic acid and its derivatives. The acid is extracted with chloroform or chloroform/ethanol and the solvent is carefully removed. The acid residue is taken up in 1 ml. ether and chromatograms are run on aliquots of the solution on Whatman No. 3 paper. The solvent system employed consists of n-butanol (40 pts.), ethanol (11 pts.) and ammonia-ammonium carbonate buffer (19 pts.), as described by Fewster and Hall (*Nature*, 1951, **168**, 78). The developed chromatogram is dried at 150° C. and sprayed with 0·025 per cent. 2·6 dichlorophenol-indophenol in a pH 7·2 buffer. Parkinson (*ibid.*) obtained the following R_F values: p-chlorobenzoic acid 0·59, salicylic acid 0·59, benzoic acid 0·50 and p-hydroxybenzoic acid 0·31. Salicylic acid shows a blue fluorescence when the unsprayed paper is examined under u.v. light (*cf. J. Ass. Off. Agric. Chem.*, 1957, **40**, 592).

Determination of Benzoic Acid

Direct extraction. With some foods, *e.g.*, soft drinks and wines, benzoic acid can be shaken out directly with ether/light petroleum after acidifying the sample with dilute sulphuric acid.

The acid is purified by shaking the bulked ether extracts with dilute sodium hydroxide and then treating the alkaline solution at 60° C. with permanganate to remove interfering acids. The excess permanganate is then reduced with sulphur dioxide and the benzoic acid extracted with ether after acidifying the aqueous solution. The solvent must be removed with not more than gentle heat before weighing.

It is advisable to confirm the result by dissolving the residue in 2 ml. acetone, adding 2 ml. water and titrating with N/20 alkali. 1 ml. N/20 alkali ≡ 0·0061 g. benzoic acid.

It should be borne in mind that cinnamic acid if present is oxidised to benzoic acid by permanganate. A method for selectively shaking out benzoic acid at pH 4 in the presence of saccharin is described on p. **217**.

Very small amounts of benzoic acid can be estimated in ether solution at 227 mμ using the spectrophotometer.

Distillation Method. With viscous foods such as coffee extracts and sauces it is advisable to employ the procedure proposed by Monier-Williams (*Analyst*, 1927, **52**, 153 and 229). The volatile acids are steam distilled from phosphoric acid and then purified by permanganate treatment and sublimation of the extracted material:—

To a 500 ml. steam distillation flask add 30–100 g. of sample, 200 ml. of water and saturate with salt (40 g./100 ml. water). Make distinctly acid with phosphoric acid and rapidly steam distil 500 ml. into a large porcelain basin containing 10 ml. of normal alkali. Wash down the condenser with 25 ml. of approx. N/10 sodium hydroxide and evaporate down the alkaline distillate to about 25 ml. on a water bath. Allow to cool to about 45° C. and add potassium permanganate solution (5 per cent) until a pink colour persists after stirring. Decolorise with sulphur dioxide and add sufficient dilute sulphuric acid to dissolve the precipitated manganese dioxide and to make the liquid acid. Transfer to a stoppered 100 ml. cylinder, saturate with salt (33 g./100 ml. water) and extract the acid with four successive portions of 15 ml. of ether/light petroleum. Transfer each extract to a 6 in. × $\frac{5}{8}$ in. test-tube* and evaporate off the solvent by means of a current of air passing through the tube which is held in water at 30° C. Wash down any deposit remaining on the sides of the tube into the bottom with a little ether and evaporate off the solvent. Remove the tube from the bath. For most purposes the residue is

* See footnote on p. **86**

sufficiently pure to be weighed at this stage without further purification. To purify, add enough clean dry sand to cover the deposit and heat the tube for one-and-a-half hours at 160° C. with the upper three-quarters of it above the top of the air oven used. In this way pure benzoic acid sublimes onto the cooler upper part of the tube. Remove the tube, break off the bottom part* containing the sand, etc. and weigh the benzoic acid on the upper part of the tube. If the crystals are dissolved off the tube with a neutral solvent the result can be confirmed by titration. Alternatively the crystals can be used for qualitative tests.

Monier-Williams (*Analyst,* 1927, **52**, 572) has described a further method for estimating benzoic acid in which the acid is vaporised and allowed to react with magnesium turnings to form magnesium benzoate using a special apparatus.

Other Methods. The amount of benzoic acid obtained after steam distillation can be estimated colorimetrically. Nicholls (*Analyst,* 1928, **53**, 19) has described a modification of Jonescu's test, in which the benzoic acid is partially converted to salicylic acid and the violet colour formed with a ferric reagent is compared with standards (see also *Analyst,* 1937, **62**, 172).

Illing (*Analyst,* 1932, **57**, 224) has described a further colorimetric method for determining benzoic acid which is based on Mohler's nitration test. The spectrophotometric method for determining benzoic acid in ether (at 227 mμ) is discussed by Englis *et al.* (*J. Agric. Food Chem.,* 1955, **3**, 964).

PARA-HYDROXYBENZOIC ACID

The esters of p-hydroxybenzoic acid and their sodium salts appear to be of value as food preservatives. It is interesting to compare this with the action of the benzoates, where the sodium salt is normally used but has no preservative value unless the pH is lowered sufficiently for the free acid to be produced. The anti-microbial activity of the p-hydroxybenzoates increases with the length of the alkyl chain up to the amyl ester. The F.S.C. recommendations provide only for the use of the methyl and propyl esters and the Committee propose that it be permitted in most instances as an alternative to benzoic acid. Esters of p-hydroxybenzoic acid are permitted as food preservatives in several European countries.

Qualitative Test for *p*-Hydroxybenzoic Acid

Millon's reagent. Dissolve 3 ml. mercury in 27 ml. cold fuming nitric acid and dilute with an equal volume of water.

Millon's reagent gives a rose-red colour with p-hydroxybenzoic acid. The free

* N.B. It is convenient to use a specially divided tube, the two parts being separated by a ground glass joint. Such tubes are obtainable from the usual suppliers. Whichever type of tube is used a disc of filter paper (larger than the diameter of the tube) should be pushed down the tube to a point above the sand prior to the sublimation.

acid is not steam volatile but the esters* are partially so. The test should be applied to the neutral ammonium salt of the acid after extraction with ether from acid solution (esters must be hydrolysed first—see the quantitative test below). As many other aromatic substances with a hydroxyl group attached to the benzene nucleus give colours with Millon's reagent (*e.g.*, salicylic acid gives an orange-red), the test cannot be considered as specific for p-hydroxybenzoic acid. Salicylic acid can be distinguished, however, by the intense violet colour given with ferric chloride (*cf.* Edwards *et al.*, *Analyst*, 1937, **62**, 178 and Stevenson and Resuggan, *ibid.*, 1938, **63**, 152; see also p. 84).

Determination of p-Hydroxybenzoic Acid

Edwards, Nanji and Hassan (*Analyst*, 1937, **62**, 178) have also proposed that the reaction with Millon's reagent can be employed for the colorimetric estimation of p-hydroxybenzoic acid. As mentioned above the acid is usually incorporated as an ester, which must be saponified before the reagent is added. The following method is applicable to cordials:—

Saturate the sample with ammonium sulphate, acidify with dilute sulphuric acid and extract with three portions of 15 ml. of ether. Remove the solvent, and to the residue add 60 ml. alcohol and 5 ml. of N/2 alcoholic potassium hydroxide. Boil under reflux for two hours. Then remove most of the alcohol, dilute the residue with water, and acidify with dilute sulphuric acid. Saturate with ammonium sulphate and extract the acid with three portions of 15 ml. of ether. Then shake the bulked ether extracts with three portions of 10 ml. of dilute ammonia. Boil the bulked ammoniacal extracts until neutral and dilute if necessary to a definite volume. Add 20 ml. of test solution containing not more than 2 mg. of the acid to a boiling tube. Also prepare a series of standards in boiling tubes containing 0·1–2·0 mg. of p-hydroxybenzoic acid each diluted to 20 ml. with water. Add 2 ml. of Millon's reagent to each tube and place them in boiling water for at least five minutes. Dilute each to 50 ml. in Nessler cylinders and compare the colours produced.

SALICYLIC ACID

Salicylic acid and its sodium salt were fairly extensively used as a food preservative until its use was prohibited by the 1925 Preservative Regulations.

Salicylates can be detected by the violet coloration produced with ferric chloride (see benzoic acid, p. 84). It should be borne in mind, however, that salicylic acid is destroyed by alkaline permanganate. The use of ferric chloride for estimating salicylic acid has been discussed by Nicholls (*Analyst*, 1928, **53**, 19) and by Edwards *et al.* (*ibid.*, 1937, **62**, 172). Salicylates can also be detected by Jorissen's test: To 20 ml. of a neutral solution containing about 0·02 g. of the extracted acid add 1 ml. of copper sulphate solution (1 per cent.), 0·01 g. of potassium nitrite and acidify with acetic acid (50 per cent.). Place in boiling water for some minutes. In the presence of salicy-

* The methyl, ethyl and propyl esters of p-hydroxybenzoic acid are sometimes referred to as Nipagin M (or Solbrol), Nipagin A and Nipasol M respectively.

late a red colour is produced. A quantitative modification of the test has been described by Nicholls (*loc. cit*).

Englis *et al.* (*J. Agric. Food Chem.*, 1955, **3**, 964) have proposed a spectrophotometric method for determining salicylic acid using ether as solvent (absorption peak 236 mμ; secondary peak at 306 mμ).

PARACHLOROBENZOIC ACID

Parachlorobenzoic acid appears to be effective for controlling mould and rope in bread and is permitted on the Continent. The acid can be separated by steam-distillation and extraction. Weiss (*Analyst*, 1934, **59**, 196) applies one of the forms of Mohler's test (*cf.* p. 84) for the detection of p-chlorobenzoic acid. The material is nitrated with sulphuric acid and potassium nitrate, diluted and made alkaline with ammonia. Then hydroxylamine hydrochloride is "layered" on top of the liquid in a test tube so that a green zone appears. The low solubility of p-chlorobenzoic acid in water (1 in 5200 at 20° C.) is said to enable a separation to be made from benzoic acid (1 in 350) by treating with water. The acids can be separated by the paper chromatography method of Parkinson (*loc. cit.*).

NITRITES

Sodium (or potassium) nitrite is used with the salt in wet brines or in dry curing mixtures in the manufacture of pickled meat products. Nitrite is also produced by reduction of nitrate due to reducing organisms in the brine. Nitrites combine with the meat pigment myoglobin in curing to form nitrosohæmoglobin, which is responsible for the characteristic red colour of pickled meat. Nitrites have a more powerful preservative action than nitrates. Several instances of poisoning have been reported on the Continent due to the ingestion of meat products containing excessive amounts of nitrite and deaths have occurred among children.

Griess-Ilosvay reaction for the Detection and Determination of Nitrites (*Bull. Soc. Chim.*, 1889, **49**, 388).

Reagent No. 1. 0·5 g. sulphanilic acid, 30 ml. glacial acetic acid and 120 ml. distilled water. Dissolve and filter.

Reagent No. 2. Dissolve 0·1 g. α-naphthylamine in 120 ml. distilled water by boiling, cool, add 30 ml. glacial acetic acid and filter.

N.B. If either solution should become coloured on keeping shake with 0·5 g. zinc dust and filter.

The Griess-Ilosvay reaction, which is commonly employed for the detection and estimation of nitrites depends on the diazotisation of sulphanilic acid by nitrous acid and the coupling of the resultant compound with α-naphthylamine to form an azo dye. A pink colour is slowly produced and reaches a maximum intensity in about fifteen minutes. When very large amounts of nitrite are present the colour forms immediately and often disappears rapidly or may not be seen at all.

For the *qualitative detection* of nitrites in products suspected of containing cured meat (for example), macerate the sample thoroughly with water, filter and add a few drops of each of the Griess-Ilosvay reagents. If nitrite is present the pink coloration appears within a few minutes.

For the *quantitative estimation* of nitrites in cured meats (p. 307), it is preferable to add a clarifying agent such as mercuric chloride or dialysed iron before filtering. This is especially necessary when cereal filler is present. If the final estimation is carried out in 50 ml. Nessler cylinders, 2 ml. of each of the reagents should be added to each glass. The markings on the appropriate B.D.H. Nessleriser disc represent amounts of nitrite nitrogen (N) present in the glass, but the preservative is normally reported as sodium nitrite:

$$1\gamma = 0{\cdot}001 \text{ mg. nitrite N; and nitrite N} \times 4{\cdot}93 = NaNO_2$$

If standards have to be prepared from sodium nitrite the solid used should first be assayed for purity by titration. Introduce into a flask 50 ml. N/10 potassium permanganate, 100 ml. water, and 5 ml. conc. sulphuric acid, warm to 40° C. and titrate with a solution containing 0·5 per cent. of the sodium nitrite.

$$1 \text{ ml. N/10 permanganate} \equiv 0{\cdot}003450 \text{ g. } NaNO_2$$

A suitable standard stock solution contains 0·493 g. of *pure* $NaNO_2$ per litre. Dilute 10 ml. of this solution to one litre, so that 1 ml. ≡ 0·001 mg. nitrite (as N). The best range for the test is from 0·0005–0·001 mg. N.

If the colour is to be measured absorptiometrically the optical density can be read at 520 mμ (Ilford green filter No. 604).

Alternative methods for determining nitrites have been described by Giblin and Chapman (*Analyst*, 1936, **61**, 686) and by Shinn (*Ind. Eng. Chem., Anal. Edn.*, 1941, **13**, 33).

NITRATES

The determination of nitrate is most usually carried out on cured meats.

BORIC ACID

Boric acid and borax were commonly used as preservatives prior to their prohibition by the 1925 Preservative Regulations. It is of interest to note, however, that the Regulations were relaxed during World War II to permit borax in bacon and margarine. This relaxation has now been withdrawn and the F.S.C.'s Sub-Committee endorse the views of the 1924 Departmental Committee, who considered boron preservatives as less desirable substances in view of their cumulative nature and their possible use to mask incipient putrefaction.

QUALITATIVE TEST FOR BORIC ACID

Turmeric test. Evaporate the sample with excess of lime water on a water bath and ignite at low redness. Cool, add a few mls. of water and acidify by the dropwise addition of 5 N hydrochloric acid (four drops in excess). Add four drops of saturated oxalic acid solution and soak a piece of turmeric paper in the liquid. Dry the paper at a low temperature. In the presence of borates the dried paper acquires a cherry-red colour which changes to greenish-black on the addition of dilute alkali (ammonium or sodium hydroxide). With liquid foods such as milk or the aqueous phase of butter the test can be performed " direct " (*i.e.*, without ashing), commencing by adding four drops of 5 N hydrochloric acid, etc. It should be borne in mind that the test is sufficiently sensitive to detect the traces of boron which are naturally present in fruits, etc. Robinson (*Analyst*, 1939, **64**, 324) has reported that by using turmeric-coated sand it is possible to put the test on a quantitative basis.

Determination of Boric Acid

Most of the methods which have been proposed for the determination of boric acid involve neutralisation to methyl red followed by the addition of mannitol, glycerol or invert sugar. This converts the boric acid to a relatively strong monobasic acid, which can then be titrated to phenolphthalein.

Prior to the titration the food may be ashed with alkali (as in the A.O.A.C. method). With butter or margarine the extracted aqueous phase may be used (p. 387) and with milk the titration can be carried out on the filtrate obtained by boiling with copper sulphate solution (p. 351). Alcock (*Analyst*, 1937, **62**, 522) dissolves the ash and distils the boric acid with methyl alcohol vapour. Quinalizarin (Ripley-Duggan, *Analyst*, 1953, **78**, 183) and carmine (Hatcher and Wilcox, *Anal. Chem.*, 1950, **22**, 567) may be employed as reagents for the colorimetric determination of boric acid.

Scott Dodd (*Analyst*, 1929, **54**, 645, 715; 1930, **55**, 23) has critically surveyed some of the methods which have been proposed for determining boric acid.

DEHYDROACETIC ACID

Dehydroacetic acid (DHA) and its sodium salt (DHA-S) appear to be very effective agents for controlling mould growth. In view of the available evidence on toxicity the U.S.F.D.A. regard DHA as a deleterious substance. The F.S.C.'s Preservatives Sub-Committee stated that "The available pharmacological evidence tends to be contradictory but indicates that the margin between the possible toxic level and the amounts likely to be used commercially is too small to justify the use of D.H.A. or its sodium salt as a food preservative."

Detection and Determination of Dehydroacetic Acid

The A.O.A.C. (1960, 9th Edn., p. 387) describes a test for the detection of dehydroacetic acid in cheese in which the acid is extracted with ether from the prepared food. The clear ether extract is evaporated to dryness, the residue is dissolved in 1 ml. of N/2 sodium hydroxide and the alkaline solution is transferred to a test-tube. Then 0·5 ml. of salicylaldehyde solution (20 per cent. v/v in alcohol) and 1 ml. of sodium hydroxide solution (1 + 1) are added, the tube is placed in boiling water for five minutes and then diluted with 2 ml. of water. A red or orange colour indicates the presence of dehydroacetic acid. The intensity of the colour is approximately proportional to the quantity present.

The quantitative method described in the A.O.A.C. involves the removal of the acid from a chloroform extract of the cheese with alkali. The alkaline extract is then acidified with hydrochloric acid and the absorbance of the acid extract is measured in a spectrophotometer at 307 mμ after thorough aeration to effect complete removal of the dissolved chloroform.

FORMALDEHYDE

Formalin (36 per cent. formaldehyde), a powerful preservative, is not permitted to be present in foods. It was formerly found in a number of samples (particularly in milk and meat).

Qualitative Tests for Formaldehyde

In the case of milk, tests for formaldehyde may sometimes be applied directly to the sample. Solid and semi-solid samples should be diluted with water and distilled slowly after acidifying with phosphoric acid. The first few mls. of distillate should be tested by one of the following tests:

Hehner's Test

Mix 2 ml. of distillate with 2 ml. of milk in a test-tube. Pour sulphuric acid (90 per cent. containing a trace of ferric chloride) down the side of the tube. In the presence of formaldehyde a violet zone forms at the junction of the layers. The violet coloration does not appear usually when relatively large quantities of formaldehyde are present. The milk supplies the tryptophane which must be present for the test to operate.

Chromotropic Acid Test

The reagent used is a saturated solution of 1·8-dihydroxynaphthalene-3.6-disulphonic acid in 72 per cent. sulphuric acid. For the test, add 1 ml. of distillate to 5 ml. of the pale straw-coloured reagent in a test-tube, mix and place in boiling water for fifteen minutes. In the presence of formaldehyde a purplish colour is produced (see also *A.O.A.C.* and *Analyst*, 1949, **74**, 610).

SCHRYVER'S TEST

Mix 10 ml. of distillate with 2 ml. of freshly prepared (and filtered) 1 per cent. phenylhydrazine hydrochloride solution and add 1 ml. of freshly prepared 5 per cent. potassium ferricyanide and 4 ml. of concentrated hydrochloric acid. A red colour is formed in the presence of formaldehyde. A positive reaction is also given by certain methylamines (*cf. Analyst*, 1909, **34**, 470). These latter substances, however, do not give the phloroglucinol reaction which is obtained by adding to an aqueous extraction a mixture of 1 per cent. phloroglucinol and an equal volume of 25 per cent. sodium hydroxide solution. In the presence of formaldehyde a rose-red colour appears.

SHREWSBURY AND KNAPP'S TEST (*Analyst*, 1909, **34**, 12)

The reagent used is prepared by mixing 100 ml. of concentrated hydrochloric acid with 0·1 ml. of concentrated nitric acid. With milk, mix 5 ml. of sample with 10 ml. of reagent and heat in water at 50° C. for ten minutes. In the presence of formaldehyde a purple colour is produced.

Determination of Formaldehyde

An approximate measure of the amount of preservative present can be obtained by comparing the colour obtained in the above methods with those given by known amounts of formaldehyde. It should be borne in mind that the results obtained tend to be rather lower than the amount originally added due to decomposition after incorporation into the food and to low recoveries in distillates due to polymerisation. It is advisable, therefore, to return the figure as " Recovered formalin," but in view of its prohibition samples containing it should be condemned anyway.

Hexamine ($C_6H_{12}N_4$), which is permitted to be used in some European countries is converted on distillation from acid solution to formaldehyde, which can be approximately estimated by the colorimetric methods described above.

Formic acid can be estimated by the following process which is based upon that due to Fincke (*Biochem. Z.*, 1913, **51**, 253):—

The apparatus is fitted up consisting of a steam generator, a large flask to contain the sample, a 500 ml. flask, which can be heated and which contains a suspension of about 3 g. of barium carbonate in water, and a condenser. The tube leading from the sample to the barium carbonate should have a spray trap, and preferably a bulb with side holes at the end so as to distribute the incoming vapour in the barium carbonate suspension. About 50 g. of the material are acidified with tartaric acid, placed in the distillation flask, heated, and steam is blown in through the sample and into the barium carbonate suspension (which is also heated), then condensed and collected in a 1-litre flask. About 1 litre of distillate should be collected, charring must be avoided, and there must be excess of barium carbonate at the end of the distillation. The barium suspension is now filtered, the insoluble carbonate being washed with hot water. The filtrate and washing should measure about 150 ml. or be evaporated to that volume. Add to the filtrate 10 ml. of 50 per cent. sodium acetate solution, 2 ml. of 10 per cent. hydrochloric acid, and 25 ml. of mercuric chloride solution (10 g. $HgCl_2$ + 15 g. NaCl in 100 ml.). Heat this mixture on a water bath for two hours, collect the precipitate of mercurous chloride on a Gooch crucible, wash, dry and weigh it. The weight × 0·0975 gives the weight of formic acid. A blank test should be made on the reagents.

PROPIONIC ACID

Propionic acid has been shown to be an effective agent in the control of moulds and rope in bread. It is usually incorporated as the calcium or sodium salt and as a preservative it appears to be more effective in acid solution. The antimicrobial action is, therefore, probably due to the undissociated acid. Propionic acid appears to be relatively non-toxic and the F.S.C.'s Preservatives Sub-Committee have proposed that it be permitted to be used in bread up to a maximum of 0·3 per cent. of the flour weight.

The A.O.A.C. (1960, p. 171), describes a method for propionic and acetic acids in bread in which the steam distilled acids are separated on a column of silicic acid.

SORBIC ACID

Sorbic acid is more effective against yeasts and moulds than bacteria. It is most effective under acid conditions and it is said to inhibit the growth of yeasts and moulds without interfering with the "beneficial" organisms present, *e.g.*, sorbic acid controls mould growth in cheese but does not impede the lactobacilli. In their Report, the F.S.C. *recommend* that sorbic acid (or its sodium salt) should be permitted in flour confectionery, marzipan, cheese, solutions of colours and in silicone anti-foam emulsion.

For its estimation sorbic acid can be steam-distilled and extracted as in the Monier-Williams procedure (p. 85), but the permanganate treatment must be omitted. The sorbic acid can be removed from its ethereal solution with sodium hydroxide and the optical density measured on the aqueous solution at 256 mμ ($E^{1\%}_{1\,cm}$ 2200 at pH 5·9). For further information the reader should refer to the papers in Food Research (1954, **19**, 20; 1955, **20**, 649; 1958, **23**, 338).

As sorbic acid is said to give a violet colour with ferric chloride, it may be confused in qualitative work with salicylic acid. Eeckhaut (*Anal. Abstr.*, 1956, **3**, No. **1156**) has, however, described a method which is said to be specific for sorbic acid. The distilled acid is extracted with alkali from its solution in ether and the aqueous extract is then evaporated to low bulk. Five ml. of sulphuric acid (25 per cent. v/v) and 1 ml. of N/10 permanganate are then added to the cooled solution. Decoloration occurs within one minute in the presence of both acids, but sorbic acid can be distinguished from salicylic acid as it yields acetaldehyde on oxidation, which can be detected by Schiff's reagent. A slight positive reaction may be obtained with cinnamic acid.

DIPHENYL, ORTHOPHENYLPHENOL

Diphenyl is effective in preventing mould growth, but it is not very active against yeasts and bacteria. It is normally incorporated into the packaging material of citrus fruits so that the vapour acts over the surface. Orthophenylphenol and its sodium salt are effective bactericides and fungicides. Certain fruits are treated with a solution of the sodium salt used in the form of a dip or a spray. Further details, including a review of the methods which have been proposed for the determination of these preservatives, are given on pages 201–203.

ANTIBIOTICS

Tetracyclines. The antibiotics chlortetracycline (aureomycin) and oxytetracycline (terramycin) have been shown to be effective preservatives especially when applied to flesh foods, provided the animal, fowl or fish is treated as soon after death as possible (*cf. Food Sci. Abs.*, 1956, **28**, 121) and standards of hygiene are adequate.

Work carried out at the National College of Food Technology (Wight and Pearson, unpublished) showed that with minced beef treated with chlortetracycline there was a marked retardation in bacterial growth (especially of Gram-negative rods), which was reflected in the lower volatile nitrogen figures obtained (when compared with controls similarly treated). Concurrently, however, the treated meat showed an increase in the F.F.A. figure of the extracted fat due to yeasts which resembled *Candida lipolytica*.

The Antibiotics Panel of the F.S.C. Preservatives Sub-Committee considered the proposals put forward that poultry, fish and whales should be permitted to be treated with chlortetracyclines. In the light of available evidence the panel were not prepared to recommend such treatment for poultry and fish largely on the grounds that (a) hygiene standards might be inadequate, (b) relatively large residues are sometimes found in the skin (particularly of fish) even after cooking, and (c) the use of tetracyclines as food preservatives might adversely affect their value when used therapeutically or might have other undesirable consequences. In the U.S.A., however, up to 7 p.p.m. of tetracycline is permitted in raw poultry.

Nisin is a " medically rejected antibiotic " which is produced by a number of strains of a cheese starter organism, *S. lactis,* and is naturally present in certain cheeses. Its value as a preservative lies mainly in the control of clostridial spores which multiply in certain cheeses and in other products leading to blowing or " hard swells." It does not enable the manufacturer to produce a satisfactory product from unwholesome cheese and nisin appears to be harmless to man. Also when nisin is added in sufficient amount to control spoilage by thermophilic organisms, the residual nisin in the product after processing is very small.

After considering this and other evidence the F.S.C.'s Antibiotics Panel concluded that " from the point of view of possible hazards, a distinction must be drawn between those canned foods where heat processing (whether the use of nisin is permitted or not) will be sufficient to destroy *Cl. Botulinum* spores or where the pH of the product is such that there is no risk of botulism, and products where these conditions may not obtain. In the latter cases, they feel that the use of nisin could increase the risk of food poisoning, but see no objection on medical grounds to the use of nisin where there is no risk of botulism. We therefore recommend that provision should be made for the addition of nisin to canned foods, defined as ' foods in hermetically sealed containers which have been sufficiently heat processed to destroy *Clostridium Botulinum*, or the pH of which is less than 4·5.' With such a definition, we do not think it necessary or practicable to specify the particular canned foods which may be so treated. It will automatically exclude those such as canned clotted cream, olives in brine, marinated fish and pasteurised canned hams about which the Antibiotics Panel have reservations."

In consequence the F.S.C. have recommended that foods as defined above should be permitted to contain nisin. No statutory limit is recommended.

Determination of Antibiotics

The amounts of antibiotic residues in foods are extremely small and for their determination microbiological assay methods must be employed. For each antibiotic there is an assay organism which is considered to be the most efficient for the attainment of reproducible results, *e.g.*, *B. cereus* is suitable for chlortetracycline. Also the extractant has to be chosen in relation to the particular food and organism. If the extractant contains a substance which itself inhibits the growth of the assay organism a " pad-plate " method is used in which small filter paper discs are dipped in the solution and dried before placing on the assay plate. The pad-plate method is described by Kohler and Broquist in the Reports of the Lederle (Cyanamid) Laboratories. Where the extractant is not itself inhibiting, the solution can be applied directly to the assay plate and methods of the " cylinder-plate " type can be employed (Brownlee *et al.*, *J. Gen. Microbiol.*, 1948, **2**, 40; 1949, **3**, 347).

Other useful references are: Grady and Williams (Antibiotics and Chemotherapy, 1953, **3**, 158); " Antibiotics Annual " (1953/54), Medical Encyclopædia, Inc. N.Y., p. 409–415.

ANTIOXIDANTS

Legislation

In 1953/4 the Food Standards Committee adopted two reports which it had received from the Preservatives Sub-Committee on the subject of antioxidants in food. The Committee stated that oxidative rancidity (p. 409) was the chief cause of rancidity in fats. Antioxidants only control this type of rancidity. Care in processing and the selection of storage conditions minimises the trouble but the more or less rapid increase in " peroxides " following an induction period causes considerable trouble. Further, antioxidants serve only to prolong the storage period of a well prepared fat and they do not mask the decomposition of food or render a rancid fat palatable. The Committee did not consider it necessary to legislate in relation to " natural " antioxidants, *e.g.*, sugar, cereal, oils, flours (such as oat flour), herbs and spices.

Subsequently The Antioxidant in Food Regulations, 1958 (S.I. 1958 No. 1454) were introduced and for the purpose of the regulations:—

> " ' antioxidant ' means any substance which delays, retards or prevents the development in food of rancidity or other flavour deterioration due to oxidation but does not include lecithin, ascorbic acid, tocopherols, citric acid, tartaric acid, phosphoric acid or any permitted preservative."

The following table summarises the requirements stated in The First Schedule to the regulations:—

Specified food	*n*-Propyl gallate or *n*-octyl gallate or *n*-dodecyl gallate or any mixture thereof	B.H.A. or B.H.T. or any mixture thereof
	Maxima (p.p.m. by weight)	Maxima (p.p.m. by weight)
(a) Anhydrous edible oils and fats, whether hardened or not and vitamin oils and concentrates	100	200
(b) Butter for manufacturing purposes	80	160
(c) Essential oils, including their flavouring constituents — isolates and concentrates	1000	1000
B.H.A. = butylated hydroxyanisole		
B.H.T. = butylated hydroxytoluene		

Also B.H.A. or B.H.T. or mixtures thereof within the limits specified may be used in conjunction with one or more of the gallates within the limits specified, provided that the total amount of antioxidant shall not exceed 300,240 and 1000 p.p.m. for a, b and c respectively.

The Second Schedule relates to the labelling of antioxidants. Each container shall bear a statement of the ingredients present in a prescribed form. The Third Schedule relates to the "Labelling of Food containing Added Antioxidant." In certain instances, *but not including sale by retail*, the presence of antioxidants in foods has to be declared on the label together with the maximum amount present (as p.p.m. w/w).

The Antioxidant in Food Regulations do not apply to foods or antioxidants intended for export.

Detection and Determination of Antioxidants in Fats

For the extraction of antioxidants use can be made of the fact that methyl, ethyl and propyl gallate are soluble in water, whilst higher gallates, BHA, etc., are soluble in 72 per cent. ethanol. For general purposes most antioxidants can be extracted by dissolving 20–40 g. of the fat in 50–100 ml. light petroleum and shaking out with ethanol (72 per cent.) or methanol. If concentration of the ethanolic extract is necessary it should preferably be carried out under reduced pressure. For the detection of *propyl gallate* (*PG*), Mahon and Chapman (*Analyt. Chem.*, 1951, **23**, 1116) add ten drops of 0·88 ammonia to 5 ml. of the alcoholic extract. The appearance of a rose colour indicates the presence of PG. For the detection of *butylated hydroxyanisole* (*BHA*), add 1 ml. borax solution (1 per cent. aqueous) and a few small crystals of 2·6 dichloroquinonechloroimide to 5 ml. of the alcoholic extract. The appearance of a blue colour indicates the presence of BHA. These authors also describe a test for *nordihydroguaiaretic acid* (*NDGA*): add 2 ml. of ferric chloride reagent (0·2 per cent. in absolute alcohol) and 2 ml. of $\alpha\alpha'$ dipyridyl reagent (0·2 per cent. in absolute alcohol) to 5 ml. of the alcoholic extract. The appearance of a red colour indicates the presence of NDGA. Butylated hydroxytoluene (BHT) also forms a red ferrous-dipyridyl complex but is more difficult to extract from the fat (see Anglin *et al.*, *J. Agric. Food Chem.*, 1956, **4**, 1019).

DETECTION OF ANTIOXIDANTS BY MEANS OF PAPER CHROMATOGRAPHY

Gander (*Analyt. Abs.*, 1956, **3**, No. 562) has described a method for the separation and detection of several antioxidants: Dissolve 40 g. lard or margarine in light petroleum (42–68° C.) and shake with ethanol (96 per cent). Place five 0·02 ml. drops of the separated alcoholic extract on Whatman No. 1 paper and develop the chromatogram with one of the following solvents: (i) methanol-*iso* amyl alcohol-benzene-water (2 : 1 : 1 : 1), (ii) water-ethyl acetate (97·5 : 2·5), (iii) methanol, (iv) *n*-butanol-acetic acid-water (4 : 1 : 5). Develop the spots by spraying with (a) silver nitrate solution (1 per cent.) followed by dilute ammonia (1 per cent.) or (b) borax solution (2 per cent.) followed by an ethanolic solution of 2 : 6 dichloroquinonechloroimide (0·01 per cent.). In employing this technique it should be borne in mind that some oils contain natural antioxidants which have similar R_F values to those of the added antioxidants (see *Analyt. Abs.*, 1957, **4**, No. 2807). Other procedures employing paper chromatography for the detection of antioxidants have been described by Mitchell (*J. Ass. Off. Agric. Chem.*, 1957, **40**, 909), Kawata and Hosogai (*Analyt. Abs.*, 1958, **5**, No. 1995) and by ter Heide (*ibid.*, 1959, **6**, No. 1536).

DETERMINATION OF GALLATES IN FATS

Cassidy and Fisher (*Analyst*, 1960, **85**, 295) have recommended the following colorimetric procedure for the determination of gallates. The antioxidant is extracted with methanol at 40–45° C. and solid ferrous ammonium sulphate is added to the resultant solution.

METHOD

Shake 10 g. of the warm liquid sample (or 5 g. solid fat + 5 ml. liquid paraffin) with 25 ml. methanol (95 per cent.) for one minute in a stoppered boiling tube with a bulb of capacity 10–12 ml. blown in its side at the bottom to effect separation of an upper layer by tilting the tube. Place in a water bath at 40–45° C. for fifteen minutes. Pour the upper layer into a 50 ml. graduated flask, repeat the extraction with 20 ml. methanol, transfer the upper layer to the flask and make up to the mark. Add 1 g. calcium carbonate to the flask, shake for thirty seconds and filter, rejecting the first runnings. To 10 ml. filtrate add exactly 1 ml. acetone and 10 mg. of *powdered* ferrous ammonium sulphate and shake for one minute. Allow to stand for thirty minutes and measure the optical density at 580 mμ in a spectrophotometer using a 1 cm. cell.

$$\text{Gallate (mg. per 11 ml. final soln.)} = dK$$

where d = optical density and using a Unicam SP500, K = 0·622 (n-propyl gallate), 0·785 (*n*-octyl) or 0·952 (*n*-dodecyl). Such factors should be determined for each batch of reagents.

A further method for the determination of gallates has been described by Vos, Wessels and Six (*Analyst*, 1957, **82**, 362). After dissolving the fat in light petroleum, propyl gallate is extracted with water and the higher gallates with absolute methanol. The determination is carried out absorptiometrically at 550 mμ in a sodium acetate buffer solution using ferrous tartrate which is specific for gallates, as reagent.

DETERMINATION OF BUTYLATED HYDROXYANISOLE (BHA)

Mahon and Chapman (*Analyt. Chem.*, 1951, **23**, 1120) proposed the following control procedure for the determination of BHA in lard and shortening:—

METHOD

Dissolve 10 g. of fat in 50 ml. light petroleum (Skellysolve H) and extract with 72 per cent. alcohol (three times with 25 ml. quantities and then once with 60 ml.). Dilute the combined alcoholic extracts to 200 ml. with alcohol (72 per cent.) and filter. Dilute a suitable aliquot to 12 ml. with alcohol (72 per cent.), add 2 ml. of 2·6 dichloroquinonechloroimide (0·1 per cent. in absolute alcohol), mix and add 2 ml. borax solution (2 per cent. aqueous). Mix and measure the optical density at 620 mμ. Prepare the standard curve by adding the reagents to quantities of BHA over the range 10–15 μg. (each dissolved in 12 ml. alcohol).

GENERAL METHODS FOR DETERMINING ANTIOXIDANTS

For the determination of antioxidants, Anglin, Mahon and Chapman (*J. Agric. Food Chem.*, 1956, **4**, 1018) separate BHA and BHT from the fat by distillation with superheated steam. The distillate is analysed for the sum of the BHA and BHT by forming the ferrous-dipyridyl complex. Then by determining the BHA with 2·6 dichloroquinonechloroimide, the amount of BHT can be determined by difference. NDGA and PG are extracted from a carbon tetrachloride solution of the fat using 50 per cent. alcohol and are determined using ferrous sulphate.

Berger, Sylvester and Haines (*Analyst*, 1960, **85**, 341) separate BHA and the gallate esters from fat by means of reverse-phase partition chromatography (using synthetic rubber as the inert support). Gallates and BHA are determined colorimetrically using ferrous tartrate and 2·6 dichloroquinonechloroimide respectively. The authors also describe a method for determining BHT in ethanol from the u.v. absorption spectra under alkaline conditions.

Whetsel *et al.* (*J. Amer. Oil Chem. Soc.*, 1955, **32**, 493) determine PG and BHA by means of equations relating to the extinctions of solutions in *iso*-propyl alcohol obtained over a series of wavelengths.

COLOURING MATTERS

LEGISLATION

In 1954/5 the Food Standards Committee adopted two reports which it had received from the Preservatives Sub-Committee on the subject of colouring matter in food. The Committee stated that soon after the discovery of the coal-tar dyes about 1850 they rapidly gained favour for colouring foods in place of the mineral pigments and natural colours which had hitherto been the only ones available. The main advantages to the manufacturer of using the coal-tar dyes are that they give him a much wider range of colours with a higher and more uniform tinctorial power and brilliance. The F.S.C. accepted that the colouring of foods was a legitimate practice. Over ninety coal-tar dyes were put forward and seventy-five of these were studied by the Committee's Pharmacology Panel from the point of view of possible harmful effects if used in foods.

The F.S.C. recommended that the parts of the Preservatives, etc. Regulations, 1925/27 which provided that only a comparatively small number of mineral, vegetable and coal-tar colours should not be allowed in foods should be replaced by a list of permitted colours. The F.S.C.'s recommendations, with a few minor modifications, were subsequently incorporated into The Colouring Matter in Food Regulations, 1957 (S.I. 1957 No. 1066), which name thirty coal-tar colours on the permitted list. The regulations prohibit the sale or importation of food containing any added colouring matter other than that specified in the First Schedule (see below). Also raw meat, game, poultry, fish, fruit and vegetables are not permitted to contain any added colouring matter at all otherwise than for marking purposes (except that citrus fruit may contain added colouring matter provided that the skin is conspicuously marked using a permitted colour, with the words " colour added "). The permitted coal-tar colours which are listed in Part I of the First Schedule to the Regulations are as follows: Ponceau MX, Ponceau 4R, Carmoisine, Amaranth, Red 10 B, Erythrosine BS, Red 2G, Red 6B, Ponceau SX, Ponceau 3R, Fast Red E, Orange G, Orange RN, Oil Yellow GG, Tartrazine, Naphthol, Yellow S, Yellow 2G, Yellow RFS, Yellow RY, Sunset Yellow FCF, Oil Yellow XP, Green S, Blue VRS, Indigo Carmine, Violet BNP, Brown FK, Chocolate Brown FB, Chocolate Brown HT, Black PN.

Part II of the First Schedule lists the other permitted colours:—

" 1. Caramel and the colour obtained from cochineal.

2. The following colouring matters of vegetable origin—any colouring matter natural to edible fruits or vegetables and alkannet, annatto, carotene, chlorophyll, flavine, indigo, orchil, osage orange, persian berry, safflower, saffron, sandalwood, turmeric; or their pure colouring principles whether isolated from such natural colours or produced synthetically.

3. Bole or iron oxide, carbon black, titanium dioxide, ultramarine and, solely for the external colouring of dragees and the decoration of sugar-coated flour confectionery, silver or aluminium in leaf or powder form.

4. The aluminium or calcium salts (lakes) of any of the scheduled water-soluble colours."

The Second Schedule relates to the " Labelling of Colouring Matter and Colouring and Flavouring Compounds." Each container shall bear a distinctly and legibly printed label in the form of one of the following declarations:—

Colouring matter—

THIS FOOD COLOUR CONFORMS TO THE STATUTORY REQUIREMENTS OF THE UNITED KINGDOM

Colouring and flavouring compounds—

THE FOOD COLOUR IN THIS COMPOUND CONFORMS TO THE STATUTORY REQUIREMENTS OF THE UNITED KINGDOM

Identification of Water-Soluble Coal-Tar Dyes Present in Foodstuffs

The general scheme for identifying coal-tar dyes present in foods normally involves:

1. Preliminary treatment of the food.
2. Extraction of the colour from the prepared solution of the food.

3. Separation of mixed colours if more than one is present.
4. Identification of the separated colours.

1. Preliminary Treatment of the Food

Assuming that an acidic colour is present, the preliminary treatment involves removing interfering substances and obtaining the dye in acid solution preparatory to boiling with wool. Suggested methods of treating various types of food are given below (*cf.* " Separation and Identification of Food Colours Permitted by the Colouring Matters in Food Regulations, 1957," Assoc. Publ. Analysts, 1960):—

(a) *Non-alcoholic liquids, e.g.*, soft drinks: As most of the foods in this group are acidic they can usually be treated directly with wool. Otherwise, acidify slightly with acetic acid or potassium hydrogen sulphate.
(b) *Alcoholic liquids, e.g.*, wines: Boil to remove the alcohol and acidify (if necessary) as for (a).
(c) *Soluble foods, e.g.*, jams, sweets, icings: Dissolve in 30 ml. water and treat as in (a).
(d) *Starch-based foods, e.g.*, cakes, custard powder, golden raising powder: Grind 10 g. of sample very thoroughly with 50 ml. of 2 per cent. ammonia in 70 per cent. alcohol, allow to stand for some hours and centrifuge. Pour the separated liquid into a dish and evaporate on the water bath. Take up the residue in 30 ml. water containing acid and treat as in (a).
(e) *Candied fruits:* Treat as in (d).
(f) *Products with a high fat content, e.g.*, sausages, meat and fish pastes: De-fat with light petroleum and obtain the colour in aqueous solution by treatment with hot water (acidify, etc., as usual). Note that oil-soluble colours tend to give coloured solutions in organic solvents.

In difficult cases treatment with warm, 50–90 per cent. acetone or alcohol (which precipitates starch) containing 2 per cent. ammonia is often helpful. The organic solvent should be removed before acidifying (see (d) above). Note that basic colours dye the wool in alkaline solution and therefore require treatment with acid where the use of ammonia is called for above.

2. Extraction of the Colour From the Food

N.B. The wool should be prepared by boiling pure white knitting wool in very dilute sodium hydroxide solution and then in water (*cf.* Thaler and Sommer, *Chem. Abstr.*, 1954, **48**, 4714).

Add a 20 cm. strip of wool to about 35 ml. of the prepared, slightly acidified solution and boil. Continue boiling, wash the wool under the cold tap, transfer it to a small beaker and boil gently with dilute ammonia. If the colour is stripped by the alkali, the presence of an acidic coal-tar dye is indicated. Remove the wool, make the liquid slightly acid and boil. Add a little wool and continue boiling until all the colour is removed. Extract the dye from the wool again with a small volume of dilute ammonia, filter through a small plug of cotton wool and evaporate to low bulk. This double stripping technique usually gives a purer product, but it is not always necessary. Natural colours may also dye the wool during the first treatment, but the colour is not usually removed by ammonia.

Basic dyes can be separated by making the food alkaline with ammonia, boiling with wool and then stripping with acetic acid. As all the 28 permitted water-soluble colours are acidic, an indication of the presence of a basic dye at this early stage suggests that a non-permitted colour is present (*N.B.* Basic rhodamine dyes were quite frequently used prior to the 1957 Regulations).

3. Separation and Identification of the Extracted Colour(s)

General tests. Although it is always advisable to run chromatograms for the final detection of the colours, much work can often be saved by applying simple qualitative tests to fragments of the dyed wool and comparing the results with those obtained from known dyes.* It is useful for instance to apply common reagents (*e.g.*, conc. HCl) which may produce characteristic colour changes. Some xanthene dyes are extracted by ether from aqueous solution. It is sometimes possible to detect the presence of mixed colours by spotting on filter paper and adding a few drops of water so that separate zones appear.

Chromatographic separations. Colours can be separated using any of the usual paper chromatographic techniques. The A.P.A. booklet recommends the use of Whatman CRL 1 paper in large beakers (see Hunt *et al.*, *Analyst*, 1955, **80**, 172) for general work and the " circular zone " plate method of Rutter (*Analyst*, 1950, **75**, 37) for the final identification.

Solvents (*A.P.A.*)

No. 1 1 ml. 0·88 ammonia + 99 ml. water.
No. 2 2·5 per cent. aqueous sodium chloride.
No. 3 2 per cent. sodium chloride in 50 per cent. ethanol.
No. 4 *iso*-butanol (1 vol.), ethanol (2 vols.), water (1 vol.).
No. 5 *n*-butanol (20 vols.), water (12 vols.), glacial acetic acid (5 vols.).
No. 6 *iso*-butanol (3 vols.), ethanol (2 vols.), water (2 vols.); then to 99 ml. add 1 ml. of 0·88 ammonia.
No. 7 80 g. phenol + 20 g. water.

Separation of components of mixed colours. Paint a horizontal line of the prepared dye solution about 2 cm. from the bottom edge of a piece of Whatman No. 1 chromatographic paper and dry. Then elute upwards in a covered beaker using the most effective solvent to effect a separation (solvent No. **5** is often useful for general purposes, but it is advisable to study the R_F values in the table first in relation to the probable colours present). If more than one colour appears to be present, cut out the separate lines of colour, extract with water or aqueous-acetone, evaporate to dryness and re-dissolve in a few drops of water. If the separation is indefinite, repeat the process using different solvents.

4. Identification of the Separated Colours

Paper chromatography. Cut a piece of Whatman No. 1 chromatographic paper so that it can be rolled into a cylinder which fits inside a litre beaker without touching the sides. Draw a line parallel to the bottom of the paper and about 2 cm. away from it. Place a spot of the concentrated solution of the unknown dye on the line together with a series of spots (about 2 cm. apart) of aqueous solutions of known dyes of similar colour and dry. Roll the paper in a cylinder and hold the top corners together with a paper clip so that the vertical edges do not touch. Into the beaker place the rolled paper (not touching the sides) together with a 1 cm. layer of selected solvent (see R_F values in the table). Cover with a clock glass and allow the solvent front to rise 10–15 cm. Identify the colour by comparison with the spots obtained from the known dyes. Run similar chromatograms using different solvents, chosen in relation to their ability to differentiate between any which have similar R_F values on the first " run " (see also *J. Ass. Off. Agric. Chem.*, 1952, **35**, 423; 1953, **36**, 802; 1954, **37**, 812). The appearance of the spots under u.v. light may also be useful for identification purposes.

* Sets of the permitted coal-tar colours and of many non-permitted ones are obtainable from Solmedia Ltd., London, E.17.

TABLE SHOWING R_F VALUES AND SPECTR

	Class
REDS	
Ponceau MX	Azo
Ponceau 4R	Azo
Carmoisine	Azo
Amaranth	Azo
Red 10B	Azo
Erythrosine BS	Xanthene
Red 2G	Azo
Red 6B	Azo
Red FB	Azo
Ponceau SX	Azo
Ponceau 3R	Azo
Fast Red E	Azo
ORANGE	
Orange G	Azo
Orange RN	Azo
Sunset Yellow FCF	Azo
YELLOWS†	
Tartrazine	Pyrazolone
Naphthol Yellow S	Nitro
Yellow 2G	Pyrazolone
Yellow RFS	Azo
Yellow RY	Azo
GREENS, BLUES & VIOLETS	
Green S	Triphenylmethane
Blue VRS	Triphenylmethane
Indigo Carmine	Indigoid
Violet BNP	Triphenylmethane
BROWNS, BLACKS	
Brown FK	Azo
Chocolate Brown FB	Disazo
Chocolate Brown HT	Disazo
Black PN	Disazo

* "Separation and Identification of Food
Association of Public Analysts (London), 19(

† The permitted list also includes 2 oil-soluble

PHOTOMETRIC PEAKS OF PERMITTED WATER-SOLUBLE COLOURS*

R_F Values						Main Spectrophotometric Peaks mμ
Solvent No.:						
1	2	3	4	5	6	
0·47	0·05	0·31	0·40	0·53	0·69	500 in acid
0·95	0·36	0·42	0·29	0·33	0·29	505 in acid
0·61	0·04	0·56	0·51	0·56	0·28	515 in acid
0·77	0·06	0·20	0·24	0·19	0·20	520 in acid
0·59	0·03	0·27	0·32	0·36	0·31	525 in acid
0·23	0·00	0·70	1·00	1·00	0·68	525 in alkali
0·85	0·17	0·36	0·35	0·45	0·39	530 in acid
0·66	0·04	0·18	0·27	0·25	0·23	515 in acid
0·10	0·00	0·06	0·32	0·38	0·15	385 and 495 in acid
0·76	0·04	0·42	0·44	0·56	0·40	500 in acid
0·49	0·03	0·32	0·39	0·59	0·50	505 in acid
0·45	0·05	0·60	0·45	0·54	0·60	505 in acid
1·00	0·59	0·70	0·53	0·52	0·58	475 in acid
0·33	0·05	0·82	0·84	0·68	0·78	485 in acid
0·78	0·26	0·65	0·49	0·45	0·56	480 in acid
1·00	0·26	0·30	0·26	0·18	0·22	430 in acid
0·78	0·31	0·68	0·52	0·53	0·56	370 in acid
1·00	0·83	1·00	0·74	0·52	0·48	400 in acid
0·94	0·60	0·74	0·47	0·40	0·48	510 in acid
0·85	0·21	0·17	0·19	0·11	0·04	435 in acid
1·00	0·88	1·00	0·73	0·57	0·50	635 in acid
1·00	0·85	1·00	0·84	0·62	0·75	415 in acid
	0·07	0·30	0·28	0·21	0·27	285 and 615 in acid
0·40	0·06	1·00	0·80	0·59	0·80	590 in neutral
0·24	0·00	0·56	0·45	0·50	0·56	455 in acid
						470 in acid
						465 in acid
0·31	0·00	0·06	0·10	0·07	0·09	570 in acid

Colours Permitted by The Colouring Matters in Food Regulations, 1957," see also J. Chromatography, 1960, **3**, D6–D13).

Colours, viz. Oil Yellow GG (azo) and Oil Yellow XP (pyrazolone).

Confirmation of the identity of dyes using the spectrophotometer

Dilute the pure neutral dye solution to a suitable intensity of colour and determine the absorption curve on the spectrophotometer in neutral, acid and alkaline solution and compare the maxima with those obtained with the known dye (see table). The A.P.A. booklet gives the absorption curves of all the 28 permitted water soluble colours obtained with a Unicam SP 500 spectrophotometer. Each dye was examined in neutral (0·02 per cent. ammonium acetate), acid (N/10 HCl) and alkaline (N/10 NaOH) solution. The more important peaks are quoted in the table.

It should be borne in mind, however, that the dyeing and stripping process sometimes causes some decomposition of colours and this consequently affects the chromatograms and spectrophotometric curves obtained.

If the dye appears to be a non-permitted one, it is particularly important to apply chemical tests. In this connection the following references should be helpful:—

Woodman ("Food Analysis", 1941, McGraw-Hill, N.Y.).
Green ("The Analysis of Dyestuffs", Griffin, London).
Clayton (J. Soc. Dy. Col., 1937, **53**, 178).

Oil-Soluble Colours

Oil Yellow GG and Oil Yellow XP are the only oil-soluble coal-tar colours on the permitted list. Oil Yellow GG can be extracted by sodium hydroxide from its ethereal solution, but XP and many of the non-permitted colours cannot. For the isolation of the purified colour, the oil should first be extracted from the food with light petroleum. Then the colour can be removed with fairly concentrated acid (*A.O.A.C.*, 1960, p. 575), or by solvent partition using N,N-dimethylformamide (DMF) and light petroleum. The latter method due to Mark and McKeown (*J. Ass. Off. Agric. Chem.*, 1958, **41**, 817) is described below:—

METHOD

Dissolve 10 g. of oil in 50 ml. light petroleum, filter and transfer to a separator. Extract with three 20 ml. portions of DMF, discarding the petrol layer. Extract the combined DMF solution with four 25 ml. portions of light petroleum, back-extracting each time with 5 ml. DMF. Discard the petrol extracts. Dilute the combined DMF solution with an equal volume of water and extract with one 30 ml. and one 10 ml. portion of chloroform. Centrifuge and discard the aqueous DMF layer. Wash the combined chloroform solution with water to remove dissolved DMF. Evaporate the chloroform solution to dryness at room temperature under reduced pressure (if the residual fat content is low the chromatographic or spectrophotometric examination may commence at this point). Dissolve the residue in 25 ml. DMF, transfer to a separator, add 25 ml. water and extract with four 25 ml. portions of light petroleum. Discard the aqueous DMF layer, wash the combined light petroleum solution with water to remove the dissolved DMF and evaporate the light petroleum solution under reduced pressure at room temperature.

Oil-soluble colours can be separated by reverse-phase paper chromatography. The papers used are previously soaked in a 10 per cent. solution of liquid paraffin in mixed ethers, drained and then dried under vacuum. The paper is spotted with the colour after dissolving in a suitable solvent (*e.g.*, ether, mixed ethers or alcohol). Suitable chromatographic solvents are (1) 70 vols. acetone + 30 vols. water and (2) 60 vols. dioxan + 40 vols. water (*cf.* Silk, *J. Ass. Off. Agric. Chem.*, 1959, **42**, 427).

Oil-soluble colours can also be identified by dissolving the extracted dye in chloroform or acetone and obtaining the absorption curve by means of the spectrophotometer. The absorption peaks for the permitted oil-soluble colours are as follows : Oil Yellow GG at 380 mμ and Oil Yellow XP at 405 mμ.

Chapter 4

SUGAR AND PRESERVES

Sugars—Syrup—Liquid Glucose—Honey—Jam—Marmalade—Fruit Curd—Mincemeat

General Tests

Molisch's General Test for Carbohydrates. Add 2 drops of 20 per cent. alcoholic solution of α-naphthol to 2 ml. of a 0·1 per cent. solution of the substance and pour 2 ml. of concentrated sulphuric acid down the side of the tube so that a layer forms below the aqueous solution. Carbohydrates, if present, produce a violet colour which is discharged on the addition of excess alkali. The reaction appears to be due to the formation of a condensation product between the α-naphthol and the hydroxy-methyl-furfural produced by decomposition of the carbohydrate.

Production of Osazones by sugars. In common with other aldehydes and ketones, many sugars react with phenylhydrazine, giving rise to osazones, many of which have a characteristic micro-crystalline appearance and melting point:—

Dissolve 0·3 g. anhydrous sodium acetate and 0·2 g. phenylhydrazine hydrochloride in 2 ml. of the sugar solution (1 per cent.) by shaking together in a stoppered bottle (1 oz. size), heat in boiling water for thirty minutes and allow to cool naturally to room temperature. Examine the crystals microscopically and compare the crystals with those produced from known sugars. With monosaccharides osazones are produced much more quickly than with maltose or lactose. Sucrose may produce an osazone due to inversion during the heating.

Detection of Reducing Sugars with Alkaline Solutions of Copper or Silver Salts

When a solution of a reducing sugar is heated with an equal volume of mixed *Fehling's solution* (p. 116) in boiling water the cupric salt is reduced to red cuprous oxide. The reaction is given by lactose and maltose as well as the monosaccharides. Sucrose also gives a precipitate if the solution has been previously hydrolysed by heating with acid. Reducing sugars give a silver mirror when heated with ammoniacal silver solutions.

Barfoed's reagent is reduced by monosaccharides, but not appreciably by lactose and maltose. The reagent is prepared by dissolving 6·5 g. of crystallised copper acetate in 100 ml. of 1 per cent. acetic acid solution.

For the test, heat 5 ml. of the sugar solution with 5 ml. of Barfoed's reagent in boiling water for three and a half minutes. Production of a red precipitate of cuprous oxide indicates the presence of a monosaccharide.

Properties of Individual Carbohydrates

The following table gives some of the properties of the more important carbohydrates:—

PHYSICAL PROPERTIES OF CARBOHYDRATES

Name	Formula	*Specific rotation $[\alpha]_D^{20}$	Melting point °C.	Melting point of osazone °C.	Reaction with Fehling's solution
MONOSACCHARIDES					
Arabinose (pentose) .	$C_5H_{10}O_5$	+190·6 → **+104·5**	160	160	+
Dextrose (hexose) .	$C_6H_{12}O_6$	+112·2 → **+52·7**	146	205	+
Fructose (hexose) .	$C_6H_{12}O_6$	−132·2 → **−92·4**	103	205	+
Galactose (hexose) .	$C_6H_{12}O_6$	+151 → **+80·2**	167	200	+
Mannose (hexose) .	$C_6H_{12}O_6$	→ **+14·2**	132	198	+
DISACCHARIDES					
Sucrose . . .	$C_{12}H_{22}O_{11}$	**+66·53**	160	—	—
Lactose (anhydrous) .	$C_{12}H_{22}O_{11}$	+34·9 → **+55·4**	252	211	+
Lactose (hydrated) .	$C_{12}H_{24}O_{12}$	+85·0 → **+52·6**	203	211	+
Maltose (hydrated) .	$C_{12}H_{24}O_{12}$	+111·7 → **+130·4**	102	206	+
TRISACCHARIDE					
Raffinose (hydrated) .	$C_{18}H_{42}O_{21}$	**+105·2**	118	—	—
POLYSACCHARIDES					
Starch . . .		**+195**	—	—	—
Cellulose . . .		+	—	—	—
Inulin . . .		**−33 to −40**	178	—	—
Glycogen . . .		**+196**	—	—	—
Dextrin . . .		**+195**	—	—	—

Arabinose is a pentose which is widely distributed in plants and occurs to some extent in animal tissues, usually in the form of its anhydrides or pentosans, which yield arabinose (with xyloses) on hydrolysis. When distilled with dilute hydrochloric acid it yields furfural which gives characteristic colour reactions with resorcinol or phloroglucinol. If other sugars are present arabinose can be estimated in the same manner as pentosans (p. 200), but if no other reducing sugar is present, by copper reduction methods.

Dextrose. Dextrose, grape sugar or glucose occurs in a variety of animal and vegetable foodstuffs. *d*-Glucose exists in the anhydrous state (m.pt. 146° C.) and in the hydrated form (m.pt. 86° C.) which contains one molecule of water. It is produced by the hydrolysis of starch and disaccharides such as sucrose and is, therefore, frequently present in foodstuffs which contain sugar. Dextrose readily reduces Fehling's and Barfoed's solutions and is fermented by yeast.

Fructose. Fructose, lævulose or fruit sugar is commonly present in fruits and is lævorotatory. It exists in considerable quantities in

* Changes due to mutarotation (p. 113) are also shown.

honey and invert sugar. The specific rotation of fructose is considerably influenced by temperature. At 20° C., $[\alpha]_D^{20}$ is—92° but the rotation decreases with rise in temperature and at 87° C. a solution of invert sugar exhibits no rotation as the dextro-rotation due to dextrose is then neutralised by the lævo-rotation due to the fructose. It also exhibits the phenomenon of mutarotation (p. 113) to a marked extent.

Galactose occurs in a polymerised form in many gums. Commercially it is a product of the hydrolysis of lactose. It readily crystallises with one molecule of water ($C_6H_{12}O_6 . H_2O$) in the form of a white powder having a m.pt. of 120° C. (anhydrous form 167° C.). It exhibits mutarotation and has a high temperature coefficient.

Mannose occurs in complex polysaccharides called mannosans.

Sucrose, saccharose, cane sugar or beet sugar, is present in a large number of foods. It does not reduce Fehling's solution and does not form an osazone. In common with other disaccharides it is inverted by acid. The resulting "invert sugar" contains equivalent amounts of dextrose and fructose.

Lactose, or milk sugar is present in milk and has only a slightly sweet taste. It melts at 203° C. with decomposition and forms a monohydrate which is stable at 100° C. but decomposes at 130° C. This is of interest in connection with the total solids of milk. Thus, although the hydrate $C_{12}H_{22}O_{11}$, H_2O is not completely dehydrated until 130° C. is reached, it seems probable the total solids obtained by drying at 100°C. includes anhydrous lactose. Lactose reduces Fehling's solution, forms an osazone and is inverted by acids yielding equal quantities of dextrose and galactose. The specific rotation of 55·4° relates to the equilibrium or γ form of the α-hydrate and β-anhydride which exist in the freshly prepared solution.

Qualitative Tests for Lactose and Maltose

Fearon's test (*Analyst,* 1942, **67**, 130). To a dilute neutral solution of the sugar add 3 drops of methylamine hydrochloride solution (5 per cent.), boil for a few seconds and then add 5 drops of sodium hydroxide solution (20 per cent.). A yellow colour changing to carmine indicates the presence of lactose or maltose.

Rubner's test (*modified*). To 2 ml. of an aqueous solution of the sugar add 0·2 g. lead acetate, boil and add twice the volume of dilute ammonia. Lactose gives a dense white precipitate, but maltose only a slight precipitate. Then boil gently for one minute. Lactose shows a salmon pink colour and maltose an orange yellow.

Maltose occurs naturally in plants, leaves and seeds, and in malt. It is produced by the decomposition of starch and is therefore present in liquid glucose. It exists as a slightly sweet sugar forming a hydrate which is decomposed at or below 100° C. Maltose is hydrolysed by acid at about 80° C. yielding two molecules of dextrose.

Raffinose is one of the few trisaccharides of interest to the food analyst. It occurs in beet sugar, molasses and cereals. It does not reduce Fehling's solution nor does it form an osazone. On hydrolysis with strong acids it yields dextrose, fructose and galactose, but with weak acids fructose and the disaccharide melibiose are produced. Raffinose can be estimated reasonably accurately by inversion with top and bottom yeasts (see *A.O.A.C.*, 1960, 9th Edn., p. 424).

The relative sweetness of some sugars is given on p. 126.

Detection of Sugars by means of Paper Chromatography

Particularly in instances where the amount of carbohydrate is small, paper chromatography represents a useful method for identifying sugars. The following table gives the composition of some suitable solvents:—

SOLVENTS FOR PAPER CHROMATOGRAPHY OF THE SUGARS

Solvent Constituent	Percentage by Volume						
	1	2	3	4	5	6	7
Ethanol	10	—	—	—	—	—	—
n-Propanol	—	70	—	—	—	—	—
iso-Propanol	—	—	75	70	65	—	—
n-Butanol	40	—	—	10	—	45	—
Ethyl acetate	—	10	—	—	—	—	40
Pyridine	—	—	—	—	15	35	20
Ammonia, 0·88	1	—	—	—	—	—	—
Water	49	20	25	20	20	20	40

In general the order of the R_F values is: pentoses (highest), hexoses, disaccharides, trisaccharides. The R_F values increase if the proportion of pyridine or water is increased, but butanol and ethyl acetate have the opposite effect.

One-way ascending chromatography using Whatman No. 4 paper is usually adequate. The solution used should preferably contain about 0·5 per cent. of the sugar and be free of protein, which may be removed with trichloracetic acid. After running the chromatograms and marking the solvent front, the paper should be dried in the air, dipped rapidly through an appropriate reagent and warmed gently at 95–110° C. until the colour development is maximal. Many dipping and spraying reagents have been described, *e.g.*, silver nitrate (M/10 $AgNO_3$ in 5M ammonia), aniline phthalate (2 per cent. in moist butanol) and naphthoresorcinol (0·1 per cent. in 0·4N HCl in 80 per cent. ethanol). Such reagents have been tabulated in the extensive review of the subject due to Lederer and Lederer (" Chromatography," Elsevier, Amsterdam).

For the quantitative determination of sugars, Wilson (*Anal. Chem.*, 1959, **31**, 1199) uses butanol pyridine/water (6 : 4 : 3) as solvent. After treatment with aniline phthalate, the coloured spots are eluted from the paper with 0·7 N hydrochloric acid in 80 per cent. ethanol (v/v) and the concentration of the sugar is determined by spectrophotometry (*cf.* Lederer and Lederer, *loc. cit.*).

TABLE OF SPECIFIC GRAVITIES[1] AND REFRACTIVE INDICES[2] AT 20° C. OF SOLUTIONS OF SUCROSE

Sucrose per cent. w/w	Specific gravity at 20/20° C.	Refractive index n_D^{20}	Sucrose per cent. w/w	Specific gravity at 20/20° C.	Refractive index n_D^{20}
0	1·00000	1·33299	51	1·23727	1·4221
1	1·00389	1·33443	52	1·24284	1·4242
2	1·00779	1·33588	53	1·24844	1·4264
3	1·01172	1·33733	54	1·25408	1·4285
4	1·01567	1·33880	55	1·25976	1·4307
5	1·01965	1·34027	56	1·26548	1·4329
6	1·02366	1·34176	57	1·27123	1·4351
7	1·02770	1·34326	58	1·27703	1·4373
8	1·03176	1·34477	59	1·28286	1·4396
9	1·03586	1·34629	60	1·28873	1·4418
10	1·03998	1·34783	61	1·29464	1·4441
11	1·04413	1·34937	62	1·30059	1·4464
12	1·04831	1·35093	63	1·30657	1·4486
13	1·05252	1·35250	64	1·31260	1·4509
14	1·05677	1·35408	65	1·31866	1·4532
15	1·06104	1·35567	66	1·32476	1·4555
16	1·06534	1·35728	67	1·33090	1·4579
17	1·06968	1·35890	68	1·33708	1·4603
18	1·07404	1·36053	69	1·34330	1·4627
19	1·07844	1·36218	70	1·34956	1·4651
20	1·08287	1·36384	71	1·35585	1·4676
21	1·08733	1·36551	72	1·36218	1·4700
22	1·09183	1·36719	73	1·36856	1·4725
23	1·09636	1·36888	74	1·37496	1·4749
24	1·10092	1·37059	75	1·38141	1·4774
25	1·10551	1·3723	76	1·38790	1·4799
26	1·11014	1·3740	77	1·39442	1·4825
27	1·11480	1·3758	78	1·40098	1·4850
28	1·11949	1·3775	79	1·40758	1·4876
29	1·12422	1·3793	80	1·41421	1·4901
30	1·12898	1·3811	81	1·42088	1·4927
31	1·13378	1·3829	82	1·42759	1·4954
32	1·13861	1·3847	83	1·43434	1·4980
33	1·14347	1·3865	84	1·44112	1·5007
34	1·14837	1·3883	85	1·44794	1·5033
35	1·15331	1·3902	86	1·45480	
36	1·15828	1·3920	87	1·46170	
37	1·16329	1·3939	88	1·46862	
38	1·16833	1·3958	89	1·47559	
39	1·17341	1·3978	90	1·48259	
40	1·17853	1·3997	91	1·48963	
41	1·18368	1·4016	92	1·49671	
42	1·18887	1·4036	93	1·50381	
43	1·19410	1·4056	94	1·51096	
44	1·19936	1·4076	95	1·51814	
45	1·20467	1·4096	96	1·52535	
46	1·21001	1·4117	97	1·53260	
47	1·21538	1·4137	98	1·53988	
48	1·22080	1·4158	99	1·54719	
49	1·22625	1·4179	100	1·55454	
50	1·23174	1·4200			

[1] Bur. Standards Circ. C440 (U.S.A.).

[2] International Scale (1936).

CORRECTIONS FOR DETERMINING THE PERCENTAGE OF SUCROSE IN SUGAR SOLUTIONS BY REFRACTOMETER WHEN READINGS ARE MADE AT TEMPERATURES OTHER THAN 20° C.
(International Temperature Correction Table, 1936; *Intern. Sugar J.*, 1937, **39**, 24s)

Temperature °C.	PER CENT. SUCROSE										
	0	5	10	15	20	25	30	40	50	60	70
	Subtract from the per cent. sucrose										
10	0·50	0·54	0·58	0·61	0·64	0·66	0·68	0·72	0·74	0·76	0·79
11	·46	·49	·53	·55	·58	·60	·62	·65	·67	·69	·71
12	·42	·45	·48	·50	·52	·54	·56	·58	·60	·61	·63
13	·37	·40	·42	·44	·46	·48	·49	·51	·53	·54	·55
14	·33	·35	·37	·39	·40	·41	·42	·44	·45	·46	·48
15	·27	·29	·31	·33	·34	·34	·35	·37	·38	·39	·40
16	·22	·24	·25	·26	·27	·28	·28	·30	·30	·31	·32
17	·17	·18	·19	·20	·21	·21	·21	·22	·23	·23	·24
18	·12	·13	·13	·14	·14	·14	·14	·15	·15	·16	·16
19	·06	·06	·06	·07	·07	·07	·07	·08	·08	·08	·08
	Add to the per cent. sucrose										
21	0·06	0·07	0·07	0·07	0·07	0·08	0·08	0·08	0·08	0·08	0·08
22	·13	·13	·14	·14	·15	·15	·15	·15	·16	·16	·16
23	·19	·20	·21	·22	·22	·23	·23	·23	·24	·24	·24
24	·26	·27	·28	·29	·30	·30	·31	·31	·31	·32	·32
25	·33	·35	·36	·37	·38	·38	·39	·40	·40	·40	·40
26	·40	·42	·43	·44	·45	·46	·47	·48	·48	·48	·48
27	·48	·50	·52	·53	·54	·55	·55	·56	·56	·56	·56
28	·56	·57	·60	·61	·62	·63	·63	·64	·64	·64	·64
29	·64	·66	·68	·69	·71	·72	·72	·73	·73	·73	·73
30	·72	·74	·77	·78	·79	·80	·80	·81	·81	·81	·81

DETERMINATION OF SUGARS

When pure syrups of known composition are being examined, the refractometer affords a rapid and fairly accurate method of estimating the sugar content. The specific gravity is also of value, but tends to be less accurate. The table on p. 109 shows the specific gravities and refractive indices of solutions of sucrose. The soluble solids as determined from the refractive index and calculated as sucrose (irrespective of the actual soluble substances present) is most useful industrially for routine control purposes when a rapid comparative measure is the major consideration. Macara (*Analyst,* 1931, **56**, 394) gives the following corrections to allow for other substances present in solutions of jam, etc.

Substance	*Percentage in sample*	*Corrections to be applied to the Total Soluble Solids from the direct Refractometric readings* Per cent.
Invert Sugar	20	+0·45
	40	+0·90
	60	+1·35
Glucose Solids . . .	20	−0·25
	40	−0·5
Citric Acid	1	+0·1
	2	+0·2
	3	+0·3

For the accurate determination of sugar, it is preferable to use polarimetry. Copper reduction methods (volumetric and gravimetric) also give accurate results provided the prescribed technique is followed exactly. For accurate results (particularly in polarimetry) it is often necessary to clarify the solution using one of the following reagents:—

Alumina cream is prepared by adding a slight excess of ammonia to a saturated solution of alum. It is either used unwashed or washed by repeated decantation.

Basic lead acetate is efficient, but it should be used in moderation, and it is preferable to add alumina cream afterwards in order to reduce the tendency to absorb sugars (especially fructose).

Neutral lead acetate is less efficient, as a clearing agent, but it does not appear to absorb sugars.

Zinc ferrocyanide is a good general clearing agent. Firstly zinc acetate solution (21·9 g. crystalline zinc acid + 3 ml. glacial acetic acid in 100 ml. water) is added to the sugar solution. Then an equal volume of potassium ferrocyanide (10·6 per cent. in water) is added and the liquid is mixed, made up to the mark and filtered (after allowing the precipitate to settle).

Animal charcoal tends to absorb sugars and it is always advisable to reject the first runnings.

Other clearing agents which are used include phosphotungstic acid and sodium tungstate.

It should be borne in mind that a correction for the volume of precipitate may be necessary with some samples and when greater accuracy is required.

Methods Prescribed for the Inversion of Sucrose in the Fertilisers and Feeding Stuffs Regulations (*cf.* pp. 115 and 373)

1955 *Regulations.* Pipette an aliquot of the clarified solution into a 100 ml. graduated flask, dilute to 75 ml. and add 10 ml. of 6·34 N hydrochloric acid (≡ 6·5 ml. conc. HCl). Mix. Place the flask in a 70° C. water bath and when the temperature of the solution reaches 67° C. (which should take two and a half to three minutes), continue the heating for a further five minutes (the solution should then be at about 69° C.). Cool immediately, exactly neutralise to litmus paper, make up to 100 ml. and filter.

1960 *Regulations.* Pipette an aliquot of the clarified solution into a 300 ml. beaker, add 15 ml. of N/1 hydrochloric acid, dilute to 150 ml. with water, cover with a clock glass and heat to boiling point. Continue to boil for two minutes, cool, add phenolphthalein, just neutralise with sodium hydroxide solution (10 per cent.), transfer to a 200 ml. graduated flask and dilute to 200 ml.

Polarimetric Methods for Determining Sugars

The specific rotation (or specific rotatory power) measured using the sodium D line can be expressed by the formula:—

$$\text{Specific rotation at } 20^\circ \text{ C.} = [\alpha]^{20}_{D} = \frac{100\alpha}{lc}$$

where α = observed angular rotation

l = length of tube in decimetres

c = concentration of solution in g./100 ml.

The following table gives the formulæ for calculating the specific rotations for various sugars:—

Sugar	Formula
Sucrose	$[a]^{20}_{D} = 66{\cdot}462 + 0{\cdot}0087c - 0{\cdot}000235c^2$
Dextrose	$[a]^{20}_{D} = 52{\cdot}50 + 0{\cdot}0188p + 0{\cdot}000517p^2$
Dextrose	$[a]^{20}_{5461 A} = 62{\cdot}032 + 0{\cdot}04257c$
Dextrose	$[a]^{20}_{5461 A} = 62{\cdot}032 + 0{\cdot}04220p + 0{\cdot}0001897p^2$
Fructose	$[a]^{20}_{D} = -113{\cdot}96 + 0{\cdot}258q$
Maltose	$[a]^{20}_{D} = 138{\cdot}475 - 0{\cdot}01837p$
Lactose	$[a]^{2}_{D} = 52{\cdot}53$
Invert sugar	$[a]^{2}_{D} = -(19{\cdot}415 + 0{\cdot}07065c - 0{\cdot}00054c^2)$
Invert sugar temperature correction	$[a]_{D} = [a]^{20}_{D} + (0{\cdot}283 + 0{\cdot}0014c)(t - 20^\circ \text{ C.})$

c = concentration in grams per 100 ml.

p = percentage by weight.

q = percentage of water.

t = temperature (°C.).

(Reproduced from Circular 440, Polarimetry Saccharimetry and the Sugars, Nat. Bur. Stds. (1942).)

The specific rotations at 20° C. for various sugars when in solution at various concentrations are given in the following table:—

Sugar	Specific rotations $[\alpha]_D^{20}$ at the following concentrations			
	10%	15%	20%	25%
Dextrose	+52·7°	+52·9°	+53·1°	+53·5°
Fructose	−90·7°	−92·0°	−93·3°	−94·6°
Invert sugar*	−19·8°	−20·2°	−20·4°	−20·6°
Lactose (hydrated)	+52·5°	+52·5°	+52·5°	+52·5°
Maltose	+138·3°	+138·2°	+138·1°	+138·0°
Sucrose	+66·5°	+66·5°	+66·5°	+66·5°

* Invert sugar has a zero rotation at 87° C.

The following specific rotations of other carbohydrates when in solution of about 10–15 per cent. concentration at 20° C. using sodium light are sufficiently accurate for most purposes: arabinose +104·5°, galactose +80·2°, raffinose +105·2°, xylose +18·8°, starch and dextrin +195°.

Mutarotation. When a sugar solution is freshly prepared it has a different rotation from that of the same solution after standing some hours. This is due to the gradual attainment of equilibrium between the pyranose and furanose isomeric forms which individually have different rotations. Errors due to mutarotation, as this phenomenon is known, can be eliminated by allowing the solution to stand for several hours before making the observations or by adding a few drops of ammonia which establishes the optical equilibrium at once.

Saccharimeters. Saccharimeters are virtually ordinary polarimeters which are calibrated in percentages of sugar, *i.e.*, a stated " normal weight " of sucrose in 100 ml. of solution gives a rotation of 100 divisions on the sugar scale. As various sugar scales had been proposed using different normal weights,* International Commissions in 1900, 1912 and 1932 formulated certain proposals. At a meeting of the Amsterdam Commission in 1932 it was recommended that new saccharimeters be calibrated as degrees sugar (°S.) on an " International Sugar Scale ":—

" It is recommended that the polarisation of the normal solution (26·000 g. of pure sucrose dissolved in 100 ml. and polarised at 20° C. in a 200-mm. tube, using white light and the dichromate

* *Normal weights* (*classical*): Ventzke and Schmidt-Hænsch scales 26·05 g. (1° = 0·3466° angular); Laurent and Soleil-Duboscq scales 16·3 g. (1° = 0·2167° angular).

filter as defined by the Commission) be accepted as the basis of calibration of the 100° point on the International Sugar Scale" (one degree = 0·34620 angular degrees).

The optical path in the dichromate solution is equivalent to a layer of liquid 15 mm. thick for a 6 per cent. solution (*i.e.*, for sucrose, thickness in mm. × % concentration = 90).

Saccharimeter readings are sometimes referred to as Degrees Ventzke (V). In using the relevant formulæ, readings must be calculated back so that they are equivalent to those given by a 26 per cent. solution using a 2 dcm. tube.

Polarimetric determination of a single sugar. If sucrose is the only optically active substance present, the reading on the saccharimeter of a 26 per cent. solution in a 2 dcm. tube at 20° C. gives the percentage present in the sample. For other samples containing another single sugar the formula is:—

$$\text{Sugar A}(\%) = V \times \frac{[\alpha]_D^{20} \text{ for sucrose}}{[\alpha]_D^{20} \text{ for sugar A}}$$

where V = saccharimeter reading.

Using a polarimeter reading in angular degrees, prepare a 20 per cent. clarified solution and measure the angular rotation (α) at 20° C. in a 2 dcm. tube:—

$$\text{Sugar A}(\%) = \frac{\alpha}{[\alpha_D^{20}] \text{ for sugar A}} \times \frac{100}{2} \times \frac{100}{20}$$

Polarimetric determination of sucrose in the presence of other sugars. For the estimation of sucrose in the presence of invert sugar as in molasses, syrups or similar products the Clerget-Herzfeld double polarisation method is often employed. This depends on the change from dextro-rotation to lævo-rotation where sucrose is hydrolysed by acids or enzymes to a mixture of equal parts of dextrose and fructose.

The A.O.A.C. procedure using the saccharimeter is as follows: Prepare 100 ml. of a solution (clarified if necessary) containing 26 g. of the material and 2·315 g. of sodium chloride and take the reading (D) in a saccharimeter using a 2 dcm. tube at 20° C.

Pipette 50 ml. of the 26 per cent. sugar solution into a 100 ml. flask and add 25 ml. of water. Then add slowly (while rotating the flask) 10 ml. of 6·34 N hydrochloric acid and place in a water bath maintained at 60° C. for nine minutes, agitating the flask for the first three minutes. Then immerse the flask in water at 20° C. When the temperature of the contents has fallen to 35° C. dilute nearly to the mark and leave the flask in the water for a further thirty minutes before making up to the mark. Then double the saccharimeter reading of this solution (at 20° C. in a 2 dcm. tube) to give that for the normal weight (I). The A.O.A.C. formula is based on the change in the saccharimeter reading at 20° C. for solutions containing the normal weight (26 per cent.) from +100° to −32·56° when sucrose is hydrolysed by the above method

$$\text{Sucrose } (\%) = S = \frac{100\,(D - I)}{132{\cdot}56 + 0{\cdot}0794\,(m - 13) - 0{\cdot}53\,(t - 20)}$$

where t = temperature at which the readings are made.
m = T.S. from original sample of inverted solution.

If the original solution contains exactly 26 per cent. of the sample and the readings are made at 20° C., the formula becomes:—

$$S = \frac{100\,(D - I)}{132{\cdot}56}$$

N.B.—The divisor of 132·56 used in this formula applies to the acid inversion method described above, but needs to be modified for different conditions of hydrolysis (see also the discussion of the Inversion Divisor Factor below).

Clerget-Herzfeld Calculation Using Angular Degrees

If the specific rotations (100α/lc) of the sample solutions before and after inversion are $[\alpha_D]^{20}_D$ and $[\alpha_I]^{20}_D$ respectively, the Clerget-Herzfeld formula using the results from angular degrees is:—

$$\text{Sucrose } (\%) = S = \frac{[\alpha_D]^{20}_D - [\alpha_I]^{20}_D}{Q}$$

The Inversion Divisor Factor (Q) is the change in the specific rotation of sucrose on inversion divided by 100. As one saccharimeter degree equals 0·34620 × angular degrees and the specific rotations are based on a theoretical 100 per cent. solution in a 1 dcm. tube, Q can be calculated from the divisor (132·56) in the A.O.A.C. formula:—

$$Q = \frac{132{\cdot}56 \times 0{\cdot}34620}{2 \times 26} = 0{\cdot}8825$$

This value for Q applies for acid inversion at 60° C. and when the polarimetric readings are measured at 20° C. using sodium light. Other factors are used for different conditions, however. It has been shown that the Inversion Divisor Factor varies according to (a) the presence of other substances, including the precipitant used, (b) the method of hydrolysis employed, *i.e.*, the figure is influenced by whether acid or invertase is employed, the concentration and temperature and (c) the nature of the light. Jackson and Gillis (see Circ. 440, Nat. Bur. Stds., 1942) found that, using acid inversion at 70° C. there was a greater likelihood of destruction of fructose than at 60° C. The latter temperature was adopted by the Milk Products Sub-Committee of the S.P.A. (*Analyst*, 1930, **55**, 115), who calculated the following divisors for readings at 20° C.:—

	Zinc ferrocyanide precipitant Q	Phosphotungstic acid precipitant Q
Sodium light	0·8825	0·8865
Mercury green line . . .	1·0392	1·0439
International sugar scale (j) light .	2·549	2·561

The Sub-Committee (*loc. cit.*) also quote the necessary correction factors which should be applied to Q for variations in the concentration and the temperature (*cf. Analyst*, 1928, **53**, 569).

The method may be extended to the estimation of another sugar (X) such as invert sugar or lactose in the presence of sucrose. Using saccharimeter readings, if S is the percentage of sucrose calculated from the Clerget-Herzfeld formula:—

$$X\ (\%) = \frac{(D - S)\ 66{\cdot}5}{[\alpha_D]^{20}_D \text{ for sugar X}}$$

For lactose hydrate and invert sugar, the specific rotations $[\alpha]^{20}_D$ are $+52{\cdot}5$ and $-20{\cdot}0$ respectively.

Using angular rotations:—

$$X\ (\%) = \frac{([\alpha_D]^{20}_D - 0{\cdot}665 \times S)\ 100}{[\alpha]^{20}_D \text{ for sugar X}}$$

where $[\alpha_D]^{20}_D$ is the specific rotation of the sample before inversion at 20° C.

These formulæ do not apply when a third optically active substance is present such as in a material containing liquid glucose as well as sucrose and invert sugar (see p. 130). It should also be borne in mind that other carbohydrates are hydrolysed to a greater or lesser extent according to the experimental conditions. The normal treatment with acid does not affect lactose, dextrose, maltose or dextrins appreciably, however, so that the Clerget process is reasonably accurate in their presence.

Raffinose can be approximately estimated by Clerget's acid inversion method (using the appropriate rotations). For greater accuracy, however, it is preferable to use the A.O.A.C. procedure which describes polarimetric estimations before and after treatment with two enzyme preparations. Invertase from "top" baker's yeast inverts sucrose to invert sugar and hydrolyses raffinose to a mixture of fructose and melibiose. "Bottom" brewer's yeast also contains melibiase and has the additional power to hydrolyse melibiose to a mixture of dextrose and galactose.

Fehling's Solution

No. 1. Dissolve 69·278 g. copper sulphate ($CuSO_4.5H_2O$) in water and make up to one litre.

No. 2. Dissolve 100 g. sodium hydroxide and 346 g. sodium potassium tartrate in water and make up to one litre.

Gravimetric Copper Reduction Methods for Determining Sugars

Reducing sugars can be determined by weighing the cuprous oxide produced by reduction of Fehling's solution (p. 116) under standardised conditions. Alternatively, the precipitate can be ignited and weighed as cupric oxide. The extent of the reduction of Fehling's solution varies with the experimental conditions, so that it is most important to calculate the results from a table of figures which apply to the particular technique employed.

Mark a 400 ml. squat form beaker at 100 ml., add to it 50 ml. of Fehling's solution and cover with a clock glass. Place in a boiling water bath, so that the boiling water surface is level with the 100 ml. mark, but the beaker should not rest on the bottom of the bath. After five minutes, add 50 ml. of sugar solution, replace the cover and continue the heating for exactly twelve minutes. Filter at once through a prepared Gooch crucible, wash with hot water and finally with alcohol. Place the crucible inside a platinum dish and ignite to constant weight in the air, so that the precipitated cuprous oxide is oxidised to cupric oxide. The amount of sugar equivalent to the quantity of copper oxide formed may be obtained from the table on page 106 (Elsdon, *Analyst*, 1923, **48**, 436).

Alternatively, the copper in the cuprous oxide can be determined by dissolving the precipitate in nitric acid and titrating with thiosulphate. Further, the precipitate can be dissolved in a known quantity of acid ferric sulphate solution and the iron so reduced can be titrated with permanganate. Both of these volumetric methods are described in the A.O.A.C. (1960, p. 427), which, however, describes Munson and Walker's method for the production of the original cuprous oxide. For this, the Fehling's and sugar solutions are heated on a gauze with a flame, so that the liquid boils in four minutes and the boiling is continued for a further two minutes. For the gravimetric process, the cuprous oxide precipitate is dried at 100° C. and weighed as such.

In the gravimetric processes, a blank should be determined, and this should include the same amount of sucrose present as in the test sample. The effect of the presence of sucrose is allowed for in the A.O.A.C. tables (*cf. J. Amer. Chem. Soc.*, 1921, **43**, 1503). Starch, if present, should be filtered off before the determination.

Copper Reducing Power (" K " Value)

The " copper reducing power " is the copper oxide reduced by 100 parts of the substance calculated as dextrose. It is often referred to as the " K Value". Thus, a syrup which contains 45 per cent. of reducing sugars expressed as dextrose has a K value of 45.

QUANTITIES OF COPPER OXIDE PRODUCED UNDER STANDARD CONDITIONS BY VARIOUS CARBOHYDRATES

Quantities expressed in Milligrams in all cases

Cupric oxide	Cuprous oxide	Copper	Dextrose	Starch	Lævulose	Hydrated lactose, $C_{12}H_{22}O_{11}.H_2O$	Anhydrous lactose	Maltose	Invert sugar	Cane sugar
100	89·9	79·9	—	—	—	59·2	56·2	72·5	—	—
110	98·9	87·9	—	—	—	65·2	61·9	79·8	45·3	43·0
120	107·9	95·9	46·5	41·8	51·7	71·2	67·6	87·2	49·2	46·7
130	116·9	103·9	50·4	45·4	55·5	77·2	73·3	95·3	53·1	50·4
140	125·9	111·9	54·2	48·8	59·5	83·2	79·0	102·0	57·0	54·2
150	134·9	119·8	58·0	52·2	63·9	89·3	84·8	109·2	61·0	58·0
160	143·9	127·8	61·8	55·6	68·1	95·4	90·6	116·8	65·1	61·8
170	152·9	135·8	65·7	59·1	72·4	101·4	96·3	124·2	69·2	65·7
180	161·9	143·8	69·6	62·6	76·7	107·4	102·0	131·5	73·4	69·7
190	170·9	151·8	73·6	66·2	80·9	113·5	107·9	138·8	77·5	73·6
200	179·9	159·8	77·6	69·8	85·3	119·7	113·7	146·3	81·5	77·4
210	188·9	167·8	81·6	73·4	89·5	126·0	119·7	153·6	85·7	81·4
220	197·9	175·8	85·5	77·0	94·0	132·2	125·6	161·0	90·0	85·5
230	206·9	183·8	90·0	81·0	98·6	138·4	131·5	168·3	94·4	89·7
240	215·9	191·7	94·2	84·8	103·0	144·6	137·4	175·7	98·7	93·8
250	224·9	199·7	98·3	88·5	107·3	150·7	143·2	183·1	102·9	97·8
260	233·9	207·7	102·5	92·3	111·8	157·0	149·2	190·5	107·1	101·7
270	242·9	215·7	106·7	96·0	116·4	163·0	154·9	197·9	111·4	105·8
280	251·9	223·7	110·8	99·7	121·0	169·2	160·7	205·2	115·7	109·9
290	260·8	231·7	115·0	103·5	125·6	175·5	166·7	212·6	120·1	114·1
300	269·8	239·7	119·4	107·5	130·1	182·3	173·2	220·0	124·6	118·4
310	278·8	247·7	123·7	111·3	134·7	188·9	179·5	227·3	129·1	122·6
320	287·8	255·7	128·2	115·4	139·5	195·4	185·6	234·7	133·7	127·0
330	296·8	263·6	132·6	119·3	144·5	202·0	191·9	242·1	138·2	131·3
340	305·8	271·6	137·1	123·4	149·5	208·7	198·3	249·5	142·8	135·7
350	314·8	279·6	141·7	127·5	154·2	215·3	204·5	256·9	147·7	140·3
360	323·8	287·6	146·4	131·8	158·9	222·0	210·9	264·3	152·7	145·1
370	332·8	295·6	151·2	136·1	163·8	228·6	217·2	271·6	157·5	149·6
380	341·8	303·6	155·6	140·1	168·5	235·2	223·4	279·0	161·9	153·8
390	350·8	311·6	160·5	144·5	173·5	241·9	229·8	286·4	166·5	158·2
400	359·8	319·6	165·2	148·7	178·4	248·6	236·2	293·7	171·4	162·8
410	368·8	327·6	170·1	153·1	183·3	254·7	242·0	301·1	176·2	167·2
420	377·8	335·5	175·0	157·5	188·3	260·9	247·9	—	181·2	172·1
430	386·8	343·5	179·9	161·9	193·2	267·2	253·8	—	186·2	176·9
440	395·8	351·5	185·0	166·5	198·4	273·8	260·1	—	191·2	181·7
450	404·8	359·5	190·0	171·0	203·6	280·4	266·4	—	196·3	186·5
460	413·8	367·5	195·0	175·5	208·8	287·2	272·3	—	201·6	191·5

Volumetric Copper Reduction Methods for Determining Sugars

The volumetric method of Lane and Eynon (*Analyst*, 1923, **48**, 220) is often the most convenient method for determining reducing sugars. The volume of sugar solution required to reduce 10 or

25 ml. Fehling's Solution completely is determined using methylene blue as the redox indicator for assessing the end point. A minute excess of reducing sugar solution is necessary for the reduction of the indicator itself.

Lane and Eynon's Method

The sugar solution should be neutral and preferably be clarified (lead and calcium salts should be removed with potassium oxalate). The concentration of the sugar solution should be such that the titration is 15–50 ml. (this can usually be achieved with 0·1–0·3 gm. sugar present per 100 ml. for 10 ml. Fehling's solution and 0·25–0·8 g. for 25 ml.). Place the sugar solution in a 50 ml. burette fitted with a pinch cock (instead of a tap) and with the glass jet bent twice so that the graduations are not directly over the flask.

Preliminary titration. Pipette 10 or 25 ml. of Fehling's solution (p. 104) into a 300 ml. conical flask, and add, from the burette, 15 ml. of the sugar solution. Boil the liquid on an asbestos-covered gauze and add further quantities of the sugar solution (1 ml. at a time) at 10–15 second intervals to the boiling liquid

MALTOSE TABLE

Ml. of sugar solution required.	For 10 ml. of Fehling's solution.				For 25 ml. of Fehling's solution			
	Hydrated maltose $C_{12}H_{22}O_{11}\,H_2O$.		Anhydrous maltose $C_{12}H_{22}O_{11}$		Hydrated maltose $C_{12}H_{22}O_{11},H_2O$.		Anhydrous maltose $C_{12}H_{22}O_{11}$.	
	Factor.*	Mg. per 100 ml.	Factor.*	Mg. per 100 ml.	Factor.†	Mg. per 100 ml.	Factor.†	Mg. per 100 ml.
15	**81·3**	**542**	**77·2**	**515**	**208·2**	**1388**	**197·8**	1319
16	81·2	507	77·1	482	207·8	1298	197·4	1233
17	81·1	477	77·0	453	207·4	1220	197·0	1159
18	81·0	450	77·0	427	207·1	1151	196·7	1093
19	80·9	426	76·9	405	206·8	1088	196·5	1034
20	**80·8**	**404·0**	**76·8**	**383·8**	**206·5**	**1032·3**	**196·2**	**980·7**
21	80·7	384·3	76·7	365·1	206·1	981·6	195·8	932·5
22	80·6	366·4	76·6	348·1	205·8	935·5	195·5	888·7
23	80·5	350·0	76·5	332·5	205·4	893·2	195·1	848·5
24	80·4	335·0	76·4	318·3	205·1	854·5	194·8	811·8
25	**80·4**	**321·5**	**76·4**	**305·4**	**204·8**	**819·0**	**194·5**	**778·1**
26	80·3	308·8	76·3	293·4	204·4	786·3	194·2	747·0
27	80·2	297·0	76·2	282·2	204·1	756·0	193·9	718·2
28	80·1	286·1	76·1	271·8	203·8	727·9	193·6	691·5
29	80·0	276·0	76·0	262·2	203·5	701·7	193·3	666·6
30	**80·0**	**266·6**	**76·0**	**253·3**	**203·2**	**677·3**	**193·0**	**643·4**
31	79·9	257·8	75·9	244·9	202·9	654·3	192·8	621·6
32	79·9	249·7	75·9	237·2	202·6	633·1	192·5	601·4
33	79·8	241·9	75·8	229·8	202·3	613·0	192·2	582·4
34	79·8	234·6	75·8	222·9	202·0	594·3	191·9	564·6
35	**79·7**	**227·6**	**75·7**	**216·2**	**201·8**	**576·5**	**191·7**	**547·7**
36	79·6	221·1	75·6	210·0	201·5	559·7	191·4	531·7
37	79·6	215·0	75·6	204·3	201·2	543·9	191·2	516·7
38	79·5	209·2	75·5	198·7	201·0	528·9	191·0	502·5
39	79·5	203·8	75·5	193·6	200·8	514·7	190·8	489·0
40	**79·4**	**198·5**	**75·4**	**188·6**	**200·5**	**501·3**	**190·5**	**476·2**
41	79·4	193·7	75·4	184·3	200·3	488·5	190·3	464·1
42	79·3	188·8	75·3	179·4	200·1	476·3	190·1	452·5
43	79·3	184·3	75·3	175·1	199·8	464·7	189·8	441·5
44	79·2	180·0	75·2	171·0	199·6	453·6	189·6	430·9
45	**79·2**	**175·9**	**75·2**	**167·1**	**199·4**	**443·0**	**189·4**	**420·9**
46	79·1	172·0	75·1	163·4	199·2	433·1	189·2	411·4
47	79·1	168·3	75·1	159·9	199·0	423·6	189·0	402·4
48	79·1	164·7	75·1	156·5	198·9	414·4	188·9	393·7
49	79·0	161·2	75·0	153·1	198·7	405·5	188·8	385·2
50	**79·0**	**158·0**	**75·0**	**150·1**	**198·6**	**397·2**	**188·7**	**377·3**

* Mg. of maltose corresponding to 10 ml. of Fehling's solution
† „ „ „ „ „ 25 ml. „ „ „

until the blue colour is nearly discharged. Then add 3–5 drops of aqueous methylene blue solution (1 per cent.) and continue titrating the boiling liquid until the indicator is completely decolorised.

Accurate titration. Repeat the titration, adding before heating, almost all of the sugar solution required to effect reduction of the copper. Boil gently for two minutes, add 3–5 drops of the methylene blue indicator and complete the titration within a total boiling time of three minutes. At the end point all the blue colour should be discharged and the liquid should be orange. The proportions of the various sugars, equivalent to 10 or 25 ml. of Fehling's solution are given in the tables (see also *Analyst*, 1943, **68**, 147).

The method is applicable to a mixture containing sucrose and reducing sugars. In order to obtain the proportion of sucrose present, carry out the titrations before and after inversion and calculate each as invert sugar. Then the sucrose figure can be obtained by multiplying the difference by 0·95.

Iodometric Determination of Aldose Sugars

Methods for determining dextrose (and other sugars) have been proposed which are based on the oxidation of the sugar to gluconic

LACTOSE TABLE

Ml. of sugar solution required.	For 10 ml. of Fehling's solution.				For 25 ml. of Fehling's solution.			
	Hydrated lactose. $C_{12}H_{22}O_{11}, H_2O$,		Anhydrous lactose. $C_{12}H_{22}O_{11}$.		Hydrated lactose $C_{12}H_{22}O_{11}, H_2O$.		Anhydrous lactose. $C_{12}H_{22}O_{11}$.	
	Factor.*	Mg. per 100 ml.	Factor.*	Mg. per 100 ml.	Factor.†	Mg. per 100 ml.	Factor.†	Mg. per 100 ml.
15	68·3	455	64·9	432	172·5	1150	163·9	1093
16	68·2	426	64·8	405	172·1	1076	163·5	1022
17	68·2	401	64·8	381	171·7	1010	163·1	960
18	68·1	378	64·7	359	171·4	952	162·8	906
19	68·1	358	64·7	340	171·1	900	162·5	855
20	68·0	340·0	64·6	323·0	170·9	854·5	162·3	811·8
21	68·0	323·8	64·6	307·6	170·6	812·4	162·0	772·3
22	68·0	309·1	64·6	293·6	170·4	774·5	161·8	735·8
23	67·9	295·4	64·5	280·6	170·2	740·0	161·6	703·0
24	67·9	282·9	64·5	268·8	170·0	708·5	161·5	673·1
25	67·9	271·6	64·5	258·0	169·9	679·5	161·4	645·5
26	67·9	261·0	64·5	248·0	169·7	652·7	161·2	620·1
27	67·8	251·1	64·4	238·5	169·5	627·9	161·0	596·5
28	67·8	242·1	64·4	230·0	169·3	604·8	160·8	574·6
29	67·8	233·8	64·4	222·2	169·2	583·3	160·7	554·1
30	67·8	226·0	64·4	214·7	169·0	563·3	160·6	535·1
31	67·8	218·7	64·4	207·8	168·9	544·8	160·5	517·6
32	67·8	211·9	64·4	201·3	168·8	527·4	160·4	501·0
33	67·8	205·6	64·4	195·3	168·6	511·0	160·2	485·5
34	67·9	199·7	64·5	189·7	168·5	495·6	160·1	470·8
35	67·9	194·0	64·5	184·3	168·4	481·1	160·0	457·0
36	67·9	188·6	64·5	179·2	168·2	467·3	159·8	443·9
37	67·9	183·5	64·5	174·3	168·1	454·3	159·7	431·6
38	67·9	178·7	64·5	169·8	168·0	442·1	159·6	420·0
39	67·9	174·1	64·5	165·4	167·9	430·5	159·5	409·0
40	67·9	169·7	64·5	161·2	167·8	419·5	159·4	398·5
41	68·0	165·9	64·6	157·6	167·7	409·0	159·3	388·6
42	68·0	161·9	64·6	153·8	167·6	399·1	159·2	379·1
43	68·0	158·1	64·6	150·2	167·6	389·7	159·2	370·2
44	68·0	154·7	64·6	147·0	167·5	380·7	159·1	361·7
45	68·1	151·3	64·7	143·7	167·4	372·1	159·0	353·5
46	68·1	148·0	64·7	140·6	167·4	363·9	159·0	345·7
47	68·2	145·1	64·8	137·8	167·3	356·0	158·9	338·2
48	68·2	142·1	64·8	135·0	167·2	348·3	158·8	330·9
49	68·2	139·2	64·8	132·2	167·2	341·0	158·8	324·0
50	68·3	136·6	64·9	129·8	167·1	334·2	158·7	317·5

* Mg. of lactose corresponding to 10 ml. of Fehling's solution.
† ,, ,, ,, ,, 25 ml. ,, ,,

acid by the action of iodine as described by Willstäter and Schudel (*Ber. Deut. Chem. Ges.*, 1918, **51**, 780):

$$C_6H_{12}O_6 + I_2 + 3NaOH \rightarrow HOCH_2(CHOH)_4CO_2Na + 2NaI + 2H_2O$$

Iodine oxidises aldoses (*e.g.*, glucose, lactose) but it has little or no effect on ketoses (*e.g.*, sucrose, fructose), so the former sugars can be determined in the presence of the latter (Hinton and Macara, *Analyst*, 1924, **49**, 2):—

To a 250 ml. bottle add 25 ml. of the neutral clarified solution containing 0·05–0·08 g. dextrose (or 0·05–0·15 g. lactose), a considerable excess of iodine (20 ml. N/10) and 5 ml. of N/2 sodium hydroxide solution. Allow the stoppered bottle to stand in the dark for fifteen minutes, then acidify with 5 ml. of 2 N sulphuric acid and immediately titrate the excess iodine with N/10 thiosulphate using starch. Carry out a blank on 25 ml. water at the same time. The difference between the two titrations represents the amount of N/10 iodine required by the sample:—

1 ml. N/10 Iodine ≡ 0·00901 g. dextrose ≡ 0·018 g. lactose hydrate
≡ 0·017 g. maltose

Invert Sugar Table for 10ml. Fehling Solution

Ml. of sugar solution required	Solutions containing besides invert sugar:—									
	No sucrose		1 g. sucrose per 100 ml.		5 g. sucrose per 100 ml.		10g. sucrose per 100 ml.		25g. sucrose per 100 ml.	
	Invert sugar factor*	Mg. invert sugar per 100 ml.	Invert sugar factor*	Mg. invert sugar per 100 ml.	Invert sugar factor*	Mg. invert sugar per 100 ml.	Invert sugar factor*	Mg. invert sugar per 100 ml.	Invert sugar factor*	Mg. invert sugar per 100 ml
15	**50·5**	**336**	**49·9**	**333**	**47·6**	**317**	**46·1**	**307**	**43·4**	**289**
16	50·6	316	50·0	312	47·6	297	46·1	288	43·4	271
17	50·7	298	50·1	295	47·6	280	46·1	271	43·4	255
18	50·8	282	50·1	278	47·6	264	46·1	256	43·3	240
19	50·8	267	50·2	264	47·6	250	46·1	243	43·3	227
20	**50·9**	**254·5**	**50·2**	**251·0**	**47·6**	**238·0**	**46·1**	**230·5**	**43·2**	**216**
21	51·0	242·9	50·2	239·0	47·6	226·7	46·1	219·5	43·2	206
22	51·0	231·8	50·3	228·2	47·6	216·4	46·1	209·5	43·1	196
23	51·1	222·2	50·3	218·7	47·6	207·0	46·1	200·4	43·0	187
24	51·2	213·3	50·3	209·8	47·6	198·3	46·1	192·1	42·9	179
25	**51·2**	**204·8**	**50·4**	**201·6**	**47·6**	**190·4**	**46·0**	**184·0**	**42·8**	**171**
26	51·3	197·4	50·4	193·8	47·6	183·1	46·0	176·9	42·8	164
27	51·4	190·4	50·4	186·7	47·6	176·4	46·0	170·4	42·7	158
28	51·4	183·7	50·5	180·2	47·7	170·3	46·0	164·3	42·7	152
29	51·5	177·6	50·5	174·1	47·7	164·5	46·0	158·6	42·6	147
30	**51·5**	**171·7**	**50·5**	**168·3**	**47·7**	**159·0**	**46·0**	**153·3**	**42·5**	**142**
31	51·6	166·3	50·6	163·1	47·7	153·9	45·9	148·1	42·5	137
32	51·6	161·2	50·6	158·1	47·7	149·1	45·9	143·4	42·4	132
33	51·7	156·6	50·6	153·3	47·7	144·5	45·9	139·1	42·3	128
34	51·7	152·2	50·6	148·9	47·7	140·3	45·8	134·9	42·2	124
35	**51·8**	**147·9**	**50·7**	**144·7**	**47·7**	**136·3**	**45·8**	**130·9**	**42·2**	**121**
36	51·8	143·9	50·7	140·7	47·7	132·5	45·8	127·1	42·1	117
37	51·9	140·2	50·7	137·0	47·7	128·9	45·7	123·5	42·0	114
38	51·9	136·6	50·7	133·5	47·7	125·5	45·7	120·3	42·0	111
39	52·0	133·3	50·8	130·2	47·7	122·3	45·7	117·1	41·9	107
40	**52·0**	**130·1**	**50·8**	**127·0**	**47·7**	**119·2**	**45·6**	**114·1**	**41·8**	**104**
41	52·1	127·1	50·8	123·9	47·7	116·3	45·6	111·2	41·8	102
42	52·1	124·2	50·8	121·0	47·7	113·5	45·6	108·5	41·7	99
43	52·2	121·4	50·8	118·2	47·7	110·9	45·5	105·8	41·6	97
44	52·2	118·7	50·9	115·6	47·7	108·4	45·5	103·4	41·5	94
45	**52·3**	**116·1**	**50·9**	**113·1**	**47·7**	**106·0**	**45·4**	**101·0**	**41·4**	**92**
46	52·3	113·7	50·9	110·6	47·7	103·7	45·4	98·7	41·4	90
47	52·4	111·4	50·9	108·2	47·7	101·5	45·3	96·4	41·3	88
48	52·4	109·2	50·9	106·0	47·7	99·4	45·3	94·3	41·2	86
49	52·5	107·1	51·0	104·0	47·7	97·4	45·2	92·3	41·1	84
50	**52 5**	**105·1**	**51·0**	**102.0**	**47·7**	**95·4**	**45·2**	**90·4**	**41·0**	**82**

* Mg. of invert sugar corresponding to 10 ml. of Fehling's solution.

Sucrose can also be estimated by titrating before and after inversion (1 ml. N/10 ≡ 0·017 g. sucrose).

Hinton and Macara (*Analyst*, 1927, **52**, 668) described a modified method for determining aldose sugars employing chloramine-T. Using this reagent the oxidation is slower and more easily controlled, and the possible secondary action on sucrose and fructose is more likely to be reduced to negligible proportions:—

To a 250 ml. bottle, add 25 ml. of the neutral clarified solution containing about 0·080 g. dextrose (0·15 g. lactose or invert sugar), 20 ml. of potassium iodide solution (10 per cent.) and then 50 ml. N/20 chloramine-T (7·04 g./litre). Allow the stoppered bottle to stand in the dark for one and a half hours, then acidify with 10 ml. of 2N hydrochloric acid and immediately titrate with N/20 thiosulphate using starch. Carry out a blank on 25 ml. water at the same time. The difference between the two titrations represents the amount of N/20 chloramine-T required by the sample.

1 ml. N/20 ≡ 0·0045 g. dextrose ≡ 0·009 g. lactose hydrate
≡ 0·0085 g. inverted sucrose

Invert Sugar Table. 25ml.

Ml. of sugar solution required	Solutions containing besides invert sugar :— No sucrose.		1 g. sucrose per 100 ml.	
	Invert sugar factor.*	Mg. invert sugar per 100 ml.	Invert sugar factor.*	Mg. invert sugar per 100 ml.
15	**123·6**	**824**	**122·6**	**817**
16	123·6	772	122·7	767
17	123·6	727	122·7	721
18	123·7	687	122·7	682
19	123·7	651	122·8	646
20	**123·8**	**619·0**	**122·8**	**614·0**
21	123·8	589·5	122·8	584·8
22	123·9	563·2	122·9	558·2
23	123·9	538·7	122·9	534·0
24	124·0	516·7	122·9	512·1
25	**124·0**	**496·0**	**123·0**	**492·0**
26	124·1	477·3	123·0	473·1
27	124·1	459·7	123·0	455·6
28	124·2	443·6	123·1	439·6
29	124·2	428·3	123·1	424·4
30	**124·3**	**414·3**	**123·1**	**410·4**
31	124·3	401·0	123·2	397·4
32	124·4	388·7	123·2	385·0
33	124·4	377·0	123·2	373·4
34	124·5	366·2	123·3	362·6
35	**124·5**	**355·8**	**123·3**	**352·3**
36	124·6	346·1	123·3	342·5
37	124·6	336·8	123·4	333·5
38	124·7	328·1	123·4	324·7
39	124·7	319·7	123·4	316·4
40	**124·8**	**311·9**	**123·4**	**308·6**
41	124·8	304·4	123·5	301·2
42	124·9	297·3	123·5	294·1
43	124·9	290·5	123·5	287·3
44	125·0	284·1	123·6	280·9
45	**125·0**	**277·9**	**123·6**	**274·7**
46	125·1	272·0	123·6	268·7
47	125·1	266·3	123·7	263·1
48	125·2	260·8	123·7	257·7
49	125·2	255·5	123·7	252·5
50	**125·3**	**250·6**	**123·8**	**247·6**

* Mg. of invert sugar corresponding to 25 ml. of Fehling's solution.

Dextrose Table

Ml. of sugar solution required	For 10 ml. of Fehling's solution.		For 25 ml. of Fehling's solution.	
	Dextrose factor.*	Mg. dextrose per 100 ml.	Dextrose factor.†	Mg. dextrose per 100 ml.
15	**49·1**	**327**	**120·2**	**801**
16	49·2	307	120·2	751
17	49·3	289	120·2	707
18	49·3	274	120·2	668
19	49·4	260	120·3	633
20	**49·5**	**247·4**	**120·3**	**601·5**
21	49·5	235·8	120·3	572·9
22	49·6	225·5	120·4	547·3
23	49·7	216·1	120·4	523·6
24	49·8	207·4	120·5	501·9
25	**49·8**	**199·3**	**120·5**	**482·0**
26	49·9	191·8	120·6	463·7
27	49·9	184·9	120·6	446·8
28	50·0	178·5	120·7	431·1
29	50·0	172·5	120·7	416·4
30	**50·1**	**167·0**	**120·8**	**402·7**
31	50·2	161·8	120·8	389·7
32	50·2	156·9	120·8	377·6
33	50·3	152·4	120·9	366·3
34	50·3	148·0	120·9	355·6
35	**50·4**	**143·9**	**121·0**	**345·6**
36	50·4	140·0	121·0	336·3
37	50·5	136·4	121·1	327·4
38	50·5	132·9	121·2	318·8
39	50·6	129·6	121·2	310·7
40	**50·6**	**126·5**	**121·2**	**303·1**
41	50·7	123·6	121·3	295·9
42	50·7	120·8	121·4	289·0
43	50·8	118·1	121·4	282·4
44	50·8	115·5	121·5	276·1
45	**50·9**	**113·0**	**121·5**	**270·1**
46	50·9	110·6	121·6	264·3
47	51·0	108·4	121·6	258·8
48	51·0	106·2	121·7	253·5
49	51·0	104·1	121·7	248·4
50	**51·1**	**102·2**	**121·8**	**243·6**

* Mg. of dextrose corresponding to 10 ml. of Fehling's solution.
† " " " " 25 ml. " "

LÆVULOSE TABLE

Ml. of sugar solution required	For 10 ml. of Fehling's solution		For 25 ml. of Fehling's solution.	
	Laevulose factor.*	Mg. laevulose per 100 ml.	Laevulose factor.†	Mg laevulose per 100 ml.
15	**52·2**	**348**	**127·4**	**849**
16	52·3	327	127·4	796
17	52·3	308	127·5	750
18	52·4	291	127·5	708
19	52·5	276	127·6	672
20	**52·5**	**262·5**	**127·6**	**638·0**
21	52·6	250·6	127·7	608·1
22	52·7	239·6	127·7	580·6
23	52·7	229·1	127·8	555·5
24	52·8	220·0	127·8	532·5
25	**52·8**	**211·3**	**127·9**	**511·5**
26	52·9	203·3	127·9	491·9
27	52·9	196·0	128·0	474·0
28	53·0	189·3	128·0	457·2
29	53·1	183·1	128·1	441·6
30	**53·2**	**177·2**	**128·1**	**427·0**
31	53·2	171·7	128·1	413·3
32	53·3	166·5	128·2	400·5
33	53·3	161·6	128·2	388·5
34	53·4	157·0	128·3	377·3
35	**53·4**	**152·6**	**128·3**	**366·7**
36	53·5	148·6	128·4	356·6
37	53·5	144·7	128·4	347·0
38	53·6	140·9	128·5	338·1
39	53·6	137·3	128·5	329·6
40	**53·6**	**134·0**	**128·6**	**321·5**
41	53·7	130·9	128·6	313·7
42	53·7	127·9	128·6	306·2
43	53·8	125·1	128·7	299·2
44	53·8	122·4	128·7	292·5
45	**53·9**	**119·8**	**128·8**	**286·2**
46	53·9	117·2	128·8	280·0
47	53·9	114·7	128·9	274·2
48	54·0	112·4	128·9	268·6
49	54·0	110·2	129·0	263·2
50	**54·0**	**108·0**	**129·0**	**258·0**

* Mg. of laevulose corresponding to 10 ml. of Fehling's solution.
† " " " 25 ml. " "

Hinton and Macara (*loc. cit.*) give tables which state the small corrections which are necessary when sucrose is present in the solution.

Somogyi's Semi-Micro Methods for Determining Sugars

Somogyi (*J. Biol. Chem.*, 1926, **70**, 599; 1945, **160**, 61) has described semi-micro methods for the determination of sugars. In the latter paper he describes the use of an alkaline copper tartrate reagent which is stable, is sufficiently alkaline to permit the determination of slowly reacting sugars (*e.g.*, maltose) and can be employed for both colorimetric and volumetric determinations.

Somogyi's Modified Alkaline Copper Tartrate Stock Reagent. Dissolve 28 g. anhydrous disodium phosphate and 40 g. sodium potassium tartrate in 700 ml. water, add 100 ml. N/1 sodium hydroxide and stir in 80 ml. of 10 per cent. solution of crystalline copper sulphate. Then add 180 g. anhydrous sodium sulphate, dissolve and make up to a litre. Allow to stand for a few days. Decant or filter.

METHOD. Add potassium iodate solution to the Somogyi stock reagent according to the expected sugar content (see table). Mix 5 ml. of this freshly prepared reagent and 5 ml. of the sugar solution in a Pyrex boiling tube, cover with a glass bulb and heat in boiling water for the requisite time (ten minutes for dextrose, twenty minutes for maltose). Cool and add the appropriate volume (see table) of 2·5 per cent. potassium iodide solution from a pipette down the wall of the tube, without stirring or shaking. Acidify rapidly with 1·5 ml. of 2 N sulphuric acid, mix and titrate with N/200 thiosulphate using starch (titration = a. ml.). Carry out a blank on 5 ml. of water at the same time (b. ml.). Calculate the sugar concentration from (b − a).

1 ml. N/200 thiosulphate = 0·135 mg. dextrose = 0·26 mg. maltose.

Expected sugar content in 5 ml. sample solution.	Volume of N/1 potassium iodate to be added to one litre of stock reagent.	Corresponding volume of 2·5% potassium iodide to be added after the reduction.
Up to 0·5 mg.	5 ml.	0·5 ml.
Up to 1 mg.	10 ml.	1·0 ml.
Up to 3 mg.	25 ml.	2·0 ml.

N/1 potassium iodate contains 35·67 g. KIO_3 per litre.

In Somogyi's colorimetric method, the arsenomolybdate reagent of Nelson (*J. Biol Chem.*, 1944, **153,** 375) is used. This is prepared by dissolving 25 g. of ammonium molybdate in 450 ml. water, adding 21 ml. of concentrated sulphuric acid and 3 g. $Na_2HAsO_4 . 7H_2O$ (dissolved in 25 ml. water) and incubating at 37° C. for 2 days. For the determination of sugars, mix 2 ml. of the stock Somogyi copper reagent and 2 ml. of clear sugar solution in a test-tube, cover with a glass bulb and heat in boiling water for ten minutes. Cool, add the arsenomolybdate reagent, dilute to 10 ml. or 25 ml. and measure the optical density at 500 mμ and compare with standards prepared similarly.

CANE SUGAR

Sugar is obtained from sugar cane (*Saccharum officinarum*), the juices of which contain 8–23 per cent. of sugar and from sugar beet (*Beta alba*), the juices of which contain 13–15 per cent. of sugar.

Sugar refining. The raw sugar is first mixed with some raw syrup and centrifuged. After washing off much of the molasses, the " affined sugar " is dissolved in water and the solution is treated with milk of lime. Carbon dioxide is then bubbled through and impurities (gums, waxes, etc.) are carried down by the precipitate of calcium carbonate which is produced. The cloudy liquid is then filtered and decolorised with animal charcoal. The colourless liquid produced is evaporated under reduced pressure, centrifuged to separate the massecuite from the syrup and the wet crystals are then dried in a current of hot air to produce *granulated sugar*. Alternatively the residue in the vacuum pan may be cast into slabs, centrifuged and the moist slabs are then dried and cut into cubes (*cube sugar*). The practice of adding blue dye (ultramarine) to sugar to make it appear whiter has apparently now been discarded in Britain.

Icing sugar is prepared by reducing crystalline sugar to a fine powder. For trade purposes a little tricalcium phosphate or starch is mixed in to reduce caking.

Demerara sugar is a brown sugar with a characteristic taste. It was originally produced in Demerara, but by a decision of the King's Bench any sugar resembling it which is a product of the West Indies or Mauritius may be so described. The practice of adding tin chloride to stabilise the colour of demerara sugar has now been largely discontinued. Artificial colouring matter and titanous chloride are, however, used now for the same purpose.

The following table shows the usual composition of different kinds of sucrose (all figures as percentages):—

	Sucrose	*Invert Sugar*	*Ash*	*Organic Non-Sugars*	*Water*
RAW SUGARS					
Cuba Cane	97·41	0·83	0·45	0·73	0·58
West Indies Cane	96·93	1·00	0·58	0·71	0·78
San Domingo Cane	97·59	0·76	0·40	0·63	0·62
Australia Cane	98·55	0·41	0·30	0·44	0·30
Natal Cane	98·35	0·48	0·37	0·44	0·36
Java Cane	97·48	0·92	0·49	0·45	0·66
Mauritius Cane	98·08	0·60	0·35	0·48	0·51
British Beet	97·12	0·02	0·69	1·03	1·14
COLOURED SUGARS FOR DIRECT CONSUMPTION					
Demerara	97·50	0·73	0·50	0·76	0·51
Pieces Sugar 1	99·14	0·09	0·05	0·10	0·62
,, ,, 2	97·25	0·40	0·33	0·53	1·49
,, ,, 3	89·03	2·36	2·17	3·54	2·90
,, ,, 4	87·23	2·83	2·58	4·21	3·15
Muscovado	94·70	1·80	1·00	1·50	1·00
REFINED WHITE SUGARS	99·965	0·010	0·007	0·010	0·018

Sucrose B.P. The 1958 B.P. requires refined sugar to comply with the following: specific rotation of a 10 per cent. w/v solution +66° to +66·7°; maxima are stated for sulphated ash (0·02 per cent.), arsenic (1 p.p.m.), lead (2 p.p.m.) and sulphur dioxide (70 p.p.m.); also the powder must comply with the limit tests for acidity or alkalinity, barium and strontium, reducing sugars and ultramarine.

Sweetness. There are no instruments which measure sweetness. In any attempt to compare the sweetness of various materials it is necessary to rely upon the subjective reactions of individual taste senses. These taste senses depend on whether the body is in need of sugar and become more acute when the body is short of sugar. Reactions may also vary with the state of health. The taste senses are also influenced by the rate at which a material dissolves on the tongue and this is probably the reason that many people consider icing sugar to be sweeter than granulated sugar and the latter to be sweeter than cube sugar. Comparative sweetness is best measured when the sugar is at a low concentration so that the solution is only just sweet.

The apparently greater sweetness of impure sugars has often been attributed to the presence of organic acids or salt. It now seems more probable that the extra sweetness is due partly to fructose, which is over 1½ times as sweet as sucrose, and partly to the illusion created by the " treacly " flavour of the sugar breakdown products.

The *relative sweetness* of some of the more common sugars has been stated as follows: fructose 173, invert sugar 130, sucrose 100, dextrose 74, galactose 32, maltose 32, lactose 16.

Analysis of Sugar

Solubility. Refined sugar should dissolve in 0·5 part of water at 20° C.

Moisture. Dry 5 g. at 100° C. to constant weight.

Ash. Moisten 10 g. with sulphuric acid (conc.), ignite gently, moisten again with sulphuric acid and re-ignite to constant weight. The sulphated ash of refined sugar should not exceed 0·02 per cent. Zerban *et al.* (*Ind. Eng. Chem., Anal. Ed.*, 1930, **2**, 32, 322; 1931, **3**, 38; 1932, **4**, 435) have shown, using dilute sugar solutions, that there is a relationship between the electrical conductivity and the ash.

Sucrose. Determine the specific rotation of a 10 per cent. w/v solution or the saccharimeter reading using a solution containing the normal weight (see p. 101).

Reducing sugars. The 1958 B.P. requires refined sugar to comply with the following limit test: To 4·0 g. in a test-tube, add 3·7 ml. of water, 1·0 ml. of a 0·1 per cent. w/v solution of methylene blue, and 0·3 ml. of sodium hydroxide solution (20 per cent. w/v), close the tube with a rubber stopper, shake vigorously for fifteen seconds, and place in boiling water. Examine the tube after two minutes without removal from the boiling water; the blue colour is not completely discharged (to comply with the B.P. test).

Paper chromatography of the sugars. The best method of differentiating between cane and beet products is to establish the presence or absence of raffinose. Raffinose is present in all beet products, but absent in products derived from sugar cane. It can be detected and estimated by paper chromatography preferably using solvent 2 (p.108).

Colouring matter. Most artificial dyes will be absorbed by wool on boiling in the presence of a trace of acid (see p.100). Ultramarine can be detected by adding dilute hypophosphorous acid to a concentrated solution of the sugar and allowing to stand. If ultramarine is present an unpleasant odour develops within an hour. A further test involves the preparation of a concentrated solution of the sugar in a tall cylinder followed by microscopical examination of any dark specks which fall to the bottom.

Trace metals. Examine for arsenic and lead. Tin was formerly determined in samples of Demerara sugar (see p. 125).

Sulphur dioxide. See p. 80 (maximum 70 p.p.m.).

SYRUP, TREACLE, MOLASSES

These materials are viscous sweet liquids containing uncrystallisable sucrose, reducing sugars, dextrin, organic acids, nitrogenous matter and inorganic constituents. Golden syrup, which contains the highest proportion of sucrose and invert sugar, is prepared from residues obtained during the production of refined sugar. After removing the crystallisable sugar, the residue is boiled with dilute sulphuric acid. After neutralising with chalk, the syrup is filtered through charcoal and concentrated under reduced pressure.

The following tables show the percentage composition of some of these products:—

	Cane Molasses	Beet Molasses	
Sucrose	30·5	48·0	48·5
Reducing sugars	22·0	0·5	2·5
Organic non-sugars	12·0	14·5	14·5
Ash	10·5	11·0	9·5
Water	25·0	26·0	25·0

	Golden Syrup			
	1	2	3	4
Sucrose	31·7	26·1	33·9	34·5
Invert sugar	45·3	47·5	39·8	45·6
Ash	1·6	3·5	1·2	2·2
Water + organic matter (not sugar)	21·4	22·9	25·1	17·7

Syrup, B.P. contains 66·7 per cent. sucrose (w/w) and should comply with the following: specific rotation at 20° C. + 56° to + 60° and weight per ml. at 20° C. 1·315–1·333 g.

Analysis of Syrups

Water. Dry on sand at 70° C. in a vacuum oven for at least five hours. As fructose tends to decompose at 75° C. in the presence of water, drying at 100–105° C. tends to produce high results.

In refineries it is more usual to measure the refractive index, calculate as pure sucrose and apply corrections for invert sugar and ash:—

$$\text{True solids} = \frac{\%\ \text{Refractometric Solids}}{1 - 0{\cdot}00025 \times \%\ \text{Invert} + \text{k} \times \%\ \text{Ash}}$$

For cane syrups and molasses, k = 0·0043.
For beet syrups and molasses, k = −0·0008.

N.B.—The formula above only applies to syrups containing 20–30 per cent. of water.

Ash. Determine the sulphated ash.

Sugars. Sucrose and invert sugar can be determined by double polarization. Liquid glucose, if present, gives a high positive direct rotation and the rotation after inversion is appreciably positive also. If the K value is also determined, the equations given in the section on Liquid Glucose can be applied (see below).

Formerly several samples of treacle and golden syrup were found to contain liquid glucose and several successful prosecutions were reported. There appears, however, to have been no objection taken to similar articles, which have been described by other names such as " crystal syrup " and " amber syrup."

As mentioned previously the presence of raffinose in syrup serves to establish the presence of products derived from sugar beet.

Trace metals. Examine for arsenic and lead.

LIQUID GLUCOSE

Nowadays liquid glucose is usually produced from maize grains. The starch is hydrolysed with hydrochloric acid and the resultant liquid is neutralised with sodium carbonate, filtered, decolorised with animal charcoal and concentrated under reduced pressure. Solid commercial glucose is produced by allowing the concentrate to solidify spontaneously. Commercial dextrose monohydrate is prepared by using more dilute solutions, filtering and seeding with pure dextrose.

Liquid glucose, B.P. (1953). The B.P. (1953) required liquid glucose to comply with the following: acidity such that 5 g. does not require more than 0·6 ml. N/10 alkali; R.I. not less than 1·490 at 20° C.; maxima for sulphated ash (0·6 per cent.), sulphur dioxide (450 p.p.m.), arsenic (1 p.p.m.) and lead (2 p.p.m.).

Dextrose B.P. The 1958 B.P. requires the anhydrous powder to comply with the following: specific rotation of a 10 per cent. w/v solution of the substance dried at 105° C. +52·5° to +53·0°; maxima are stated for loss on drying at 105° C. (2·0 per cent.), sulphated ash (0·1 per cent.) sulphur dioxide (70 p.p.m.), arsenic (1 p.p.m.) and lead (2 p.p.m.); also the powder must comply with the limit tests for acidity, chloride, sulphate, less soluble sugars and dextrins.

The tests in the B.P. are a useful guide to the procedures which can be applied to liquid and solid commercial samples.

Various methods have been described for the determination of the individual carbohydrates in liquid glucose. One relatively simple procedure is Steinhoff's volumetric method, which is based on the assumption that dextrins do not reduce Fehling's solution (see U.S. Nat. Bur. Stds., Circ. C 440, *loc. cit.*). Patterson *et al.* (*Analyst*, 1957, **82**, 812; 1960, **85**, 75) have described a procedure in which the mono- and di-saccharides are estimated by a modification of Somogyi's method after separation on a carbon column. The purity of the eluates were confirmed by paper chromatography. The following results on 11 samples of commercial liquid glucose have been extracted from the latter paper:—

Density	41·6–44·9° Baumé
*Dextrose equivalent (% w/w)	34·0–65·0
Total sugar solids (% w/w)	78·9–85·1
Dextrose (% w/w)	10·4–33·2
Maltose (% w/w)	8·0–24·3
Higher saccharides (by diff.) (% w/w)	24·8–64·3

From other work liquid glucose contains 14–21 per cent. of water and yields 0·3–0·5 per cent. of ash.

The proportions of dextrose, maltose and dextrin can also be assessed by Wiley's method, which is based on the assumption that the specific rotations $[\alpha]_D$ of dextrose, maltose and dextrin are +52·7, +139·2 and +198 respectively (*cf.* p. 94):—

O(%) = Total organic solids = Total solids—Ash

K = Cupric reducing power (p. 105)

S = Specific rotation

Maltose + dextrose + dextrin = O . . (i)

Maltose = 3·195S + 4·642K − 6·326O . (ii)

Dextrose = K − 0·62 × % maltose . . (iii)

Dextrin = O − (dextrose + maltose) . . (iv)

If D is the direct saccharimeter reading (Ventzke degrees) for a solution containing the normal weight, equation (ii) becomes:—

Maltose = 2·125D + 4·632K − 6·326O . (v)

Dextrin has a rather high specific rotation and the amount present has a considerable bearing on the rotation of different samples of liquid glucose. In previous editions of this volume the mean saccharimeter reading of liquid glucose was quoted as 170° V. and the mean K value as 42 (figures which are accepted by the Customs authorities). The following table quotes the polarimetric values which are now becoming accepted more generally (*cf.* Chatt, *Analyst*, 1952, **77**, 335):—

Sample	K Value	Saccharimeter reading for normal weight V°	Specific rotation $[\alpha]^{20}_D$
Liquid glucose	36·5	179	119
Dry solids†	43	210	140

* Percentage of reducing sugars calculated as dextrose on the dry basis.

† Calculated on the basis of 85 per cent. total solids.

Determination of Liquid Glucose in Admixture with Sucrose and Invert Sugar

(1) Estimate the sucrose content by Clerget's method (p. 115):—

$$\text{Sucrose } (\%) = S = \frac{(D - I)\ 100}{133{\cdot}6} \qquad \text{(i)}$$

(2) Determine the K value (p. 105) by the Lane and Eynon method and then using the figures in the above table apply the following equations:—

$$0{\cdot}365G + 1{\cdot}0Y = K \qquad \text{(ii)}$$

$$1{\cdot}79G - 0{\cdot}34Y = D - S \qquad \text{(iii)}$$

where G and Y represent the percentages of liquid glucose and invert sugar respectively and D is the direct saccharimeter reading. The percentage of liquid glucose in the sample can then be calculated algebraically from these equations.

Using angular rotations, S can be calculated from the specific rotations before and after inversion (p. 115) to give equation (i), equation (ii) is the same as above and equation (iii) becomes:—

$$1{\cdot}19G - 0{\cdot}21Y = [\alpha_D]^{20}_{D} - 0{\cdot}665S$$

For calculations of the *dry glucose solids* (G^1), equation (ii) then becomes:—

$$0{\cdot}43G^1 + 1{\cdot}0Y = K$$

Using saccharimeter readings, equation (iii) becomes:—

$$2{\cdot}10G^1 - 0{\cdot}34Y = D - S$$

Using angular rotations, equation (iii) becomes:—

$$1{\cdot}40G^1 - 0{\cdot}21Y = [\alpha_D]^{20}_{D} - 0{\cdot}665S$$

At 87° C., the positive rotation of dextrose equals the negative rotation of fructose. Invert sugar, therefore, shows a zero rotation at 87° C. The rotation at 87° C. of the solution after inversion is entirely due to liquid glucose. Owing to the expansion of the solution due to the higher temperature employed readings taken at 87° C. should be multiplied by 1·0315.

HONEY

Honey is the saccharine product gathered by bees from the nectar of flowers. The colour of honey varies from almost colourless to nearly black according to the flower from which the nectar is taken. Storage at cool temperatures causes honey to "granulate," *i.e.*, the sugar crystallises.

The main constituents of honey are dextrose, fructose, sucrose, dextrins, mineral matter, proteins and wax. Pollen is invariably present in comb honey, but may be absent in products which have been finely strained. Enzymes present cause changes in the proportions of the original sugars present and the sucrose may disappear completely on storage. The presence of over about 5 per cent. of sucrose may be due to artificial feeding of the bees with the sugar during the winter or during periods of drought.

Most natural honey has a negative optical rotation, but sucrose and liquid glucose both make the rotation " more positive ". Technical invert sugar is invariably present in artificial honeys.

The B.P.C. 1959 requires honey to comply with the following standards: (a) optical rotation of a 20 per cent. solution in a 1 dcm. tube +0·6° to −3·0°; (b) weight per ml. at 20° C. from 1·352–1·358 g.; (c) ash not more than 0·3 per cent.; and (d) Fiehe's test for technical invert sugar to be negative.

Extensive tables giving composition figures and other analytical results for various honeys have been presented by Mitchell *et al.* (*Analyst*, 1954, **79**, 435; 1960, **85**, 412).

Analysis of honey

Sampling. Granulated honeys should be melted by heating in a water bath at 50° C.

Total solids and water. The total solids can be estimated by drying in a vacuum oven at 70° C. Alternatively the figure can be calculated from the specific gravity of a 20 per cent. w/v solution:—

$$\text{Total solids} = \frac{\text{S.G. of 20\% solution} - 1}{0{\cdot}00386}$$

Chataway (*Can. J. Res.*, 1932, **6**, 352) has presented figures relating the refractive index of honey with the water content:—

R. I. (20° C.)	1·5041	1·5015	1·4990	1·4965	1·4940
Water (%)	13·0	14·0	15·0	16·0	17·0
R. I. (20° C.)	1·4150,	1·4890,	1·4866,	1·4844,	1·4822
Water (%)	18·0	19·0	20·0	21·0	22·0

Temperature correction = −0·00023 per degree.

Most genuine honeys contain 13–23 per cent. of water.

Ash. Ignite the sample preferably using an Argand burner. The ash of genuine honey seldom exceeds 0·35 per cent. Only a minute trace of sulphate should be present in pure honey.

Total nitrogen. Determine by the macro Kjeldahl method (p. 22). The nitrogen in the dry material of genuine honey is usually less than 0·25 per cent.

Free acid. Dilute 10 g. with 75 ml. water and titrate with N/10 alkali to phenolphthalein. Mitchell *et al.* (*Analyst*, 1954, **79**, 435; 1960, **85**, 412) found that the figures given by numerous honeys ranged from 13–65 ml. N/10 alkali per 100 g. Alternatively the acidity may be reported in terms of formic acid. Most honeys have a pH value of 3·7–5·0.

Detection of Technical Invert Sugar

During the manufacture of invert syrup by acid hydrolysis of sucrose, 5-hydroxymethylfurfuraldehyde is also formed.

Fiehe's test for the detection of invert sugar in honey depends on the formation of a red colour when the aldehyde reacts with resorcinol. The modified test due to Lampitt *et al.* (*Analyst*, 1929, **54**, 381) is as follows:—

Dissolve 20 g. honey in 20 ml. cold water and shake gently with 40 ml. ether. Decant off the ether and evaporate at room temperature. Dissolve the residue in 10 ml. ether and to 2 ml. of this ether extract add 2 ml. of a fresh 1 per cent. solution of resorcinol in concentrated hydrochloric acid. A positive reaction is indicated by the immediate appearance of a pink colour in the acid layer. This darkens and after twenty minutes a deep cherry-red colour appears at the junction of the layers. Then for a confirmatory test, allow the rest of the ether extract from the honey (8 ml.) to evaporate in a porcelain dish at room temperature and to the residue add 2 ml. of fresh aniline acetate solution (1 ml. redist. aniline + 4 ml. glacial acetic acid). A positive reaction is indicated if a pink to orange colour appears within fifteen minutes. Lampitt *et al.* state that both the resorcinol and aniline tests will give distinct positive reactions with honey containing 5 per cent. of commercial invert sugar. Invert syrups prepared by enzymic inversion do not contain the aldehyde and, therefore, do not give a reaction.

Lampitt *et al.* (*loc. cit.*) also showed that heated natural honeys give a positive reaction to Fiehe's test. This was confirmed by Franzke and Iwainsky (*Anal. Abstr.*, 1957, **4**, No. 3115), who by means of ascending paper chromatography detected the aldehyde in natural honeys which had been heated at 100° C. for one hour. They employed butanol-ethanol-water (2 : 4 : 1) as solvent and resorcinol-trichloracetic acid as spraying reagent. They also detected the 5-hydroxymethylfurfuraldehyde by means of the absorption spectrum of the aqueous solution from 240–320 mμ (Max. 282·5 mμ; $E^{1\%}_{1\,cm.} = 1300$).

Determination of Sugars

The principal sugars present in honey are fructose and dextrose together with a little sucrose (see p. 131) and a little dextrin. Auerbach and Bodlander (*Analyst*, 1924, **49**, 389) have stated that the essential difference between genuine and artificial honey lies in the ratio of fructose to dextrose present:—

	Fructose/Dextrose Ratio
Genuine honey	106–119 : 100
Artificial honey	Less than 90 : 100

These authors describe the following method for the calculation of the ratio:—

Dilute 2 g. to 250 ml. and estimate the total reducing sugars gravimetrically on 25 ml. Estimate the dextrose iodimetrically by adding excess N/10 iodine (at least twice the iodine used) to 25 ml. of the honey solution and 100 ml. of equal parts of M/5 sodium bicarbonate and carbonate solutions. Allow to stand in the dark for two hours, acidify with 12 ml. of dilute sulphuric acid (25 per cent.) and titrate with N/10 thiosulphate. Carry out a blank at the same time. The difference between the two titrations represents the dextrose (1 ml. N/10 iodine $\equiv$ 0·009005 g. dextrose). Fructose = total reducing sugars − dextrose.

The polarimetric readings are useful for assessing the genuineness of honey. Formerly it was usual to regard with suspicion any honey which gave a positive optical rotation, as such would be occasioned by the addition either of liquid glucose or sucrose. It is usually assumed nowadays, however, that if the specific rotation falls between +5° and −15° and if these figures are only slightly altered after inversion, the sample is probably genuine. If the specific rotation does not fall within these limits (or if a more complete examination is necessary), the copper reducing power should also be determined and calculated as dextrose.

Cane sugar (S) can be calculated from the Clerget-Herzfeld formula (p. 114). When the percentage of dextrin is small the percentages of dextrose (d) and fructose (f) can be calculated from the polarisation and the K value:—

$$d + 0{\cdot}915 f = \mathrm{K}$$

$$f = \frac{0{\cdot}793\ \mathrm{K} + \mathrm{S} - \mathrm{D}}{2{\cdot}08}$$

where D = direct polarisation on Ventzke scale at 20° C.

When an appreciable quantity of dextrin is present, the composition may be calculated in terms of sucrose, invert sugar and dextrin, as Clerget's formula gives the cane sugar, and the K value divided by 0·915 gives the invert sugar (or the weight of reduced copper may be calculated directly to invert sugar) (*cf. Analyst*, 1938, **63**, 315).

Liquid glucose can be estimated from Browne's formula, which is based on the difference in the invert polarisations at 20° and 87°. The difference is due to the lowering of the rotation of fructose with increasing temperature and, therefore, depends on the amount of invert sugar present. Liquid glucose if present decreases the difference in the polarisation figures irrespective of the specific rotation. Browne's formula assumes that the average percentage of invert sugar after inversion in honey is 77 per cent:—

$$\text{Commercial glucose (\%)} = \frac{100 - 7700\,(\mathrm{V}_{87} - \mathrm{V}_{20})}{26{\cdot}7 \times \mathrm{I}}$$

where V_{87} and V_{20} are the saccharimeter polarisation figures for a solution containing the normal weight when measured in a 2 dcm. tube at 87° and 20° C. respectively; and I is the percentage of reducing sugars in the sample after inversion, calculated as invert sugar.

Dextrin. Transfer 8 g. of honey into a 100 ml. graduated flask using 4 ml. of water, make up to the mark with absolute alcohol, shake thoroughly and allow to stand overnight. Filter off the precipitate and wash with alcohol. Then dissolve the precipitate in hot water, dry at 70° C. under reduced pressure and weigh. Dissolve the crude dextrin in a definite volume of water (approx. 0·5 g. ppt./50 ml.) and determine the reducing sugars before and after inversion:—

$$\text{Dextrin} = \text{Crude dextrin} - (\text{sucrose} + \text{reducing sugars}).$$

Microscopical examination. A microscopical examination of honey should always be made. Genuine honey almost invariably shows some pollen grains and traces of wax besides crystals of sugar. Starch granules should be absent. The honey is diluted with about twice its volume of water and allowed to settle overnight in a conical glass. The sediment, which will include the pollen grains, is removed with a pipette and mounted on microscope slides. Separation with the centrifuge is not so satisfactory.

Honeydew honey. During periods of prolonged drought bees may have to supplement their normal supplies of nectar from plants by honeydew, which is a sweet, sticky fluid which is excreted on foliage by leaf-sucking insects (*e.g.*, aphids). Honeydew represents a nuisance as the resultant honey tends to be dark and have an undesirable taste. Kirkwood *et al.* (*Analyst*, 1960, **85**, 412) have shown that compared with floral honey, honeydew honey on the average contains more dextrin and less reducing sugars and has a higher ash, a higher pH and a less negative optical rotation.

PRESERVES

Jam, Marmalade and Jelly

The basic principle of jam and marmalade manufacture is the boiling together of fruit (suitably prepared), sugar and water. During the boiling the proportion of solid matter in the mix increases (due to the evaporation of water), a proportion of the sucrose is converted to invert sugar and a gel is produced on cooling. It has been shown that three separate components, viz., sugar, pectin and acid, all play a part in forming the gel. Where the fruit is low in pectin strength or acid, therefore, the deficiency is made up by adding solutions of such ingredients towards the end of the boiling. Jelly manufacture involves boiling the fruit with water and then the extract (after filtration) is boiled with the sugar (see *Principles of Fruit Preservation* by Morris, 1951, p. 34, Chapman and Hall, London).

Legal standards. The Preservatives Regulations state limits for sulphur dioxide in fruit and fruit pulp for conversion into jam ranging from 1500–3000 p.p.m. Much sulphur dioxide is lost during the boiling and the limit for jam, marmalade and fruit jelly is 100 p.p.m.

Quality standards for jam and other preserves are stated in **The Food Standards (Preserves) Order, 1953** (S.I. 1953 No. 691, as amended by S.I. 1953 No. 1307):—

THE SCHEDULE

1. For the purposes of this Schedule:

   (a) " fruit " includes ginger, rhubarb and marrow.

   (b) " fruit content " means as respects any jam or marmalade the total quantity of fruit of the variety or varieties used in the manufacture of the jam or marmalade, being, where such variety or varieties are specified in the description thereof on sale, offer or exposure for sale, the total quantity of the variety or varieties so specified. The said quantity is the total quantity of the prepared fruit used in the manufacture of the jam or marmalade or in the case where fruit pulp is used the total quantity of the prepared fruit used in the manufacture of the pulp (no account being taken in any case of any added fruit juice or added pectin not derived from the prepared fruit) expressed as a percentage based on the number of pounds of prepared fruit required to be used in making 100 lb. of finished jam or marmalade.

   (c) " hermetically sealed container " means a container sealed while the contents are hot with a closure capable of maintaining the partial vacuum resulting from cooling.

   (d) " jam " includes jelly jam.

   (e) " marmalade " includes jelly marmalade.

   (f) " prepared fruit ":

      (i) as respects ginger, rhubarb, melon and marrow means respectively peeled ginger root or stem, trimmed rhubarb, and melon and marrow without seeds, stem or rind;

      (ii) in all other respects means sound fresh fruit from which stalks and calices have been removed.

(g) " percentage of soluble solids " means as respects any product the percentage by weight of soluble solids ascertained, when the product is at a temperature of 20° C., by means of a refractometer, no correction being made for insoluble solids.

2. *Standard for Jam*

(1) Jam shall contain a percentage of soluble solids of not less than 68½ per cent. unless packed in hermetically sealed containers when it shall contain not less than 65 per cent.

(2) The fruit content of jam shall be not less than as specified in the following table:

Minimum Fruit Content

Description of Jam	Fruit content (per cent.)
Blackberry or Bramble	38
Blackcurrant	25
Damson	35
Fig and Lemon	40 (8)
Gooseberry	30
Gooseberry and Raspberry	30
Gooseberry and Strawberry	30
Greengage	38
Loganberry	30
Melon and Lemon	40 (8)
Melon and Pineapple	40 (5)
Melon and Ginger	40 (1)
Raspberry	30
Raspberry and Gooseberry	30
Raspberry and Loganberry	30
Raspberry and Redcurrant	30
Redcurrant	35
Rhubarb and Ginger	40 (1)
Strawberry	38
Strawberry and Gooseberry	35
Youngberry	38
All other varieties	40

(3) Except as provided in paragraph (4) below first named fruit in all mixed jams should amount to not less than 50 per cent. and not more than 75 per cent. of the fruit content.

(4) Where figures appear in brackets in the second column of the table above, the figure in each case denotes the minimum quantity by weight of the second named fruit to be contained in the finished jam expressed as a percentage based upon the number of parts by weight of the second named fruit required to be present in 100 parts by weight of finished jam.

3. *Standard for Marmalade*

Marmalade shall contain a percentage of soluble solids of not less than 68½ per cent. unless packed in hermetically sealed containers when it shall contain not less than 65 per cent., and:

(a) in the case of marmalade other than ginger marmalade the fruit content shall be not less than 20 per cent. citrus fruit;

(b) in the case of ginger marmalade the fruit content shall be not less than 25 per cent. drained ginger only or 25 per cent. drained ginger and citrus fruit but so that not less than 15 per cent. shall be drained ginger. " Drained ginger " means ginger (in syrup) drained free from syrup.

4. *General*

No jam or marmalade shall contain any added acid other than citric, tartaric or malic acid.

Analysis

The analysis of jam should include the determination of the **soluble solids, fruit content, sulphur dioxide** and **trace metals** and an examination for the types of **dyes** and **fruit** present. The determination of **pectin** is occasionally required and industrially the proportion of **reducing sugars** and the **pH value** are of some considerable importance.

Soluble solids. The Preserves Order requires jam sold in hermetically sealed containers to contain at least 65 per cent. of soluble solids (minimum of 68·5 per cent. for an " open " container). Such amounts are considered to be necessary to ensure the inhibition of the growth of moulds and yeasts. The Order requires the soluble solids to be calculated from the refractive index at 20° C. (uncorrected for insoluble solids). The sucrose conversion table is given on p.109. Macara (*Analyst*, 1931, **56**, 393) has given the appropriate conversions from butyro-refractometer readings and the corrections which can be used for reducing sugars and citric acid if the true soluble solids are required.

The individual sugars present can be determined on the cleared solution, both before and after inversion by either a volumetric, polarimetric or a gravimetric method. Industrially the determination of reducing sugars is of some considerable importance. *Reducing sugars* alone are best determined by Lane and Eynon's method by titrating 10 ml. of Fehling's solution with a 1·5 per cent. solution of the jam (cleared with zinc ferrocyanide) using methylene blue as indicator. Manufacturers prefer the reducing sugar content to fall within the range 20–40 (calculated as a percentage of the preserve) in order to prevent the separation of crystals during storage. Crystals of sucrose may separate if the amount present is high (reducing sugars below 20 per cent.) and invert sugar may form a honey-like mass if the reducing sugar content exceeds 40 per cent. The invert sugar content increases with increased acidity and boiling time. Under inversion is sometimes corrected by replacing some of the cane sugar with invert syrup.

The presence of **liquid glucose** has sometimes been objected to when present in jam and marmalade. It is indicated from the polarimetric readings if the rotation after inversion is positive or only slightly negative. In this case it is necessary to determine the cupric reducing power. Then the proportions of cane sugar, invert sugar and liquid glucose are given by the formulæ on p. 130. Dry glucose may be deduced with fair accuracy by assuming 20 per cent. of water in liquid glucose. The amount of glucose found by the writer in a number of genuine jams did not exceed 3 per cent.

Dextrin may be detected by alcohol precipitation as described under honey, and its presence is suggestive of added starch syrup, although certain fruits contain dextrins. These may be distinguished by dissolving the crude dextrin in water and adding iodine solution. Erythrodextrin, which is characteristic of corn syrup, gives a red colour.

Fruit content. The minima for fruit in jams, jellies and marmalade are prescribed in the Preserves Order (see p. 135). The fruit content can be assessed from:—

$$\text{Fruit (\%)} = \frac{\text{Percentage of A in the jam}}{\text{Average percentage of A in the fruit}} \times 100$$

Where A is any suitable constituent, *e.g.*, *insoluble solids, non-sugar solids, seed count.* As *acid* and *pectin* are sometimes added during the boiling to make up for deficiencies in the fruit, such values are of doubtful value for calculating the fruit content. Ranges of figures for various determinations on fruits have been the subject of numerous papers, *e.g.*, *Analyst*, 1928, **53**, 32; 1931, **56**, 35; 1934, **59**, 231; 1935, **60**, 592; 1936, **61**, 251; 1938, **63**, 262; 1940, **65**, 540; *J. Sci. Fd. Agric.*, 1950, **1**, 8 (abstracted in the table); 1958, **9**, 18. The last mentioned paper gives the figures for imported fruits such as banana, grapefruit, lemon, melon, sweet orange and pineapple and pH values for many British fruits.

From the table it will be seen that the figures vary widely for each constituent and the calculation of the fruit content using mean values can only be regarded as a reasonable approximation. Although an explanation should be sought from the manufacturers if the calculated fruit content should fall much below the prescribed minimum based on average figures, it is inadvisable to take legal action unless it is still low when the calculation is based on the minimum figure for A in the fruit. Some of the appropriate analytical methods are described below.

N.B.—Exclude stones when weighing out for the determinations.

ANALYTICAL DATA OF THE EDIBLE PORTIO[N]
(Money and Christian, *J. Sci. Fd. Agric.*, 1950, **1**,

		Insoluble solids %	No. of seeds per 100 g. %	Refract-ometer solids %	Total solids %
Apples	max.	2·54		16·4	18·1
	min.	1·05		6·8	10·8
	average	1·50		12·0	14·0
Apricots	max.	2·49		18·5	18·35
	min.	1·17		7·7	7·3
	average	1·60		11·6	12·6
Blackberries	max.	10·00		11·4	21·1
	min.	4·05		6·0	12·1
	average	6·33	3180	8·9	15·4
Blackcurrants	max.	9·72		17·6	24·4
	min.	3·28		8·0	13·7
	average	5·92	4450	13·3	19·7
Cherries (Black)	max.	3·92		22·2	25·0
	min.	1·30		12·7	12·9
	average	1·95		17·4	18·7
Damsons	max.	2·45		27·3	29·1
	min.	1·46		11·0	13·4
	average	1·91		16·8	18·4
Gooseberries	max.	4·38		13·6	15·1
	min.	1·51		4·0	7·9
	average	2·21	600	7·7	11·0
Greengages	max.	1·99		25·0	26·7
	min.	1·14		10·2	10·95
	average	1·45		16·7	16·7
Loganberries	max.	8·00		14·9	23·3
	min.	2·15		7·8	13·2
	average	6·35		9·6	16·3
Plums (Red)	max.	1·99		18·0	21·5
	min.	0·66		7·4	8·1
	average	1·17		11·7	13·0
Raspberries	max.	7·93		12·6	24·8
	min.	1·40		5·0	9·4
	average	5·12	4190	8·6	13·9
Strawberries	max.	3·40		12·0	13·8
	min.	0·86		4·5	6·8
	average	2·41		8·2	10·2

F SOME COMMON FRUITS
acara, *Analyst*, 1931, **56**, 43)

Total sugars %	Non-sugar solids %	Acidity No. of ml. N/10 per 100 g. %	Pectin as calcium pectate %
13·05	6·5	270	0·79
5·3	2·5	20	0·19
9·6	4·3	130	0·55
11·85	10·4	400	1·32
1·6	4·1	81	0·42
6·05	6·5	242	0·99
10·4	15·8	438	1·19
2·4	8·35	75	0·40
4·5	10·9	227	0·63
10·6	17·7	650	1·79
1·6	9·1	121	0·55
6·3	13·4	503	1·13
17·3	9·8	169	0·80
7·7	3·7	70	0·06
12·4	6·3	109	0·32
13·8	15·6	453	1·35
5·6	7·5	226	0·39
8·5	9·9	320	1·03
10·25	10·3	415	1·20
2·0	3·7	173	0·27
4·6	6·4	284	0·62
14·5	13·9	435	1·42
4·05	5·1	80	0·66
8·5	8·2	192	1·11
7·3	22·2	488	0·68
1·1	7·2	151	0·50
4·3	12·0	375	0·59
13·4	10·0	380	1·43
2·3	3·6	25	0·34
7·4	5·6	225	0·79
8·7	17·7	518	0·88
1·7	5·8	106	0·10
4·5	9·5	283	0·40
9·8	8·1	323	0·90
2·8	2·6	81	0·13
5·65	4·6	136	0·54

Insoluble solids. The insoluble solids figure is probably the most commonly used for the calculation of the fruit content. For the determination, boil 20 g. of sample gently with 200 ml. of water for half an hour and then filter through a 13·5 cm. paper (previously dried at 100° C. in a weighing bottle and weighed, etc.). Wash the filter thoroughly with hot water, return the paper to the weighing bottle, dry at 100° C., cool and weigh. From the increase in weight calculate the percentage of insoluble solids in the sample.

Seed count. Heat 50 g. of sample in a porcelain basin with a little water on a water bath and when the insoluble matter has sufficiently disintegrated, count the seeds. Calculate the number of seeds present in 100 g. of jam. The seed count is of assistance for mixed jams, where one fruit contains seeds and the other does not. The individual seeds should also be examined under the lens to confirm that they are genuine. A set of genuine seeds should be kept for comparison purposes.

The non-sugar solids (Hughes and Maunsell, *Analyst,* 1934, **59,** 231) represent the difference between the total solids (after drying at 100° C. in a vacuum oven) and the total sugars (as invert sugar).

The lead number (Hinton, *Analyst,* 1934, **59,** 248) which depends on the acid constituents of the fruit, has also been suggested for determining the fruit content of jams (see also Moir, *Analyst,* 1944, **69,** 343).

The potassium (K_2O) content is another constituent which may be employed for assessing the fruit content. Potassium can be determined by flame photometry, by the A.O.A.C. gravimetric chloroplatinate method or by the perchlorate method (also gravimetric—*e.g.*, *Analyst,* 1948, **73,** 449).

Pectin. An important constituent of many fruits and therefore of jams, is pectin; on the strength and amount of this substance the solidifying power of the conserve largely depends. It seems that the term " pectin " is really generic and covers a range of molecular compositions.[1] Pectic acid itself may have the formula $C_{17}H_{24}O_{16}$, and forms a calcium salt which is insoluble in acetic acid. Pectin includes a range of esters of this acid and it appears from the work of Tutin, Tollens and others that the pectin in all the common fruits and vegetables is substantially the same—at least from the analytical point of view. In the presence of sugar and acid it forms a gel; the gelatinising power is much influenced by the acidity,

[1] Hinton, *Food Investigation Report,* No. 48. H.M.S.O.

quite a small increase in which is said to be equivalent to a considerable quantity of sugar.

Several methods have been published for the estimation of pectin, the more important of which are those of Carré and Haynes (*Biochem. J.*, 1922, **16**, 60) and King (*Analyst*, 1925, **50**, 371). King heats 50 g. of the jam on the water bath with hot water and disintegrates the tissue with the aid of a glass rod, then adds hot alcohol little by little, with constant stirring until the volume reaches 300 ml. The mass at the bottom is frequently stirred for about two hours, or until no gelatinous particles are visible; then the whole is filtered on a coarse paper, using the pump if necessary towards the end of the filtration. The filter and its contents are washed back into a beaker with warm alcohol, treated again with 300 ml. of alcohol, and refiltered. The pectin is dissolved from the residue with boiling water and the solution filtered and cooled, a small excess (about 0·02 N) of sodium hydroxide is added, and the solution is allowed to stand for an hour to complete the hydrolysis. After this time acetic acid is added to give a concentration of free acid of about 0·1 N, then 20 ml. of 10 per cent. calcium chloride solution. The mixture is allowed to stand for an hour, then boiled and filtered through a tared paper. The gelatinous precipitate is washed with boiling water, then transferred again to the beaker, boiled with 300 ml. of water, filtered once more through the same paper, dried and weighed as calcium pectate (*cf.* A.O.A.C.).

King gives the following percentages of calcium pectate for some of the common varieties of jam:—

Plum and apple	0·13
Blackberry and apple	0·25
Plum	0·46
Damson	0·53
Raspberry	0·31
Gooseberry	0·45

The Pectin Sub-Committee of the Jam Panel of the B.F.M.I.R.A. (*Analyst*, 1951, **76**, 536) have reported on the determination of grade strength of pectins as commercially supplied for jam manufacture.

Agar was formerly added to jams in order to stiffen the preserve. It can be identified by microscopical examination—the presence of diatoms indicates the presence of agar (see also *Analyst*, 1921, **46**, 239; 1923, **48**, 433; 1925, **50**, 371).

Acidity. Determine by titrating 10 g. of sample with N/10 sodium hydroxide using phenolphthalein and calculate the total acidity as either a percentage of citric acid (or malic acid if apple predominates) or as the number of mls. of N/10 alkali required per 100 g. The result may, however, be of little value for the calculation of the fruit content, as pH adjustment is often practised to improve the set. The substances which are added for this purpose include citric acid, the so-called " buffer salts " such as sodium bicarbonate and sodium citrate. Industrially the pH value itself is usually determined also. A high pH often indicates unsatisfactory setting properties and a low pH suggests that the preserve will " weep " on storage. Most jams have a pH value between 3·0 and 3·5.

Examination for the type of fruit. The microscopic examination of the pulp must be made on the wet insoluble residue obtained after heating a quantity of the sample with water and decanting off the liquid. The microscopic appearance of the common fruit pulps is described in Winton's "*The Structure and Composition of Foods*" (Vol. II, Wiley, N.Y.). Jakloviv has described a more elaborate method in which the type of fruit is identified from the bands produced on alumina (see Dickinson and Goose, "*Laboratory Inspection of Canned and Bottled Foods,*" 1955, p. 135, Blackie, Lond.).

On the addition of iodine, apple pulp usually shows blue patches due to the presence of starch. Partridge (*Analyst*, 1926, **51**, 346) has made the method more sensitive by preparing a 10 per cent. extract of the jam with warm water (60° C.) and then straining and centrifuging. The iodine solution is added to the residue, which contains any starch present in a more concentrated form.

Preservatives. Most jam is manufactured from sulphited pulp containing up to 3000 p.p.m. of sulphur dioxide but most of the preservative is lost during boiling. Here the maximum of 100 p.p.m. only represents a tolerated residue (sugar is the preservative of jam). *Sulphur dioxide* can be conveniently determined on 50 g. by the iodimetric method.

Formerly *salicylic acid* was frequently added to preserves. It can be extracted from the acidified sample with ether and its presence can be detected with ferric chloride (*cf. Analyst*, 1905, **30**, 124). Under such conditions *benzoic acid* and *saccharin* would also be extracted. As a little boric acid is often present in fruit traces may be found in some jams.

Trace metals. Jams should be examined for arsenic, copper, lead and, if canned, tin.

Added colour. Add a tuft of white wool to about 20 g. of jam diluted with 50 ml. of water, acidify with dilute acetic acid and boil for five minutes. Then remove the wool, wash it in running cold water and after extracting the colour by boiling with dilute ammonia continue by means of the method described on p. 100.

LEMON CURD

Lemon curd is manufactured by boiling together sugar, cornflour (or wheat flour), margarine or butter, egg, citric acid, oil of lemon (introduced as essence), colour and water. Precautions are usually taken in the manufacture to reduce the loss of the oil of lemon by volatilisation.

Legal Standards

The Food Standards (Preserves) Order, 1953, prescribes:—

" 5. *Standard for Fruit Curd (including fruit flavour curd)*

(1) Each 100 parts of fruit curd shall contain not less than:—
(a) 4 parts of fat;
(b) 0·33 part of citric acid;
(c) 0·125 part of oil of lemon (or 0·25 part of oil of orange);
(d) 1 part of dried whole egg
or $1\frac{1}{2}$ parts of sugar dried whole egg
or $3\frac{3}{4}$ parts of liquid or frozen whole egg
or $4\frac{1}{2}$ parts of shell egg.

(2) The percentage of soluble solids contained in fruit curd shall be not less than 65 per cent."

Analysis

The analysis of fruit curd (lemon or orange) should include the determination of the **soluble solids, acidity, fat, volatile oil** and **egg content** (from the alcohol soluble phosphorus) in addition to the examination for **preservatives, trace elements** and **added colour.**

Soluble solids. Determine by refractometer (as for jam). Industrially the proportion of reducing sugars is also of importance as it indicates whether the preserve may crystallize on storage (see Jam).

Acidity. Titrate 10 g. of sample with standard alkali using phenolphthalein and calculate the total acidity as citric acid.

Fat. Either employ the Rose-Gottlieb method (use 1–2 g. + 9 ml. water) or dilute the sample with water and extract the fat several times with ether.

Volatile oil. Distil the oil of lemon (or orange) from sodium chloride solution in the B.P. volatile oil apparatus (p. 244) preferably using 300 g. of sample. From the volatile oil figure obtained as v/w (per cent.) and the s.g. of the oil (0·85), convert the result to w/w (per cent.).

Egg content. The method for determining egg in lemon curd based on the organic phosphorus content (*cf.* salad cream) has been described by Bagnall and Smith (*Analyst,* 1945, **70**, 211):—

Reflux 20 g. of sample with 100 ml. of alcohol (95 per cent.) in a boiling water bath for six hours. Filter on a Buchner funnel. Retain the filtrate and re-extract the residue as before with more alcohol and filter. Evaporate the combined filtrates to dryness in a basin. Wet oxidise the residue in the basin and determine the phosphate in the resultant solution either gravimetrically or colorimetrically.

Using the above method, the authors found that the following percentages of phosphate (as P_2O_5) were extracted: dried egg—1·20, wheat flour—0·077[1], rye flour—0·064.

Starch. Bagnall and Smith (*loc. cit.*) also describe the following procedure for determining starch in fruit curd, which is modified from the method of Illing and Whittle (*Food,* 1944, **13**, 32):—

To the residue on the Buchner funnel obtained in the method above add 75 ml. of 0·7 per cent. aqueous potassium hydroxide solution and gelatinise the starch by simmering gently for thirty minutes. Transfer the hot liquid to a 200 ml. graduated flask, cool and make up to volume. Filter. Neutralise 20 ml. of filtrate with 5 per cent. acetic acid using phenolphthalein. Add 8 ml. of N/10 iodine solution and 4 ml. of *either* 10 per cent. potassium acetate solution *or* alcohol and allow to stand until the precipitate settles and centrifuge. Decant the supernatant liquid. Rub the residue with a rod and treat it with 12 ml. of a mixture containing 10 ml. of 95 per cent. alcohol and 2 ml. N/10 thiosulphate, adding in several small quantities. When the particles are thoroughly broken up with the rod, add 25 ml. of 80 per cent. alcohol and filter through a weighed Gooch crucible, wash with 95 per cent. alcohol, dry and weigh the starch. Multiply the starch figure by $\frac{4}{3}$ to give the proportion of wheat flour. The starch content is usually about 3–4 per cent. Insufficient starch or incompletely gelatinised starch tends to cause fat separation.

Preservatives. Nowadays preservatives are seldom met with in lemon curd. Formerly salicylic and benzoic acids were occasionally found.

Trace elements. Fruit curd should be examined for arsenic, copper and lead.

Added colour. See p. 100. The starch should be removed with alcohol before treating the sample with the wool.

[1] Present day wheat flour usually gives a somewhat lower figure.

MINCEMEAT

Mincemeat is manufactured by mixing together apples, dried fruit (currants, sultanas and raisins), mixed peel, sugar, suet, acetic acid and flavourings such as salt, cinnamon, cloves, nutmegs and essence of lemon. The mixture should contain sufficient soluble solids (about 68 per cent.) to prevent growth of micro-organisms. The mixture is not heated during the manufacturing process.

Legal Standards

The Food Standards (Preserves) Order, 1953, prescribes:—

" 6. *Standard for Mincemeat*

(1) Each 100 parts of mincemeat shall contain:

(a) not more than 0·5 part acetic acid (80 per cent. or glacial);
(b) not less than 30 parts of added sugar;
(c) not less than 30 parts of dried fruit and peel;
(d) not less than 2·5 parts of suet or equivalent fat.

(2) The percentage of soluble solids contained in mincemeat shall be not less than 65 per cent."

Analysis of Mincemeat

The routine analysis of mincemeat should include the determination of **soluble solids** (by refractometer), **fat, sugars** (before and after inversion by Lane and Eynon's method), and the **fixed and volatile acidity** and examination for **preservatives** and **trace metals.** The industrial control of mincemeat has been discussed by Devey (*Food*, 1959, **28**, 205; *cf.* Grange, *ibid.*, p. 323).

Sampling

The sample should be passed through a mincer prior to the analysis.

Fat

The fat can be determined by the Werner-Schmid method using 1–2 g. sample plus 9 ml. water.

Fixed and volatile acidity. The acetic acid content of mincemeat can be estimated by steam distilling 5 g. of sample and titrating the distillate with N/10 alkali using phenolphthalein as indicator (1 ml N/10 ≡ 0·006 g. acetic acid). Alternatively the volatile acidity can be obtained from the total and fixed acidities. The total acidity is obtained by direct titration of a 5 g. sample, and the fixed acidity by titrating a further 5 g. sample after evaporating it down several times with water. The difference between the two titrations is the volatile acidity which should be calculated as acetic acid. The fixed acidity is usually reported as citric acid (1 m. N/10 ≡ 0·007 g.).

Preservatives. Sulphur dioxide may be present in mincemeat derived from the fruit used (*e.g.*, apple). It should be borne in mind also that traces of boric acid may be found in mincemeat as it is present naturally in fruits.

Trace metals. Mincemeat should be examined for arsenic, copper and lead. Canned samples should also be examined for tin.

The dried fruit content of mincemeat cannot be determined with any degree of accuracy, but Devey (*loc. cit.*) has claimed some partial success with a method which involves (a) removal of fat, sugar and water by treatment with light petroleum, warm water and acetone respectively, and (b) picking out with forceps the particles of dried fruit and peel, which are then dried, weighed and calculated back in relation to the composition of the fruit.

Chapter 5

CEREALS AND STARCH PRODUCTS

Starches—Cereals—Flour—Bread—Cake—Custard and Blancmange Powders

STARCHES

Properties. The starches enter more largely into the composition of human food than any component except water. The chemical constitution of the group is not yet understood, but all have the general formula $(C_6H_{11}O_5)_n$, where n is probably not less than 1000. Starch is readily convertible into sugar by hydrolysis either by an enzyme such as diastase or by heating with an acid and its estimation usually depends upon this reaction. When heated with water, the grains swell up and burst at about 70° C, forming a paste. When iodine solution is added a characteristic blue colour is developed, due to starch iodide, which is decomposed on heating, but is re-formed on cooling. Starch has no melting point. It gradually swells and chars on heating, with the formation of caramel and its derivatives. It does not reduce Fehling's solution until after hydrolysis and its optical rotation is variable, depending upon the variety of cereal from which it has been obtained. The specific rotations $[\alpha]^{20}_{D}$ of some of the commoner forms, according to Ewers, are given on p. 161, but they vary somewhat according to the exact conditions of hydrolysis and purification.

Starch has a density of approximately 1·50—1·53, varying a little with the type. It is insoluble in water and in organic solvents, but on grinding with water sufficient is dissolved or passes through a filter to give a blue colour with iodine. It is usually slightly acid in its reaction.

Microscopical examination. The starches or cereals, when ground, are usually identified by their microscopic characteristics. To a small drop of water on a slide is added a minute quantity of the starch on the point of a knife. Then the powder is thoroughly mixed with the water, taking care to break any air bubbles, and covered with a coverslip. Excess of water may be removed by means of a piece of filter paper applied without pressure. The examination should first be made with the lower power and then with the $\frac{1}{6}$-inch objective. A better idea of the shape of the granules can often be obtained by rolling them by gently moving the coverslip, and the striations and hilum can be brought into relief by modifying the illumination. When there is any doubt as to whether or not particular grains are starch, a little very dilute iodine solution run under the coverslip enables the point to be easily decided.

Some of the commoner starches, and a few of the less common ones which are met sometimes as adulterants of cocoa, ginger, pepper, and other foods, are illustrated here. Details of others may be found in Greenish and Collin's " Anatomical Atlas of Vegetable Powders " (Churchill, Lond.), but in all cases of doubt it is advisable to compare the sample with known specimens.

Most of the various proposals which have been put forward from time to time for the detection and estimation of foreign starch in mixtures have not been altogether satisfactory. It is best to rely upon the microscope, which will readily show any adulteration by a starch of another group, such, for example, as maize in rice or in wheat, but when two starches of the same group are present it is exceedingly difficult to detect a small quantity with certainty. Wheat and barley, for instance, are very similar. Wallis points out that the large grains of wheat starch reach as much as 45 μ, whereas those of barley do not exceed 40 μ. He bases a counting method with lycopodium on this fact (*Pharm. J.*, 1922, **109**, 82).

In mixed cereals, fragments of the husk are often to be found by sifting, and examination of them will yield information. An alternative procedure is to boil a quantity of the material with acid and alkali as for a fibre determination (p. 41) and examine the residue on the filter with the microscope. Vogt (*Zeit. Unters. Nahr. Genussm.*, 1921, **42**, 145) gives a method which also may help in these cases, based upon the differential staining of flours by Congo red and the alkalinity of the ash. König and Bartschat (*Zeit. Unters. Nahr. Genussm.*, 1923, **46**, 321) give an empirical method for estimating rye flour in admixture with wheat flour, which is accurate to about 5 per cent., provided that the acidity is not abnormally high. The method is based upon the observation that, of the total protein of wheat, 29·1 per cent., and of rye 51·5 per cent. is soluble in a saturated solution of calcium sulphate:—

Ten grams of flour are moistened in a 500 ml. flask with the calcium sulphate solution at room temperature, made up to the 500 ml. mark with the solution, shaken in a machine for an hour, the nitrogen then estimated by Kjeldahl's method on 100 ml. of the clear filtered solution, and the result expressed as a percentage of the total nitrogen. The following tables may be used for interpolating the composition of the mixture:—

Wheat	*Rye*	
100	—	29·1
90	10	31·34
80	20	33·58
70	30	35·82
60	40	38·06
50	50	40·30
40	60	42·54
30	70	44·78
20	80	47·02
10	90	49·26
—	100	51·50

When two starches (which can be readily identified microscopically) are present in admixture their relative proportions can be approximated by comparison with known mixtures, say, of 10, 20, 30 or 50 per cent. It is necessary to ensure thorough mixing by trituration, say, of 0·5 g. with 20 ml. of water, then removing 1 drop with a platinum loop and comparing it with an exactly similar quantity of the unknown starch in a like quantity of water. A mixture of glycerine and water 50: 50 is preferable when dextrin or gum is present. A more elaborate method capable of yielding accurate results depends on mixing the powder with lycopodium and counting the number of granules. For details, see Wallis (*Analyst*, 1916, **41**, 357).

The more important diagnostic characters of the starches are as follows:—

Wheat starch. Mostly round grains with very faint striation, some are small, but others are of large size with maximum diameter 35–45 μ.[1] The larger grains are oval or lenticular when rolled, and some have a longitudinal furrow, but there is no hilum. With polarised light they exhibit a cross.

Barley starch. Closely resembles wheat, but the grains are smaller with diameter not greater than 35 μ. There are not so many intermediate sized grains, the longitudinal furrow is more frequent and on rolling their form is seen to be lemon shaped. Under polarised light there is a cross.

Rye starch. This starch also resembles wheat, but the grains are distinctly larger, measuring in some cases as much as 60 μ. There are three- or four-rayed fissures in many of the larger granules, and with polarised light the cross is more distinct than with wheat or barley.

Rice starch. These are closely packed angular grains without hilum, fairly uniform in size, measuring from 6 to 9 μ. Compound grains are numerous, but are broken by slight pressure. The usual cross is seen under polarised light.

Oat starch. Oat is very similar to rice, but the grains are somewhat larger, averaging 10–11 μ. Compound grains are numerous and are not so readily fractured as are those of rice. Oat starch is peculiar in not exhibiting the usual cross when examined in polarised light, a feature which provides a means of differentiation from rice.

[1] $1\ \mu = 0{\cdot}001$ mm.

Fig. 9. Wheat starch. × 125.

Fig. 10. Barley starch. × 125.

Fig. 11. Rye starch. × 125.

Fig. 12. Rice starch. × 125.

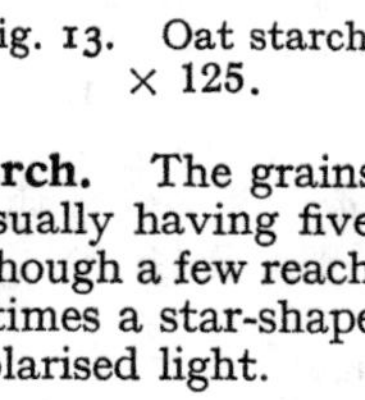

Fig. 13. Oat starch. × 125.

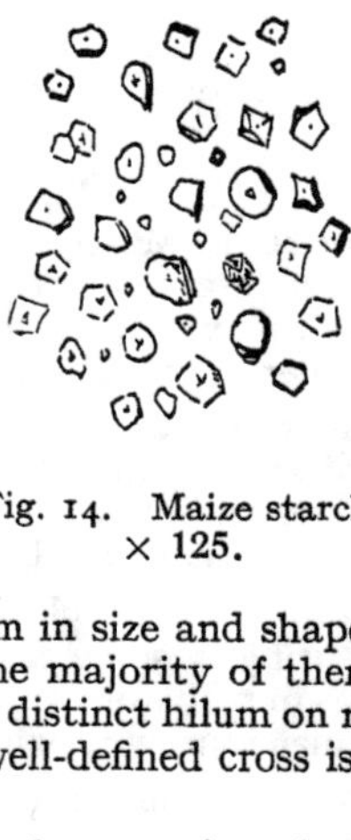

Fig. 14. Maize starch. × 125.

Maize starch. The grains are fairly uniform in size and shape, they are polygonal, usually having five to six sides. The majority of them measure about 15 μ, though a few reach 20 μ. There is a distinct hilum on most of the grains, sometimes a star-shaped fissure, and a well-defined cross is shown by the aid of polarised light.

Potato starch. Potato is composed of quite large grains of oval or conchoidal shape with oyster-shell markings of length from 100 μ downwards. A few are rounded or flattened and are much smaller (15 μ). A distinct hilum is visible in all grains, usually near one end. Under polarised light there is a cross which is centred at the hilum.

Tous-les-mois, or Queensland arrowroot. The largest of the starches. The grains are generally similar to those of potato, but are larger, some reaching as much as 135 μ. The concentric striations are more close than those of potato and the hilum nearer to the end of the grains.

Curcuma starch, or East Indian arrowroot. This is generally similar to potato but smaller. The grains average about 50 μ in length and have a tapering projection at the larger end. The hilum is not very distinct, but the concentric striations are well marked.

St. Vincent arrowroot. This also has a similarity to potato starch but is smaller and of much more uniform size. The grains have an average length of 45 μ. Neither the striations nor hilum are very distinct.

Manihot starch, or Brazilian arrowroot. The grains of this starch are variable in shape, some circular with a few polygonal or muller-shaped.

Fig. 15. Potato starch. × 125.

Fig. 16. Tous-les-mois. × 125.

Fig. 17. East Indian arrowroot. × 125.

Fig. 18. St. Vincent arrowroot × 125.

Fig. 19. Brazilian arrowroot. × 125.

Fig. 20. Sago. × 125.

The average diameter is about 20 μ, but some large grains are seen (up to 30–33 μ). The hilum is visible on the circular grains and is in the form of a cross. There are faint concentric striations.

Maranta starch, or ordinary West Indian arrowroot. Similar to St. Vincent arrowroot both in size and shape. A distinctive feature is the slit or dark line, frequently transverse, at the hilum. J. R. Stubbs (*Analyst*, 1926, **51**, 400) calls attention to and illustrates sweet potato starch as an adulterant of arrowroot and cornflour. So-called marble arrowroot is not a true arrowroot but is occasionally used as an adulterant (see Jones, *Analyst*, 1934, **59**, 493).

Arum starch, or Portland arrowroot. One of the smaller starches, its grains being usually 7–12 μ in diameter (occasionally as much as 14 μ). Some are circular with a dot-like hilum of small angular or truncated shape, and others are angular. There is a general resemblance to manihot starch, but on a smaller scale.

Fig. 21. West Indian arrowroot. $\times$ 125.

Fig. 22. Portland arrowroot. $\times$ 125.

Fig. 23. Tapioca. Unaltered. Heated.

Fig. 24. Pea starch. $\times$ 125.

Fig. 25. Bean starch. $\times$ 125.

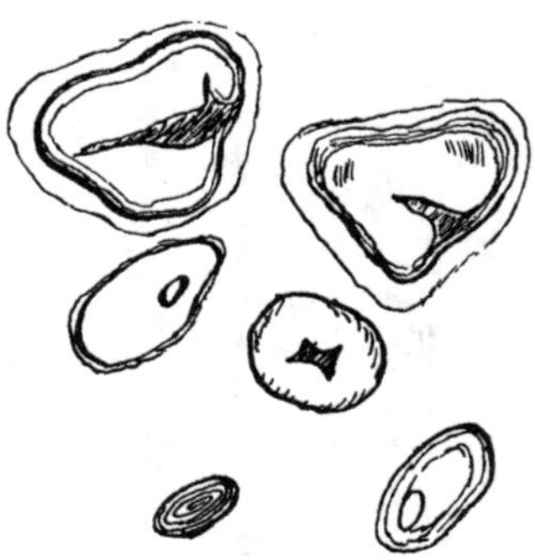

Fig. 26. Dextrin.

Sago. Sago starch which is obtained from the stem of the sago palm, *Metroxylon rumphii* and other species, is usually seen in a swollen form as rounded or ovoid granules, some angular and of unequal size, varying from 20–60 μ or more. Some granules are saucer-like in appearance, and have a dark patch in the centre.

Tapioca starch. This is prepared from manihot starch by a process of heating the moist mass. In appearance it is similar to sago but smaller. The unswollen grains are roughly circular in shape with concentric rings and usually a hilum. The size is from 15 to 25 μ in diameter. In the gelatinised tapioca of commerce a few unswollen grains are often visible, but the majority are about three times the above size, and have irregular and sometimes saucer-like shapes with no regular markings. The centre is usually dark.

Fig. 27. Banana starch. × 125.

Fig. 28. Chestnut starch. × 125.

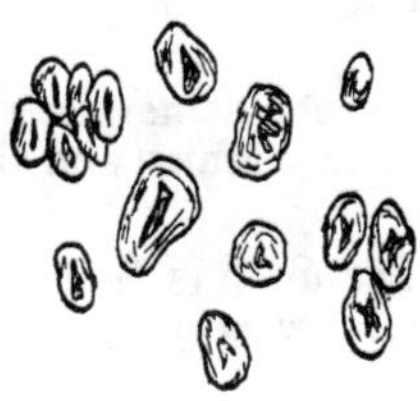

Fig. 29. Acorn starch. × 125.

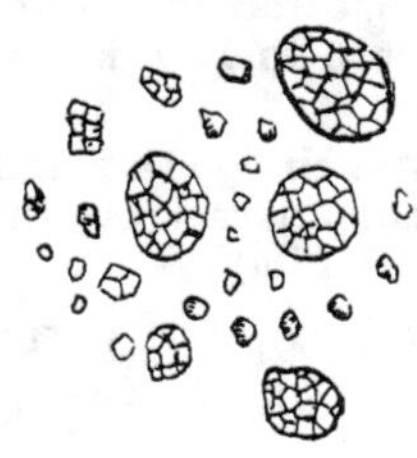

Fig. 30. Darnel starch. × 125.

Fig. 31. Buckwheat starch. × 125.

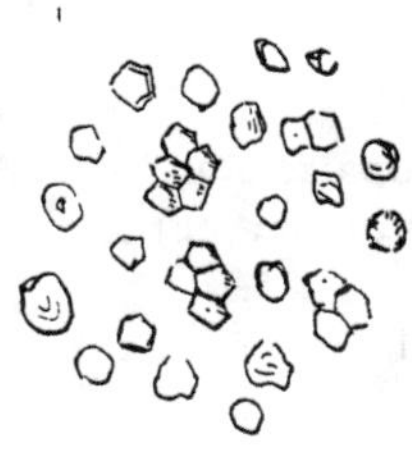

Fig. 32. Millet starch. × 125.

Pea, bean and lentil starches. These three starches are closely similar in form. They are of irregular bean-shaped or elliptical form, and most of the grains show concentric markings. Bean starch grains are the largest, being up to 57 μ in length. Pea starch is intermediate, having grains from 15 to 47 μ and lentils are mostly from 20 to 40 μ with a few as large as 45 μ. The hilum and markings are more distinct in beans than in lentils and there is with each of these two forms a slit at the hilum. With pea starch the markings and hilum are less pronounced, and there is no slit in most of the grains.

Dextrin. This is seen under the microscope as large swollen grains of irregular size and shape, which give a brown or reddish-brown colour with weak solution of iodine. Oyster-shell markings are visible on some of the particles. Dextrin should be mounted in alcohol.

The following starches are of less frequent occurrence:—

Banana starch. This has a general resemblance to curcuma starch, but is rather smaller. The grains are about 30 μ in length, narrow, and somewhat irregular in shape, with an eccentric hilum and circular striations.

Chestnut starch. Is very irregular, both in shape and size. Amongst the more characteristic types are pear-shaped grains, about 16–18 μ in length, with a hilum near the centre. There are many small grains with no visible markings.

Acorn starch. Is similar in appearance to bean starch, but the grains are more regular both in their oval shape, large slit in the hilum, and length, which is about 70 μ.

Darnel starch. Is polygonal and angular like rice, with a hilum in many grains, the average diameter being 4 μ. It usually exhibits numerous compound ovoid lumps which are not easily broken up by pressure on the cover-glass.

Buckwheat starch. Exhibits angular grains of irregular size and shape, usually from 6 to 12 μ in diameter, and having curved sides. There is no hilum.

Millet starch. The starch of *Panicum* is of the same general shape as buckwheat and darnel, but there is a hilum on many grains and the diameter is greater (10–15 μ).

The microscopy of various damaged starches has been studied by McMasters *et al.* (*J. Agric. Food Chem.*, 1957, **5**, 455).

The average percentage composition of a number of flours, starches, etc., is given below (the remainder consists mainly of starch):

Cereal	Moisture	Protein	Fat	Fibre	Ash
Arrowroot	14	0·4	0·1	Trace	0·1
Barley Meal	12	10·5	2·2	4·5	2·1
Cornflour (maize starch)	12	0·5	0·1	Trace	0·1
Farina (potato starch)	17	0·4	0·1	—	0·2
Lentil	9	24·0	1·8	3·2	1·8
Maize meal	11	9·3	3·5	1·8	1·5
Oatmeal	8	14·0	6·8	1·5	1·7
Pea flour	10	24·0	1·7	2·3	2·7
Pearl barley	12	8·6	1·1	0·4	0·9
Potatoes	77	1·5	0·5	0·5	0·8
Rice	11	6·9	1·1	0·4	0·3
Rye flour	13	8·0	1·8	0·4	0·8
Sago	14	0·2	0·2	0·1	0·2
Tapioca	13	0·4	0·1	0·1	0·1
Wheat rusk	4	13·9	0·5	—	1·7

WHEAT FLOUR

Composition. The wheat grain has the following average percentage composition: endosperm 85, bran 12·5, germ 2·5. The composition of wheat flour, however, varies considerably according to the class of wheat, its country of origin, and the proportion of outer parts removed by the particular milling process. As the outer portions contain more protein, fat, fibre and ash than the starchy endosperm, the proportion of each of these constituents decreases as the extraction percentage gets less (see composition table). Also, wholemeal flour is brownish in colour, whereas the finest grades (containing the least fibre, etc.) are white. Wheat flour differs from other flours in containing a considerable proportion of gluten and to this its special suitability for breadmaking is due. The composition of the gluten present has a bearing on the " strength " and water-holding properties of the flour. The two proteins forming the greater part of the gluten are glutenin and gliadin, and, while the latter appears to be identical in strong and weak wheats, the former exists in different varieties. The figures given in the composition table (below) have been extracted from various sources including the " Report of the Panel on Composition and Nutritive Value of Flour " (H.M.S.O., 1956).

COMPOSITION OF UNFORTIFIED WHEAT FLOURS OF VARIOUS EXTRACTION RATES AND WHEAT GERM AND WHEAT BRAN

	Flour (72%)	Flour (80%)	Wholemeal (95–100%)	Wheat Germ	Wheat Bran
Moisture (%)	13–15	13–14·5	13–14	9–12	14
Protein (%N × 5·7)	8–13	8–14	10–15	25–30	12–16
Fat (%)	0·9–1·4	1·0–1·6	1·5–2·5	8·5–11·0	3·0–4·0
Carbohydrates (%)	65–70	64–70	60–68	39–45	—
Fibre (%)	0·1–0·3	0·2–0·4	1·8–2·5	2·0–2·5	9–12
Ash (%)*	0·3–0·5	0·6–0·9	1·2–2·0	4·0–4·5	4·0–6·0
Calcium (mg. Ca/100 g.)*	15	20	30		
Iron (mg. Fe/100 g.)	1·2	1·7	2·5		
Vitamin B_1 (mg./100 g.)	0·10	0·25	0·40		
Nicotinic Acid (mg./100 g.)	0·8	1·3	6		

N.B. " Strong " flours are richer in protein than " weak " flours.

* Most flour (except wholemeal) is fortified with creta (see Regulations), which increases the ash and introduces the equivalent of approx. 125 mg. Ca per 100 g.

Statutory Standards for Flour

The statutory standards for flour are prescribed in The Flour (Composition) Regulations, 1956:—

Description of Flour	Compositional Requirement
All flour other than flour containing the whole of the products derived from the milling of wheat	To contain *creta præparata* as follows:— (i) Not less than 235 milligrams per 100 grams of flour, and (ii) Not more than 390 milligrams per 100 grams of flour
All flour	To contain quantities of the undermentioned *nutrients* as follows:— Iron . . Not less than 1·65 milligrams; Vitamin B_1 . Not less than 0·24 milligrams; Nicotinic acid or nicotinamide Not less than 1·60 milligrams per 100 grams of flour; such nutrients to be added (where addition is necessary), in the case of iron, in the form of reduced iron (ferrum redactum)or ferric ammonium citrate and, in the case of vitamin B_1, nicotinic acid and nicotinamide, in a form conforming to the standards of the British Pharmacopoeia or the British Pharmaceutical Codex.*

In the regulations "creta præparata" means the substance so described in the British Pharmacopoeia or the British Pharmaceutical Codex,* which conforms to the standards laid down therein and which is of granularity standard:

(i) not more than 0·1 per centum to remain on a 100 mesh sieve, and
(ii) not more than 0·2 per centum to remain on a 240 mesh sieve;

"Flour" means the product derived from or separated during the milling or grinding of wheat (other than wheat offals and wheat germ separated for disposal as such) and includes meal and any flour derived wholly or partly from wheat that, before being ground or milled, has been malted or subjected to any process but does not include semolina."

Food Standards Committee Report on Bread and Flour, 1960

In 1960 the F.S.C. published a Report on Bread and Flour. The report reviews the evidence which had been put forward since the late thirties which

* The B.P. prescribes the following standards for the added materials included in the regulations:—

"Creta"—minimum of 97 per cent. of $CaCO_3$ with reference to the substance dried at 100° C.
"Reduced iron"—minimum of 80 per cent. of metallic iron.
"Ferric ammonium citrate"—to contain 20·5–22·5 per cent. of Fe.
"Vitamin B_1"—the official substance is aneurine hydrochloride which is required to contain 20·6–21·2 per cent. total Cl, 10·4–10·7 per cent. of Cl present as hydrochloride and not less than 98·0 per cent. of $C_{12}H_{17}ON_4SCl,HCl$ (all calculated with reference to the substance dried at 105° C.).
"Nicotinic acid"—minimum of 99·0 per cent. of $C_6H_5O_2N$ with reference to the substance dried at 105° C.
"Nicotinamide"—minimum of 98·5 per cent. of $C_6H_6ON_2$ with reference to the substance dried under vacuum over P_2O_5.

led to the introduction of regulations which prescribed that minimum quantities of certain nutrients should be present in flour.

The following is a summary of the recommendations contained in the Report in relation to **FLOUR**:—

(1) The definition of flour in the Flour (Composition) Regulations, 1956, should be replaced by the following:—" Flour means the product derived from or separated during the milling or grinding of wheat, whether or not any of the wheat has been malted or subjected to any process, and includes meal but does not include separated wheat offals, separated wheat germ or semolina."

(2) Only flour containing not less than 0·6 per cent. fibre calculated on the dry matter should be described as " wheatmeal."

(3) The term " wholemeal " should only be used to describe flour containing the whole of the product derived from the milling of cleaned wheat. Wholemeal should not be required to contain added nutrients.

(4) Wholemeal and wheat malt flour should not be required to contain added creta.

(5) Further investigation should be made into the possibility of using alternative forms of calcium (dicalcium phosphate, tricalcium phosphate, calcium sulphate, or calcium lactate) and into methods of improving the distribution of creta. If alternative forms of calcium are permitted the limits should be expressed as calcium. It should not be compulsory to add creta to flour used for self-raising flour by the addition of not less than 4½ lb. A.C.P. per 280 lb. sack.

(6) Methods of analysis are recommended in relation to enforcement of the regulations.

(7) Sampling of flour for the purpose of the compositional regulations should be confined to mills and docks. Provision should be made for analyses of the master mix (containing the three added nutrients) and the arrangements for adding it (and creta) should be checked.

(8) Enforcement could be carried out by a central inspectorate (using centralised laboratories or the Government Laboratory).

(9) With certain reservations the following additions should be allowed: (a) certain bleachers and improvers (see p. 165), (b) enzyme preparations, (c) caramel (but no other colouring matter), (d) emulsifying agents (only in pre-mixes for bread, if their presence is disclosed).

Analysis of Flour

The routine analysis of flour may include the determination of **moisture, ash, added creta** (except wholemeal), **oil, protein, acidity, iron, aneurine** and **nicotinic acid,** an examination for **improvers** and **bleaching agents** and a **microscopical examination.**

Industrially certain other types of analysis are of some importance, viz., examination of the gluten, physical tests on the dough produced from the flour, determination of the maltose, colour and grade figures and the " filth " test (p. 47). The methods which are described below are essentially those which are in the opinion of the writer of the greatest interest to the official analyst. For further information on these and the other methods of flour analysis the reader should refer to " Modern Cereal Chemistry " by Kent-Jones and Amos (Northern Publishing Co. Ltd., Liverpool, Eng.).

Some standard analytical methods for flour are described in the Wheat Commission's " Examination and Analyses Byelaws, 1939 " (S.R. & O. 1939 No. 850).

Moisture. The Wheat Commission method for determining moisture involves drying 5 g. of sample at 100° C. in a dish having a diameter of 1½–2½ inches and a height not more than 2 inches (and provided with a lid). Industrially, more rapid methods are employed using higher drying temperatures, *e.g.*, the Carter-Simon method (which involves drying for fifteen minutes at 155° C.), vacuum ovens, infra-red heating, etc. Such methods may give results over 1 per cent. higher than those obtained using the Wheat Commission method, but they are often sufficiently accurate when considered on a comparative basis and can usually be correlated with the standard method. The results obtained from drying methods represent the amount of " free " moisture produced at the drying temperature, *i.e.*, some of the water present remains " bound " and is probably associated with the proteins. Distillation methods using 25 g. of sample and toluene or xylene as the immiscible solvent are said to give a figure which is equivalent to the total water present. In view of the different results obtained according to the technique employed, it is especially important to consider the method of determination when interpreting figures for the moisture content of flour.

Moist flour with a moisture content in excess of about 13 per cent. (Wheat Commission method) is liable to attack by micro-organisms, mites and insects.

Ash. The outer portions of the wheat grain contain considerably more mineral matter (about 5 per cent.) than the inner parts, so the ash figure of white flour is considerably less than that of wholemeal flour. In the absence of added mineral matter, therefore, the ash figure gives an indication of the grade—" Patent " gives an ash of about 0·3–0·4 per cent., " Straight-run " (72 per cent. extraction) about 0·45 per cent., lower grades 0·5–0·65 per cent. or more and " Wholemeal " 1·2–1·8 per cent. If added creta is present as required by The Flour (Composition) Regulations for most flour, it is somewhat difficult, however, to assess the grade from the ash figure.

The Wheat Commission method for determining the ash involves igniting 3–5 g. of the flour at 600° C. The ash of wheat flour consists principally of the phosphates of potassium and magnesium. On analysis the ash of flour gives mean percentages of 49 (as P_2O_5), 37 (as K_2O), 6 (as MgO) and 5·5 (as CaO). Such figures obviously do not apply if the flour contains added creta. Further, it should be borne in mind that, as the bran contains more magnesium and less potassium and calcium than the centre of the grain, such proportions vary with the extraction rate.

The Wheat Commission also lays down the procedure for determining the **acid-insoluble ash**:

Add 3–4 ml. of 10 N hydrochloric acid to the ash and evaporate on the steam bath and heat for a further hour. Add 25 ml. of 2 N hydrochloric acid, heat for five minutes in the steam bath and decant the liquid through an ashless filter paper. Re-extract the residue in the dish with more of the dilute acid. Transfer the residue to the filter paper and wash thoroughly with hot water. Then place the paper and residue in the original dish and ignite, cool, weigh, etc.

Oil or fat. The fat in flour is usually determined either by direct extraction with petroleum ether or by extraction with mixed ethers after acid hydrolysis. The varying amounts of fat extracted by different methods have been compared by Herd and Amos (*Cereal Chem.*, 1930, 7, 257), who obtained the following results:—

Type of Wheat	Fat Contents Calculated on Dry Matter (%)			
	Soxhlet Methods		Hydrolysis Methods	
	Petroleum Spirit	Di-ethyl Ether	Acid	Alcohol ("lipoids")
Straight Run Flour	—	1·5	2·0	—
Patent Flour .	1·2	1·2	1·7	2·0
Bran . . .	4·2	4·4	6·7	5·0
Germ . . .	7·8	8·3	9·6	10·3

Herd and Amos concluded (a) that direct extraction measures the free fat only, (b) that acid hydrolysis methods give the total fat content (free + combined) and (c) the usual alcohol method for estimating lipoids gives the fat + unhydrolysed lipoids.

For the determination of oil by the direct method, it is preferable to use a Soxhlet or Bailey-Walker extractor with petroleum spirit (B. Pt. 40–60° C.) as solvent. The material should be removed from the apparatus during the extraction and then finely ground with sand before returning to the extractor.

For the *acid hydrolysis method*, 2 g. of flour is stirred with 2 ml. of alcohol and then 7 ml. of concentrated hydrochloric acid and 3 ml. of water are added. The material is heated at about 80° C. for about half an hour. Then 10 ml. of alcohol is added to the cooled hydrolysed mixture and the fat, etc., extracted three times with 25 ml. of ether, followed by 25 ml. of petroleum spirit as is employed for dairy products (p. 330).

N.B. The determination of lipoids is described in the A.O.A.C. (*cf.* Herd and Amos, *loc. cit.*).

Protein. The total nitrogen is determined on 1·4 g. by the Kjeldahl method (p. 22). The crude protein is calculated using the factor N × 5·7 (*cf. Cereal Chem.*, 1926, **3**, 194). Wholemeal flours contain more protein than white flour, prepared from the same grist.

Gluten. The method usually employed for the estimation of gluten is not altogether satisfactory, and consists in adding to 20 g. of flour about 15 ml. of water so as to make a stiff dough. This is allowed to stand in a beaker of water and is then squeezed in the fingers and gently kneaded under a stream of running water until all the starch is washed away and water squeezed out runs away quite clear. No loss of gluten need occur in this operation. The residue of moist gluten is squeezed as dry as possible and weighed, or it is better to tear it into tiny pieces and dry in the water oven till the weight is nearly constant, and to return the result as dry gluten. The "strength" of the flour depends upon both the amount and the physical qualities of the gluten. Crude gluten obtained in this way contains albumin, globulin, glutenin, gliadin and proteose.

Fibre. The general method for determining crude fibre as laid down in the Fertilisers and Feeding Stuffs Regulations, 1960, is described on p. 41. In 1943, the S.P.A. (*Analyst*, 1943, **68**, 276) suggested for the determination of fibre in National Flour that the somewhat similar procedure described in the (then current) 1932 Regulations should be modified as follows: (a) omission of the initial extraction with petroleum spirit, (b) the initial "creaming" of 3–5 g. of flour by adding it slowly to 20 ml. of the cold sulphuric acid (1·25 per cent.) before the rest (180 ml.) of the (boiling) acid is added, (c) the prescribed use of a reflux condenser for the maintenance of constant volume during the digestion, (d) transference of the fibre, after the washing with ether, from the filter paper to a dish (previously ignited and weighed) with boiling water. Then the water is evaporated off on the water bath and the dish dried at 100° C., weighed and ignited.

The percentage of fibre in the flour varies from about 0·1 (white of 72 per cent. extraction) to about 2 (wholemeal).

A method for the determination of *pentosans* is described in the A.O.A.C. (*cf.* p. 200).

Acidity. Three different procedures have been proposed for the determination of the acidity of flour, *viz.* titration of (a) a water extract, (b) an alcoholic extract and (c) the F.F.A. of the extracted fat.

Water extract method. Shake 18 g. of flour with 200 ml. of CO_2-free water in a conical flask and place in water bath at 40° C. for one hour (with the flask loosely stoppered). Filter and titrate 100 ml. of clear filtrate with N/20 sodium hydroxide solution to phenolphthalein. The acidity of the water extract is usually calculated as lactic acid. The acidity of the water extract of brown flour is higher than that of white flour and increases during storage.

Alcohol extract method. Add 10 g. of flour to 100 ml. of neutral 90 per cent. alcohol and allow to stand for twenty-four hours with occasional shaking. Then titrate 50 ml. of the supernatant liquid (or filtrate) with N/20 alcoholic sodium hydroxide solution to phenolphthalein. The acidity of the alcohol extract is usually calculated as sulphuric acid (*cf. Cereal Chem.*, 1929, **6**, 515; 1931, **8**, 29; 1932, **9**, 128; 1938, **15**, 580).

The F.F.A. or Acid Value as determined on the petroleum ether extract of the flour tends to rise more rapidly as the flour deteriorates during storage than the acidity of the aqueous or alcoholic extracts.

pH. Shake 10 g. or 20 g. of flour with 100 ml. of water and allow to stand for at least thirty minutes. Filter and determine the pH of the filtrate either visually or with a pH meter. The pH of flour usually falls within the range 6·0–6·8. Bleaching of flour with chlorine gas causes a fall in the pH value.

Starch. A number of methods are available for the determination of starch—(a) Lintner's method involving acid dispersion of the starch followed by its polarimetric estimation, (b) O'Sullivan's method involving enzymic hydrolysis followed by estimation of the sugar using Fehling's solution, (c) miscellaneous methods. Most of these methods can be employed, with slight modifications, for the determination of starch in other foods.

(a) Lintner's Method

About 5 g. of the material are triturated with 20 ml. of water, and 40 ml. of hydrochloric acid (sp. gr. 1·15) are added in small portions at a time. Then the mixture is washed into a 200-ml. flask with hydrochloric acid (12 per cent. w/w HCl), 10 ml. of phosphotungstic acid solution (5 per cent.) are added to precipitate proteins, and the volume is made up with the precipitate to 200 ml. by the addition of more 12 per cent. hydrochloric acid. The mixture is well shaken, filtered, and the rotation of the filtrate is observed in a 200-mm. tube. The mean specific rotation of starch is taken as + 200°, but for special starches Ewers (*Z. Offent. Chem.*, 1908, **41**, 8) gives the following values at 20°, which strictly are only applicable under the special conditions set out in his paper, using potassium ferrocyanide as the clearing agent:—

Wheat . . .	183·6	Potato . . .	186·5
Barley . . .	181·5	Rye . . .	181·0
Rice . . .	186·0	Maize . . .	184·2
Oats . . .	181·3		

When the reading is made on Ventzke's scale and 5 g. in 100 ml. have been used, the reading, multiplied by 1·912, gives the percentage directly.

(b) O'Sullivan's Diastase Method

Three grams. of the finely powdered material are extracted several times with ether in a Soxhlet extractor, afterwards washed on a filter, first with dilute (10 per cent.) alcohol, then with strong alcohol; the residue is drained and washed into a flask with about 50 ml. of water, and heated for fifteen minutes in a bath of boiling water with constant stirring, so that all the starch is gelatinised and a homogeneous mixture is obtained. The solution is now cooled to 55°, about 0·03 g. of diastase dissolved in a little water is added, and the mixture kept at 55°–60° for an hour or more (a drop of the solution should give no blue colour with iodine). Then the temperature is raised to 100°, the solution is filtered and the filtrate and washings are cooled and made up to 250 ml. Two hundred ml. of the filtrate are taken for the acid treatment, 20 ml. of hydrochloric acid are added, and the mixture heated in a boiling water bath for two and a half hours. The solution is cooled, then nearly neutralised with sodium carbonate, diluted to 500 ml., and the dextrose is estimated by the volumetric or gravimetric Fehling method. Dextrose multiplied by 0·90 gives the equivalent of starch.

(c) MISCELLANEOUS METHODS

Clegg (*J. Sci. Food Agric.*, 1956, **7**, 40) has described a rapid method for estimating starch in cereals, etc. which is based on the colorimetric reaction between anthrone and sugars to produce a green colour.

Although many methods have been proposed for the *direct determination of starch*, the results obtained are not necessarily more accurate than the estimation by difference. Care must be taken, however, if the " difference " method is used if materials such as dried milk or soya are present. Other methods for determining starch are included in the following references: *J.A.O.A.C.*, 1927, **10**, 108, 473; *J.S.C.I.*, 1931, **50**, 20 T; *Analyst*, 1934, **59**, 673; 1938, **63**, 328, 421).

Maltose figure. The " maltose figure " is a measure of the sugar produced by enzymic action on the starch present when the flour is incubated with water (the " **diastatic activity** "). Industrially the maltose figure is of considerable interest as it gives an indication of the gas producing powers of the corresponding dough. For its determination, Blish and Sandstedt (*Cereal Chem.*, 1933, **10**, 359) incubate the flour in the presence of an acetate buffer (pH 4·7) at 30° C. for one hour and the sugar produced is then determined volumetrically using alkaline ferricyanide. The result is expressed as mg. maltose produced by 10 g. of flour. Unlike the original procedure of Rumsey, in which the flour was incubated with water at 27° C. and the sugar determined using Fehling's solution, the Blish-Sandstedt method seems to be little affected by any creta which may be present. A modification of the Blish-Sandstedt method has been proposed by Hildebrand and McClellan (*Cereal Chem.*, 1938, **15**, 107).

Iron. The Flour (Composition) Regulations require all flour to contain at least 1·65 mg. of iron per 100 g., such nutrient to be added (if necessary) in the form of Reduced Iron, B.P. (required to contain at least 80 per cent. Fe) or ferric ammonium citrate, B.P. (required to contain 20·5–22·5 per cent. Fe).

Added iron can be detected *qualitatively* by wetting the flour with potassium thiocyanate (5 per cent. in N/1 HCl) and comparing the red colour after fifteen minutes with that produced by a non-enriched flour similarly treated.

For the *determination* of iron in flour, the F.S.C. (Analytical Panel) Report on Bread and Flour states that the method due to Pringle (*Analyst*, 1946, **71**, 491) using *o*-phenanthroline is satisfactory:—

Weigh out a suitable quantity of the sample (containing 0·05–0·4 mg. Fe) into a silica basin, add 10 ml. of glycerol-alcohol mixture (1 : 1) and, after a preliminary heating at the mouth of the muffle furnace, ash overnight at 600° C. Cool, add exactly 1·0 ml. concentrated nitric acid, evaporate at the muffle entrance and then replace the basin in the furnace for an hour. Cool, add 5 ml. of 5 N hydrochloric acid to the ash, place on a steam bath for fifteen minutes and filter through a hardened filter paper into a 100 ml. graduated flask. Then add 3 ml. of diluted hydrochloric acid (1 in 100) to the basin, bring to the boil on a hot plate and pass through the filter. Repeat this process four times and then wash the basin and filter with hot water. Make the filtrate up to 100 ml., mix and pipette 10 ml. into a 25 ml. graduated flask. Add 1 ml. of 2 per cent. sulphur dioxide solution and titrate with 2 M sodium acetate using a small piece of congo red paper (colour change from blue to pink). Then add 2 ml. of 0·25 per cent. aqueous *o*-phenanthroline solution and make up to the mark. After allowing the colour to develop overnight, read the optical density in a 4 cm. cell at 520 mμ against a blank prepared in the same manner as the test solution.

STANDARD GRAPH

Dissolve 0·7024 g. ferrous ammonium sulphate, $Fe(NH_4)_2 (SO_4)_2 6H_2O$, in water, add two drops of hydrochloric acid and dilute to 1000 ml. Dilute 50 ml. to 1000 ml. so that 1 ml. $\equiv$ 0·005 mg. Fe. For the standard graph pipette aliquot amounts of this solution (over the range 0–0·05 mg. Fe) into a series of 25 ml. flasks and add the appropriate reagents (see above).

More rapid results can be obtained using $\alpha : \alpha'$ dipyridyl as complexing agent (see A.O.A.C.; and *Cer. Chem.*, 1951, **18**, 825), but Pringle (*loc. cit.*) has shown that it is less sensitive than *o*-phenanthroline.

Mineral additions can be separated by shaking the flour with carbon tetrachloride (see section on " Self-Raising Flour ").

Creta. The phytic acid present in flour renders some of the calcium unavailable to the body. To counteract this, most types of flour are required to contain 235–390 mg. of *creta præparata*, B.P. (see p. 156). Wholemeal flour is exempted from this requirement (this is somewhat anomalous as a higher extraction rate results in an increase in the phytic acid content—see p. 5 of F.S.C. Report). Two difficulties arise in connection with the estimation of the added *creta*: (a) the uneven distribution in the flour, and (b) the uncertainty as to the accuracy of the blank.

1. Determination of the Calcium Content by Titration of the Oxalate Using Permanganate

The first of three methods for determining creta in flour proposed by Greer, Mounfield and Pringle (*Analyst*, 1942, **67**, 352) involves ashing 10 g. of flour and taking up the ash in hydrochloric acid. The calcium is then precipitated as the oxalate, which is dissolved in hot 4 N sulphuric acid and the solution titrated at 80° C. with N/20 permanganate. For exact details the original paper should be consulted.

$$1 \text{ ml. N/20 permanganate} \equiv 1 \text{ mg. Ca} \equiv 2{\cdot}50 \text{ mg. } CaCO_3$$
$$\equiv 2{\cdot}58 \text{ mg. Creta, B.P. containing 97 per cent. } CaCO_3$$

A subtraction must be made from the figure obtained to allow for the natural calcium present in the flour, which may vary from 15 (Patents) to 30 (Wholemeal) when expressed as mg. Ca per 100 g.

2. Determination of Creta by Back Titration

Greer and his co-workers (*loc. cit.*) also proposed the following rapid titration method:

Stir 25 g. of flour with 50 ml. of 15 per cent. magnesium sulphate solution to a smooth suspension, add 50 ml. N/10 hydrochloric acid and stir again. After two minutes, filter through a fluted 18·5 cm. No. 1 Whatman paper and re-filter the first runnings. To 25 ml. of filtrate add 1·5 ml. of bromocresol green indicator and titrate in a boiling tube with N/10 sodium hydroxide solution until the colour matches a colour standard (consisting of 30 ml. of Clark and Lubs phthalate buffer of pH 4·4 and 1·5 ml. of the indicator in a boiling tube). The magnesium salt used acts as a protein precipitant and using an end point at pH 4·4 interference from dissolved carbon dioxide is reduced.

The titration is not theoretical (the reaction is not complete) and it is necessary to calculate the factors for known mixtures, including that for untreated flour. The authors state that 1 ml. of N/10 alkali per 25 ml. filtrate is equivalent to 4 oz. creta/280 lb. sack (0·09 per cent.).

3. Determination of the Carbon Dioxide Produced on Addition of Acid by Gravimetric, Titrimetric and Gasometric Methods

Greer and his co-workers (*loc. cit.*) have also described a method in which 20 g. of flour is mechanically shaken with 70 ml. of water and 20 ml. 5 N hydrochloric acid and the carbon dioxide is absorbed and weighed in a soda-lime tube using a gravimetric train. The method gives accurate results even in the absence of a control, but in view of the special apparatus required, it is rather more suitable where large numbers of samples are analysed regularly.

The time taken for the determination can be reduced considerably by absorbing the carbon dioxide produced in baryta and back titrating (Hartley and Green, *Analyst*, 1943, **68**, 142) or by measuring the volume of gas produced in a Chittick apparatus (Fraser and Weston, *Analyst*, 1950, **75**, 402). The average blank for the flour in these two methods is approximately equivalent to 5 mg. $CaCO_3$ per 100 g. of flour.

4. Determination of the Calcium Content Colorimetrically Using Chloranilic Acid

The colorimetric determination of calcium using chloranilic acid (2 : 5 dichloro-3 : 6-dihydroxy-1 : 4-benzoquinone) has been applied to flour by Sawyer, Tyler and Weston (*Analyst*, **1956**, **81**, 362). Although the reaction is not specific for calcium, the authors found that the concentration of interfering ions contributed by the flour is constant and is relatively small compared with the total calcium content when added creta is present. The method is most suitable for the routine determination of creta when a number of samples are being examined. The method does not involve ashing of the sample or filtration.

Special Reagents

Standard calcium solution. Dissolve 8·3626 g. of calcium carbonate (A.R.) in the minimum quantity of acetic acid and dilute to 250 ml. (0·1 ml. of the solution = 5 oz. $CaCO_3$ per sack if 3 g. flour are taken).

Buffer solution (*pH* 4·0). Equal volumes of 0·2 M sodium acetate and 0·8 M acetic acid solution.

Chloranilic acid solution. Dissolve 1 g. in one litre of hot buffer solution. Cool and filter.

Method

To a series of 100 ml. conical flasks (ten for standards and one for each sample) add 40 ml. of chloranilic acid solution (0·1 per cent.). For the standards add 0, 0·1, 0·2, 0·3 and 0·4 ml. of the standard calcium solution (each in duplicate). For each sample, add 3·00 g. of the flour. Stopper each flask and shake them, first vigorously and then gently, in a mechanical shaker for thirty minutes (loosen the bungs at intervals). Allow the flasks to stand overnight in a cool place. Using a pipette with a rubber bulb, transfer 10 ml. of the supernatant liquor to a ½ in. colorimeter tube and read the extinction, E_1, at 540 mμ, using water as reference solution.

Decolorise the residual chloranilic acid by adding one drop of a 15 per cent. solution of sodium dithionite ($Na_2S_2O_4$) and re-determine the extinction, E_2. The second reading allows for the turbidity of the extract and E_1–E_2 is the extinction due to the colour of the solution. Construct a graph relating E_1–E_2 (for the standard extracts) and the added calcium.

Tyler and Weston (*Analyst*, **1956**, **81**, 375) have pointed out that as creta contains about 230 p.p.m. of *fluorine*, its addition at the rate of 14 oz. per sack contributes about 0·7 p.p.m. to the flour.

Improvers and Bleaching Agents

Improvers. During storage flour increases in " strength " and there is a consequent improvement in the baking qualities. The addition of " improvers," however, produces somewhat similar changes in the properties of the gluten (in the dough) in a matter of hours. The changes which take place during ageing (natural or artificial) have not been fully established, but in practice there is some bleaching and increased water absorption. Numerous substances have been added as flour improvers such as acid phosphates, persulphates (of ammonium or potassium), bromate and various bleaching agents (see below). They are seldom used in amounts in excess of about 1 in 5000.

Bleaching agents. The demand for a white flour has encouraged the use of bleaching agents which convert the xanthophylls to colourless compounds. Such treatment has been said to be approximately equivalent to the conversion of a lower grade flour to a higher ("patent") grade. It should be borne in mind, however, that the chemicals employed do not bleach any branny particles present. Further, the grade figure as measured by the Kent-Jones and Martin Flour Colour Grader (*Analyst,* 1950, **75**, 127;133 and *Chem. & Ind.*, 1956, No. 50, 1490) shows little alteration due to the artificial bleaching process.

The bleaching substances which have been added to flour include oxides of nitrogen, chlorine, nitrogen trichloride, chlorine dioxide and benzoyl peroxide.

The Preservatives Sub-Committee Report on Flour Improvers (contained in the F.S.C. Report on Bread and Flour, 1960) stated that, "at present, the use of some form of chemical improver is a technological necessity," but considered "that further study should be given to alternative physical methods of bread making with the ultimate object of rendering chemical improvers unnecessary." The report added, however, that "There is no strong justification for the use of bleaching agents." It was *recommended* that all bleaching and improving agents in use should be subject to long-term chronic toxicity studies. Meanwhile only treatment with the following substances should be permitted:—

Maturing agents—ascorbic acid, potassium bromate, ammonium or potassium persulphate, chlorine dioxide.

Bleaching agents, etc.—benzoyl peroxide (up to 50 p.p.m.), chlorine (for cake flour), sulphur dioxide (for biscuit flour).

Examination of Flour for Improvers and Bleaching Agents

For the detection of improvers it is sometimes possible to effect separation from the flour by shaking it with carbon tetrachloride (p. 174). **Phosphates,** for instance, can be detected by the molybdate reaction on the sediment obtained.

Persulphates of ammonium or potassium are usually used as improvers for flour at a concentration of about 110 p.p.m. (calcium sulphate being used as diluent). Bread made from flour containing ammonium persulphate is likely to contain about 90 p.p.m. of ammonium sulphate.

Persulphates can be detected by making 20 g. of flour into a paste with 20 ml. water and pouring an alcoholic solution of benzidine (1 per cent.) over it. Particles of persulphate in contact with the benzidine reagent are revealed as deep blue specks (*cf. Analyst,* 1912, **37**, 90). Other reagents such as methylene blue and potassium iodide have also been proposed for the detection of persulphates. A quantitative method for determining ammonium persulphate in flour has been described by Auerbach *et al.* (*Cereal Chem.*, 1949, **26**, 490).

Potassium bromate is usually used at a concentration of 10–15 p.p.m. The " improving " action does not take place until the flour is made into dough and its use results in the bread containing 10 p.p.m. of potassium bromide.

Bromates can be detected by adding to the wetted flour a solution containing 0·5 per cent. potassium iodide in 2 N hydrochloric acid and mixing thoroughly. The appearance of blackish spots suggests the presence of bromate. A method for the determination of bromate has been described by Armstrong (*Analyst,* 1952, **77**, 460).

Vitamin C is said to act similarly and be as effective as bromate when used at a concentration of 20–80 p.p.m. (in 80 per cent extraction flour). Vitamin C can be detected by adding a 0·1 per cent. solution of 2 : 6 dichlorophenolindophenol to the wetted flour. White spots appear in a few minutes in the presence of the vitamin.

Chlorine, Chlorine dioxide and Nitrogen Trichloride

Chlorine gas acts as both a bleaching agent and an improver, but it is no longer used for bread flour. Flour intended for the manufacture of certain cakes (*e.g.*, Angel food, layer cake), however, may be treated with up to 900 p.p.m. With flour, chlorine combines with the colouring matter, the starch and the protein (with reduction of nutritive value of the gluten). Treatment with chlorine also causes the pH of the flour to fall.

Chlorine dioxide (**Dyox**) also acts as both an improver and a bleaching agent and about 80 per cent. of the flour produced in the U.K. is treated with it. The level of use for flour is about 15–30 p.p.m. When flour is treated with chlorine dioxide there is some destruction of vitamin E, but there is no evidence that any other harmful effect takes place.

Nitrogen trichloride (**agene**) was formerly used as a bleaching and maturing agent for much of the flour produced in the U.K. Work by Mellanby (1946), however, showed that dogs fed on a diet containing a large proportion of bread made from " agenised " flour developed hysteria. Subsequently the toxic factor responsible was isolated and identified as methionine sulphoximine. Then in 1950 the Jameson Committee (comprised of representatives of the M.O.F., M.O.H., M.R.C. and the milling industry) recommended that the treatment of flour with agene should be discontinued and replaced by chlorine dioxide. This was accepted by the Ministers concerned and by the milling industry.

Tests for chlorine and related substances rely on the fact that flour treated with them contains chlorine in the fat. The methods used, however, are seldom sensitive enough to detect the small amounts present in flour fat due to treatment with the dioxide or agene.

Qualitative test. Extract 30 g. of flour with 50 ml. of light petroleum, filter or centrifuge and remove the solvent from the clear solution by allowing it to evaporate. Heat a piece of copper wire in a bunsen flame until it is black, dip it into the extracted fat and return it to the flame. A green colour at the edge of the flame suggests chlorine treatment of the flour (Beilstein's test).

For the *quantitative estimation of chlorine,* shake 500 g. of flour with 700 ml. of light petroleum and filter through a Buchner funnel to obtain a clear filtrate. Measure the volume of the filtrate to obtain the sample weight equivalent, evaporate to low bulk and transfer to a platinum dish. Add 20 ml. alcoholic potash (4 per cent.), evaporate and ignite gently. Extract the ash twice with 20 ml. of dilute nitric acid and filter through a chloride-free paper. Wash through with water, ignite the paper in the platinum dish and extract again with dilute acid. Evaporate the combined extracts on the water-bath to about 50 ml. and add 5 ml. of N/50 sodium chloride and 10 ml. of N/50 silver nitrate. Cool, filter into a flask, wash through with cold water and titrate with N/50 thiocyanate using ferric alum as indicator. Express the results as chlorine in p.p.m., after deducting that due to the N/50 sodium chloride added. Untreated flour gives a figure of about 1 p.p.m. (*cf. A.O.A.C.* and *J.S.C.I.*, 1930, **49**, 223 T).

Benzoyl peroxide (**Novadelox**) is used at levels of about 30 p.p.m. to supplement the bleaching effect of chlorine dioxide in about half of the bread-flour produced in the U.K. Benzoyl peroxide oxidises the colouring matter of the flour and causes some loss of vitamin E.

Flour treated with this compound may contain benzoic acid in addition to unchanged benzoyl peroxide. The peroxide can be converted into benzoic acid with sodium hydroxide (in the presence of acetone). The total benzoic acid can then be converted to salicylic acid and determined colorimetrically. For further information reference should be made to the following: *Analyst,* 1933, **58**, 4; 1953, **78**, 467 and the *A.O.A.C.* Benzoyl peroxide does not appear to affect the baking quality of the flour.

Oxides of nitrogen (**the electric bleacher**). Nitrogen peroxide is now seldom used as a bleaching agent for flour. As little as 4 p.p.m. is said to be sufficient and the flour so treated may contain 3 p.p.m. of residual sodium nitrite, the presence of which can be detected by the Griess-Ilosvay reaction:—

Shake 20 g. of flour with 200 ml. nitrite-free water (previously warmed to 40° C.) and keep at that temperature for one hour, shaking as often as possible.

Filter. Add 2 ml. of each of the Griess-Ilosvay reagents (page 76) to 50 ml. of the clear filtrate and compare the pink colour obtained using either a nessleriser disc or standard solutions of sodium nitrite similarly treated. Calculate the nitrite (free and combined) as N or $NaNO_2$ in p.p.m. Untreated flour contains up to 2 p.p.m. (as $NaNO_2$). Oxides of nitrogen do not appear to affect the baking quality of the flour.

Sulphur dioxide is only used (as the gas or as sodium bisulphite) for the treatment of biscuit flour. The normal level of use is about 20 p.p.m. (as SO_2).

Pesticide residues. For methods which are suitable for the determination of pesticide residues in flour and other foods the reader should consult the following references: for DDT see *Analyst,* 1960, **85**, 600; for BHC (hexachlorocyclohexane) see *ibid.*, 1960, **85**, 21; for malathion see *ibid.*, 1960, **85**, 915.

Vitamins in Flour

Aneurine.* The Flour (Composition) Regulations, 1956, prescribe that all flour shall contain at least 0·24 mg. of vitamin B_1 per 100 g. (see p. 156). Many of the methods which have been proposed for the determination of aneurine are based on the thiochrome method of Jansen (*Rec. Trav. Chim.*, 1936, **55**, 1046). Under standard conditions, the net fluorescence of the thiochrome produced by the oxidation of the vitamin (using alkaline ferricyanide) is directly proportional to its concentration over a given range. The fluorescence can be measured (a) against a quinine standard (null-point fluorimeter), or, (b) as direct deflections (deflection type fluorimeter), or, (c) visually. Methods which are suitable for the determination of vitamin B_1 in cereals, malt, yeast and meat are described in the report of the Sub-Committee on Vitamin Estimations (Aneurine Panel) of the S.P.A.'s Analytical Methods Committee (*Analyst,* 1951, **76**, 127). The F.S.C. (Analytical Panel) Report on Bread and Flour states that the procedure described by Ridyard (*Analyst,* 1949, **74**, 18),† using a modified Spekker fluorimeter, is satisfactory. The following simplified *fluorimetric procedure* for determining aneurine in flour has been described by the American Association of Cereal Chemists (" *Cereal Laboratory Methods,*" 1957, 6th Edn., p. 510):

Special reagents

Potassium chloride solution: 250 g. dissolved in 2 per cent. acetic acid to produce one litre.

Quinine standard solution: 50 mg. quinine sulphate dissolved in N/10 sulphuric acid to produce 500 ml. Dilute 10 ml. of this solution and 2·6 ml. sulphuric acid (conc.) to one litre.

* Referred to also as " Thiamine."

† See also *Analyst,* 1950, **75**, 634.

Standard aneurine solution (B): Prepare a 0·01 per cent. aqueous solution of aneurine hydrochloride. Dilute 2 ml. to one litre with the acidified potassium chloride solution (above).

METHOD

Place a quantity of flour containing 3–5 µg. of aneurine (*e.g.*, 2 g. of sample) in a conical flask, add 50 ml. of the acid potassium chloride solution and place in a water-bath maintained at 70° C. for thirty minutes, shaking as often as possible. Centrifuge or filter. Pipette 5 ml. of the clear extract into a 50 ml. centrifuge tube, quickly add 3 ml. of sodium hydroxide solution (15 per cent.), mix, add two drops of fresh potassium ferricyanide solution (1 per cent.) and again mix rapidly. After one minute add 15 ml. of *iso*-butyl alcohol, shake vigorously for one minute, centrifuge and remove the water layer. Clarify the *iso*-butyl alcohol layer by adding 1 ml. of ethyl alcohol and measure the fluorescence after adjusting the fluorimeter to a suitable value using the quinine solution or a glass standard. Also oxidise 5 ml. of standard aneurine solution B. Determine the blank on 5 ml. of extract by adding *iso*-butyl alcohol and alkali, but omitting the ferricyanide. Then if:

R_X = fluorimeter reading with unknown;

R_{Xb} = blank fluorimeter reading with unknown;

R_S = fluorimeter reading with B_1 standard;

R_{Sb} = blank fluorimeter reading with B_1 standard;

and S = sample weight in g.

$$\text{Then mg. aneurine/100 g.} = \frac{R_X - R_{Xb}}{R_S - R_{Sb}} \times \frac{1}{S}$$

The rapid *visual method* for determining vitamin B_1 in flour given below summarises the procedure described by a sub-committee of the M.R.C. (*Biochem. J.*, 1943, **37**, 433):—

Cream 2 g. of the flour with 25 ml. of 2 per cent. hydrochloric acid, allow it to stand overnight and, after shaking, centrifuge. Transfer 5 ml. of the clear liquor to a 15 ml. centrifuge tube, add one drop of 10 per cent. hydrochloric acid and 5 ml. of *iso*-butyl alcohol. Shake the tube for at least one minute, centrifuge, and note the volume of the aqueous layer.

Into a 25 ml. stoppered bottle (A) place 1·5 ml. of the aqueous layer, add 2 ml. of methyl alcohol, 1 ml. of 20 per cent. sodium hydroxide solution, then within thirty seconds 0·15 ml. of 5 per cent. solution of potassium ferricyanide, shaking thoroughly after each addition. Then add 0·25 ml. of hydrogen peroxide (20 vol.) and allow to stand for one minute. Add 10 ml. of *iso*-butyl alcohol, shake vigorously for two minutes and allow to separate. In a second bottle (B) put 1·5 ml. of the aqueous layer from the centrifuge tube and follow the same procedure but omitting the ferricyanide. Into a third bottle (C) put 1·5 ml. of dilute standard aneurine solution and follow the same procedure as with bottle A.

To compare the fluorescence of the thiochrome produced, draw off the aqueous layer from each bottle with a pipette, wash the *iso*-butyl alcohol with 3 ml. of water, remove the water, and add to each 2 ml. of ethyl alcohol. Measure 10 ml. of the clear extract from bottles A and B into separate tubes. Compare the fluorescence in a darkened room in front of the window of the ultra-violet lamp. Add quantities of the standard C to the blank B until the fluorescence observed matches that of A. At the last matching add a volume of *iso*-butyl alcohol to the sample A equal to that of the standard added to the blank B. To calculate the aneurine content: if x is the volume in ml. of standard *iso*-butyl alcohol extract (C) required and r_1 and r_2 the volumes in ml. of the extracts before and after the preliminary washing with *iso*-butyl alcohol, then the aneurine content of the sample is:—

$$\frac{4x}{10} \times \frac{25}{1{\cdot}5 \times 2} \times \frac{r_2}{r_1} = \frac{10\,xr_2}{3r_1}\ \mu g./g. \equiv \frac{xr_2}{3r_1}\ mg./100\ g.$$

The standard stock solution contains 20 µg. of aneurine hydrochloride per ml. of 0·01 N HCl containing 25 per cent. of ethyl alcohol. Dilutions of this may be made on the day of test with 0·01 N HCl so that 1·5 ml. = 4 µg. of aneurine hydrochloride. One International unit of vitamin B_1 = 3 µg. of aneurine hydrochloride.

N.B. For the purposes of The Flour (Composition) Regulations, 1956, the results should be reported in terms of vitamin B_1 B.P. (see p. 144). The F.S.C. has recommended, however, that the result should be expressed as 100 per cent. pure anhydrous vitamin B_1.

Nicotinic acid, nicotinamide and riboflavin may be assayed microbiologically. The general principle involved in such determinations is as follows: Firstly, a micro-organism is chosen which requires the vitamin being assayed as an essential nutrient. Secondly, a basal medium is used which represents an adequate diet for the selected organism in all respects except for the vitamin. The procedure is to add to the tubes of the basal medium (a) differing known amounts of an extract of the material under examination, and (b) differing known amounts of the vitamin being determined. After incubation, the extent to which the organism has grown is assessed (*e.g.*, by titration of the acidity produced). Then the amount of the vitamin in the original sample is obtained, either statistically or from a standard curve by comparing the results from both sets of tubes.

The Flour (Composition) Regulations, 1956, prescribe that all flour shall contain at least 1·60 mg. of nicotinic acid per 100 g. The F.S.C. (Analytical Panel) Report on Bread and Flour states that the determination of *nicotinic acid* is satisfactory if the recommended method of the S.A.C.* (*Analyst*, 1946, **71**, 401), using *Lactobacillus arabinosus* is employed and " any chemical method or alternative microbiological method selected should be shown to give results strictly comparable with this method."

A chemical method for determining nicotinic acid in foods has been proposed by Dennis and Rees (*Analyst*, 1949, **74**, 481), which is based on the yellow colour developed when pyridine is treated with cyanogen bromide in the presence of an aromatic amine.

The S.A.C.* (*Analyst*, 1946, **71**, 398) has also described a microbiological assay method for determining **riboflavin** using *Lactobacillus helveticus*. A fluorimetric method for determining riboflavin has been described by Hodson and Norris (*J. Biol. Chem.*, 1939, **131**, 621; *cf.* " Cereal Laboratory Methods," 1957, 6th Edn., pp. 500–503).

* Formerly the S.P.A.

Vitamin E, which is present in wheat germ oil, can be determined by the chemical method described on p. 441.

SELF-RAISING FLOUR

Most self-raising flours consist essentially of a mixture of flour, sodium bicarbonate (about 1·15 per cent.) and an acid phosphate (about 1·5 per cent. A.C.P. or A.S.P.). The flour used is usually derived from a soft wheat (the F.S.C. Report on Bread and Flour states, however, that there is no reason why any type of flour should not be used, provided that the appropriate description is used). The Food Standards (Self-Raising Flour) Order, 1946 (S.R. & O. 1946 No. 157) prescribes a minimum of 0·40 per cent. of available carbon dioxide as determined in a specified manner (see below). No maximum is prescribed for residual carbon dioxide. In addition, a maximum of 3 p.p.m. of fluorine is prescribed in the Fluorine in Food Regulations.

Analysis of Self-Raising Flour

The analysis of self-raising flour should include (a) a **microscopical examination**, (b) the determination of **the total and residual carbon dioxide, ash** and the possible contaminants, *viz.* **arsenic, lead** and **fluorine** and (c) the qualitative detection of the acid ingredient. In addition, the quantitative estimation of the ingredients may be required (on the same lines as described for baking powder).

Available carbon dioxide. Although the actual method for the final determination of the carbon dioxide is left to the individual analyst, the Food Standards (Self-Raising Flour) Order prescribes the following methods of analysis:—

"2. The available carbon dioxide shall be determined by ascertaining the difference between the total carbon dioxide and the residual carbon dioxide; and the total carbon dioxide and the residual carbon dioxide shall respectively be determined in the following manner:—

(a) Total Carbon Dioxide: Shall be determined by ascertaining the weight thereof evolved when the self-raising flour is treated with excess of dilute sulphuric acid, the evolution being completed either by boiling for five minutes or by means of reduced pressure.

(b) Residual Carbon Dioxide: Shall be determined by taking not less than five grams of the self-raising flour, which shall be mixed to a smooth paste with distilled water, and a further quantity of distilled water amounting in all to not less than twenty times the weight of the flour shall then be incorporated. The liquid shall be heated in a boiling water bath for thirty minutes, being vigorously stirred for the first five minutes and thereafter for approximately half a minute at intervals of approximately five minutes. The liquid shall forthwith be boiled for three minutes, being vigorously stirred during the whole of such period, and then transferred to an apparatus for determining carbon dioxide, through which carbon dioxide-free air shall be passed for not less than ten minutes. The residual carbon dioxide is the weight thereof evolved when the self-raising flour so treated is further treated with excess of dilute sulphuric acid, the evolution being completed either by boiling for five minutes or by means of reduced pressure."

The carbon dioxide can be determined gravimetrically, volumetrically (by back titration after absorption in barium hydroxide) or gasometrically (Chittick method), *i.e.*, by similar methods to those employed for baking powder (pp. 186–187), but more of the sample should be taken for the test. In view of the relatively large bulk of flour used, the addition of a few large glass beads is useful for shaking the flour with the acid and thereby ensuring completion of the reaction.

The Determination of Sodium Bicarbonate in Self-Raising Flours containing Creta Præparata, B.P.

In view of the relatively low bicarbonate content of self-raising flour, the presence of creta makes an appreciable contribution to the total carbon dioxide as determined in the usual way. Stephenson and Hartley (*Analyst*, 1955, **80**, 461) have shown, however, that if a dilute solution of acid sodium pyrophosphate (A.S.P.) is added to a self-raising flour containing creta, the carbon dioxide is released from the bicarbonate but not from the calcium carbonate. In the method, 17 g. of the self-raising flour is placed in the flask of a Chittick gasometric apparatus (*A.O.A.C.*, 1960, 9th Edn., p. 97) and the solution in the gas burette is brought to the zero mark. Then 45 ml. of fresh A.S.P. solution (2·5 per cent w/v) is added to the flour and the mixture is shaken for two minutes and allowed to stand for five minutes before equalising the pressure in the gas burette using the levelling bulb. Then the volume of gas produced is read off and corrected for temperature and pressure. After the reagent blank has been determined for the method, the carbon dioxide produced by the bicarbonate can be calculated. The " total " carbon dioxide due to both the bicarbonate and the creta can be determined in the apparatus in the usual way using 17 g. of flour and 45 ml. of dilute sulphuric acid (1 + 5). As the increase in the volume of gas produced by the creta is relatively small, both determinations should be performed a number of times in order to obtain a reasonably accurate difference figure. For the modified calculations which are necessary for these methods the original paper should be consulted.

NOTE

The F.S.C. has *recommended* that it should not be compulsory to add creta to flour used for self-raising flour by the addition of A.C.P. at the rate of not less than 4½ lb. per 280 lb. sack.

Mineral additions. A useful method for detecting or estimating the added mineral matter is to shake 50 g. of the flour with 200 ml. of *dry* carbon tetrachloride in a separator having a short delivery tube and fitted with a tap containing a hole of wide bore. On allowing the mixture to stand, the flour floats on the surface and the added bicarbonate and acid ingredients fall to the bottom and may be tapped off into a basin. After evaporating off the solvent the residue can be examined qualitatively for added ingredients by the usual tests, *e.g.*, bicarbonate, phosphate, tartrate and calcium. This method enables such ingredients to be examined in their "original" state, *i.e.*, without any interaction of the ingredients (unlike the ash). For the estimation of the additions, the residue in the separator should be given two further shakings with carbon tetrachloride.

Contaminants of the acid phosphate. High amounts of arsenic, lead and fluorine have on occasion been found in acid phosphates which are used in self-raising flour. The methods and limits for these elements are given in Chapter 3. A method for determining lead* in flour has been proposed by Hart (*Analyst*, 1951, **76**, 692). The amount of calcium sulphate permissible in acid phosphate was the subject of a report to the L.G.B. (1911, No. 13). In the report, Hamill recommended that no more than 10 per cent. of $CaSO_4$ should be allowed in acid phosphate, and that the ratio of SO_3/P_2O_5 should not exceed 1/7. If the estimation of both sulphate and phosphate is carried out on the whole self-raising flour, an allowance must be made for that naturally present in the flour (SO_3 0·0028 per cent.; P_2O_5 0·34 per cent.). No such correction is necessary if the analysis is carried out on the mineral matter separated with carbon tetrachloride.

BREAD

The F.S.C. Report on Bread and Flour (1960) stated: **Bread** may be defined as the product obtained by the doughing, fermenting and baking of flour with or without other ingredients. Yeast is the main raising agent used, but in a very few cases fermentation is replaced by, for example, a chemical liberation of carbon dioxide as in soda bread. The main ingredients introduced by the baker are water and salt, but small quantities of fat, sugar, milk powder, malt, other cereals and soya flour and substances such as emulsifying agents, yeast foods and caramel are also variously added.

White bread should be made by baking a fermented dough made from white flour conforming to the standard for flour at present laid down in The Flour (Composition) Regulations, 1956, yeast and water, and could contain the following additional ingredients: (a) salt, edible oils and fats, milk or milk products, sugars; (b) enzyme active preparations (*e.g.*, malt extract, malt flour); (c) soya flour (not more than two parts by weight to 100 parts of flour); (d) wheat gluten, wheat germ; (e) poppy seeds, caraway seeds,

* Relatively large amounts of phosphates are often present in self-raising flour.

cracked wheat or oat grain and oatmeal in or on the bread up to two parts by weight to 100 parts of flour; (f) yeast stimulants; (g) emulsifying agents (super-glycerinated fats, stearyl tartrate, lecithin): (h) propionates (as recommended in F.S.C. Report on Preservatives—see p. 454); (i) acetic acid, vinegar, A.C.P. and A.S.P. (to inhibit " rope ").

Brown bread should be made chiefly from wheatmeal and have a fibre content of not less than 0·6 per cent calculated on the dry matter. The F.S.C. prefer the term " brown bread " to " wheatmeal bread."

Wholemeal bread should be bread made from wholemeal, without any addition of white flour or wheatmeal, and could contain any of the additional ingredients of white bread except milk and milk products, soya bean flour, prepared wheat gluten, wheat germ, oat grain and oatmeal.

Soda bread should be bread leavened with sodium bicarbonate and a suitable acid constituent instead of yeast and could contain any of the additional ingredients of white bread except (f).

Malt, fruit, nut and egg bread should be bread containing malt, fruit, nut and egg respectively, in amounts sufficient to characterise the type of bread sold.

Milk bread should be required to contain not less than 3·6 per cent. by weight of whole milk solids or skimmed milk solids, calculated upon the weight of the loaf. Labels should be required to bear a declaration: " contains whole milk solids " or " contains milk solids not fat."

The F.S.C. Report also recommended that no special claims should be allowed for bread containing less than 16 per cent. protein (on dry basis); breads containing 16–22 per cent. protein should be allowed to be described as "**gluten bread,**" but no reference to protein should be allowed. Breads containing more than 22 per cent. protein should be allowed to be described as "**high protein**" **bread,** but there should be no reference to protein except in a statement "contains X per cent. protein"; the term " **starch reduced** " should only be applied if the carbohydrate content on a dry basis is less than 50 per cent.; claims that breads have a specific weight-reducing property should not be allowed; no bread should be described as " wheat germ bread " unless it contains 10 per cent. added processed wheat germ calculated on a dry basis on the bread.

The acidity changes from about 0·1 to 0·5 per cent. (as lactic acid) as the bread changes from fresh to stale (see also Bechtel, *Trans. Amer. Ass. Cer. Chem.*, 1955, **13**, (2), 108). Surface active agents such as G.M.S. and polyoxyethylene stearate may possibly be added as anti-staling agents (see Carson *et al.*, *Cer. Chem.*, 1950, **27**, 438 and Appendix II).

Although micro-organisms seldom cause trouble with flour, certain bacteria and moulds are liable to grow in bread. The spores of *B. mesentericus* may survive the baking process and, particularly if the bread is cooled slowly or wrapped before cooling, are liable to cause "**rope**" (Amos and Kent-Jones, *Analyst,* 1931, **56**, 572). If this occurs, sticky yellowish-brown spots appear in the crumb and an odour resembling decaying fruit is apparent. In extreme cases the middle of the loaf becomes brown and, on breaking, the crumb can be pulled out into threads (hence the name). "**Bleeding bread**" is the name given to a disease due to *B. prodigiosus*, which causes a bright red colour to appear in the crumb, the formation of which is sometimes very rapid. The spores of most moulds are destroyed during baking. Mould in bread usually commences from the outside of the loaf and its growth is encouraged by storage in conditions of high humidity.

Analysis of Bread

The general analysis of bread should include the determination of **moisture** (dry at 100° C.), **fat** (acid hydrolysis p. 151), **protein** (p. 34), **ash**, **salt** and **fibre** (p. 41) and a microscopical examination. It is preferable to carry out most of the analyses on the sample after drying at 100° C. A separate minced portion should be used for the moisture determination so that the results can be calculated back to the original sample. The composition figures for various types of bread, quoted by McCance and Widdowson, are as follows:—

	White	Wholemeal	Hovis	Malt	Currant
	%	%	%	%	%
Water . .	32	37	39	24	38
Protein (N × 6·25) .	7·9	8·4	11·4	9·1	7·0
Fat . .	0·7	1·6	3·7	3·3	3·4
Sugar .	tr.	tr.	tr.	19	13
Starch .	54	44	41	31	33

Analyses by König show that the ash varies from 0·88 to 1·52 per cent. and the fibre from 0·31 to 1·12 per cent.

Improvers and bleachers undergo changes during the baking process but in some instances the derived product can be detected. For instance the presence of traces of benzoic acid in the bread usually indicates that the flour used had been treated with benzoyl peroxide. During baking, persulphates are converted to the sulphate and bromates to the corresponding bromide. The presence of these anions does not necessarily prove that improvers have been used, however, as they may be derived from the wheat or other ingredients. The losses and changes which occur during baking make the accurate estimation of improvers and bleachers a matter of some difficulty.

In former times, **alum** was said to have been used to improve the colour of white bread. For its detection the bread is soaked in a mixture containing 1 ml. of freshly prepared 5 per cent. tincture of logwood, 5 ml. of 10 per cent. ammonium carbonate solution and 20 ml. water. Then the reagent is squeezed out and the bread dried in the 100° C. oven. In the presence of alum a violet tint remains on drying, whereas uncontaminated bread gives a buff colour.

Iron can be determined in the same way as is described for flour.

Vitamin B_1. The F.S.C. (Analytical Panel) Report on Bread and Flour recommends that vitamin B_1 should be determined by the thiochrome method of Ridyard (*Analyst*, 1949, **74**, 18). During preparation of the extract it is necessary to treat bread with phosphatase (*ibid.*, 1951, **76**, 127) in order to free combined thiamine. The loss of vitamin B_1 during bread-making is about 15 per cent.

Detection and Determination of Milk in Bread

The F.S.C. has recommended that milk bread should contain not less than 3·6 per cent. by weight of milk solids (see p. 175). The F.S.C. Report also states that " For the purpose of formal samples it may be necessary to take three loaves in order to obtain a representative part in each of the three parts of the sample."

The following table from F.S.C. Report shows that milk bread contains slightly greater amounts of protein, fat and calcium than ordinary bread:—

Nutrient		No addition of milk	Bread containing:— Whole milk solids			Skim milk solids	
			5·5%	4·2%	3·0%	4·2%	2·1%
Protein	g.	7·8	8·8	8·5	8·5	9·2	8·5
Fat*	g.	0·7	2·1	1·8	1·4	0·7	0·7
Carbohydrate	g.	54·7	53·6	54·0	54·0	54·3	54·3
Calcium	mg.	102·0	145·0	134·0	127·0	148·0	127·0
Iron	mg.	1·45	1·42	1·43	1·44	1·44	1·45
Thiamine	mg.	0·162	0·169	0·166	0·162	0·173	0·166
Riboflavin	mg.	0·025	0·085	0·071	0·056	0·088	0·056
Nicotinic acid	mg.	1·49	1·45	1·46	1·47	1·47	1·48

* No allowance is made in these calculations for the contribution of any other fats that might be added to the bread. Each figure shows the amount present per 100 g. of bread.

The fact that milk contains lactose, but the other constituents do not, is more usually used for the tentative detection and determination of milk in bread. Lactose can be *detected* in bread by means of paper chromatography.

Lactose can be *determined* in bread after removing the other sugars present by fermentation with yeast. The following method due to Choi *et. al.* (*Cer. Chem.*, 1950, **27**, 398), involves extracting the sugars from bread with dilute alcohol, destroying the non-lactose reducing sugars by fermenting the alcohol-free extract with yeast and determining the lactose remaining by Somogyi's method:—

SPECIAL REAGENTS

Yeast suspension. Wash baker's yeast by centrifuging with four times its volume of distilled water. Dilute to a 25 per cent. suspension (see also *J. Dairy Res.*, 1936, **7**, 41).

Yeast nutrient solution. Containing 1·7 per cent. bacto-peptone, 0·5 per cent. dipotassium phosphate and 0·33 per cent magnesium sulphate.

Protein precipitant. Dissolve 50 g. sodium tungstate and 6 g. disodium phosphate in 200 ml. water. Add slowly 220 ml. of 2 N HCl, mix and dilute to 500 ml.

Somogyi's reagent. See p. **111** and *J. Biol. Chem.*, 1926, **70**, 599.

METHOD

Weigh 15 g. of air-dried bread into a 200 ml. graduated flask containing 60 ml. water. Mix, add 35 ml. ethanol (95 per cent.) and immerse in a boiling water-bath for fifteen minutes. Cool, dilute to the mark with more ethanol and centrifuge. Evaporate 150 ml. of the supernatant liquor to 40 ml. and dilute to 100 ml. in a graduated flask. Transfer 10 ml. of this bread extract to a 50 ml. conical flask, add 6 ml. yeast suspension and 5 ml. yeast nutrient solution. Also set up a blank commencing with 10 ml. water. Stopper the flask with a stopper through which passes 6 mm. glass tubing (about 10 cm. long) and incubate at 30° C. for two and a half hours with shaking. Then centrifuge for ten minutes at 1000 r.p.m. Transfer the supernatant liquor into a 50 ml. graduated flask and wash the residue with two 10 ml. portions of water. To the liquor + washings add 2·5 ml. of protein precipitant, make up to 50 ml., mix and filter. For the determination of lactose, pipette 5 ml. of the filtrate into a tube, neutralise to phenol red with N/2 sodium hydroxide, add 5 ml. Somogyi's reagent and two drops of benzene. Then immerse the tube in boiling water for fifteen minutes, cool, add 2·5 ml. 2 N sulphuric acid and titrate the excess iodine with N/200 thiosulphate (see p. 112).

$$\text{Non-fatty milk solids (\%) (in air-dried bread)} = 2 \times \frac{8{\cdot}33 \times L \times 10^4}{100 - M}$$

where L = grams of lactose in the 5 ml. aliquot.
M = moisture in air-dried bread (%).

Zentner (*Analyst*, 1956, **81**, 483) has devised a method for determining lactose in bread by means of descending paper chromatography with a *n*-butanol-pyridine-water system containing silver nitrate. The concentration of lactose is determined from the optical densities of the silver spots in a densitometer.

The proportion of butter fat present in bread can be assessed by carrying out the semi-micro Reichert process on the extracted fat (p. 380; *cf. J. Ass. Off. Agric. Chem.* 1956, **39**, 469).

CAKE AND FLOUR CONFECTIONERY

The analysis of flour confectionery seldom presents difficulties to the " official " analyst, but samples may require to be examined for **added colour** and **preservatives.** The preservatives which might be present on occasion are sorbic acid, propionic acid or boric acid. The general quality can be assessed by determining the **moisture** (dry at 100° C.), **oil** and **sugar** (see below), **ash, salt, protein** and **egg** (p. 197). Further, the oil (extracted directly from the dried material) should be examined for identity and edibility (see Chapter 14). Chocolate products should be examined for cocoa.

Emergency legislation during World War II stated *maxima* for oils and fats and sugar in cakes (because of shortages of raw materials) and methods for their determination were subsequently published by Nicholls *et al.* (*Analyst*, 1943, **68**, 48):

Oil. Weigh 2 g. of minced sample into an extraction tube, add 2 ml. alcohol and 10 ml. dilute hydrochloric acid (25 conc. : 11 water by vol.) and maintain at 65° C. for thirty minutes. Cool, add 10 ml. alcohol and shake with 25 ml. ether. Then add 25 ml. light petroleum, shake again, blow the fat solution into a weighed flask, etc., and continue as usual.

Sugars. Grind 5 g. of minced sample with warm water (45° C.) and transfer into a 250 ml. graduated flask using a total volume of 200 ml. of warm water. Maintain at 45° C. for one hour, mixing frequently. Cool, add dialysed iron, make the aqueous portion up to the mark, shake and filter. Invert 50 ml. of the filtrate in a 100 ml. flask at 70° C. (p. 112) in the usual way and determine the sugars before and after inversion by Lane and Eynon's method (p. 119). Insoluble matter is allowed for by multiplying by 0·99.

DETERMINATION OF COCOA (Moir and Hinks, *Analyst*, 1935, **60**, 446)

Moisten 20 g. or other convenient quantity of cake in a mortar with alcohol (80 per cent. v/v) and rub to a smooth paste. Transfer to a flask with the addition of a further 200 ml. of alcohol (80 per cent.), add 2·5 g. of freshly-ignited magnesium oxide and digest for one and a half hours under reflux air-condenser in a boiling water-bath. The flask should be shaken frequently to prevent the cake from clotting. Filter through a Buchner funnel and re-digest the residue twice with 100 ml. quantities of 80 per cent. alcohol. If fat is to be determined, collect the alcoholic filtrates in a large measuring flask, make up to the mark with alcohol, and shake vigorously to maintain the fat in an emulsified state throughout the liquid. Remove an aliquot portion for the determination of fat. Evaporate the remainder on the water-bath to about 100 ml., replacing the alcohol by water. Add two to three drops of dilute hydrochloric acid. Transfer with hot water to a 150 ml. measuring flask, and clarify with zinc ferrocyanide as described in the method for cocoa, making up to 150 ml. with water. A larger flask and more zinc acetate and potassium ferrocyanide may be used if the amount of fat is large. Filter, evaporate a measured quantity of the filtrate to small bulk (10 to 20 ml.) and extract with chloroform. If the amount of alkaloid is small (2 to 4 mg.) three 50 ml. quantities of chloroform are sufficient. Otherwise, extract as described on p. 236. Reserve a portion of the chloroform extract for the qualitative murexide test.* Distil the chloroform from the reminder, dissolve the residue in hot water, and determine the nitrogen either by digestion and distillation into N/10 acid, as with cocoa, or by digestion and nesslerisation. In the latter case add 0·1 g. of sucrose and 3 ml. of concentrated sulphuric acid to the residue in a small Kjeldahl flask. Heat over a small flame till frothing ceases, add 0·01 g. of selenium, and continue heating till colourless and then for half an hour longer. Cool, dilute, neutralise with caustic soda and make up to 100 ml. Determine the ammonia- or nitrogen-content of

* See also *Analyst*, 1935, **60**, 200.

this solution by nesslerisation. Correct the result obtained for the ammonia-content of the digestion reagents. Use the factors of 3·26 and 100/3·15 for conversion of nitrogen into alkaloid, and alkaloid into dry fat-free cocoa-matter, respectively.

It is of interest to note that Circular FSL/18/45 was issued by the M.O.F. in 1945 in connection with the licensing of chocolate sponge, cake and pudding mixtures. A minimum of 10 per cent. cocoa was stated (corresponding to 4 per cent. of fat-free cocoa in the product).

Blancmange powders usually consist of a cornflour base with added flavour and colour. Occasionally other starches such as farina (from potato) have been mixed with cornflour for these products. Chocolate blancmange usually contains some 5–10 per cent. of cocoa, together with chocolate colouring and vanilla essence. For the determination of theobromine + caffeine (and hence the cocoa) in chocolate blancmange powder the method of Moir and Hinks (*Analyst*, 1935, **60**, 439), which is given on p. 179, is available.

Custard powders are usually composed of cornflour, flavouring (salt, vanilla extract) and a yellow dye. On occasion the cornflour has been replaced by other starches such as potato, sago or arrow-root. Wheat flour, which has also been used, tends to produce a gel in the final product with a less desirable consistency. As a result of a case heard in 1925 it would appear that an offence is committed if a starch preparation containing no egg is sold as " custard " not qualified by the word " powder." Egg is not now present in any of the brands of custard powder on the market. If egg is present, the protein cells and oil will be revealed in the microscopical examination and it can be estimated by means of the organically combined phosphorus (p. 197). Sweetened custard powders containing dextrose, sucrose or saccharin have been sold on occasion. Samples of custard powder should be examined for *ash, sulphur dioxide* and *added colour*.

Industrially, determinations of the *moisture* and *vanillin* contents of custard powders are also of some importance:—

Moisture is conveniently determined by a drying method. If the Carter-Simon rapid method of drying for fifteen minutes at 155° C. is employed the moisture content should not exceed 15 per cent.

Vanillin. Mix 20 g. of sample with 198 ml. of 50 per cent. isopropyl alcohol to produce a 10 per cent. extract (approx.). Filter, evaporate 100 ml. of filtrate to about 25 ml., transfer to a 100 ml. graduated flask and determine the vanillin content by the A.O.A.C. method for vanilla extracts as follows:—

Add 4 ml. of lead acetate solution (5 per cent. of each of neutral and basic and make up to the mark. Filter and pipette a suitable aliquot of the filtrate (say 5 ml.) into a 50 ml. graduated flask, add 5 ml. Folin-Denis reagent (10 per cent. sodium tungstate, 2 per cent. phosphomolybdic acid and 5 per cent. orthophosphoric acid/water), make up to the mark with saturated sodium carbonate solution and allow to stand for exactly ten minutes. Filter. Measure the blue colour of the filtrate against standard vanillin solutions (similarly prepared) in a 1 in. cell in a tintometer or photometrically at 610 mμ. A blank should be performed using 20 g. of cornflour or the equivalent of the farinaceous matter present. Most custard powders contain 0·02–0·06 per cent. of vanillin. **Coumarin,** a flavour also used on occasion in custard, can be estimated on a prepared solution by the A.O.A.C. method which employs a diazonium reagent. It would appear that coumarin may be carcinogenic.

RICE

The " facing " and polishing of rice during the preparation for sale is described in Hamill's report to the Local Government Board, No. 8 (1909). Small quantities of blue pigment and of oil have been employed to improve the colour and translucency. Talc, when used for facing rice, is used in conjunction with glycerin. In the early part of the century the practice had become general in the trade and a maximum of 0·5 per cent. was recommended for extraneous mineral matter. The use of such mineral additions has, however, now largely disappeared. In the case of ground rice the ash figure gives some indication of such additions. The ash of ground rice is usually between 0·20 and 0·30 per cent. and samples which have a figure in excess of this range should be regarded with suspicion and the acid insoluble ash should be determined also. An unfaced rice usually gives a readily fusible ash, *i.e.*, it gives a glassy residue when heated over an Argand burner, whereas the ash of faced rice, containing more silica, is not readily fusible and is flaky. Ground rice should also be examined microscopically.

Talc in whole rice can be determined by the following method due to Krźiźan (Z. *Unters. Nahr. Genussm.*, 1906, **11**, 564): Shake 20 g. of sample with dilute ammonia and dilute hydrogen peroxide. Heat to about 60° C. so that the gas formed causes the particles of talc to come away from the surface. Decant off the liquid containing the talc, wash the grains several times with water and add these washings to the decanted liquid. Heat the liquor with hydrochloric and chromic acids to oxidise suspended meal, filter off the talc, wash, ignite and weigh. In unpolished rice the residue does not normally exceed 0·025 per cent.

Compared with milled rice, hulled rice is rich in phosphoprotein and lecithin, which is valuable nutritionally. The differences in the phosphorus and potassium contents of whole and polished rice can be seen from the following figures:—

	Phosphorus (%P_2O_5)	Potassium (%K_2O)
Whole rice	0·27	0·25
Polished rice	0·09	0·07

Particularly in countries where rice is the staple diet, the presence of vitamin B_1 in the outer coating of the grain is of considerable importance physiologically. It has been established that a diet consisting of polished rice is liable to cause beri-beri, due to the consequent low intake of the antineuritic vitamin.

RYE FLOUR

Rye can be detected by microscopical examination, which reveals the large round starch grains in addition to smaller sizes which resemble those of wheat (also some rye grains show a stellate hilum). Rye flour is occasionally contaminated with **ergot**, a parasite which is the sclerotium of *Claviceps purpurea* and originates in the ovary of *Secale cereale*. Ergot is slightly curved, hard and dark purple in colour. Its microscopical characters are somewhat ill-defined, but chemical tests have been devised which are based on the fact that ergot tends to give a coloured extract. In employing such tests it is essential that comparisons are made with the uncontaminated material. The following method is included in "*Cereal Laboratory Methods*, 1957" (*A.A.C.C., Minnesota*):—

Shake 10 g. flour with 20–30 ml. ether and 1·2 ml. of sulphuric acid (5 per cent.) and allow to stand for six hours in a closed flask. Filter and wash the residue with ether until 40 ml. of filtrate is obtained. Add 1·8 ml. saturated sodium bicarbonate. If ergot is present the lower layer becomes violet (*cf. Analyst*, 1906, **31**, 363).

Ergot can also be detected by means of the spectroscope: About 50 g. of the flour are extracted with warm alcohol for several hours and the solution is filtered, concentrated to about 5 ml. and examined spectroscopically. In the presence of ergot the solution has a reddish brown colour and characteristic absorption bands are seen in the blue and in the green. Two bands at 538 and 499 mμ (and a faint one at 467 mμ) are characteristic of ergot.

PEARL BARLEY

Pearl barley consists of the whole barley grain with the husk removed. It is sometimes faced with talc, which can be estimated by Kržižan's process described under rice (see also Liverseege and Hawley, *J. Soc. Chem. Ind.*, 1915, **34**, 203). A microscopical examination should always be carried out, particularly in any dusty portions, which frequently contain meal mites (*Acarus farinæ*).* Sulphur dioxide used to be found occasionally in pearl barley.

* General reference: "The Mites of Stored Food", Technical Bulletin No. 9 by A. M. Hughes (H.M.S.O., 1961).

SOYA MEAL

Soya is a pulse—not a cereal. During World War II soya flour was frequently present in sausages. The percentage composition of soya beans, meal and flour is shown in the following table:—

	Soya beans	Soya meal	Soya flour
Water	10·5	8·1	6·5
Oil	18·0	7·5	20·5
Protein	35·0	42·1	42·5
Fibre	4·6	5·0	3·5
Ash	4·8	5·5	5·0
Carbohydrate	27·1	31·8	22·0
Lecithin (as P_2O_5)	0·26	0·30	0·28

cf. Illing and Whittle, *Food,* 1944, **13**, 32.

Sometimes the oil is partially extracted and this results in an increase in the proportion of the other constituents in many samples of soya flour. The pulse proteins have high biological values (*cf.* Drummond, *J. Roy. Soc. Arts.*, 1942, **90**, 431).

Soya is almost devoid of starch, but it can be identified microscopically by the presence of " hour-glass " cells. These are best seen on the fibrous residue separated by boiling the sample with dilute acid and alkali as in the determination of fibre. Provided the product being examined has not been heated during manufacture, soya can be detected by the **urease test.** The test depends on the fact that soya contains the enzyme urease which liberates ammonia from urea:—

Shake 3–5 g. of the material with 5 ml. of 2 per cent. urea solution and wipe the top part of the tube dry with a cloth. Put a piece of moistened red litmus paper in the top of the tube, cork lightly and incubate at 37° C. for two hours. If soya is present the litmus paper turns blue, due to the ammonia liberated by the urease.

In the absence of lecithin from egg or other products, soya can be approximately determined from the organic phosphorus.

The oil extracted from soya has a characteristic taste and a tendency to rancidity. The values obtained from soya oil are:—

Weight per ml. (20° C.)	0·918–0·924
Refractive index at 40° C.	1·4675–1·4685
Iodine value	129–137
Saponification value	190–195
Reichert value	4–5
Acid value	Up to 3·5
Unsaponifiable matter	0·7–1·6 per cent

Chapter 6

BAKING POWDERS: EGGS: SALAD CREAM

BAKING POWDER

BAKING powder consists essentially of an acid ingredient mixed with an approximately equivalent amount of sodium bicarbonate and diluted with starch. The acidic ingredients which are most commonly employed are cream of tartar, tartaric acid, acid sodium pyrophosphate (A.S.P.) and acid calcium phosphate (A.C.P). Other ingredients which have been suggested for baking powder are adipic acid, sodium aluminium sulphate, citrates and calcium lactate.

Although alum has been used in the U.S.A. the idea has met with opposition in Britain. Of the acid ingredients used, cream of tartar and A.S.P. react slowly but tartaric acid and A.C.P. act rapidly. Some powders contain mixtures of A.S.P. and A.C.P. and others of cream of tartar and tartaric acid. The presence of ingredients which react more quickly also means a more rapid loss of available carbon dioxide during storage. In food manufacture, however, the separate ingredients are often weighed out just prior to the process. The acid phosphates which are then used sometimes have the acidity reduced by the addition of starch and such mixtures are often referred to as **" Cream Powders."** A common mixture of this type contains 64 per cent. of A.S.P., so that the acidity is equivalent to that of cream of tartar (2 parts acid to 1 part bicarbonate). The " strength " or " Neutralising Value " of an acid ingredient is expressed as the number of parts of sodium bicarbonate which will neutralise 100 parts of the acid phosphate.

Golden raising powder is essentially baking powder but containing in addition a yellow dye.

Analysis of Acid Phosphates

The analysis of acid phosphates should include the determination of **acidity, arsenic, lead** and **fluorine.**

The **acidity** can be determined by boiling 0·84 g. of the acid phosphate or cream powder with 20 g. of salt and 50 ml. water and titrating with N/20 sodium hydroxide. The titration multiplied by 5 gives the number of parts of sodium bicarbonate equivalent to 100 parts of acid phosphate (the neutralising value).

For routine purposes, arsenic and lead can be determined direct on the solution of the material in hydrochloric acid. Fluorine should be determined by the S.P.A. method.

Cream powders (see above) should also be examined microscopically in order to identify the starch present.

Legal Requirements for Baking Powder and Golden Raising Powder

The Food Standards (Baking Powder and Golden Raising Powder) Order, 1944 (S.R. and O, 1944 No. 46; 1946 No. 157) prescribes the following standards:—

	* Available carbon dioxide (minimum) per cent.	* Residual carbon dioxide (maximum) per cent.
Baking Powder . . .	8	1·5
Golden Raising Powder .	6	1·5

* As determined by the prescribed method.

In addition a maximum of 15 p.p.m. of fluorine is prescribed in the Fluorine in Food Regulations.

A sample which contains an excessive amount of residual carbon dioxide tends to produce a " soda-taste " in the products in which the baking powder is used.

Analysis of Baking Powder and Golden Raising Powder

The analysis of baking powder and golden raising powder should include (a) a **microscopical examination**, (b) the determination of **the total and residual carbon dioxide,** the **ash** and the possible contaminants, *viz.*, **arsenic, lead** and **fluorine** and (c) the qualitative detection of the acid ingredient. In addition the original composition may be required and it is then necessary to determine the alkalinity of the ash, phosphate, tartrate, etc. Golden raising powder should also be examined for the added colour present.

Available carbon dioxide. Although the actual method for the final determination of the carbon dioxide is left to the individual analyst, the Food Standards (Baking Powder and Golden Raising Powder) Order prescribes the following methods of analysis:—

" 3. (1) **The residual carbon dioxide** shall be determined in the following manner:

A sample of 2 grammes of baking powder or golden raising powder, as the case may be, shall be treated with 25 millilitres of water and evaporated to dryness on a boiling water-bath and subsequently treated with a further 25 millilitres of water and evaporated in like manner. The residual carbon dioxide is the weight thereof evolved when the sample so treated is further treated with excess of dilute sulphuric acid at room temperature, the evolution being completed either by boiling or by means of reduced pressure

(2) **The available carbon dioxide** shall be determined by ascertaining the difference between the total carbon dioxide and the residual carbon dioxide; and the total carbon dioxide shall be determined by ascertaining the weight thereof evolved when the baking powder or golden raising powder, as the case may be, is treated with excess of dilute sulphuric acid at room temperature, the evolution being completed either by boiling for five minutes or by means of reduced pressure."

The carbon dioxide can be determined gravimetrically, volumetrically or gasometrically. The latter method is often preferred in industry for rapid routine work.

(A) For the *gravimetric method*, a " train " is set up which consists of (1) a tower containing stick potash for removing carbon dioxide from the air, (2) a bolt-head flask containing a few large glass beads and fitted with a tap funnel, (3) a small reflux condenser, (4) concentrated sulphuric acid contained in tubes containing pumice, or in Drechsel bottles (connected to the top of the reflux condenser), (5) absorption tubes with taps containing Sofnolite or Carbosorb, (6) a guard sulphuric acid tube or bottle and (7) an aspirator or controlled vacuum pump. If Drechsel bottles are used in the apparatus, the type containing a sinter is more efficient than those with a plain tube. The method is as follows:—

Pass a current of air through the apparatus for ten minutes and weigh the absorption tubes. Then add the sample (for the total carbon dioxide use 0·5 g. of baking powder or 5 g. of self-raising flour; for the residual carbon dioxide use the prescribed amount after the appropriate treatment) to the dry bolt-head flask. Apply gentle suction and slowly introduce 20–50 ml. of water (CO_2-free) into the flask via the tap funnel and shake. Then add 25 ml. of 5 N sulphuric acid slowly to the flask and gradually bring up to the boiling point and boil for five minutes. Then close the taps in the absorption tubes and weigh the absorbed carbon dioxide. As some water may already be present from the preliminary treatment in the determination of the residual carbon dioxide, less water is added to the flask (*e.g.*, self-raising flour). A blank should also be performed.

$$\text{Available } CO_2 = \text{Total } CO_2 - \text{Residual } CO_2$$

(B, 1) The *volumetric method* can be carried out using a similar train to that employed in the gravimetric method except that the absorption tubes (5) are replaced by two Drechsel bottles containing an excess of N/10 barium hydroxide solution (usually 25 ml. in each bottle is sufficient). At the end of the process the combined baryta solutions are back titrated with N/10 oxalic acid using phenolphthalein as indicator (see calculation in B, 2 below).

(B, 2) The *volumetric method* can also be carried out using a simpler apparatus consisting of a large Buchner-type filter flask containing a test-tube of convenient size:—

Weigh out 0·25 g. of baking powder into a dry test-tube. Pipette 50 ml. of N/10 barium hydroxide solution into a filter flask (500–700 ml.) and insert the tube containing the sample. Attach a tap funnel to the neck of the flask by means of a rubber bung so that the delivery end of the funnel fits inside the neck of the test-tube. Evacuate the flask via a short piece of rubber pressure tubing attached to the side tube of the flask. Close the evacuated flask by means of a screw clip fitted to the piece of tubing. Then run excess of 5 N sulphuric acid drop by drop from the tap funnel into the test-tube containing the sample. Close the tap and allow to stand overnight so that the carbon dioxide produced reacts completely with the baryta. In the morning remove the funnel, wash the liquid on the *outside* of the test tube into the flask with a little water and back titrate the unused baryta with N/10 oxalic acid using phenolphthalein as indicator (a ml. N/10). Carry out a blank (titration = b ml.N/10).

$$\text{Total } CO_2\ (\%) = \frac{(b - a) \times 0{\cdot}0022 \times 100}{\text{weight taken}}$$

For the residual carbon dioxide treat 2 g. of the sample in the manner prescribed in the Order (see above). Transfer the residue to a suitable test-tube with the assistance of a little water, insert the tube in a filter flask and determine the " residual " in the same way as described for the " total." Again the:—

$$\text{Total } CO_2 - \text{Residual } CO_2 = \text{Available } CO_2$$

This method is less suitable for the analysis of self-raising flour.

(C) In the *Chittick gasometric method* (*A.O.A.C.*, 1960, 9th Edn., p. 97) the volume of the carbon dioxide produced is measured in gas-measuring burettes. After correcting the volume to N.T.P. the weight of gas evolved from the baking powder is calculated. The Chittick method gives rapid results with both baking powder and self-raising flour and is especially useful in industrial control laboratories. The apparatus should be periodically checked using a known weight of sodium bicarbonate (*cf. J. Ass. Off. Agric. Chem.*, 1923, **6**, 453; 1927, **10**, 36).

Examination of Acid Ingredients of Baking Powder

QUALITATIVE TESTS

Shake about 4 g. of sample in a separator, with 60 ml. of carbon tetrachloride and transfer the deposit of baking chemicals to a small dish and evaporate on the water-bath. The dry residue can then be examined qualitatively for the acid ingredients present:—

(a) FLAME TEST

(b) PHOSPHATES. Heat a portion of the residue gently with dilute nitric acid and ammonium molybdate solution in a test-tube. The presence of phosphate is indicated by the formation of a yellow precipitate.

If the molybdate test is positive, shake a further portion of the material with water and filter. Add silver nitrate solution to the neutral filtrate. A yellow precipitate indicates the presence of an orthophosphate (A.C.P.) and a white amorphous precipitate indicates pyrophosphate (A.S.P.). Salt, if present, interferes with the test and its absence should be confirmed by adding nitric acid which dissolves the phosphates but not the chloride of silver.

(c) CALCIUM. Dissolve a portion of the residue in dilute hydrochloric acid, neutralise with ammonia, make just acid with acetic acid and add ammonium oxalate solution. Calcium is indicated by the formation of a white precipitate of the oxalate.

(d) TARTRATE. Acidify a portion of the residue with acetic acid, add a drop of ferrous sulphate solution, a few drops of hydrogen peroxide and excess of sodium hydroxide solution. A purple colour indicates the presence of tartrate.

Confirm by warming some of the residue with a few drops of potassium bromide solution (10 per cent.), a few drops of resorcinol solution (2 per cent.) and sulphuric acid (conc.). Tartrate gives a blue colour, which after cooling, gives a red colour when poured into water (*cf. J.S.C.I.*, 1943, **62**, 176*T*).

Reactions of Baking Powder Residues containing the commoner acid ingredients

Test	*A.C.P.*	*A.S.P.*	*Cream of Tartar*	*Tartaric acid*
(a) Flame test .	Na + Ca	Na	Na + K	Na
(b) Phosphate .	+ (*)	+ (*)	—	—
(c) Calcium . .	+	—	—	—
(d) Tartrate . .	—	—	+	+

* Use $AgNO_3$ test in neutral solution for differentiating between the two types of phosphate.

QUANTITATIVE EXAMINATION

DETERMINATION OF CREAM OF TARTAR AND TARTARIC ACID (A.O.A.C.)

To 2·5 g. sample in a 250 ml. graduated flask add 100 ml. warm water (about 50° C.) and allow to stand for thirty minutes at room temperature, shaking occasionally. Cool, make up to the mark with water, shake vigorously and filter through a large fluted paper. Pipette two portions of 100 ml. each of *clear* filtrate into 250 ml. beakers (A and B) and evaporate to about 20 ml. To beaker A add 3·5 ml. N/1 potassium hydroxide, mix well, and add 2 ml. acetic acid. Treat beaker B in a similar manner but use N/1 sodium hydroxide instead of potassium hydroxide. Then treat each mixture separately as follows: Cool to 15° C., stir vigorously for one minute and place in a refrigerator overnight. Filter the precipitate on a Gooch crucible containing a thin layer of asbestos and wash the precipitate remaining on the sides of beaker through the filter with 75 ml. ice-cold alcohol (80 per cent.). Finally wash the sides of the crucible with 25 ml. alcohol and suck dry. Transfer the contents of the crucible back into the original beaker with 100 ml. of hot water and titrate with N/10 alkali using phenolphthalein as indicator. If the titrations are " A " and " B " respectively:—

$$\text{Total tartaric acid (\%)} = 1{\cdot}5\,(A + 0{\cdot}6)$$
$$\text{Cream of tartar (\%)} = 1{\cdot}88\,(B + 0{\cdot}6)$$
$$\text{Tartaric acid (free) (\%)} = 1{\cdot}5\,(A - B)$$

The figure 0·6 represents the solubility of cream of tartar in terms of N/10 alkali.

PHOSPHATE AND CALCIUM

Dissolve 1 g. of sample in hot dilute nitric acid, boil for thirty minutes to convert pyrophosphate into orthophosphate and filter. Divide the filtrate into portions C and D. To aliquot C add molybdic acid solution, warm to 68° C., dissolve the washed phosphomolybdate precipitate in slight excess of N/2 sodium hydroxide and titrate the excess with N/2 acid.

Alternatively dissolve the phosphomolybdate precipitate in dilute ammonia, add magnesia mixture, filter off the magnesium ammonium phosphate, wash with ammonia, dry, ignite to magnesium pyrophosphate and weigh (see p. 45). Calculate the phosphate as P_2O_5 (and to the equivalent of acid phosphate—see p. 190).

To determine the *calcium* in aliquot D, make the cold solution just neutral with ammonia, add ammonium acetate (considerable excess) and acetic acid, boil and add ammonium oxalate. Filter off the precipitate of calcium oxalate (which may contain traces of iron and aluminium), wash, dissolve in acid and titrate hot with N/10 permanganate (see p. 42). 1 ml. N/10 permanganate $\equiv$ 0·002804 g. CaO. Also calculate the calcium as A.C.P.

SULPHATE

Boil 3 g. of sample with dilute hydrochloric acid to hydrolyse the starch and precipitate the sulphate as barium sulphate in the usual way. Calculate the calcium (as CaO) equivalent to the calcium sulphate present and deduct it from the total calcium (as CaO). Then the phosphate can be calculated:—

$$0{\cdot}607x + 0{\cdot}458y = A$$
$$0{\cdot}234x + 0{\cdot}543y = B$$

where x and y are the percentages of $CaH_4P_2O_7$ and $Ca_3P_2O_8$ respectively and A and B are the percentages of P_2O_5 and CaO (after deducting the CaO present as $CaSO_4$).

STARCH in baking powder is commonly estimated by difference, but may be determined by dissolving the raising ingredients in the minimum quantity of cold dilute hydrochloric acid, filtering off the crude starchy matter and weighing it directly. Alternatively the starch can be determined polarimetrically or by the diastase method.

Calculation of the Original Composition of Baking Powders

Although the amount of available carbon dioxide falls during storage, the original composition of a baking powder (when manufactured) can be assessed reasonably accurately. The main figures required for the calculation are the residual carbon dioxide and the amount of the acid ingredient(s) present, *e.g.*, phosphate as P_2O_5, calcium, cream of tartar, tartaric acid. The figures for ash and the alkalinity of the ash are often useful also for checking the calculation. Having determined the amounts of the acid constituents present, the bicarbonate is given by:—

$$\left\{\begin{array}{l}\text{Total } NaHCO_3 \equiv NaHCO_3 \text{ equivalent to the acid present} \\ \text{(available)} + NaHCO_3 \text{ equivalent to the residual } CO_2.\end{array}\right.$$

The amount of filler is usually obtained by difference:—

Filler (%) = 100 — Total $NaHCO_3$ (%) — Total acid ingredients (%)

Salt is occasionally present and this must be allowed for in the calculation of filler and also when using the ash figure for confirmation of the general results. The moisture content of the filler can also be calculated by assuming that all the moisture in the baking powder is associated with the starch present. As decomposition takes place on heating in the oven, the moisture content must be determined by distillation from toluene, by drying over sulphuric acid in a vacuum desiccator or by means of the Karl Fischer technique.

For calculating the amount of bicarbonate equivalent to cream of tartar or tartaric acid, the theoretical equations can be used:—

CREAM OF TARTAR:—

$$\underset{84\text{ g.}}{NaHCO_3} + \underset{188\text{ g.}}{HOOC(CHOH)_2COOK} = NaOOC(CHOH)_2COOK + \underset{44\text{ g.}}{CO_2} + H_2O$$

Therefore, 1 lb. bicarbonate ≡ 2·24 lb. cream of tartar.

TARTARIC ACID:—

$$\underset{168\text{ g.}}{2\,NaHCO_3} + \underset{150\text{ g.}}{HOOC(CHOH)_2COOH} = NaOOC(CHOH)_2COONa + $$

$$+ \underset{88\text{ g.}}{2CO_2} + 2H_2O$$

Therefore, 1 lb. bicarbonate ≡ 0·893 lb. tartaric acid.

For phosphatic powders the theoretical equations are of value, but the average practical equivalents of the commercial materials used often give a better idea of the original composition:—

ACID SODIUM PYROPHOSPHATE:—

$$\underset{168\text{ g.}}{2NaHCO_3} + \underset{222\text{ g.}}{Na_2H_2P_2O_7} = Na_4P_2O_7 + \underset{88\text{ g.}}{2CO_2} + 2H_2O$$

Therefore *theoretically*, 1 lb. bicarbonate ≡ 1·32 lb. Pure A.S.P.
But, 1 lb. bicarbonate ≡ 1·36 lb. Commercial A.S.P.
Commercial A.S.P. contains approximately 63·9 per cent. P_2O_5.

ACID CALCIUM PHOSPHATE:—

$$\underset{672\text{ g.}}{8\,NaHCO_3} + \underset{702\cdot 3\text{ g.}}{3CaH_4P_2O_8} = Ca_3P_2O_8 + 4Na_2HPO_4 + \underset{352\text{ g.}}{8CO_2} + 8H_2O$$

Therefore *theoretically*, 1 lb. bicarbonate = 1·045 lb. Pure A.C.P.
But, 1 lb. bicarbonate ≡ 1·25 lb. Commercial A.C.P.
Commercial A.C.P. contains approximately 57·0 per cent. P_2O_5 and 18·4 per cent. Ca.

For an alum baking powder the equation is:—

$$6\,NaHCO_3 + Na_2Al_2(SO_4)_4 = Al_2(OH)_6 + 4Na_2SO_4 + 6CO_2$$

EXAMPLE 1

Acid ingredient—A.S.P.

Total CO_2 in sample = 14·1%

Residual CO_2 in sample = 0·2 % ≡ 0·38 % $NaHCO_3$

Total ash = 44·2%

$$P_2O_5 = 25\cdot4\% = \frac{100}{63\cdot9} \times 25\cdot4 = 39\cdot7\%\text{ A.S.P.}$$

$$= \frac{1}{1\cdot36} \times 39\cdot7 = 29\cdot2\%\ NaHCO_3\text{ (available)}$$

Total sodium bicarbonate = 29·2 + 0·4 = 29·6%

Moisture = 3·5%

COMPOSITION OF VARIOUS BAKING POWDERS (B) AND GOLDEN RAISING POWDERS (G)

	G	B	B	B	G	B	B	B	B
% Moisture	5·4	3·7	7·6	7·4	4·3	4·7	6·5	4·2	3·7
% Acid Sodium Pyrophosphate	27	37	26·9	—	24·0	—	—	12·7	18·8
% Acid Calcium Phosphate (Coml.)	—	—	—	25·7	11·8	—	—	—	—
% Sodium bicarbonate	20	28	22·6	19·9	27·5	40·1	23·7	24·4	29·8
% Cream of tartar	—	—	—	—	—	—	3·3	—	4·8
% Tartaric acid	—	—	—	—	—	33·7	17·6	16·5	10·0
% Filler*	53	35	50·5	54·4	36·7	26·2	55·4	46·4	36·6
% Available CO_2 (when analysed)	10·0	13·7	6·1	7·1	7·2	19·7	11·0	11·9	12·5
% Available CO_2 (original)	10·5	14·6	11·5	9·9	14·2	—	—	12·3	15·2

*Filler including moisture as added.

ORIGINAL COMPOSITION:—

	Per cent.
Sodium bicarbonate	29·6*
Acid Sodium Pyrophosphate . .	39·7*
Filler (by difference)	30·7*
Total carbon dioxide	15·5
Available carbon dioxide . . .	15·3

The alkalinity of the ash of the sample was equivalent to 30·5 per cent. $NaHCO_3$ which confirms that the figure of 29·6 per cent. bicarbonate is of the right order.

* *N.B.* In reporting the results of the original composition it is common (industrially at least) to return the figures as whole numbers.

EXAMPLE 2

Acid ingredients—A.S.P. and A.C.P.

Total CO_2 in sample = 12·1%

Residual CO_2 in sample = 0·1% = 0·2% $NaHCO_3$

Total ash = 43·7%

Calcium = 2·1% (as Ca) = 11·4% A.C.P. = 6·5% P_2O_5

Total P_2O_5 = 23·5%

P_2O_5 due to A.S.P. = 23·5 − 6·5 = 17·0% = 26·6% A.S.P.

Bicarbonate equivalent to A.C.P. = 11·4/1·25 = 9·1%

Bicarbonate equivalent to A.S.P. = 26·6/1·36 = 19·6%

Bicarbonate equivalent to residual CO_2 = 0·2%

Total bicarbonate = 28·9%

Moisture = 2·8%

ORIGINAL COMPOSITION:—

	Per cent.
Sodium bicarbonate	28·9
Acid sodium pyrophosphate (coml.) .	26·6
Acid calcium phosphate (coml.) . .	11·4
Filler (by difference)	33·2
Total carbon dioxide	15·1
Available carbon dioxide . . .	15·0

N.B. With baking powders containing cream of tartar and tartaric acid the calculations follow similar lines. The theoretical equations given above can be employed and the final result can often be readily checked from the ash and the alkalinity of the ash on the assumption that the bicarbonate is converted on ignition to Na_2CO_3, that the cream of tartar goes to K_2CO_3 and the tartaric acid gives no ash.

EGGS

Hen's eggs on the average weigh 57 g. and consist of 57 per cent. of white, 32 per cent. of yolk and 11 per cent. of shell.

Average composition of egg-white, egg-yolk and whole egg (edible) from commercial fresh eggs (cf. J.A.O.A.C., 1932, **15**, 314):—

	Egg-White	*Egg-Yolk*	*Whole Egg*
Water (%)	87·8	49·5	73·95
Fatty matter (Lipids) (%)	0·03	31·9	11·55
Nitrogen (%)	1·72	2·61	2·04
Ash (%)	0·7	1·0	0·8
Dextrose (%)	0·40	0·17	0·32
Chloride (% NaCl)	0·29	0·30	0·29
Phosphorus (% P_2O_5)	0·04	1·38	0·52
pH	9·1	6·3	7·7
Specific gravity	1·055	1·026	1·048*

* Excluding shell.

The shell of the hen's egg contains about 3·5 per cent. of protein, 1·5 per cent. of water and 95 per cent. of ash (mainly calcium carbonate with traces of magnesium carbonate and calcium phosphate).

By means of precipitation with ammonium sulphate and by electrophoretic methods at least nine proteins have been identified in egg-white, *viz.* ovalbumins A_1 and A_2, conalbumin C, ovomucoid O, globulins G_1, G_2 and G_3, ovomucin and avidin. The yolk contains simple proteins (livetins) and phosphoproteins (vitellin and vitellenin). The solids of egg-yolk also contain true fats, phospholipins (of which ovolecithin comprises about 70 per cent.), sterols (*e.g.*, cholesterol) and a little carbohydrate.

The yolk is surrounded by the vitelline membrane, which is semi-permeable. During storage water diffuses slowly from the white to the yolk and the s.g. of the whole egg decreases. Eggs can be inspected by placing them in front of a bright light (candling). As the egg stales, the white ceases to be transparent and becomes darker and opaque. At the same time the yolk is less visible and the air space gets larger. Candling should also reveal blemishes such as blood spots. If a good fresh egg is cracked and poured on to a plate a large proportion of thick white adheres closely to a compact yolk; whereas a thin white surrounding a flattened yolk indicates a lower quality egg. Changes in the taste and odour must obviously be considered also. Chemically, the ammoniacal nitrogen (estimated after aeration into N/50 acid) and the acidity of the fat both increase as eggs spoil.

Useful methods for the examination of the shells of eggs suspected of being preserved (in water-glass, lime water or oil) have been described by Nicholls (*Analyst*, 1931, **56**, 383).

Analysis of Eggs

The chemical estimation of the main constituents of shell eggs are not normally carried out, but if required the following points should be borne in mind:—

PROTEIN:—

From a study of the nitrogen content of the proteins present, the following Kjeldahl factors for the total protein would seem to be appropriate: White, N × 6·70; Yolk, N × 6·62; Whole egg, N × 6·68.

FATTY MATTER:—

The fatty matter extracted from eggs depends on the particular solvent used. The extract consists principally of oil (see below) and certain phosphoproteins. Although the total amount of extract obtained may vary with the solvent used, the ratio of N/P_2O_5 appears to be constant. For the estimation of fatty matter in liquid eggs the Rose-Gottlieb process is suitable (see method described under condensed milk—p. 372). If the acidity of the fat is to be required, the determination should be carried out on the ether extract of the dried materials.

Egg oil. Characteristic figures:—

Refractive index at 20° C.	1·4655–1·4670
Saponification value	188–198
Iodine value	60–70
Unsaponifiable matter	Up to 4 per cent.
Phosphorus (as P_2O_5)	1·4 per cent.

Frozen whole egg is prepared using freezing temperatures between − 10° and − 40° F. Such temperatures, combined with air circulation, cause rapid freezing, which is important, as it reduces bacterial multiplication and produces a more uniform product. After freezing the product is stored below 15° F. The product should contain less than 75 per cent. water, at least 10 per cent. of oil (extract) and the F.F.A. of the fatty extract should not exceed 3 per cent. as oleic acid. The product imported from China occasionally contains boric acid.

Frozen egg yolk always contains some adhering white. The degree of separation affects the solids content (minimum 43 per cent.), which is controlled at the plant using a refractometer.

DRIED EGG

Dried egg is usually produced by spray-drying. The moisture content of fresh samples should not exceed 5 per cent., but it may rise during storage to as much as 9 per cent. The other main constituents present usually fall within the following percentage ranges: protein (N × 6·68) 40·4 − 50·5, chloroform extract (fat, lecithin, etc.) 40–46, ash 3·2–4·2, organic phosphorus (as P_2O_5) 1·05–1·40 (chloroform as solvent).

Organic phosphorus. The amount of organic phosphorus extracted varies with the solvent used. Manley and Lobley (*Analyst,*

1948, **73**, 30) extract 3–10 g. sample for 2–3 hours with warm chloroform in a continuous extractor. The extract is then dried and weighed to ascertain the fat content. The extract is then wet oxidised using 2–5 ml. of sulphuric acid and nitric acid, and the phosphorus precipitated as phosphomolybdate and determined gravimetrically as magnesium pyrophosphate. Alternatively, the phosphorus can be determined absorptiometrically (p. 46). For a further method employing methanol see the section on Salad Cream.

Solubility can be determined by the following method:—

One g. of the sample is weighed into a dry beaker, 10 ml. of water are added and the mixture is allowed to soak for three hours, then heated on a water-bath at 50° for half an hour. The mixture is transferred to a centrifuge tube without washing the beaker, and spun for at least a minute. The upper layer is poured off without disturbing the sediment, the beaker is then washed with 10 ml. of water at 50°. The water is poured into the centrifuge tube and, after mixing, is spun as before. The upper water is poured off and the sediment is washed with a further 10 ml. of water. After this the sediment is transferred to a tared filter or basin, dried and weighed.

A good quality fresh sample usually shows a solubility of over 90 per cent. (*cf.* Hawthorne, *J. Soc. Chem. Ind.*, 1944, **63**, 6).

Samples of dried egg should also be examined for *borate, arsenic* and *lead*.

Sugar-dried egg. At least one half of the aerating power of fresh egg is lost after the whole egg has been spray-dried. It has been shown, however, that no such loss occurs if certain sugars are dissolved in the liquid egg before spray-drying. Imported commercial sugar-dried egg usually contains 2 parts of egg-solids to 1 part of sucrose. The moisture content should preferably not exceed 3·5 per cent.

For further information relating to the technology of the above egg products the following reference should be consulted:—

" Eggs and Egg Products," by Brooks and Taylor, 1955 (D.S.I.R., Fd. Inv., Spec. Rept. No. 60, H.M.S.O.).

SALAD CREAM, MAYONNAISE, SALAD DRESSINGS

Salad cream is a wholly or partly cooked semi-solid emulsion of edible vegetable oil, egg yolk, vinegar, salt, sugar, spices (*e.g.*, mustard) and emulsifying agents such as gums and starches. Although mayonnaise only has to conform to the same legal requirements as salad cream, it used to be generally accepted that the former was a superior product containing more oil and less emulsifying agent. In the U.S.A. mayonnaise is required to contain at least 65 per cent. of oil, whereas the minimum for salad dressing is 30 per cent.

LEGAL REQUIREMENTS FOR SALAD DRESSINGS

The Food Standards (Salad Cream and Mayonnaise) Order, 1945 (S.R. & O. 1945 No. 1177 as amended by S.R. & O. 1946 No. 157) prescribes a minimum of 25 per cent. (w/w) of edible vegetable oil and a minimum of 1·35 per cent. (w/w) of egg-yolk solids for salad cream, mayonnaise and any other salad dressing. Products of a lower standard may be sold, however, provided that the label bears the following statement: " This product is not a salad cream or mayonnaise and does not comply with the statutory standard prescribed for these products." Such a statement is " to be clearly legible, to appear conspicuously and in a prominent position on the label . . ."

Code of Practice (C.P. 22/1950) states that **horseradish cream, sauce or relish** shall contain a minimum of 25 per cent. horseradish root or the equivalent of dehydrated root and essential oil. Horseradish cream shall contain at least 25 per cent. of a " cream " containing at least 25 per cent. of edible vegetable oil and 7·5 per cent. N.F.M.S. (all percentages w/w).

Analysis. The analysis of salad cream in the " official " laboratory should include the determination of the **oil, egg-yolk solids** and **trace elements.** The oil should be identified by the usual **oil values** and be examined also for possible **rancidity.** *Industrially,* some additional determinations are also called for:—

(a) total solids (dry on sand), (b) acidity (as acetic acid), (c) sugar, (d) ash, (e) salt (determine on ash), (f) efficiency of emulsification (under the microscope the bubbles of oil should be small and regular in size), (g) viscosity (time to run through a " standard " funnel), (h) colour and (i) stability. Imported samples occasionally contain added preservative.

Oil. Use a modified Werner-Schmid method (it is convenient to employ a Mojonnier tube):—

Shake 1 g. of sample with 10 ml. of 5 N hydrochloric acid and place in a beaker of water at 70° C. Then boil the water and remove the tube after boiling for half an hour. Add water, cool and shake with 25 ml. of ether followed by 25 ml. of light petroleum. Transfer the organic solvent layer to a weighed flask. Extract at least twice more, remove the solvent and dry the oil for forty minutes in the 100° C. oven.

Oil values. The acid extracted oil from the determination of oil (above) should *not* be used for assessing the edibility of the oil. For this, mix a separate portion of the sample with excess of anhydrous sodium sulphate and extract the oil by stirring in fresh ether. Filter the ether extract into a flask, remove the solvent and dry in the oven. Determine the iodine value, saponification value and rancidity values (F.F.A., peroxide value, Kreis test) on the oil so obtained.

Egg-yolk solids. The accurate estimation of egg-yolk solids in salad cream is somewhat difficult. It is probably most frequently determined from the organic phosphorus extracted by means of an organic solvent. The following method, employing methanol, appears to give reasonably satisfactory results (*cf. Analyst,* 1948, **73**, 30):—

Reflux 20 g. of salad cream with 100 ml. of absolute methyl alcohol for six hours and allow to stand overnight. Filter, re-extract the residue with methanol and wash through the filter with more solvent. Evaporate the combined filtrates containing the extracted organic phosphorus to low bulk on the water-bath. Transfer to a digestion flask, wet oxidise with sulphuric-nitric acid and wash into a 100 ml. graduated flask. Make up to 100 ml. at room temperature, mix and filter. Pipette 50 ml. of filtrate into a further 100 ml. graduated flask, neutralise with ammonia (0·88), make just acid with dilute nitric acid, add 25 ml. of vanadate-molybdate reagent and dilute to the mark. Mix and measure the optical density at 470 mμ

Alternatively, the phosphorus in the wet oxidised residue can be determined volumetrically after precipitating as phosphomolybdate, etc.

$$\text{Egg-yolk solids} = P_2O_5 \times 56$$

$$\text{Dried egg} = \text{Egg-yolk solids} \times 1{\cdot}48$$

N.B. The methanol extracts 1·20 per cent. P_2O_5 from dried egg compared with 0·15 per cent. from soya flour.

A method for determining egg in salad cream has been proposed by Daubney and Sexton (*Analyst,* 1950, **75**, 305). This is based on the colorimetric determination of the choline derived from egg lecithin by means of acid hydrolysis and conversion of the choline into its pink reineckate, which is soluble in acetone.

In a more recent method, Casson and Griffin (*Analyst,* 1959, **84**, 281) employ the enzyme lecithinase D (prepared from cabbage leaves) for the hydrolysis of the phospholipids. The choline liberated is separated by adsorption on a cation-exchange resin and, after elution, is determined colorimetrically as the reineckate at 520 mμ.

Trace elements. The most important of these are arsenic (p. 55) and lead (p. 65). With the large amount of oil present the wet oxidation process may prove troublesome and Middleton and Stuckey's " cold " method (p. 52) may be more convenient.

OTHER DETERMINATIONS

See list above (*cf.* Devey, *Food Trade Rev.*, 1959, **29** (2), 16).

Chapter 7

FRUIT AND VEGETABLE PRODUCTS

Fruits—Vegetables—Tomato Purée—Tomato Ketchup—Fruit Juices—Soft Drinks—Syrup of Blackcurrant

General Composition of Fruits and Vegetables

Fruits. The major part of the edible portion of fresh fruits consists of water (75–95 per cent. for most types). Fruits are poor sources of protein (0·2–1·3 per cent. as N × 6·25), but, in general, contain a reasonable amount of carbohydrate. The latter may include varying proportions (according to the fruit, maturity, etc.) of dextrose, fructose and sucrose and possibly starch (*e.g.*, banana, apple). The principal acids present in fruits are citric, tartaric and malic acids. The total acidity falls after picking. The pH of fruits varies from 2·5–4·5 (most types 3·0–3·5). Other constituents of fruits include cellulose and woody fibres, mineral salts, pectin, gums, tannins, colouring matters and volatile oils. The main feature of the composition from the nutritional point of view is that certain fruits, particularly blackcurrants, most citrus fruits and strawberries, are good sources of vitamin C. Some analytical data of some common fruits which are useful in connection with jam analysis are given on pp. 138–9.

Dried fruits (*e.g.*, currants, raisins, dates and figs) contain about 13–22 per cent. of water, and in consequence the other main constituents are present in proportionately greater amounts than in the corresponding fresh fruit. Some samples, therefore, contain over 60 per cent. of total sugars. Dried fruits are almost devoid of vitamin C, but prunes and dried apricots represent useful sources of vitamin A.

Nuts are especially rich in fat, protein and vitamin B_1.

Vegetables. As with fruits, the major part of the edible portion of fresh vegetables consists of water (84–96 per cent. for most types).

Peas, beans and potatoes contain reasonable amounts of protein and carbohydrate. Potatoes which have been stored at temperatures around 30° F. tend to be over-sweet due to conversion of starch to sugar, but storage at high temperatures (*e.g.*, at 60° F.) tends to produce a " mealy " potato due to the production of starch from sugar. Some vegetables are comparatively good sources of calcium (especially watercress), iron (especially spinach) and phosphorus. Vitamin C is present to the largest extent in green vegetables, and, especially in view of the comparatively large amounts eaten, other important dietary sources are potatoes, peas and turnips (also swedes). Carrots are a good source of vitamin A, which is also present in green vegetables and fresh peas. Dried peas (and beans) are almost devoid of vitamins A and C. Sulphite is added in the preparation of dehydrated vegetables (moisture content about 5 per cent.), however, in order to retain colour and vitamin C. Although much vitamin C is lost during cooking, it appears to be rather more stable in dehydrated than in fresh vegetables.

REFERENCES
" Manual of Nutrition," Min. Agric. Fish. & Food, H.M.S.O., 1959.
" The Chemical Composition of Foods," McCance and Widdowson, H.M.S.O., 1960.

Analysis of Fruit and Vegetables

The analysis of fresh fruit and vegetables in terms of water, protein, carbohydrates, mineral matter, fibre, etc., is not very frequently required, nor does the determination of such constituents present any special difficulties. Only certain special determinations are therefore described below. From the legal point of view, fresh fruits may be required to be examined for preservatives (*e.g.*, diphenyl, orthophenylphenol and possibly boric acid and thiourea) and mineral oil. The appropriate regulations are quoted on pp. 450–453. It should be borne in mind that traces of boric acid are present naturally in fruits. *Dried fruits* should be examined for sulphur dioxide and possibly mineral oil and trace metals.

Pectin. Many methods have been proposed for the estimation of pectin which depend on extraction and hydrolysis by water, followed by precipitation with alcohol. Ling *et al.* (*J.S.C.I.*, 1925, **44**, 253T) have reported favourably on the method due to Carré and Haynes (*Biochem. J.*, 1922, **16**, 60):—

A suitable quantity of the pulped vegetable or fruit is repeatedly extracted with cold water, and the mixed extract is boiled and filtered through paper. An aliquot part of the filtrate is diluted to 300 ml. and to it are added 100 ml. of 0·1 N sodium hydroxide solution. The mixture is allowed to stand overnight, then 50 ml. of N acetic acid are added and five minutes later 50 ml. of

2 N calcium chloride solution. After standing for an hour the liquid is boiled for a few minutes and filtered. The residue is washed with boiling water until free from chlorides, again boiled with water and filtered on a Gooch crucible, washed, dried and weighed as calcium pectate. To this substance Carré and Haynes ascribe the empirical formula $C_{17}H_{22}O_{16}Ca$, but King (*Analyst*, 1925, **50**, 371) gives it the formula $C_{17}H_{24}O_{16}Ca$. King's method has already been given.

Pentosans. Besides the pectins, there are other substances present in fruits which yield five-carbon sugars on hydrolysis, *viz.*, the pentosans. These appear to have the general formula $(C_5H_8O_4)_n$ and are formed in increasing quantities as the liquefaction of plant tissue proceeds. There appears to be a general relationship between the amount of crude fibre and pentosans, which is to be expected as the latter substances are so closely allied to cellulose. The estimation of pentosans is seldom more useful than that of fibre, which is much less tedious. When required it is carried out by distillation with hydrochloric acid with which they yield furfuraldehyde.

Two g. of the finely divided material are distilled from a distillation flask fitted with a tap funnel, with 100 ml. of 12 per cent. hydrochloric acid (*i.e.*, sp. gr. 1·06). When about 30 ml. have passed over, a similar volume of the same acid is run into the flask and the distillation repeated. This is continued until all furfural has been distilled off, which may be ascertained by testing a drop of the distillate with aniline acetate paper. No pink colour should appear. It is usually necessary to distil over upwards of 300 ml. To the mixed distillate containing the furfural is added a large excess of solution of phloroglucinol in dilute hydrochloric acid, and the mixture is allowed to stand for twelve hours. The precipitate is filtered off, washed with 150 ml. of cold water, dried in the oven, and weighed. The weight thus obtained, multiplied by 0·544 gives the equivalent of furfuraldehyde. In order to calculate pentosans therefrom, an allowance of 10·4 mg. is deducted (Tollens, *Z. Rubenz. Ind.*, **46**, 480) from the weight of furfural and the difference multiplied by 1·82.

A more rapid alternative method is that of Ling and Nanji (*Biochem. J.*, 1921, **15**, 466):—

The distillate obtained as already described is made up to a definite volume, say 400 ml. Then 25 ml. are carefully neutralised in a 100 ml. flask with 3 N sodium hydroxide solution, using methyl orange as indicator (it is important to keep the mixture cool at this stage). The solution is acidified with acetic acid, and 10 ml. of 2 per cent. aqueous solution of phenylhydrazine are added, the mixture is diluted nearly to 100 ml. and warmed to 50–55° C. for twenty minutes, cooled, made up to the mark and filtered. Ten ml. of filtrate are added to a like volume of 0·1 N iodine solution, diluted to 100 ml., and titrated back with 0·05 N thiosulphate solution. A similar titration is made on 10 ml. of the phenylhydrazine solution, and the difference calculated as furfuraldehyde in accordance with the equations:—

$$PhNH.NH_2 + 2I_2 \longrightarrow 3HI + C_6H_5I + N_2$$

$$C_4H_3O.CHO + PhNH.NH_2 \longrightarrow C_4H_3OCH{:}N.NHPh + H_2O.$$

Wittmann has given the following percentage figures for the pentosan content of various fruits: juniper berries 6·0, raspberries 2·7, elderberries 1·2, grapes (Japanese) 1·6, blackberries 1·2, strawberries 0·9, cranberries 0·75, bilberries 0·8, gooseberries 0·5, currants 0·4.

Vitamin C

The ascorbic acid content of fruit and vegetables can be estimated by macerating the sample (preferably mechanically) with a stabilising agent such as 5 per cent. metaphosphoric acid or trichloracetic acid and titrating the decanted or filtered extract with 2·6 dichlorophenolindophenol as described for juices on p. 211 (see also *Biochem. J.*, 1942, **36**, 155). The following procedure using the same reagent is virtually the Canada Dept. of Agriculture's modification of the absorptiometric method for determining vitamin C proposed by Loeffler and Ponting (*Ind. Eng. Chem., Anal. Ed.*, 1942, **14**, 846).

Special Reagents

Stock ascorbic acid solution. Prepare a 0·1 per cent. solution of ascorbic acid in 0·4 per cent. oxalic acid solution.

Working standards (WS). Take 5, 10, 15, 20 and 25 ml. of stock ascorbic acid solution and make up each to 500 ml. with 0·4 per cent. oxalic acid solution. These solutions numbered 1 to 5 contain 1, 2, 3, 4 and 5 mg. ascorbic acid per 100 ml. respectively.

Standard dye solution. Twelve mg. of 2 : 6 dichlorophenolindophenol per litre.

Standard curve. To four absorptiometer tubes add the following:—

D.W—10 ml. water	*S*—WS No. 1 + 9 ml. water
No. 1—1 ml. 0·4% oxalic acid	*No.* 2—1 ml. WS No. 1

Adjust the absorptiometer to zero, using water (DW) and a green filter (approx. 520 mμ). To tube No. 1 add 9 ml. standard dye solution (0·0012 per cent.), mix and record the reading (L_1) *exactly* fifteen seconds after adding the dye. Then adjust the instrument to zero with tube S in the absorptiometer. To tube No. 2 add 9 ml. dye, mix and read after fifteen seconds (L_2). Record L_1 and L_2 for each working standard and construct the standard curve with concentrations of ascorbic acid (mg/100 ml.) as abscissæ and L_1–L_2 for each working standard as ordinates.

Method

Macerate 50 g. of sample for three minutes in a blender with 350 ml. of 0·4 per cent. oxalic acid solution and filter. Obtain L_1 as described above. To tube S add 1 ml. filtrate + 9 ml. water and adjust the instrument to zero. Then to tube No. 2 add 1 ml. filtrate + 9 ml. dye and record L_2 after fifteen seconds. Calculate L_1–L_2 and obtain the concentration of ascorbic acid from the standard curve. If necessary repeat with a more suitable concentration of sample (*cf. Ind. Eng. Chem., Anal. Ed.*, 1941, **13**, 793; 1943, **15**, 182).

Diphenyl is an effective anti-mould agent which may be present in citrus fruits up to a maximum of 100 p.p.m.

Diphenyl is incorporated into the paper wrappers or the lining papers of the boxes so that the vapour inhibits mould growth on the surface of the fruit. Work carried out in the U.S.A. in the Hazleton Laboratories on orange treated in this way gives some idea of the distribution of the diphenyl over the fruit: whole fruit 20–100 p.p.m.; peel 100–300 p.p.m.; pulp or juice up to 1 p.p.m. In Britain, Tomkins and Isherwood (*Analyst*, 1945, **70**, 330) found 40 p.p.m. in the peel of some commercially treated oranges and negligible amounts in the pulp. Marmalade made with treated oranges was found to contain 0·1 p.p.m. diphenyl.

Diphenyl can be determined spectrophotometrically using cyclohexane as solvent (absorption max. at 251 mμ; $E_{1\,cm.}^{1\%} = 1200$). With wrapping materials the preservative can be readily obtained in solution by merely soaking the paper in the solvent (*cf.* Rajzman, *Analyst*, 1960, **85**, 116).

With fruits, diphenyl can be removed from the food by steam distillation, but the presence of citrus oils tends to cause interference. Baxter (*J. Ass. Off. Agric. Chem.*, 1957, **40**, 249) describes a method in which the interfering oils are destroyed by permanganate oxidation and also reports favourably on the modified " chromatostrip " procedure of Stanley *et al.* (*ibid.*, p. 294). A further method involving direct extraction with chloroform has been presented by Hoeke and Cats (*Analyt. Abstr.*, 1958, **5**, No. 3522). Other methods for determining diphenyl in fruits have been described by Tomkins and Isherwood (*loc. cit.*) and Steyn and Rosselet (*Analyst*, 1949, **74**, 89).

o-**Phenylphenol** is effective both as a bactericide and as a fungicide and may be present in some fruits up to certain prescribed maximum limits (see Preservatives Regulations, p. 451). Mould growth on citrus fruits can be suppressed by dipping in a solution of the sodium salt. This treatment appears to give a residue of about 60 p.p.m. on the whole fruit. During storage the sodium salt is converted into the free phenol. In view of its lower volatility, *o*-phenylphenol is probably less effective than diphenyl when incorporated in the wrapping papers. Experiments have shown that the amounts taken up by the peel and pulp are about 6–17 mg. and 0–0·04 mg. per orange respectively. Marmalade made from treated oranges, however, was found to contain 4·5 p.p.m. of *o*-phenylphenol, which is much higher than the amount of diphenyl found (0·1 p.p.m.) when similarly treated. Non-citrus fruits are dipped in or sprayed with an alkaline solution of the phenol and are then sometimes washed with water. Most of the residue appears to remain on the outside of the fruit and would not be ingested if the fruit is peeled before consumption. For further information on the subject the reader should refer to the F.S.C. Report on Preservatives and the paper of Tomkins and Isherwood (*Analyst*, 1945, **70**, 330).

Orthophenylphenol (OPP) can be determined spectrophotometrically using cyclohexane as solvent. The absorption peak is at 245–250 mμ, but Harvey and Penketh (*Analyst*, 1957, **82**, 498) recommend that the determination should be carried out at 284 and 295 mμ. If both diphenyl and OPP are present, the fruit should firstly be distilled from alkali so that the former comes over but the OPP remains behind. Then by acidifying and continuing the steam distillation the phenol comes over. Thomas (*Analyst*, 1960, **85**, 551) has employed gas chromatography for the simultaneous determination of diphenyl and OPP.

The method of Schiffman (*J. Ass. Off. Agric. Chem.*, 1957, **40**, 238) for the determination of OPP involves steam distillation of the acidified fruit. Then the distillate is made alkaline, extracted with light petroleum to remove oils and the aqueous solution containing the phenol is clarified with Filter-Cel. The OPP is then determined colorimetrically using 4-aminoantipyrine at a controlled pH. Harvey and Penketh (*loc. cit.*) describe methods for extracting OPP from various materials and the phenol is finally measured spectrophotometrically in cyclohexane or colorimetrically in aqueous solution using the stabilised diazo compound, Brentamine fast red GG.

Thiourea in Oranges

Thiourea has been used as a rot and mould suppressant in oranges. M.O.F. Circular MF 2/54, however, stated that its use should be considered as a contravention of the Preservatives Regulations. Thiourea can be detected by the following test (A.O.A.C.):—

Pentacyanoammonioferroate reagent. Dissolve 10 g. $Na_2Fe(CN)_5NO.2H_2O$ in 40 ml. ammonia solution and allow to stand overnight at about 0° C. so that all the nitrosoferricyanide decomposes (shown when a few drops no longer produce a red colour when added to a solution of creatinine in N/1 sodium hydroxide). Filter and add absolute alcohol to the filtrate until no further precipitate forms. Filter, wash the pentacyanoammonioferroate precipitate with absolute alcohol until free from ammonia and dry in a vacuum desiccator. Store in a desiccator containing calcium chloride, in the dark. Prepare a 1 per cent. aqueous solution, expose it for one day to the air and light and store in a dark bottle in the dark.

Qualitative test. Shake some orange juice with two-thirds its volume of ether, centrifuge and separate the lower layer. Stir in Filter-Cel and transfer to a Hirsch funnel. Keep the vacuum on for a short time and agitate the solution to remove most of the ether. Then to 5 ml. of filtrate add 5 drops of the pentacyano-ammonioferroate reagent. If no blue colour appears, add a few drops of N/10 iodine dropwise and shaking after each addition. About 5 drops are usually necessary for the development of the maximum colour (excess reduces the colour). The presence of thiourea is indicated if a bluish green colour develops.

The A.O.A.C. (1960, 9th Edn., p. 401) describes a further qualitative test using a modified reagent which is also employed for the estimation of thiourea in oranges and in frozen peaches. In the quantitative method the blue colour is measured at 610 mμ (*cf. J.A.O.A.C.*, 1948, **31**, 476 and *Anal. Abs.*, 1958, **5**, No. 3522).

CANNED FRUITS AND CANNED VEGETABLES

A Code of Practice for Canned Fruit and Vegetables was approved in 1958. This describes standards of fill, composition and size of canned fruits and vegetables which have been accepted by the Fruit and Vegetable Canners' Association of Great Britain as the recognised practice of the British canning industry. The code is given in detail in *Food Trade Review*, 1959, **29** (4), 3. The code states:—

(a) Minimum filled weights for the various fruits and vegetables in the different sizes of can used.

(b) Ranges for the proportions of the vegetables and fruits in Vegetable Macedoine (Mixed Vegetables) and Fruit Salad.

(c) Standards of size of green peas and stringless beans for canned vegetables.

(d) The designations of the types of syrup to be used on labels to describe the density of the syrup at the time of canning:—

	Class A Not less than	Class B Not less than
Light Syrup	15° Brix	15° Brix
Syrup	30° ,,	22° ,,
Heavy Syrup	40° ,,	30° ,,
Extra Heavy Syrup	50° ,,	40° ,,

N.B.—Degrees Brix refers to the reading obtained on a special hydrometer calibrated in percentage sugar, *e.g.*, a 40° Brix syrup has a s.g. at 20° C. of 1·179 and contains 40 per cent. w/w of sugar.

The fruits in Class B are apricots, peaches, pears, pineapple, fruit salad and fruit cocktail. All other fruits are in Class A.

The fruit contains much less soluble solids than the syrup. During processing, therefore, the fruit dilutes the syrup. The density of the final syrup (referred to as the "cut-out" syrup) is in consequence considerably lower than that of the added syrup, *e.g.*, an added syrup of 40° Brix will usually cut-out at 22–25°. The "cut-out" syrup strength depends on several factors—the strength of the added syrup, the particular fruit and its stage of ripeness and the ratio of fruit to syrup used. Similarly the drained weight (the weight of the fruit in the product retained on a sieve after draining for two minutes) varies considerably according to various factors including the processing conditions. The original filled weight of samples of canned fruit can be assessed approximately, however, from the ranges of drained weight given below:—

Canned Fruit (packed in 40° Brix syrup)	*Drained Weight* (as % of filled weight)
Blackberries	65–90
Cherries	80–95
Gooseberries	80–95
Loganberries	65–90
Plums	70–95
Raspberries	60–90
Strawberries	55–80

N.B.—A fall of 5° Brix in the strength of the added syrup causes a drop of about 1 per cent. in the drained weight and a fall of about 2·5–3° Brix in the cut-out syrup.

In the absence of a Brix hydrometer the syrup strength can be readily checked by means of a refractometer (see table on p. 109).

Canned fruits are normally examined for vacuum, pH, added colour and metallic contamination, *e.g.*, tin (in relation to the internal appearance of the can), lead and possibly arsenic and zinc. Complaints occasionally arise alleging that cans of grapes packed in light syrup contain glass. These fragments invariably consist of crystals of argol (cream of tartar derived from the grapes) and Manley and Alcock (*Analyst*, 1950, **75**, 280) consider that their deposition is encouraged by cold storage. Dickinson and Fowler (*Analyst*, 1955, **80**, 706) have also reported the presence of "glass-like" crystals of calcium *dl*-tartrate in cans of French cherries, but their origin would appear to be rather obscure. The amount of calcium necessary to cause the formation of these crystals suggests that the cherries had been in contact with concrete at some stage.

Canned vegetables. In addition to the provisions of the Code of Practice mentioned previously (p. 204), some additional requirements of the Canned Fruit and Vegetables Order (now revoked) are of interest:—

(e) Most vegetables to be packed in a brine containing 1·25–2·5 per cent. (w/w) of salt, but peas must contain in addition at least 2·3 per cent. of sucrose.

(f) Colouring matter was only permitted in butter beans, haricot beans, stringless or runner beans, peas and spinach.

(g) Spinach purée to contain 0·1–0·6 per cent. of salt in the canned product.

(h) Canned beans in tomato sauce were required to contain at least 5 per cent. of sugar (as invert sugar) and at least 28 per cent. of solid matter when dried for sixteen hours at 98° C.

The drained weights of canned vegetables tend on the average to be just slightly greater than the filled weights (range 95–110 per cent. for most types).

Canned vegetables are normally examined for vacuum, pH, metallic contamination (arsenic, copper, lead and tin) and added colouring matter (see also *A.O.A.C.*, 1960, *9th Edn.*, *p.* 448).

Canned peas. The Labelling of Food Order states that peas which have been dried, soaked or otherwise processed prior to canning must be described as " processed peas." Also peas so treated must not be referred to as " fresh," " garden " or " green."

TOMATO PRODUCTS

Tomato Purée, Tomato Paste

Although English tomatoes have an excellent flavour, large quantities of concentrate (purée) have to be imported from Italy, Spain, Hungary and elsewhere to meet the demand for products such as tomato ketchup, canned soups and sauces for canned beans and canned fish. Raw tomatoes contain about 5 per cent. of total solids. The so-called " Double-concentrated " purée of commerce contains about 30 per cent. of total solids and for manufacturing purposes is often packed in large " 5 kilo " cans. Most imported purées have a solids content of 25–35 per cent. Some contain up to 5 per cent. of added salt.

The average composition of raw tomato and the dry solids of tomato purée is:—

	Tomato (raw)	Tomato Purée (calculated on dry solids)
	Per cent.	Per cent.
Total solids	4·7 (4·4–6·8)	100
Ash	0·45	9·5 (8·5–12·0)
Salt	0·044	0·98*
Protein	0·68	13·8
Total Sugar (as invert sugar)	2·70	54·0
Titratable acidity (as citric acid)	0·45	7·3 (5·8–13·4)
Potassium (as K_2O)	0·26	4·8

* Figure for " natural " salt only (none added).

Analysis of tomato purée

In " official " laboratories, samples of tomato purée are examined principally for **possible metallic contamination, i.e., copper, lead, arsenic and tin** (if canned). *Industrially* the examination is usually rather more searching and includes many of the following determinations:—

Total solids. The solids content can be determined by drying under partial vacuum at 70° C. (*Analyst,* 1941, **66**, 319) or by the rapid refractometric method of Sipple (*Food Res.*, 1936, **1**, 145). The U.S.F.D.A. standard for tomato paste is a minimum of 25 per cent. salt-free solids.

Ash. Use 5 g.

Salt. Determine on the ash. Tomatoes contain about 1 per cent. of chloride (as sodium chloride) in the dry solids. Any excess represents added salt.

Sugars. Boil 20 g. of sample gently with water and make up to 100 ml. at room temperature. Filter. Invert 10 ml. of filtrate with hydrochloric acid, make up to 200 ml. and titrate against Fehling's solution (Lane and Eynon's method). Report the total sugars as invert sugar.

Acidity. The total acidity can be determined by direct titration of 2 g. with N/10 alkali using phenolphthalein as indicator. Calculate as citric acid. The volatile acidity can be determined after distillation from salt.

Colour. This is normally assessed by means of the Lovibond Tintometer. Italian purées tend to be a richer red colour and the proportion of red to yellow is higher than in products from most other countries. The presence of black specks in the purée indicates inefficient sieving of the raw materials.

Howard Mould Count. See *A.O.A.C.*, 1960, *9th Edn., p.* 626.

pH Value

Other determinations on tomato purée, *e.g.*, insoluble solids, sulphate, lead numbers and potassium have been described by Morpeth, *Analyst,* 1948, **73**, 449 (see also the section on Tomato Ketchup below).

Tomato Ketchup

Tomato ketchup usually contains tomato purée, sugar, vinegar, salt, onion and garlic or other spices. Thickeners such as carob bean gum and added colour are also often present. In view of the comparatively low acidity, tomato ketchup tends to be attacked by moulds, in particular, and it is therefore filled as hot as possible before capping.

LEGAL REQUIREMENTS

The Food Standards (Tomato Ketchup) Order, 1949 (S.I. 1949 No. 1817, as amended by S.I. 1956 No. 1167) states:—

1. The standard for tomato ketchup, catsup, sauce and relish shall be as follows:—

(a) Tomato ketchup, catsup, sauce and relish shall contain not less than six per cent. by weight of tomato solids derived from clean and wholesome tomatoes or from tomato purée, or its equivalent, made from clean and wholesome tomatoes.

(b) The tomatoes, tomato purée or its equivalent or the tomato ketchup, catsup, sauce or relish shall be so strained, with or without heating, as to exclude seeds or other coarse or hard substances.

(c) Tomato ketchup, catsup, sauce and relish shall contain no fruit or vegetables other than tomatoes except onions, garlic and spices added for flavouring purposes.

2. No tomato ketchup, catsup, sauce or relish shall contain any copper in excess of 20 parts of copper per million parts of the tomato ketchup, catsup, sauce or relish, as the case may be.

Analysis of Tomato Ketchup

In "official" laboratories, samples of tomato ketchup are examined for **metallic contamination**, *i.e.*, **arsenic** (p. 56), **lead** (p. 65) and especially **copper** (p. 62), **preservatives,** especially **benzoic acid** (p. 84) and **added colours** (p. 99), **ash, salt, total solids** and the assessment of the **tomato solids.**

Tomato solids. The amount of tomato present in tomato ketchup is usually assessed from the potassium content. Morpeth (*Analyst*, 1948, **73**, 449) determines the potassium on the ash (taking 25–30 g. sample) gravimetrically as the perchlorate and gives an average figure of 4·79 per cent. (as K_2O) of the tomato solids for tomato purée. The potassium content can also be measured using a flame photometer (see *Analyst*, 1957, **82**, 200).

The (total ash minus salt) per cent. figure which represents about 8·5 per cent. of the tomato solids is also useful as a rough check, on other results. It is not usually convenient to determine the insoluble solids (average 12·6 per cent.) because of the presence of added gums, and calculations based on the non-sugar solids (46–54 per cent.) must involve an allowance for added sucrose (the percentages quoted are based on tomato solids).

Other determinations which are made, particularly in factory control, are:—

Sugars. Determine before and after inversion using Lane and Eynon's method and calculate as invert sugar.

Acidity and volatile acidity, colour in Lovibond tintometer (see also *Food Trade Rev.*, 1958, **28** (12), 7), viscosity (time to flow through a "standard" funnel), storage tests (examine after incubation for one month at 37° C.) and Howard mould count (*A.O.A.C.*, 1960, 9th Edn., p. 626).

COMPOSITION OF TOMATO KETCHUP

Most samples fall within the following ranges:—

	Per cent.
Total solids	20–40
Tomato solids	6–17
Ash	2·9–4·0
Salt	1·6–3·6
Total sugars	10–25
Total acidity (as acetic acid)	0·8–2·3
Copper	Less than 5 p.p.m.

FRUIT JUICES

Most fruit juices which appear as such on the retail market are derived from citrus fruits. After expression in a reamer the juice is strained, flash pasteurised, filled into cans or bottles and sealed. Added sugar is sometimes present. Concentrated juice is prepared by distillation under reduced pressure or by freezing. Sulphur dioxide is invariably present in the concentrated orange juice intended for children and the process is conducted so that there is the minimum loss of vitamin C. A high proportion of the fruit juice handled in Britain is used in the manufacture of squashes and some juices which have been exported to Britain for this purpose appear to have been sophisticated with materials such as liquid glucose and potassium phosphate.

The analysis of fruit juices may include the determination of the following: total solids, acidity, ash, alkalinity of the ash, phosphate, potassium, nitrogen (and the formol titration), sugars, vitamin C, trace metals and preservatives. Industrially the refractive index, the specific gravity and the pH are usually determined and the juice is examined for enzymic activity. A simple method for identifying citrus juices employing paper chromatography has been described by Dickinson and Harris (*Analyst*, 1950, **75**, 687).

Total solids. Determine preferably by drying at 70° C. under reduced pressure.

Degree of concentration is usually assessed industrially from the refractive index (or specific gravity). The following figures are extracted from the Reports of the war-time Citrus Fruit Juices Control:—

Degree of Concentration	*Grapefruit* R.I. (20° C.)	*Lemon* R.I. (20° C.)	*Orange* R.I. (20° C.)
3×	1·3756–1·3790	1·3681–1·3710	1·3816–1·3860
4×	1·3926–1·3970	1·3796–1·3820	1·3996–1·4040
5×	1·4086–1·4120	1·3911–1·3940	1·4181–1·4230
6×	1·4236–1·4270	1·4041–1·4070	1·4381–1·4430
7×	1·4391–1·4430	1·4149–1·4175	1·4566–1·4610

(*cf. Ind. Eng. Chem., Anal. Ed.*, 1939, **11**, 447.)

COMPOSITION OF FRUIT JUICES

The following table gives the composition figures obtained from various sources (*e.g.*, Stern, *Analyst*, 1943, **68**, 44 and Morgan, *Food*, 1954, **23**, 286):—

Juice	Specific Gravity	Total Solids % w/w	Acidity (as citric acid) % w/v	Ash % w/v	Alkalinity of ash (as K_2CO_3) % w/v	P_2O_5 % w/v		Vitamin C mg./100 ml.
						a	b	
Orange, maximum	—	—	3·5	0·63 (b)	0·53 (b)	0·052	0·052	80
Orange, minimum	—	—	0·4	0·29 (b)	0·24 (b)	0·028	0·019	20
Orange, mean	1·045	10·8	1·4	0·43 (b)	0·355 (b)	0·038	0·032	45
Grapefruit, maximum	—	—	3·0	0·56 (b)	0·45 (b)	0·043	0·043	65
Grapefruit, minimum	—	—	0·4	0·34 (b)	0·255 (b)	0·028	0·027	35
Grapefruit, mean	1·040	10·4	1·6	0·47 (b)	0·35 (b)	0·033	0·034	—
Lemon, maximum	1·040	11·2	—	0·56 (b)	0·415 (b)	0·042	0·042	70 (c)
Lemon, minimum	1·030	9·2	—	0·29 (b)	0·13 (b)	0·020	0·020	30 (c)
Lemon, mean	1·035	10·0	—	0·37 (b)	0·29 (b)	0·027	0·027	45 (c)
Lime, maximum	—	—	—	0·44 (a)	0·37 (a)	0·030	—	40
Lime, minimum	—	—	—	0·33 (a)	0·27 (a)	0·019	—	5
Lime, mean	1·035	9·3	7·5	0·38 (a)	0·32 (a)	0·025	—	—
Apple	1·060	13·0	0·8	0·3	—	—	—	10
Blackcurrant	1·055	13·5	3·2	—	—	—	—	200

(a) Stern's figures; (b) Combination of figures of Morgan and Stern; (c) *cf.* Morgan, *Analyst*, 1953, **78**, 323.

Acidity. For routine purposes it is usually sufficient to determine the titratable acidity (using phenolphthalein). The acidity of juices is usually calculated as the predominant acid, *i.e.*, that of citrus juices as citric acid, apple as malic acid and grape juice as tartaric acid. The flavour of fruit juice is related to the ratio of soluble solids to the total acidity and as this ratio increases as the fruit ripens it is sometimes used for assessing the quality of the juice. The volatile acidity is usually calculated as acetic acid. Methods for determining the individual acids present in fruits are described in the A.O.A.C.

Ash. The ash is usually determined on the filtered juice. The **alkalinity of the ash** is then determined by boiling with a measured excess of N/10 acid, cooling and back titrating with N/10 alkali using methyl orange (1 ml. N/10 $\equiv$ 0·0069 g. K_2CO_3).

Phosphate. For the determination of phosphate in fruit juices it is preferable to make the sample alkaline and evaporate and ash. The phosphate can then be determined either volumetrically or colorimetrically (p. 32) on the solution obtained by dissolving the alkaline ash in hydrochloric acid (*cf.* Morgan, *Food*, 1954, **23**, 286).

Potassium can be determined by the well-known chloroplatinate or cobaltinitrite gravimetric methods (details are given in the A.O.A.C.). Alternatively flame photometry may be employed.

Nitrogen. The total nitrogen is determined by the macro Kjeldahl method.

Sugars. The presence of added sucrose can be detected by determining sugars before and after inversion by Lane and Eynon's method (p. 119). Liquid glucose can be detected by means of the polarimeter.

VITAMIN C—VOLUMETRIC DETERMINATION

Standard indophenol solution. Dissolve 0·05 g. 2:6-dichlorophenolindophenol in water, dilute to 100 ml. and filter. To standardise, dissolve 0·0500 g. pure ascorbic acid in 60 ml. of 20 per cent. metaphosphoric acid and dilute with water to exactly 250 ml. Then pipette 10 ml. of this solution into a small flask and titrate with the indophenol solution until a faint pink colour persists for fifteen seconds. Express the concentration as mg. ascorbic acid equivalent to 1 ml. of the dye solution. The dye solution keeps for a few weeks if stored in the refrigerator, but it should be standardised before use against a freshly prepared solution of ascorbic acid.

METHOD. Pipette 50 ml. of unconcentrated juice (or the equivalent of concentrated juice) into a 100 ml. graduated flask, add 25 ml. of 20 per cent. metaphosphoric acid as stabilising agent and make up to the mark with water. Pipette 10 ml. of the solution into a small flask, add 2·5 ml. of acetone and titrate with the indophenol solution until a faint pink colour persists for fifteen seconds. Calculate the vitamin C content in the sample as mg. per 100 ml. (or 100 g.). The acetone may be omitted if sulphur dioxide is known to be absent. Its function is to form the acetone-bisulphite complex with sulphur dioxide which otherwise interferes with the titration. Sometimes a small proportion of the ascorbic acid in foods becomes reversibly oxidised during ageing and forms dehydroascorbic acid. If this is suspected first estimate the ascorbic acid as already described, then through another proportion of the solution pass a stream of hydrogen sulphide for ten minutes. Stopper the flask and allow it to stand overnight in a refrigerator; then remove the hydrogen sulphide by bubbling nitrogen through the mixture and titrate as before. The difference between the two titrations gives a measure of the dehydroascorbic acid. For a discussion of the possible sources of error, see Harris and Olliver (*loc. cit.*). One International Unit of vitamin C = 50 μg. of ascorbic acid. A method for determining dehydroascorbic acid alone has been described by Roe *et al.*, *J. Biol. Chem.*, 1943, **147**, 399; 1944, **152**, 511.

The vitamin C content of fruit juices can also be estimated by the absorptiometric method described on p. 201. With coloured extracts the vitamin should preferably be determined by electrometric titration (see Harris *et al.*, *Biochem, J.*, 1942, **36**, 183 and Liebmann and Ayres, *Analyst*, 1945, **70**, 411).

Parkinson (*J. Sci. Food Agric.*, 1952, **3**, 555) has shown that interference by ferrous iron in the determination of ascorbic acid in canned foods can be eliminated by passing an oxalic acid extract through Zeo-Karb 215.

Trace metals and preservatives. The routine examination of fruit juices should include examination for trace elements such as copper, lead and zinc and for preservatives such as sulphur dioxide and benzoic acid. A method for the determination of the " free " and " bound " sulphur dioxide is described on p. 81.

Cation exchange capacity of citrus juices. A rapid method for the detection of adulteration of citrus juices based on the cation exchange capacity is described in " The Bulletin of the Research Council of Israel," 1959, **7C,** No. 4.

Enzymic activity. The pectic enzymes present in the raw juice should be destroyed during the pasteurisation process. If they are not destroyed, products into which the juice is incorporated (*e.g.*, squashes) tend to go clear (*i.e.*, the desirable cloudiness is not retained) due to the action of the enzymes on the pectinous substances present. The activity can be assessed by adding the juice to a sterilised pectin solution, neutralising to pH 7·0 and then maintaining at 30° C. The acid produced after thirty minutes is then titrated with dilute standard alkali. This titration is approximately proportional to the enzymic activity.

SOFT DRINKS

Legal Requirements

The standards for soft drinks are contained in The Food Standards (Soft Drinks) Order, 1953 (S.I. 1953 No. 1828, as amended by S.I. 1954 No. 1089). The Order provides for most types of liquid soft drinks but does *not* apply to sales to a caterer in containers containing more than one gallon; *nor* to any soft drink for which justifiable medicinal claims are made, because of the presence of medicaments in sufficient quantity; *nor* to a glucose beverage which contains at least 23 per cent. (w/v) of liquid glucose (or 10 per cent. dextrose monohydrate); *nor* to a soft drink which is intended for diabetics (and is so described) and contains no added sugar.

The numerical standards for soft drinks are prescribed in The Schedule to the Order:—

THE SCHEDULE

PART I

Standard for soft drinks for consumption without dilution

Column 1	Column 2	Column 3	Column 4	Column 5
Description of soft drink	Minimum quantity of fruit juice (expressed in terms of juice of natural strength) per 10 gallons	Minimum quantity of added sugar per 10 gallons	Maximum quantity of saccharin* per 10 gallons	Other requirements as to composition
Any citrus fruit juice and barley drink	48 fluid oz.	18 oz.	82 grains	—
Lime juice and soda	48 fluid oz.	18 oz.	82 grains	—
Drinks made from whole fresh oranges	—	18 oz.	82 grains	To be made from not less than 5½ lb. of comminuted fresh oranges per 10 gallons.
Any other description of soft drink containing fruit juice	80 fluid oz.	18 oz.	82 grains	—
Indian or quinine tonic-water	—	18 oz.	82 grains	To contain not less than ½ grain of quinine (calculated as quinine sulphate B.P.) per pint.
Soda-water	—	—	—	To contain not less than 5 grains of sodium bicarbonate per pint.
Ginger beer and herbal and botanical beers and any other description of soft drink (except those mentioned in Part II of this Schedule)	—	18 oz.	82 grains	—

1. Where any soft drink referred to in the above table contains not less than 7½ lb. of sugar per 10 gallons the maximum saccharin content specified in column 4 shall not apply.

2. In the case of any soft drink referred to in the above table which is subjected to brewing in the course of its manufacture allowance shall be made, in assessing the minimum quantity of added sugar therein, for any change in sugar content due to chemical action during such brewing.

* For the purposes of the Soft Drinks Order, " Saccharin " means the substance defined as such in the B.P. 1953, which required saccharin to contain not less than 97·0 per cent. of $C_7H_5O_3NS$.

Part II

Standard for soft drinks for consumption after dilution

Column 1	Column 2	Column 3	Column 4	Column 5
Description of soft drink	Minimum quantity of fruit juice (expressed in terms of juice of natural strength) per 10 gallons	Minimum quantity of added sugar per 10 gallons	Maximum quantity of saccharin per 10 gallons	Other requirements as to composition
Any citrus fruit juice and barley drink	$1\frac{1}{2}$ gallons	$7\frac{1}{2}$ lb.	$\frac{7}{8}$ oz.	—
Drinks made from whole fresh oranges	—	$7\frac{1}{2}$ lb.	$\frac{7}{8}$ oz.	To be made from not less than $27\frac{1}{2}$ lb. of comminuted fresh oranges per 10 gallons.
Any other squash, crush, cordial or concentrate containing citrus fruit juice	$2\frac{1}{2}$ gallons	$7\frac{1}{2}$ lb.	$\frac{7}{8}$ oz.	—
Any squash, crush, cordial or concentrate containing any other fruit juice	1 gallon	$7\frac{1}{2}$ lb.	$\frac{7}{8}$ oz.	—
Any other description of soft drink (except those mentioned in Part I of this Schedule)	—	$7\frac{1}{2}$ lb.	$\frac{7}{8}$ oz.	—

FOOD STANDARDS COMMITTEE REPORT ON SOFT DRINKS

In 1959, the F.S.C. published a Report on Soft Drinks which states that " There is a strong case for continued regulations to control the description and composition of soft drinks." The Committee's principal recommendations are contained in Paragraph 68, which states:—

68. " We therefore recommend that the Food Standards (Soft Drinks) Order, 1953, should be replaced by regulations to give effect to the above conclusions and in particular:—

(a) to provide that all soft drinks containing fruit juice, with or without other constituents of fruit (including ' bitter orange ' etc., but excluding flavoured beverages) should bear on the label a declaration in prescribed form of the percentage fruit or juice content of the drink;

(b) to provide the following minimum standards of composition:—

Soft Drinks	Ingredient to be controlled	Standard for drink for consumption without dilution	Standard for drink for consumption after dilution
1. Citrus squashes	Fruit juice (per cent. vol./vol.)	5	25
2. Citrus fruit drinks	Potable fruit content (per cent. wt./vol.).	2	10
3. Citrus juice and barley drinks	Fruit juice (per cent. vol./vol.)	3	15
4. Citrus fruit and barley drinks	Potable fruit content (per cent. wt./vol.).	—	7
5. Lime juice and soda	Lime juice (per cent. vol./vol.)	3	—
6. Non-citrus fruit squashes	Fruit juice (per cent. vol./vol)	—	10
7. Indian or quinine tonic water	(a) Quinine (parts per million calculated as quinine sulphate B.P.).	57	—
	(b) Added sugar (per cent. wt./vol.).	4½	—
	(c) Saccharin (parts per million)	80 (maximum)	—
8. Soda water	Sodium bicarbonate (parts per million).	570	—

(c) to provide that the above soft drinks (but no others) should be exempt from the requirements of the Labelling of Food Order as respects declaration of ingredients;

(d) to prohibit the use of saccharin or other artificial sweetening agent in soft drinks other than diabetic soft drinks and tonic water appropriately labelled;

(e) to prohibit the use in the labelling and advertising of soft drinks of any form of testimonial or nutritional claim based on the carbohydrate content; and to prohibit the use of the term ' glucose ' in the labelling and advertising of soft drinks;

(f) to restrict the acids used for acidifying and flavouring purposes in drinks containing fruit juice, with or without other constituents of fruit, to citric, tartaric and malic acids, and in other soft drinks to those acids with the addition of lactic, acetic and phosphoric acids; and to allow the addition of ascorbic acid to drinks containing fruit juice, with or without other constituents of fruit, to restore the natural vitamin content of the relevant fruit ingredient;

(g) to apply the appropriate minimum standards to all sales of soft drinks by a caterer, and the compulsory declaration of fruit or juice content to sales by a caterer of soft drinks pre-packed in bottles or other containers;

(h) to define ' soft drinks ' and the scope of the regulations as in the present Order except as respects fruit juices, glucose beverages, catering sales and other consequential amendments."

Appendix B in the F.S.C. Report discusses the relative merits of sucrose and glucose as sources of energy. This implies that in their view the facts do not support the contention that glucose has any special advantages over other carbohydrates as a source of energy. Nine references are quoted.

LABELLING OF SOFT DRINKS

Code of Practice (C.P. 5/2) relates to the labelling of soft drinks. Its main provisions are that if a soft drink does not contain the prescribed minimum amount of fruit juice, no illustrations of fruit should appear on the label and the description should include the word " Flavour," or, in the case of ready-to-serve drinks, the suffix -ade (*e.g.*, " Orangeade "). Also descriptions including words such as " cream," " milk," " kola," " hop bitters," " dandelion and burdock " should not be used unless the ingredients named are present in the product. No exception is taken, however, to the descriptions " Cream Soda " and " Ice Cream Soda " for clear, vanilla-flavoured ready-to-drink beverages. Exception is taken to claims of nutritional, dietary, tonic, restorative or medicinal properties which cannot be substantiated. Descriptions which may be suggestive of alcoholic liquors should not be used (except " Ginger Beer," " Ginger Ale," " Non-Alcoholic Ginger Wine," " Non-Alcoholic Shandy," etc.). Full details of the above Code are given in " The Advertising, Labelling and Composition of Food " (M.O.F., 1948, H.M.S.O., Lond.).

Ready-to-drink beverages. Most ready-to-drink beverages are carbonated. These are usually prepared from a concentrated syrup containing sugar (and possibly saccharin), fruit juice (or flavouring essence), citric acid and preservative (usually sodium benzoate).

Added colouring matter is added to some drinks and products such as ginger-beer contain a foaming agent of the saponin-type. In the manufacture a definite volume of the syrup is measured into the bottle. The bottle is then filled with a mixture of water and liquid carbon dioxide and closed. The degree of carbonation employed varies according to the product.

Squashes, cordials, crushes. Concentrated soft drinks contain more or less similar ingredients to the ready-to-drink beverages but in greater amounts. They are, however, not carbonated. Lime juice cordial is filtered to a bright clear liquid before bottling, but most other products should remain cloudy on storage provided that pasteurised juice has been used (p. 212). Squashes containing fruit juice usually contain sulphur dioxide, but with " Comminuted drinks " the preservative present is usually benzoic acid.

" Bitter Orange " and " Bitter Lemon " have a basis of fruit juice or comminuted fruit and often contain some added bitter principles, *e.g.*, quinine. The F.S.C. Report recommends that such products should be regarded as squashes or comminuted drinks (as appropriate) for the purposes of the regulations.

Analysis of Soft Drinks

In general the analysis of soft drinks should include the determination of **total solids, ash, sugars** (**before and after inversion by Lane and Eynon's method**), **acidity, pH, preservatives** (**usually benzoic acid or sulphur dioxide**) **and trace metals** (*e.g.*, **Pb, Cu, As**). In appropriate instances such products should be examined for saccharin, vitamin C, added colouring and the fruit juice content should be assessed (see below). Most of the procedures which are required follow the normal pattern of general analyses and only certain special methods are described below.

SAMPLING. With carbonated beverages the froth tends to make accurate pipetting difficult and it is usually necessary to pour the sample backwards and forwards from one beaker to another before commencing the analysis. The A.O.A.C. (1960, 9th Edn., p. 119) describes a manometric method for the determination of the carbon dioxide.

BENZOIC ACID AND SACCHARIN. If a soft drink is made strongly acid any saccharin present can be readily extracted with ether. Any benzoic acid present, however, will be extracted also.

If the Monier-Williams' distillation and sublimation method (p. 85) is employed also, it will give the amount of benzoic acid present and the saccharin content can be deduced by subtraction from the total ether extract.

A simple procedure for extracting saccharin and benzoic acid separately from soft drinks has been devised by Jones, who firstly removes the benzoic acid at pH 4·0. Then the saccharin is extracted after making the liquid strongly acid.

BENZOIC ACID. To a suitable volume of sample add 20 ml. of pH 4·0 buffer solution (containing 10 per cent. sodium citrate and 6·5 per cent. citric acid) and extract the benzoic acid by shaking with three separate portions of 25 ml. of ether. Wash each extract with the same 4 ml. of water, filter the combined extracts and remove the solvent (remove the last few ml. of solvent at room temperature with the assistance of bellows). Dissolve the residue in 2 ml. acetone, add 2 ml. water and titrate the acid with N/20 sodium hydroxide using phenol red as indicator.

$$1 \text{ ml. N/20 NaOH} \equiv 0{\cdot}0061 \text{ benzoic acid.}$$

SACCHARIN. To the aqueous liquid in the separator add the wash water and 10 ml. of concentrated hydrochloric acid and extract the saccharin by shaking with three separate portions of 25 ml. of ether. Wash the combined extracts with two separate portions of 10 ml. of water and then shake the combined wash waters with ether. Combine this ether with the combined ether extracts, filter and remove the solvent. Add 3 ml. acetone to the residue, evaporate, add 2 ml. water and titrate with N/20 sodium hydroxide using bromthymol blue as indicator.

$$1 \text{ ml. N/20} \equiv 0{\cdot}00916 \text{ saccharin.}$$

The presence of saccharin should be confirmed by qualitative tests.

A rapid spectrophotometric method for determining benzoic acid has been described by English (*Analyst*, 1959, **84**, 465).

FRUIT JUICE CONTENT. Numerous methods have been suggested for the assessment of the fruit juice content of soft drinks, *e.g.*, determination of the ash, alkalinity of the ash, phosphorus, potassium, acidity, formol titration, total nitrogen and "albuminoid ammonia" (*cf. Anal. Abstr.*, 1959, **6**, No. 1925). The calculation of the fruit content from such determinations is, however, difficult due to the wide variation in the composition of natural juices and because there are as yet no generally recognised average figures. Stern's figures for the phosphate content quoted in the table on p. 210 have perhaps been the ones most frequently used by analysts up to the present time. The estimation of the fruit juice content of drinks may be further complicated if the juice or the product has been sophisticated, *e.g.*, with inorganic substances containing potassium, phosphate, etc. In cases of doubt it is advisable to carry out as many of the determinations mentioned above as possible and to compare the results with those obtained with authentic samples of natural juice.

Many of the methods described for the analysis of fruit juices (p. 209 can be applied to soft drinks and only certain special procedures are included below. The type of fruit present can be detected by means of the chromatographic method of Dickinson and Harris (*Analyst*, 1950, **75**, 687).

FORMOL TITRATION. The visual titration method described for vinegar (p. 300) is useful as a rapid routine check of the fruit content. The formol figure can be obtained more accurately, however, by means of a potentiometric titration as employed by Intonti *et al.* (*Anal. Abstr.*, 1960, **7**, No. **745**) for the determination of orange juice.

Concentrate 200 ml. of the beverage in a porcelain dish on a water-bath at 80° C. for two hours. Cool, neutralise to exactly pH 8 using firstly N/1 and then N/10 sodium hydroxide. Add 20 ml. of neutral 40 per cent. formaldehyde (formalin, previously neutralised to pH 8) and titrate to pH 8 using N/10 sodium hydroxide. The titre (ml.) divided by 2 represents the " formaldehyde number " (F) and the amount of orange juice in the sample (per cent. w/v) is given by the expression 1·05 F/1·4. The method can be readily carried out using the pH meter.

THE " ALBUMINOID AMMONIA VALUE." Mitra and Roy (*Analyst*, **1953**, **78**, **681**) have applied the determination of the " albuminoid ammonia value " in vinegar for assessing the fruit juice content of squashes, etc. The procedure for vinegar (p. 302) is used except for the following modifications:—

(a) The s.g. of soft drinks are considerably greater than 1·0 and results are expressed on the w/w basis. Therefore, 5 g. of sample is taken instead of 5 ml.

(b) After the first distillation, 4 g. of solid potassium permanganate is added as well as the 50 ml. of alkaline permanganate (excess permanganate is necessary to counteract the reducing action of the sugars).

(c) For the second distillation a trap is inserted between the distilling flask and the condenser.

As with vinegar, Mitra and Roy observed a broad correlation between the albuminoid ammonia and total nitrogen figures for various fruit juices and squashes:—

	No. of samples	Albuminoid ammonia p.p.m. as N (w/w)	Total nitrogen p.p.m. (w/w)
Lemon juice	15	294–456	800–1090
Orange juice	23	337–700	807–1530
Grape juice	7	250–360	600–790
Apple juice	4	130–210	365–650
Pineapple juice	14	260–520	540–1036
Mango juice	35	290–1000	750–2000
Tomato juice	15	333–570	806–1200
Lemon squash*	7	72–150	205–420
Orange squash*	7	100–220	200–616
Lime juice cordial*	7	80–100	185–315

* Containing at least 25 per cent. fruit juice (v/v).

Products containing lower amounts of fruit juice gave correspondingly smaller values.

COMMINUTED DRINKS

The F.S.C.'s Report on Soft Drinks draws attention to the increasing popularity of drinks (both concentrated and ready-to-drink types), which are made by comminuting the whole fruit as provided for in the Food Standards Order. Such products are, however, prepared by a number of different processes and there are considerable variations in the proportion of insoluble matter which is rejected and in the ratio of juice to other constituents remaining in the drink. In view of this and also because of the possible use of imported concentrates, the F.S.C. has recommended that the description "drinks made from whole fresh oranges" should be replaced by "orange drink," "lemon drink," etc. In addition, the present numerical provisions obscure the fact that the standard for comminuted drinks is effectively lower than the standard for citrus fruit squashes and the Report recommends that all soft drinks containing fruit juice should bear on the label a declaration of the percentage fruit or juice content. The problems of analytical control were considered by a Working Party (set up by the F.S.C.), which recommended that if there was to be a separate standard for comminuted drinks and one which would be enforceable by analysis, it would have to be expressed in terms of "potable fruit content" (*i.e.*, the total fruit constituents of the drink whether derived from juice, pith or peel) rather than the weight of fruit used in the manufacture.

The *analysis* of drinks containing comminuted fruit follows similar principles to those outlined previously for other types of soft drinks. It should be borne in mind, however, that the preservative present is usually benzoic acid rather than sulphur dioxide.

Born (*Chem. & Ind.*, 1957, No. 23, p. 734) has described a possible method for distinguishing between a drink made from whole oranges from one made with juice only. The method involves chromatography of a benzene extract (diluted with light petroleum) on an alumina column, followed by spectrophotometry of a well-defined fraction at 325 mμ. The peel constituent responsible for the absorption peak is probably a flavonoid.

Syrup of Black Currant

Syrup of Black Currant, B.P.C., is prepared by treating the pulped fruit with pectinase to destroy the pectin and then, after pressing and clarifying, the weight/ml. at 20° C. is adjusted by the addition of water to 1·042–1·062 g. Then 700 g. of sucrose is dissolved in 560 ml. of this juice and sulphurous acid is added. The B.P.C. requires Syrup of Black Currant to conform to the following standards: ascorbic acid—minimum of 0·055 per cent. w/w as $C_6H_8O_6$; weight/ml. at 20° C.—1·27–1·30 g.; and sulphur dioxide—maximum of 350 p.p.m. w/w

B.P.C. Method for Assay of Vitamin C

Mix 5 g. of sample with 25 ml. metaphosphoric acid solution (20 per cent.), add 20 ml. acetone and dilute to 100 ml. with water. To four portions, each of 3 ml. of the solution, add 0·4, 0·5, 0·6 and 0·7 ml. respectively of dye solution (0·1 per cent. 2:6-dichlorophenolindophenol, sodium salt and standardised* so that 1 ml. = 0·0002 g. ascorbic acid). Mix by means of a stream of carbon dioxide, add 3 ml. chloroform and agitate for a further fifteen seconds. Viewing the solutions against a white background, select the two which are on either side of the end point (one colourless containing x ml. dye; one pink containing x + 0·1 ml. dye). Then prepare a further six solutions containing the following volumes of dye: x, x + 0·02, x + 0·04, x + 0·06, x + 0·08, and x + 0·1 ml. The amount of dye which was added to the solution showing the faintest pink colour is equivalent to the amount of ascorbic acid present in 3 ml. of the sample solution (*cf.* Wokes *et al.*, *Quart. J. Pharm.*, 1942, **15**, 314; Evered, *Analyst*, 1960, **85**, 515).

* The dye solution should be standardised by dissolving 0·03–0·04 g. ascorbic acid in 10 ml. of metaphosphoric acid solution (20 per cent.) and diluting with water to 250 ml. Then 10 ml. of this solution is titrated with the 0·1 per cent. dye solution.

Chapter 8

BEVERAGES

Tea—Coffee—Coffee Essences—Chicory—Cocoa—Chocolate—Cocoa Butter

TEA

Legislation

Circular FSH 16/58 issued by the Ministry of Agriculture, Fisheries and Food stated that, after 1958, H.M. Customs and Excise would discontinue sampling tea on importation for adulteration and unfitness under their permissive powers in Section 229 of the Customs and Excise Act, 1952. The circular added that the powers and duties of Port Health Authorities, Food and Drugs Authorities and other Local Authorities to inspect food for these purposes remain unaffected.

Code of Practice C.P.10 states that exception is taken to claims that a tea is " digestive," or that it is recommended by the medical profession, or that it possesses nutritional or dietary qualities not to be found in other blends, or that it is composed of some particular portion of the leaf so that the blend is entirely free from tannin.

There are no legal analytical standards for tea in Britain, but the following overseas regulations are of interest:—

	Canada	Australia	Switzerland	U.S.A.
Total ash (%) .	8 (max.)	—	—	4–7
Water-soluble ash (%) . .	2·75 (min.)	—	—	—
Extractives (%) .	30 (min.)	30 (min.)	—	—
Stalks (%) .	—	—	22 (max.)	—

Definition and manufacture. Tea consists of the prepared leaves of various species of *Thea*, the commonest being *T. sinensis, T. bohea, T. viridis* and *T. assamica.* There are four stages in the manufacture of black tea, *viz.* withering (exposure on racks to soften the leaf), rolling (rupturing the cells to release the juice for fermentation), fermentation (at about 80° F. to produce changes in the tannin) and firing (drying, using hot air). Green tea is unfermented, so the fermentation process is omitted and Oolong tea is only partially fermented. Apart from the country of origin and the individual garden marks, teas are usually graded as Orange Pekoe, Broken Orange Pekoe, Broken Orange Pekoe Fannings, Pekoe, Broken Pekoe, Broken Pekoe Souchong, Fannings and Dust. The terms " Pekoe " and " Orange " refer to the young leaf. Good teas are usually " tippy," *i.e.*, they contain a good proportion of " Tips," which are the small golden pieces of leaf bud from the tip of the shoot. Teas as sold are blends of up to twenty different types of tea. Blending is decided upon in relation to the district of origin, flavour, colour, body, pungency, price and often the hardness of the water in the area in which the tea is likely to be infused.

Adulteration. In the past, tea was admixed with foreign leaves, spent tea, iron filings, etc., and green tea was often " faced " with prussian blue, etc., but, apart from some samples which contain excessive stalk, and others which are " musty," adulteration is virtually a relic of the past.

Composition. The chief constituents of tea are moisture, tannin, nitrogenous matter (including caffeine), oil, wax, inorganic matter (especially potassium salts) and fibre. The following table gives the ranges of analytical figures obtained with genuine teas and spent tea. The figures are due to the writer and various other sources:—

	†Genuine tea %	* Spent tea %
Moisture	3·9–9·5	—
Total ash	4·9–6·5	—
Water soluble ash	3·0–4·2	Below 1
Acid insoluble ash	0·1–0·4	—
Alkalinity of soluble ash (as K_2O)	1·2–2·0	Below 0·3
Extractives	38–50	10
Caffeine	1·9–3·6	Below 1
Tannin	7·3–15·1	—
Total nitrogen	5·0–6·2	—
Crude fibre	14–18	—
Ether extract	10–11	—

The constituents which are of most importance in relation to the flavour of tea are the caffeine, tannin and the volatile oil which is present to the extent of about 0·5–0·9 per cent. In good teas the ratio of caffeine to tannin is about 1 : 3.

* Spent tea (or exhausted tea) consists of the dried leaves after they have been infused.

† Figures covering ranges for teas from India, Ceylon, China and Japan. In general, Japan tea contains less moisture and extractives, but more ash than the other types.

Analysis of Tea

If a full analysis of tea is required the following should be carried out—an organoleptic and microscopical examination and the determination of moisture, ash, water soluble ash, alkalinity of soluble ash, acid insoluble ash, total nitrogen, extractives, tannin, caffeine, crude fibre, ether extract, stalks and trace elements (particularly lead). For the routine analysis of tea the determination of the **ash, alkalinity of soluble ash, stalks** and **lead** together with a **microscopical examination** should reveal most forms of adulteration.

Organoleptic examination. It is unfortunate that the chemistry of tea, like that of wines, has not yet reached the art of the taster, and tea is valued more accurately by its flavour, expressed in unscientific but well-known terms, than by the most elaborate chemical analysis. For the tasting or analysis of tea, infusions of different strength are commonly used. The professional tea taster pours 100 ml. of water which has just boiled on to 3 g. of the leaves and allows it to infuse for five minutes, then pours off the liquor. For the estimation of tannin or caffeine, three- and five-minute extractions with 2·5 per cent. of tea are frequently made as this strength corresponds approximately with the liquor usually consumed, but such figures, although useful for comparative purposes, do not bear any definite relation to the total quantities of the constituents concerned. The amounts so extracted vary from 40 to 80 per cent. of the total.

The odour of the infused tea should also be noted and any mustiness should be reported.

Microscopical examination. Soak the infused leaves in chloral hydrate solution (50 per cent.) for several hours and then bleach them with hypochlorite and examine under the microscope. With green tea, however, the leaves are ready for microscopical examination after heating them in chloral hydrate solution for a short time. Teas after clearing should be treated with phloroglucin + hydrochloric acid in order to stain the sclerenchymatous matter. The main diagnostic characters are the long hairs and the somewhat peculiarly shaped idioblasts (both of which stain with phloroglucin + HCl), the stomata surrounded by tangentially elongated cells and the cluster crystals of calcium oxalate (Fig. 33).

Moisture. Dry 5 g. at 100° C. for one and a half hours. Tea with a moisture content in excess of about 11 per cent. has a tendency to go mouldy and consequently produce a musty infusion.

Ash. Ignite at as low a temperature as possible so that a greyish ash (not green) is produced. The total ash should preferably not

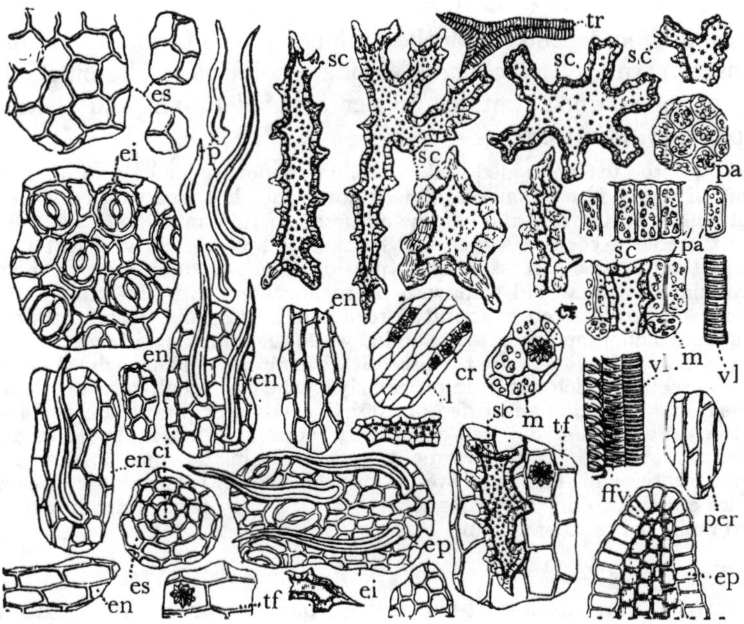

Fig. 33. Microscopical features of Tea (reproduced from Greenish and Collin's "Anatomical Atlas of Vegetable Powders," 1904, Churchill, Lond.).

cr, Crystals.
ei, Lower epidermis.
en, Neural epidermis.
ep, Apex of marginal tooth.
es, Upper epidermis.
ffv, Debris of fibrovascular bundles.
l, Bast with cluster crystals.
m, Spongy parenchyma.
p, Simple hairs.
pa, p'a', Palisade cells.
per, Pericycle, slightly lignified.
sc, Idioblasts from the mesophyll and cortical tissue.
s'c', Idioblasts from the pitch of the stem.
tf, Cortical tissue.
tr, Tracheids.
vl, Vessel.

exceed 7 per cent. The main constituents of the ash are potassium (27–36 per cent. as K_2O) and phosphorus (14–18 per cent. as P_2O_5).

Water soluble ash, alkalinity of the soluble ash, acid insoluble ash. Determine by the usual methods. It should be noted, however, that if the original ash is green (see above) the filtrate after treating with water is usually pink and this is liable to interfere with the methyl orange end-point when titrating the alkalinity. The pink colour due to manganese compounds can be removed, however, by evaporating down the green ash with water on the water-bath until no further colour is produced. The alkalinity of the soluble ash which is usually calculated as K_2O (1 ml. N/10 ≡ 0·0047 g. K_2O) is useful for the detection of the presence of exhausted leaves (see Table on page 221).

TANNIN—METHOD I (Smith, *Analyst*, 1913, **38**, 312)

This method depends upon the complete precipitation of tannin by means of saturated cinchonine sulphate solution. The cinchonine tannate so obtained contains 55 per cent. of tannin and therefore 4·3 per cent. of nitrogen:—

Ten g. of the tea are boiled under a reflux condenser for half an hour with 800 ml. of water, filtered on a Büchner funnel while hot, and washed with 200 ml. of boiling water, and the volume adjusted to 1000 ml. when cold. Fifty ml. of the cold decoction, which will be cloudy, are diluted with water, and extracted with three successive quantities of 30 ml. of chloroform to remove the caffeine, which would otherwise be precipitated with the tannin. The aqueous solution is then evaporated to about 20 ml. and 50 ml. saturated solution of cinchonine sulphate are added while the liquid is still hot and clear. After standing for some hours the precipitate is filtered off through asbestos in a Gooch crucible which has been previously washed with half-saturated cinchonine sulphate solution and dried at 100° C. The precipitate also is washed with the same solution and well drained by suction. Then it should be dried over sulphuric acid *in vacuo* overnight, and finally in the oven at 100° C. If the precipitate, while still quite wet, is put directly into the oven it fuses and becomes difficult to dry to constant weight. The dry precipitate is assumed to contain 55 per cent. of tannin.

TANNIN—METHOD II

Proctor's modification of the Lowenthal permanganate oxidation process is as follows:—

REAGENTS

(1) Standard potassium permanganate solution containing about 1 g. per litre, standardised against 0·1 N oxalic acid; (2) 2 per cent. solution of gelatin; (3) solution of 5 g. of indigo-carmine in 1 litre of water containing 50 ml. of sulphuric acid.

METHOD

Three titrations are made, working on a 1 per cent. extract of the tea; (a) dilute 4 ml. of extract to about 500 ml. with water, add 20 ml. of the indigo solution and run into the potassium permanganate solution until the indigo is bleached and a permanent pink colour remains which is visible at the margin of the yellow solution; (b) a similar titration is made on 40 ml. of the indigo solution diluted to 500 ml.; (c) 8 ml. of the tea extract are added to 25 ml. of the gelatin solution and an excess of ordinary salt is added, together with 10 ml. of dilute sulphuric acid. The mixture is diluted, shaken with a little kieselguhr, filtered, and the precipitate washed with water. To the filtrate is added 40 ml. of indigo solution; then it is titrated with the permanganate solution. For calculation of the result it is clear that $2a - b$ represents the volume of permanganate required to oxidise the tannin and other oxidisable matter in 0·08 g. of the tea, and that c represents the permanganate required to oxidise the non-tannin matters in 0·08 g. of tea and 40 ml. of the indigo solution. Hence, $2a - c$ represents the volume required to oxidise the tannin in 0·08 g. of the tea. It is important for accuracy that the amount of permanganate used to oxidise the tannin and other matters in the tea, *i.e.*, $a - b$, should not exceed one-third of b; if it does so, a less quantity of tea extract should be taken. The results are returned in terms of oxalic acid which has an equivalent of 63, and as it happens that the actual reduction equivalent of tea tannin is about 62·3, the results so expressed are almost identical with those obtained by calculation as tannins.

CAFFEINE—(A.O.A.C.)

To 5 g. of material in a 500 ml. flask add 10 g. of heavy magnesium oxide and 200 ml. of water. Boil gently under a reflux condenser for two hours; then cool, dilute to 500 ml., and filter through dry paper. Of the filtrate 300 ml. are boiled in a large flask with 10–20 ml. of 10 per cent. sulphuric acid until the volume is reduced to about 100 ml. The liquid is now filtered into a separating funnel, the flask and filter being washed with 1 per cent. sulphuric acid, and shaken out with six successive portions of chloroform, 25, 20, 15, 10, 10, 10 ml. portions being used. The combined extracts are treated with 5 ml. of 1 per cent. potassium hydroxide solution and the latter washed with a further 10 ml. of chloroform. The total chloroform extract is evaporated in a tared flask and the residue dried in the water-oven and weighed as caffeine. The residue is then transferred to a Kjeldahl flask with small portions of sulphuric acid, and the nitrogen determined in the usual way, the factor used being N $\times$ 3·464 = caffeine (*cf. Analyst,* 1910, **35**, 103).

Crude fibre. See page 41

Ether extract. Dry 2 g. of sample at 100° C. and extract with light petroleum for some hours in a continuous extractor.

Stalk (Beeson's method). Boil 5 g. for fifteen minutes with 500 ml. water. Transfer to a large basin and pick out the stalks with forceps. Dry the leaves and stalks separately in the 100° C. oven and weigh. The proportion of stalk should preferably be below 25 per cent. It is useful to examine the wet leaves and stalks for adulteration under a hand lens when the parts are being separated.

COFFEE

Legislation on pure coffee. There are no numerical standards prescribed in Britain in relation to pure coffee, but Code of Practice C.P.10 states there are no grounds for describing coffee as digestive or as having nutritional properties other than those of a stimulant.

Definition, preparation and roasting. Most commercial coffee is derived from the prepared seeds of various species of *Coffea, e.g., C. arabica, C. liberica.* Coffee is grown in East Africa, Brazil, India and other tropical countries. The fruits are often called " cherries " which they resemble and the outer layers of the fruit are removed after washing and drying. Most of the " parchment " (the hard inner pericarp which encloses the two hemispherical seeds in the fruit) is finally removed by the use of exhaust fans and sieving. After removing worthless beans, the coffee is graded. It arrives in this country as green coffee, but many types have a greenish, bluish or brownish tinge. On the flattened side there is a longitudinal cleft within which there is often a little parchment retained. The green beans are roasted by heating to 350–450° F. for 15–20 minutes and the seed increases in size, due to the production within the bean of carbon dioxide which acts as a preservative until it is released by grinding. Consequently, ground coffee is often packed under vacuum or in the presence of an inert gas.

Adulteration. Formerly, coffee was sometimes adulterated with roasted chicory, dandelion and other roots, acorns, figs, date stones and cereals. Most of the adulterants have considerably higher water extracts than coffee. Also the roots tend to yield a higher acid insoluble ash, contain more fibre and pentosans, but contain less oil.

Composition. Raw coffee contains water, protein, the alkaloids caffeine and coffearine, oil, sugar, dextrins, pentosans, cellulose, " caffetannic acid " (chlorogenic and coffalic acids), ash and various other constituents. During roasting, most of the water is lost, the carbohydrates are partially caramelised, the caffetannic acid falls to about one half and there is a slight loss of caffeine. In view of their importance in the preparation of the beverage, and in considering losses during storage, the volatile constituents are also of interest. These include pyridine, furfural, aldehydes, diacetyl, acetylmethyl carbonol, acetone, volatile phenols and hydrogen sulphide.

Analytical figures from various sources are given below for raw and roasted coffee, roasted chicory and dandelion root.

	Raw coffee			Roasted coffee			Roasted chicory			Roasted dandelion
	Min.	Max.	Aver.	Min.	Max.	Aver.	Min.	Max.	Aver.	Figs. for one sample
Moisture	8·2	13·8	10·3	0·3	5·6	2·2	2·5	12·0	5·5	10·0
Total ash	3·0	4·5	4·0	3·4	4·9	4·3	4·0	6·7	5·0	2·0
Water soluble ash	—	—	2·9	3·0	3·6	3·2	1·6	3·3	2·8	—
Ditto (as % of total ash)	—	—	—	65	85	75	—	—	55	—
Alky. of sol. ash (as K_2CO_3)	—	—	—	1·9	3·2	2·4	—	—	—	—
Acid insoluble ash (as % of total ash)	—	—	—	—	—	Ngl.	10	35	20	—
Crude fibre	—	—	—	10·5	15·3	13·0	—	—	6·9	—
Tannin	—	—	9·0	—	—	4·6	—	—	—	—
Total nitrogen	—	—	2·7	2·1	3·3	2·6	—	—	1·4	—
Caffeine	1·1	1·8	1·3	0·9	1·8	1·2	—	—	Nil	Nil
Starch	—	—	—	0·9	3·5	2·3	—	—	2·1	—
Ether extract	11·4	13·7	12·2	8·0	14·2	13·5	0·9	3·9	2·1	—
Extractives of dry sample*	25	34	—	23	33	*	70	78	*	50

* Depends to some extent on extraction method (see discussion below).

It has been shown that there is loss of caffeine on roasting amounting to about 5 per cent. of the total present, but as there is also loss of water, the amount found in the roasted coffee varies little from that in the original bean. The sugars present amount to about 4 per cent. as dextrose in the raw berries and 1·5–2 per cent. after roasting. According to Tatlock and Thomson (*J.S.C.I.*, 1910, **29**, 138) the fixed oil of coffee has the following characteristics: iodine value 99, saponification value 179·5, unsaponifiable matter (%) 5·2, s.g. 0·935.

STANDARDS FOR FRENCH COFFEE AND VIENNESE COFFEE

The Food Standards (Coffee Mixtures) Order, 1952 (S.I. 1952 No. 1697) requires coffee and chicory mixtures, including French coffee, to contain only coffee and chicory and states a minimum of 51 per cent. w/w of coffee. Coffee with fig flavour or seasoning, including Viennese coffee, is required to contain only coffee and figs, and the Order states a minimum of 85 per cent. w/w of coffee. The Labelling of Food Order requires the words " French Coffee " to be followed by the words " coffee and chicory mixture " and the words " Viennese Coffee " to be followed by the words " coffee with fig seasoning (or flavouring)."

Analysis of Coffee or Coffee and Chicory

The determinations which are necessary for the complete examination of samples of coffee and of coffee and chicory can be readily deduced from the left hand column of the table opposite. For these and for the determinations of trace metals, the usual procedures can be followed and only those which are especially applicable are included below. For routine purposes, all samples should be examined microscopically and the amount of soluble extractives should be determined.

MICROSCOPICAL EXAMINATION

Boil the sample with water, " clear " (if necessary) with chloral hydrate solution and then stain with phloroglucinol and hydrochloric acid (see p. 233). The main microscopical features of coffee, chicory and dandelion are as follows:—

Coffee. Longitudinal and transverse sclerenchymatous fibres of the pericarp.

Chicory. Exceptionally large vessels, up to 115 μ with rather short pits.

Dandelion. Concentric rings due to groups of sieve tubes; vessels have long pits and are rather more numerous than in chicory.

Moisture. Dry 5 g. at 100° C. to constant weight. The moisture content of roasted coffee should preferably not exceed 5 per cent.

Chicory. The most reliable method for detecting chicory is by means of the microscope (see above). A simple but satisfactory test may be made by gently sprinkling the powdered sample on the surface of water in a tall cylinder. The coffee floats on the surface, but the chicory begins to sink down within a few seconds. Most of the falling chicory particles leave behind them a trail of colour due to the large amount of caramel they contain.

Most of the methods which have been proposed for the determination of chicory in a mixture with coffee, are based on the fact that the hot water extract of chicory is nearly three times that of coffee. Jones' method gives reasonably good results:—

Boil 5 g. of the dried sample with 200 ml. water for fifteen minutes and pour the liquor into a 400 ml. beaker. Then re-boil the residue with 50 ml. water, mix the extracts, cool and filter into a 250 ml. graduated flask. Make the filtrate up to 250 ml., pipette 50 ml. into a weighed dish, evaporate to dryness on the water-bath, dry in the 100° C. oven and weigh.

From the weight of the residue, calculate the percentage of soluble matter (E) in the original dried sample. Then if E_1 and E_2 are the average water extracts of coffee and chicory respectively:—

$$\text{Chicory } (\%) = \frac{(E - E_1)}{(E_2 - E_1)} \times 100$$

For many years, figures of 24 and 70 were recommended as the average values of E_1 and E_2 respectively. It seems possible, however, that somewhat higher average figures would now be more appropriate. It should be borne in mind also that the proportion of soluble matter extracted depends on the particle size of the sample used and the extraction technique employed. Hughes and Wise (*J.S.C.I.*, 1934, **53**, 189 T) therefore recommended that a standardised procedure be used:—

Grind the sample so that it passes through a 40-mesh sieve (the authors found that grinding below this size did not cause any increase in the amount of material extracted). Place 5 g. of sample in a copper gauze basket (2·5 × 4 cm.) fitted with three springs at the top which grip the inner walls of a 300 ml. Pyrex boiling tube. Add a glass bead to the tube, place the basket near the top of the tube and run in from a pipette 100 ml. of water (through the sample). Then weigh the tube and push the basket down to a position rather less than 1 cm. from the water surface. Fit a reflux condenser and bring the liquid gently to the boil. After about five minutes, adjust the flame so that the froth on the surface just rises above the top of the basket. Continue with gentle, steady boiling for thirty minutes. Cool, weigh and add water, if necessary, to restore the original weight and then filter and determine the specific gravity at 15·5° C. Adjust the result to the dry weight basis, correcting for the moisture content of the sample (W) by means of the following expression:—

$$(\text{s.g.} - 1) \times \frac{100}{100 - W} + 1$$

Using the same extraction method, the authors found that the average specific gravities ($d^{15\cdot5}_{15\cdot5}$) of 5 per cent. extracts of dry roasted coffee and chicory were 1·0058 and 1·0143 respectively. The specific gravity can be calculated from the total solids of the extract by means of the expression: S.G. $= 1 + 0{\cdot}004S$, where S is the total solids of the extract (as % w/v).

In a later paper from the same laboratories (*Chem. & Ind.*, 1955, No. 20, p. 549) it has been pointed out that, on the average, 5 per cent. extracts of the types of coffee and chicory which are now being more commonly used, have higher specific gravities, *i.e.*, they yield a higher proportion of soluble matter than the types used formerly. The average figures given in the two papers for the extracts of dry coffee and chicory using the above extraction method are as follows:—

Sample	S.G. ($d^{15\cdot5}_{15\cdot5}$) of 5% dry extract	Equivalent to:— Soluble matter (as % w/w of sample)	Reference date
Coffee . .	1·0058	29	1934
,, . .	1·0064	32	1955
Chicory . .	1·0143	71·5	1934
,, . .	1·0154	77	1955

Most of the carbohydrate in chicory root is present in the form of inulin.

Caffeine

Extract 10 g. of the sample with boiling water under reflux for five hours, evaporate the extract down to fairly low bulk and transfer to a 100 ml. graduated flask (total volume 85 ml.). Add 5 ml. of each of the " zinc ferrocyanide " clearing agents (see p. 99), make up to the mark, mix and filter into a 100 ml. measuring cylinder. Measure the volume of filtrate, transfer to a separator, add ammonia and extract the caffeine as described under coffee extract.

Trace Metals

Samples of coffee may require to be examined for copper, arsenic and lead. Atkinson *et al.* (*J. Sci. Food Agric.*, 1950, **1**, 264) have studied the possible sources of the relatively high amounts of arsenic which have been found in chicory root.

Liquid Coffee Extracts

Liquid coffee extracts (or essences) consist of a mixture of a concentrated decoction of coffee with added sucrose. Caramel is often present.

Legislation

Standards for liquid coffee (and coffee and chicory) extracts are prescribed in the Food Standards (Liquid Coffee Essences) Order, 1945 (S.R. & O. 1945 No. 389 as amended by S.R. & O. 1946 No. 157):—

	Caffeine (% w/v) (minima)	Coffee (lb./gallon) * (minima)
Coffee extract . . .	0·5	4
Coffee and chicory extract .	0·25	2

* Represents a defence if the defendant proves that not less than these amounts have been used. This allows for the use of coffees with a low caffeine content.

The Preservatives Regulations permit " coffee extract " to contain up to 450 p.p.m. of benzoic acid, but its presence must be declared on the label.

Analysis of Coffee Extracts

The analysis of liquid coffee extracts may include the following: the determination of **total solids, ash, sugars, caffeine,** and **trace metals** (*e.g.*, **As, Cu, Pb**) and examination for **preservatives** (*e.g.*, **benzoic and *p*-hydroxy benzoic acids**). Industrially, the solids content is usually assessed from the specific gravity or the refractive index and the pH value is also determined.

Caffeine. Mix 20 ml. of coffee extract with 5 ml. water in a 100 ml. graduated flask. Add 5 ml. of *zinc acetate solution and then 5 ml. of *potassium ferrocyanide solution, dilute to the mark and shake. Allow to stand for five minutes and filter through a rapid filter paper. Measure the volume of filtrate, transfer to a separator, add 10 ml. of 0·88 ammonia and extract five times with 25 ml. portions of chloroform. Shake the bulked extracts first with 10 ml. N/1 sodium hydroxide solution, and then with 10 ml. of water. Shake the wash liquors with a little chloroform and add the lower layer to the bulked chloroform extracts. Filter the chloroformic solution into a weighed flask; distil off the solvent, dry at 100° C. and weigh the caffeine. If the extract is coloured, dissolve in a little hot water, cool and wash into a 100 ml. Kjeldahl digestion flask with 10 ml. of sulphuric acid (conc.). Add 2 g. sodium sulphate and 0·2 g. copper sulphate, digest, make alkaline and distil into boric acid solution in the usual way (p. 23). Titrate the ammonia with N/10 hydrochloric acid (1 ml. N/10 ≡ 4·855 mg. anhydrous caffeine). Instead of using ammonia, Harris (*J. Sci. Food Agric.*, 1953, **4**, 205) extracts the caffeine in the presence of a phosphate buffer (19 per cent. w/v $Na_3PO_4 . 12\ H_2O$) at pH 11 and the bulked chloroform extracts are washed with N/1 sulphuric acid.

Prange and Walther (*Analyt. Abstr.*, 1957, **4**, No. 1681) have suggested that the final volumetric determination can be hastened by converting the extracted caffeine to the periodide, which is dissolved in methanol and titrated with thiosulphate.

Sugars. Determine the sugars after clarifying by Lane and Eynon's method (p. 119). It should be borne in mind that roasted coffee and chicory contain, on the average, about 0·7 and 20 per cent. of reducing sugars, calculated as dextrose.

Composition of Coffee Extracts

The amounts of coffee and chicory which have been used in the preparation of liquid extracts, can be assessed (approximately) from the determinations of caffeine, sugars as dextrose, total solids and ash. Tatlock and Thomson (*J. Soc. Chem. Ind.*, 1910, **29**, 138) observed that whatever the degree of extraction of coffee, the ratio of caffeine and ash to total extract is constant. The coffee equivalent to the caffeine found is first calculated on the average basis of 1·25 per cent. Then the ash due to the coffee is taken as 4·2 per cent. of the coffee and the excess of ash calculated as due to chicory which is assumed

* See p. 111.

to have an average ash of 5·2 per cent. The following are examples of essences examined in this manner:—

	1	2.	3.	4.	5.
	Per cent.	Per cent.	Per cent.	Per cent.	Per cent.
Water	34·90	35·25	55·57	33·79	36·66
Sugars (as dextrose) . .	45·60	60·00	37·00	41·20	40·35
Organic matters not sugars	17·35	2·50	5·70	22·56	21·42
Ash	1·85	2·10	1·50	2·19	1·34
Caffeine	0·30	0·15	0·23	0·26	0·23
Coffee (dry) . . .	25	12	20	22	20
Chicory (dry) . . .	16	32	13	25	10

From the analysis of decoctions of various samples of coffee, Edwards and Nanji (*Analyst,* 1937, **62**, 841) found that the following ratios applied:—

	Ratios for genuine coffee decoctions	*Suggested limiting ratio (max.)*
Extractives/caffeine . .	12·9/1 to 20·6/1	24/1
Extractives/ash . .	about 6/1	7/1
Ash/caffeine . . .	2·57/1 to 3·04/1	4/1

The authors state that the above limiting values are generous and therefore, if they are exceeded, adulteration is probable. It should be borne in mind, however, that the presence of added sugar causes an increase in the extractives figure and in the first two ratios, but that the ash/caffeine ratio remains virtually unaltered. For the application of the use of the above ratios for the detection of adulteration and the calculation of the proportion of chicory which may have been used for preparing decoctions, the original paper should be consulted.

Chicory root contains a relatively large proportion of inulin which yields fructose on hydrolysis. It has been suggested, therefore, that this constituent may be useful for assessing the amount of chicory present. In this connection, the following references should be useful: *Analyt. Abstr.*, 1960, **7**, No. 2976; *Analyst,* 1953, **78**, 287.

Solid Coffee Extracts ("Instant Coffees")

Solid coffee extracts are prepared essentially by drying coffee decoctions. A few products are made by mixing liquid glucose with the decoction before drying and solid extracts of coffee and chicory are also prepared. In order to conserve the flavour in the cup, the temperatures used during processing are kept as low as possible to reduce the loss of volatile constituents. The product tends to be hygroscopic and is packed into airtight tins in rooms which are maintained under conditions of low humidity.

The general analysis of solid coffee extracts may include the determinations of *moisture, ash, alkalinity of the soluble ash, sugars and caffeine* and an examination for *trace metals* (*e.g.*, As, Cu, Pb) and *sulphur dioxide.* For the determination of caffeine, dissolve 5 g. in hot water, transfer to a 100 ml. graduated flask, add " zinc ferrocyanide," make up to the mark, filter, and extract as described for liquid extracts. Sugars should preferably be determined polarimetrically.

Most " Instant Coffees " give figures which fall within the following ranges: moisture 2–4 per cent., ash 7–11 per cent., alkalinity of soluble ash 8–12 ml. N/10 acid per g. and caffeine 2·8–5·0 per cent. Extracts prepared from coffee and chicory give lower figures for ash (3–4 per cent), alkalinity of the soluble ash (3·5–4·5 ml. N/10/g.) and caffeine (about 0·5 per cent.).

COCOA

Cocoa consists of the prepared seeds of *Theobroma cacao*, which is a native of tropical America. Most commercial cocoa comes from the Gold Coast, Nigeria, Brazil, the West Indies, Java and Ceylon. The mature orange to red fruit is 15–20 cm. long, has a shape resembling a pointed vegetable marrow, and contains thirty to sixty seeds embedded in a mucilaginous pulp. After fermentation in boxes, the separated seeds (" beans ") are dried to a moisture content of 6–8 per cent. The colour and taste of the final beans is influenced considerably by the method of fermentation.

Cocoa beans are 20–30 mm. long, flattened-ovoid and chocolate brown in colour. The kernel consists of two folded chocolate coloured cotyledons, which can be easily broken into angular fragments (nibs).

In the manufacture of cocoa, the beans are roasted at 200°–280° F. for a quarter of an hour to two hours and during roasting the moisture content falls to 2·5–4 per cent.

The seedcoat or shell which represents about 12 per cent. of the weight of the roasted bean, becomes loosened preparatory to its removal by winnowing. The roasted nib is ground to a semi-plastic substance which is referred to as chocolate liquor or cacao mass, containing 50–58 per cent. of fat (cocoa butter). Then the cacao liquor (heated below 200° F.) is mixed with a concentrated solution of potassium or sodium carbonate and most of the moisture is dried off at about 230° F. Then about half of the fat is pressed out from the hot liquor. Cocoa powder is produced by pulverising the resultant press cake which normally contains 20–25 per cent. of fat. If the cocoa butter is not partially removed, the resultant beverage is too rich and there is excessive separation of globules of fat. The colour of the final product varies with the fat content, *i.e.*, the higher the fat content the richer and darker is the colour of the cocoa powder. **" Alkalisation "** is the name given to the treatment with alkali. Just over one per cent. of potassium carbonate is normally used, and cocoas treated in this way are often referred to as " soluble cocoas." There does not appear to be an increase in solubility, but the particles remain in suspension longer. Advocates of this somewhat controversial treatment claim that the presence of alkali increases the digestibility of the proteins. Details of the manufacture of cocoa and chocolate are given by Jensen (*The Chemistry, Flavouring and Manufacture of Chocolate Confectionery and Cocoa*, 1931, Churchill, Lond.) and by Chatt (*Cocoa*, 1953, Intersci., N.Y. & Lond.).

Code of Practice (**C.P.1**) states that " **Cocoa** " manufactured for drinking purposes should contain at least 20 per cent. of cocoa butter, and not more than 5 per cent. of shell. It may contain alkali and small quantities of salt and/or other flavouring material.

The code also states that " **Drinking Chocolate** " in solid form consists only of cocoa and sugar, to which milk (F.C. or skimmed) and a small quantity of flavouring materials may be added. If in liquid form the sugar ingredient may be present as syrup from cane or beet sugar or invert syrup or liquid glucose.

Composition

The main constituents of cocoa are moisture, tannin (including the anthocyanin pigment " cacao red "), nitrogenous matter (proteins, theobromine, caffeine, etc.), starch and other carbohydrates, cocoa butter and inorganic matter (especially potassium and phosphorus).

The following table gives analytical figures (mostly due to Winton, Silverman and Bailey) for roasted cocoa nib, roasted cocoa shells and commercial cocoa. All figures are expressed as percentages on the whole sample:—

	Cocoa nib			Cocoa shell			Commercial cocoa		
	min.	max.	aver.	min.	max.	aver.	min.	max.	aver.
Moisture	2·3	3·2	2·7	3·7	6·6	4·9	2·7	7·8	5
Fat	48	57	52	1·7	5·9	2·8	19	28	25
Total ash	2·6	42	3·3	7·1	20·7	10·5	3·8	8·9	5·4
Water soluble ash	0·7	1·9	1·2	2	5·7	3·7	3·2	5·7	4
Alkalinity of sol. ash (as K_2CO_3)	1	2·3	1·7	3·5	4·1	3·8	1·3	4·8	2·5
Acid insoluble ash	Nil	0·07	0·02	0·05	11·2	2·5	0·01	1·4	0·24
Crude fibre	2·2	3·2	2·6	12·8	19·2	16·6	2·3	5·2	3·9
Total nitrogen	2·2	2·5	2·4	1·7	3·2	2·3	2·1	3·7	2·7
Theobromine	0·8	1·3	1	0·2	0·9	0·5	0·8	1·6	1·2
Caffeine	0·1	0·7	0·4	0·04	0·3	0·16	0·04	0·3	0·2
Starch	6·5	9	8·1	3·4	5·2	4·1	8·7	13·5	11·2
Cold water extract	9	14	11	20	25	23	15	22	19
Amount passing through 100 mesh sieve	—	—	—	—	—	—	85	97	92

Analysis of Cocoa

The routine analysis of cocoa should normally include a microscopical examination and the determination of **moisture, ash, alkalinity of the ash, fat, crude fibre, arsenic** and **lead.** Other determinations which may be necessary are those of **soluble ash, acid insoluble ash, starch, nitrogen, theobromine, caffeine** and the **cold water extract.**

Microscopical Examination

The most characteristic microscopical features of cocoa are the brown pigment, the almost colourless small starch grains (about 7 μ), the " club-shaped " hairs and the fat crystals (Fig. 34).

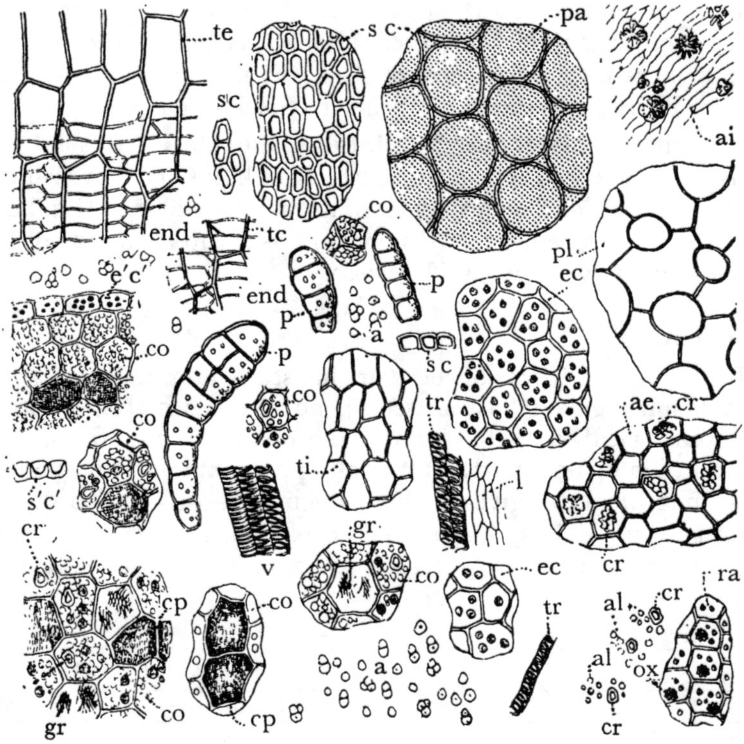

Fig. 34. Microscopical features of Cocoa (from Greenish and Collin; *loc. cit.*)

a, Starch grains.
ae, Outer layer of endosperm.
ai, Inner layer of endosperm.
al, Aleurone grains.
co, Cotyledon.
cp, Pigment cells containing cocoa-red.
cr, Crystals of fat.
ec, Epidermis of cotyledon, surface view.
e'c', Epidermis of cotyledon, profile.
end, Inner epidermis of pericarp.
gr, Crystals of fat.
l, Blast from fibro-vascular bundles.
ox, Calcium oxalate crystals.
p, Pluricellular hairs.
pa, pl, Parenchyma of seed-coat.
ra, Cells of radicle.
sc, Sclerenchymatous layer of seed-coat, surface view.
s'c', Sclerenchymatous layer of seed-coat, in profile.
te, Outer epidermis of seed-coat to which the inner epidermis of the pericarp (end) is adhering.
ti, Inner epidermis of seed-coat.
tr, v, Vessels, etc., from fibro-vascular bundle.

The fact that the husk contains considerably more mucilage, spiral vessels and sclerenchymatous cells than the nib can be used to assess approximately the proportion of shell present. The pieces of mucilage are stained pink by ruthenium red (0·02 per cent. in 10 per cent. lead acetate).

Moisture. Dry 5 g. over sand at 100° C. for four hours. The moisture content should preferably not exceed 6 per cent.

Ash. Ignite 5 g. at 550° C. The ash of unalkalised cocoa seldom exceeds 5 per cent. Alkalisation of course raises the ash, the soluble ash and the alkalinity. South Africa and Australia have adapted a limit of 8 per cent. of total ash calculated on the dry fat-free material.

Fat. Extract the fat with light petroleum in a Soxhlet extractor for twelve hours, regrind the residue and continue the extraction for a further three hours. With more efficient extractors of the Bailey-Walker or Bolton-Revis type, the total extraction time can be reduced to six to eight hours. Diethyl ether should not be used as it tends to dissolve theobromine and caffeine as well as fat.

The fat can also be determined by this modification of the Rose-Gottlieb method:—

Heat 0·5 g. of cocoa with 5 ml. alcohol for ten minutes on a water-bath at 75° C. Cool, shake well with 12·5 ml. diethyl ether, add 0·5 ml. of dilute ammonia (10 per cent. w/v NH_3) and shake again. Add 4·5 ml. water and 12·5 ml. light petroleum and shake gently. Syphon off the upper layers into a weighed flask, repeat the extraction three more times with mixed ethers, remove the solvent and weigh the fat in the usual way (p. 31 (*cf.* Kobe, *Analyt. Chem.*, 1950, **22**, 700).

The Code of Practice for cocoa states a minimum of 20 per cent. cocoa butter. The fat may be examined by the methods given on p. 227. Bywaters *et al.* (*Analyst,* 1927, **52**, 324) describe a method for detecting illipé fat.

THEOBROMINE—METHOD I

The A.O.A.C. procedure for determining theobromine using tetrachloroethane as solvent is essentially that originally due to Wadsworth (*Analyst,* 1921, **46**, 35):—

Mix 10 g. of cocoa with 2·5 g. freshly calcined magnesium oxide in a small porcelain dish, and then triturate with 9–20 ml. of water until all the particles are wetted. Place the dish on a water-bath for thirty minutes, mixing frequently to prevent any part from becoming perfectly dry. Then triturate the mixture well so that every particle is damp. Transfer to a 250 ml. flask (A), add 150 ml. tetrachloroethane, attach a reflux air condenser and boil for thirty minutes. Filter whilst almost boiling, into a larger flask (B), return the residue and filter paper to A and reflux again for twenty minutes with a further 120 ml. of solvent. Filter into B, etc., and then repeat the processes of refluxing and filtering twice more. Distil off the solvent from the combined filtrates until the volume is reduced to 3–5 ml. To the cold residue add, with rotation, 65 ml. of diethyl ether, mix well, stopper and allow to stand overnight. Filter off the precipitate on a tared filter paper, wash with several 6 ml. portions of ether, dry at 100° C., cool and weigh. Add 0·004 g. to the weight found to compensate for the solubility of the precipitate in ether and calculate the result as a percentage of theobromine in the original sample.

By employing a modified Soxhlet apparatus, Lowe (*Analyst,* 1949, **73**, 679) found that the extraction can be completed in one hour.

Theobromine—Method II

Gerritsma and Koers (*Analyst,* 1953, **78**, 201) have criticised Wadsworth's and other methods for the determination of theobromine and have described a relatively simple method for cocoa residues. The theobromine is extracted by shaking with chloroform and determined volumetrically by the method of Boie (*Pharm. Ztg.*, 1930, **75**, 968):—

To a 500 ml. bottle add 2 g. of sample, 270 ml. of chloroform and then 10 ml. of ammonia solution (10 per cent. w/w NH_3). Stopper and shake vigorously for five minutes. Add 12 g. of anhydrous sodium sulphate, shake again and allow to stand overnight. Then filter and wash through with 100 ml. of chloroform. Distil off the solvent and remove the final traces in the 100° C. oven. Add 50 ml. of water to the extracted theobromine and boil. Cool, exactly neutralise with N/10 sodium hydroxide (1 or 2 drops), using phenol red (0·5 ml. of 0·1 per cent. in alcohol) as indicator and add 20 ml. of N/10 silver nitrate (theobromine + $AgNO_3 \rightarrow$ theobromine—Ag + HNO_3). Titrate the acid which is liberated with N/10 sodium hydroxide to the red colour of the indicator (pH 7·4). 1 ml. N/10 NaOH $\equiv$ 18·0 mg. theobromine (*cf.* Holmes, *Analyst,* 1950, **75**, 457 and *Analyt. Chem.*, 1948, **20**, 40).

Theobromine and Caffeine (" Total Alkaloids ")

The " total alkaloid " content of cocoa can be determined by the method described by Moir and Hinks (*Analyst,* 1935, **60**, 439). Grind 2 g. of prepared cocoa or 2 or more g. of cocoa-nib or chocolate to a smooth paste with a little alcohol (80 per cent. v/v), and transfer to a 200-ml. flask with more alcohol of the same strength, sufficient to produce a total volume of about 100 ml. Add 1 g. of freshly-ignited magnesium oxide and digest in a boiling water-bath for one and a half hours under a reflux air-condenser, with occasional shaking. Filter while hot through a small Büchner funnel, return the residue to the flask, and re-digest for half an hour with 50 ml. of the alcohol. Filter and repeat the digestion with a further 50 ml. of the alcohol. Evaporate the combined extracts on the water-bath, adding hot water from time to time to replace the alcohol lost. When all the alcohol is removed, finally evaporate to about 100 ml., add 2 or 3 drops of hydrochloric acid (10 per cent. HCl) and transfer the liquid to a 150-ml. measuring flask. Cool, add 5 ml. of zinc acetate* solution and mix, and then add 5 ml. of potassium ferrocyanide* solution. Make up to the mark with water and mix thoroughly by shaking. Allow the flask to stand for a few minutes, and filter through a dry filter paper.

Evaporate a measured quantity (about 120 ml.) of the filtrate to about 10 ml., transfer to a separator and extract by vigorous shaking with five successive 30-ml. quantities of chloroform. Run off the chloroform after each extraction into a second separator, and wash the combined extracts with 3 to 5 ml. of water. Repeat the above process of extraction with five successive quantities of chloroform, wash the second chloroform extracts with the same wash-water, combine the whole of the extracts, and remove the chloroform by distillation. Dissolve the residue in a little hot water, transfer to a Kjeldahl flask, add 0·2 g. of sucrose and 10 ml. of concentrated sulphuric acid. Heat over a small flame till frothing ceases; add 0·02 g. of selenium and digest

* See p. 111.

till colourless. Heat for one hour longer, and determine the ammonia in the usual way by distillation into N/10 acid. Alternatively, the ammonia may be distilled into boric acid solution.

Use the factor 3·26 for conversion of nitrogen into " total alkaloid." This factor is based on the assumption that the alkaloids in cocoa nib consist of theobromine and caffeine in the ratio of 9 to 1. The authors found that roasted fat-free dry blended cocoa nibs contained on the average 3·15 per cent. of " total alkaloids." The amount of fat-free cocoa solids in a product such as chocolate can therefore be calculated by dividing the percentage of " total alkaloids " by 0·0315.

Harral (*Analyst*, 1940, **65**, 408) converts the theobromine to caffeine by shaking with methyl sulphate + alkali and both alkaloids are extracted in the form of caffeine, which is more readily extracted with chloroform than theobromine (*cf. Analyst*, 1937, **62**, 791).

Shell. The Code of Practice states that cocoa shall contain not more than 5 per cent. of shell. In assessing the amount present, use is frequently made of the fact that the shell contains a considerably greater proportion of fibre than does the nib. The general method for determining crude fibre is given on p. 41. As the result obtained depends on the particle size, Lockwood (*Analyst*, 1939, **64**, 92) recommends that the sample used for the determination should have been previously passed through a No. 100 B.S. Sieve (or No. 10 bolting silk). A modified method of determination is also described by Lockwood, who employs average fibre figures of 5·8 and 18·5 for the nib and shell respectively (both calculated on the dry fat-free material). Using these figures, the percentage of dry f.f. shell in the sample (S) can be calculated from the following equation:—

$$S = \frac{(K-5{\cdot}8)[100-(F + W)]}{12{\cdot}7},$$

where F and W are the percentages of fat and water in the sample, and K is the percentage of crude fibre in the dry fat-free sample (*cf.* Baker and Hulton, *Analyst*, 1918, **43**, 197; Knapp and McLellan, *ibid.*, 1919, **44**, 2; Fincke and Thaler, *Z. Untersuch Lebensm.*, 1942, **84**, 97; and *Food*, 1957, **26**, 43). South Africa has adopted a limit of 6·5 per cent. crude fibre, calculated on the dry fat-free material.

Added starch. The presence of added starch will have been indicated in the microscopical examination. Starch can be estimated using diastase:—

De-fat 8 g. of the sample and weigh out 5 g. of the dry residue into a flask. Add 100 ml. of 12 per cent. alcohol, shake, filter and wash with 20 ml. of industrial alcohol. Wash the extracted residue into a flask with 50 ml. of ammonia-free water, heat in boiling water for twenty minutes and cool. Add

0·1 g. of diastase (+ water) and maintain at 50–55° C. for two hours. Then cool, make up to 250 ml. and filter. Mix 200 ml. of filtrate with 20 ml. hydrochloric acid (s.g. 1·125) and heat in boiling water for three hours. Cool, neutralise (just) with sodium hydroxide solution and make up to 250 ml. Then estimate the reducing sugars using Fehling's solution in the usual way, and calculate as dextrose (dextrose × 0·90 = starch). Using this method, cocoa containing no added farinaceous matter shows a starch percentage of about 7 per cent. (*cf.* A.O.A.C.).

Sugars may be present in drinking chocolate (see p. 233.). For their estimation, prepare a 5 per cent. solution (clarified with zinc ferrocyanide or basic lead acetate) and determine the sugars present in the filtrate, before and after inversion by Lane and Eynon's method. Lactose, if present, should preferably be determined polarimetrically.

CHOCOLATE

The following table gives the usual ranges of composition figures for plain chocolate and milk chocolate together with those obtained on one sample of nut chocolate:—

	Plain chocolate Per cent.	Milk chocolate Per cent.	Nut chocolate Per cent.
Moisture	0·8–2·3	0·8–1·8	1·1
Total ash	1·0–2·5	1·8–2·4	1·5
Water soluble ash	0·7–1·1	—	—
Fat	30–40	30–40	36·6
Sucrose	37–55	35–47	45·2
Lactose	—	7–11	6·9
Starch	2–7	—	—
Total nitrogen	0·6–1·3	0·8–1·5	1·12
Theobromine	0·4–0·7	0·2–0·5	—
Crude Fibre	0·8–1·5	0·4–1·0	—
*Values on the fat**			
Reichert value	—	4·3, 5·6	4·8
Polenske value	—	1·2, 1·1	1·3
Kirschner value	—	4·0, 5·3	4·2

* Results from individual samples.

The quality of chocolate can be assessed by determining the following: **moisture, fat, sugars, nitrogen, "total alkaloids," ash, water soluble ash, crude fibre** and **trace metals** (*e.g.*, **As, Pb** and **Cu**). In addition, the oil values should be determined on the fat.

Prior to the analyses the sample should be grated on a nutmeg grater and immediately transferred to an airtight jar. The **moisture** can be determined by drying at 100° C. and the **fat** by continuous extraction using light petroleum. **Lactose** and

sucrose can be estimated, after clearing the solution with lead subacetate, using the polarimeter or by Lane and Eynon's volumetric method (*cf. Chatt, loc. cit.*, p. 237). The amount of **non-fatty milk solids** in milk chocolate can be calculated from the lactose figure and the **milk-fat** from the Reichert-Polenske-Kirschner values of the separated fat.

The amount of dry fat-free cocoa matter present in chocolate can be calculated from the " total alkaloids " figure.

COCOA BUTTER

Cocoa butter may be examined by the methods described in the chapters dealing with fats. As the analysis of chocolate may include the examination of the fat for the purpose of detecting foreign fat or estimating the amount of milk fat, the usual data for cocoa butter are shown below:—

Colour of fat	Yellowish-white
Specific gravity (15·5° C./15·5° C.)†	0·950–0·975
Refractive index at 40° C.	1·456–1·458*
Melting point (°C.)†	31–34°*
Iodine value	35–40*
Saponification value	188–195*
Unsaponifiable matter (%)	Up to 1·5
Acid value	Not more than 4·0*
Acid value (usual range)	1·1–2·8
Reichert value	0·2–1·0
Polenske value	Up to 0·5
Kirschner value	0·2
Titre (°C.)	48–51

* B.P. 1958 limits.

† When cocoa butter is melted it does not, on cooling, regain its normal s.g. and m.pt. for some time and seventy-two hours should elapse between the time of melting and observing these constants.

Some tests which are especially applicable to cocoa butter are given below.

Beef fat can be detected by Björklund's test:—

Dissolve 3 g. of fat in 10 ml. of ether in a test tube. Plug the tube with cotton wool and set aside in a cool place. Pure cocoa butter usually separates only below 10° C., but when it does, it appears as tufts, leaving a clear solution, whereas beef fat, if present, separates in flocculent masses and the supernatant liquor remains turbid. Cocoa butter crystals dissolve on warming to 18° C., whereas beef crystals do not. The crystals should be examined microscopically.

Critical temperature of dissolution of fat in acetic acid (*A.O.A.C.*)

APPARATUS

Insert a thermometer reading to 0·1° C. into a cork that fits a 6 × ¾ in. test tube and extend it far enough into the tube so the bulb is covered by 10 ml. of liquid. Place the tube in a larger tube (4 × 1¼ in.) containing glycerol and hold it firmly in place with a cork having a groove cut in its side to equalise the pressure when heat is applied.

METHOD

Filter a portion of the sample through a dry paper in an oven at 110° C. Allow the filtered sample to cool until barely warm and weigh 5 g. sample and 5 g. of 99·5 per cent. acetic acid into the test tube. Insert the cork holding the thermometer and place the test tube in the glycerol-bath. Heat and shake frequently until a clear solution is obtained. Allow to cool, with constant shaking, without removing from the bath. Note the temperature at which the first sign of turbidity appears. Carry out a control test using pure cocoa butter. As the free fatty acids lower the turbidity temperature a correction must be made for the acid value of the sample. If the concentration of the acetic acid reagent is such that the turbidity temperature of pure cocoa butter is about 90° C., one unit of acid value will cause a reduction of 1·4° in the critical temperature of dissolution (if about 100° the corresponding reduction is 1·2° C. and intermediate values can be calculated by interpolation). Also determine the acid value (p. 409) of both the sample and the pure cocoa butter. Then multiply the acid value by the correction factor and add the result to the observed turbidity temperature. The figure so obtained is the true critical temperature of dissolution. If this temperature is lower than that of pure cocoa butter by more than 3° (or 6° for the fat from milk chocolate), adulteration with coconut, palm kernel, corn, peanut, cottonseed oils, etc., or their stearins, is indicated.

Acetone-carbon tetrachloride test

REAGENT

Equal volumes of acetone and carbon tetrachloride.

METHOD

Dissolve 5 g. of the warm fat, which has been previously filtered through dry filter paper in an oven at about 110° to remove traces of moisture, in 5 ml. of the acetone carbon tetrachloride reagent in a test tube. Allow the solution to stand in iced water for twenty or thirty minutes. Put on a test on a sample of pure cacao butter at the same time for comparison. If hydrogenated oil, tallow, oleostearine or paraffin wax is present, a white flocculent precipitate will soon appear. If the water is cold enough, cacao butter may solidify. If a precipitate is formed, remove the sample from the iced water and allow it to remain at the room temperature for a time. Solidified cacao butter will rapidly melt and go into solution, but if the precipitate is due to any of the above-mentioned possible adulterants, a much longer time will be required for complete solution to take place.

For the detection of cacao butter substitutes, see Knapp, Moss and Melley (*Analyst*, 1927, **52**, 452), and Bywaters *et al.* (*ibid.*, p. 324).

Chapter 9

HERBS AND SPICES

Herbs and spices consist of the dried leaves, flowers, buds, fruits, seeds, barks or rhizomes of various plants. They are only present in foods in small amounts but they make important contributions towards the odour and flavour due to the presence of the volatile oil (essential oil) and fixed oil. Many instances of adulteration have been reported. Foreign leaf has been found in herbs due to either deliberate addition or careless gathering (grass, weeds, etc.). Spices have been adulterated with foreign spices, the exhausted spice, cereals, ground stones and shells from fruits and nuts, sand, sawdust, etc. Most adulterants can be detected microscopically, but care must be exercised in distinguishing between a deliberate sophistication and relatively small amounts derived from matter which remained on the grinding stones from the previous grinding.

Apart from the general requirements of the Food and Drugs Act, there are few British Standards for herbs and spices (only mustard and curry powder have been made the subject of Food Standards Orders). The standards for spices used for pharmaceutical purposes which are prescribed in the B.P. and B.P.C. and the numerous U.S. standards are, however, most useful as a guide to quality. Food spices which are retailed as being of B.P. quality must conform to the B.P. standards.

General Methods of Analysis

Before the main analysis, a cursory examination should be made for the presence of waste material such as excess stalks, foreign organic matter (B.P. 1958, p. 892), insects, rodent excrement, etc. In addition, whole spices must be ground thoroughly before the analysis is commenced. The *routine analysis* of most herbs and spices should include the determination of **total ash, acid insoluble ash, fixed ether extract, volatile oil, arsenic** and **lead** and a **microscopical examination.** It may be necessary also to determine **water soluble ash, crude fibre, alcohol extract, cold water extract, nitrogen** and **starch.**

Moisture should preferably be determined by distillation from an inert solvent. Drying (preferably at 70° C. in a vacuum oven) is also employed. Although volatile oil is also lost during the drying, the result obtained is often sufficiently accurate for comparison purposes.

Total ash, water soluble ash and **acid insoluble ash** are all determined by the conventional methods (see B.P. 1958, p. 884). A high ash or acid insoluble ash usually suggests inferior quality.

Crude fibre measures the amount of " unwanted " material such as husks and shell. It is particularly useful for assessing the quality of white pepper. It is determined by the usual method laid down in the Fertilisers and Feeding Stuffs Regulations.

Fixed ether extract should be determined in a continuous extraction apparatus on 2–3 g. of the powdered sample, using diethyl ether as solvent. After one hour's extraction the material should be removed, re-ground and returned to the apparatus for a further two hours' extraction. After removal of the solvent, the extract is heated at 100° C. for three quarters of an hour and then cooled and weighed. The fixed ether extract is expressed as percentage w/w.

Volatile oil. If in the determination of fixed ether extract (above) the solvent is first allowed to evaporate off at room temperature and then the extract is dried over sulphuric acid for one day, the weight obtained represents the total ether extract (fixed + volatile oil). If the total ether extract is then dried at 100° C. the weight lost represents the **volatile ether extract** (% **w/w**). Although this method may have to be used if only a small quantity of sample is available, the B.P. method (when possible) gives more satisfactory results. It involves distilling the oil over with water and measuring the volume in a graduated tube. As the excess of condensed water overflows back into the distillation flask, the level in the graduated receiving tube automatically remains constant. If the oil is heavier than water (or its s.g. is only just below 1·0), the oil is allowed to " add on " to a measured volume of xylene and the increase in volume represents the volatile oil. The general procedure is as follows:—

Place in the distillation flask (Fig. 35) a suitable quantity of the ground herb or spice (usually 20 g.), 300 ml. of water and porous pot. Connect up the still-head and run water into K until it overflows at B. Then insert the stopper K′ together with a small strip of paper, which acts as a " safety valve " by separating *slightly* the ground portions of the tubulure K and its stopper K′. The stopper K′ should be lightly held in position by means of a rubber band. Boil the liquid at a moderately fast rate on a heating mantle or a burner, whilst making sure that the lower part of the condenser G is cold. Continue the distillation for three to five hours. Then read off the volume of oil distilled over (ten minutes after removing the source of heat). Calculate the volatile oil in the spice as a percentage (v/w).

For Herbs and Spices with Oils Heavier than Water

With materials such as aniseed, cassia, cinnamon, clove, fennel, parsley, pimento and turmeric, the following modification is necessary:—

Place in the distillation flask 300 ml. of water and porous pot (no sample). To the stillhead add exactly 1 ml. of xylene in addition to the water and " close " as before. Boil the water in the flask for thirty minutes and read off the volume of xylene (Vx) in the graduated tube ten minutes after removing the source of heat. Add the sample, carry out the distillation and read off the total volume (V_T) as before. Then $V_T - V_X$ represents the volume of volatile oil which has distilled over.

Sage and Fleck (*Analyst*, 1934, **59**, 614) have described an apparatus for the determination which can be constructed from ordinary laboratory glassware.

The results obtained from the distillation method, although comparative, are probably lower than the amount actually present in the spice due to some extent to the solubility of the oil in the water. The result is also affected by the particle size (see also *Quart. J. Pharm.*, 1935, **8**, 435; 1937, **10**, 471).

Alcohol extract. Shake 5 g. of sample frequently with 100 ml. of 95 per cent. alcohol in a closed flask for six hours. Then allow to stand for eighteen hours and filter. Evaporate 25 ml. of filtrate to dryness in a tared basin, dry at 100° C., cool and weigh. Calculate the result as a percentage of the original sample.

Cold water extract (water soluble extractive). This is determined in the same way as the alcohol extract except that the alcohol is replaced by water. Exhausted spices give low alcohol and water extracts.

Nitrogen. Determine on 2 g. by the macro-Kjeldahl method.

Starch. Determine by the diastase method.

Microscopical Examination should be carried out on all samples of ground herbs and spices. The prepared slides should be examined firstly with the low power objective ($\frac{2}{3}$ in.) and then under the high power ($\frac{1}{6}$ in.). Also it is preferable to have a set of authentic samples available for comparison purposes.

A " *water-slide* " should firstly be prepared by dispersing the sample with a drop of alcohol and then adding one to two drops of glycerol solution (30 per cent. in water) before sliding on the coverslip. The water-slide is particularly useful for detecting starch. The presence of starch can be confirmed by adding a drop of very dilute iodine solution which produces the usual dark blue colour. Unless light coloured particles are seen in the cursory examination it is usually unnecessary to prepare a water-slide of leafy material (unless powdered).

A " *cleared-slide* " is prepared by gently boiling the material with chloral hydrate solution (prepared by warming 80 g. of crystals with 50 ml. of water) either on the slide or in a tube until the particles look fairly transparent. Chloral hydrate has a twofold action: (a) it removes material such as starch thereby concentrating the other tissues and (b) it removes colouring matter, etc., from the tissues so that their outlines can be seen much more clearly. Sclerenchymatous tissue can be stained red by warming the cleared material

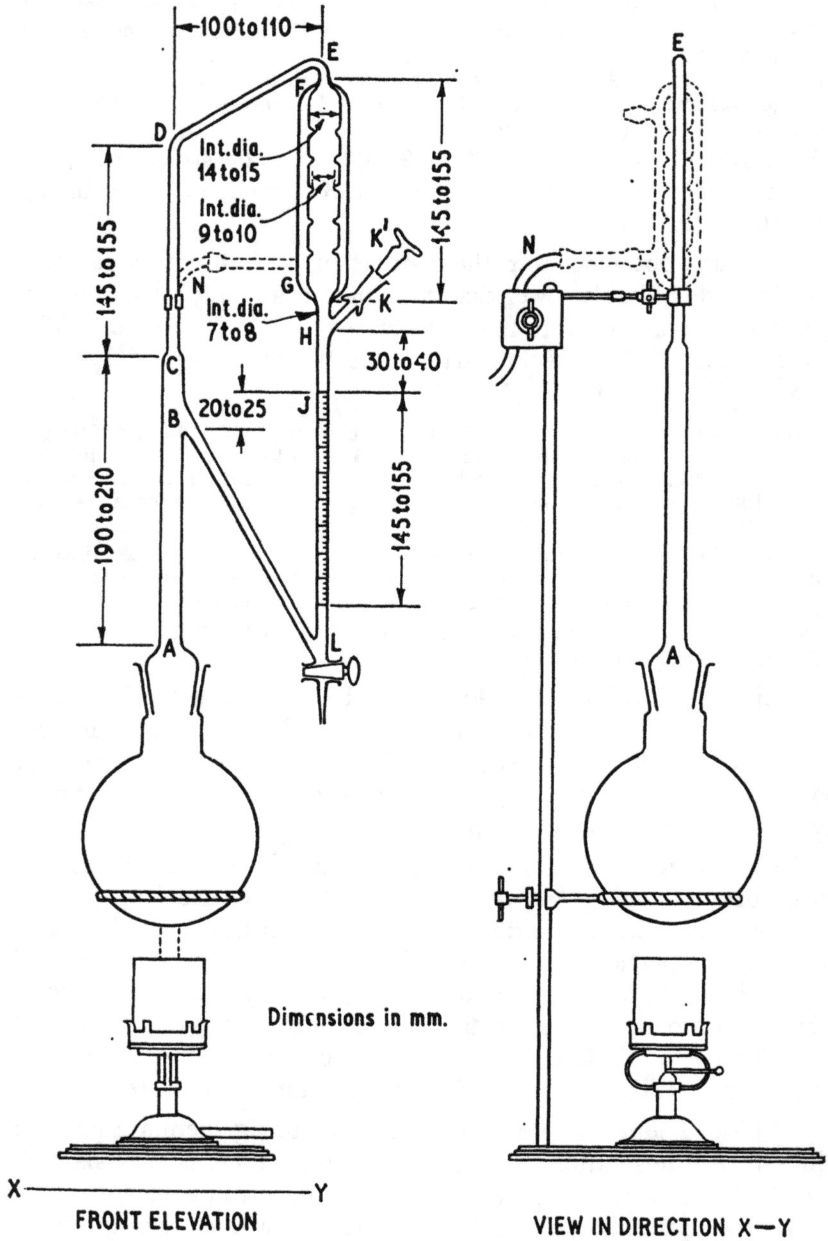

Fig. 35. B.P. Apparatus for the Determination of Volatile Oil.

with excess of phloroglucin solution (1 per cent. in 90 per cent. alcohol) followed by a drop of concentrated hydrochloric acid (*cf. Food Manuf.*, 1956, **31**, pp. 114 and 198).

The following works are useful for the study of the microscopical examination of herbs and spices:—

Greenish and Collin's " Anatomical Atlas of Vegetable Powders," Churchill, London.
Winton's " Structure and Composition of Foods," Vol. IV, Wiley, N.Y.
Wallis' " Textbook of Pharmacognosy," Churchill, London.

HERBS

Dried herbs are derived from flowering plants and consist of flowering tops and may include smaller stems, flowers and fruits, but the leaf usually predominates. Herbs are commonly employed in meat products, stuffings, soups and fish dressings. Typical analytical figures together with the U.S. standards are given in the following table:—

Herb	Total ash (%)	Acid insol. ash (%)	Crude fibre (%)	Volatile oil (%)	Stalks, etc. (%)	Number of samples
Basil . . .	11–5	0·3		1·8		1
Bay	2·8–4·1	0·2–0·8		0·7–2·3	10	9
Marjoram . . .	8·9–12·0	0·4–2·7		1·2–2·5	2–12	11
U.S. standards (maxima) . . .	16	4·5			19	
Mint . . .	10·0–14·4	0·7–2·2		0·5–3·0	4–15	12
Parsley . . .	12·4–18·4	1·0–6·0		6·6 (1 samp.)	2·3 (1 samp.)	5
Rosemary . . .				0·9		1
Sage						
*English varieties .	5·4–14·3	0·5–3·5		0·6–1·5	4–14	Numerous
U.S. standards .	10 (max.)	1 (max.)	25 (max.)	1 (min.)	12 (max.)	
Savory . . .	6·3–9·9	0·1–0·8		0·9–1·5	5–15	5
Thyme	7·0–19·2	0·8–9·8		0·4–2·5	2–8	Numerous
U.S. standards (maxima) . . .	14	4†			†	

Trace elements in herbs:—
Arsenic content—up to 2 p.p.m. As (maximum 5 p.p.m.).
Lead content—up to 10 p.p.m. (recommended maximum 10 p.p.m.).

* Sage from the Mediterranean area tends to have a low ash (*e.g.*, 3·5 per cent.) and contains more volatile oil (up to 4·5 per cent.).

† B.P.C. 1934 prescribed, for thyme, maximum for " foreign organic matter and stems over 1 mm. in diameter " of 3 per cent. and for " acid insoluble ash " of 4 per cent.

Basil (Ocimum basilicum). The diagnostic microscopical characters of basil are the warty, one- or two-celled pointed hairs with broad bases and the four-celled oil glands.

Bay (Laurus nobilis). The dried leaves may be up to 10 cm. in length, brown in colour, thick, lanceolate and tapering at both ends. The upper surface is smooth, the lower one dull. The main constituents of the volatile oil are cineole, geraniol, eugenol and terpenes. The most characteristic features in the microscopical examination are the stomata on the lower epidermis which often occur in pairs.

Marjoram (Marjorana hortensis). The dried herb may contain a proportion of flowering tops. The small ovate leaves are covered with velvety hairs. Microscopically marjoram shows wavy cells on the upper surface, very sinuous cells on the lower side, eight- to twelve-celled oil glands and two types of hairs: (a) long, pointed, multi-cellular, often warty and with broad bases, (b) glandular hairs on short stalks. The U.S. standards for marjoram are quoted in the table.

Mint (Mentha viridis). The dried herb may contain separated stalks. The volatile oil of mint contains about 50 per cent. of menthol. In the 1930's instances of the adulteration of mint with ailanthus leaves (tree of heaven) were reported. The pieces of ailanthus are a rather lighter green than the pieces of mint leaf. The principal distinguishing features in the microscopical examination are (see also *Analyst*, 1934, **59**, 33):—

Microscopical features of Mint and Ailanthus leaves

Microscopical feature	Mint leaf	Ailanthus leaf
Epidermal cells	Wavy. Not striated	Almost straight. Striated
Hairs	Two or more celled with clear outline	Single celled
Oil glands	Conspicuous, yellow centre	None
Crystals	Absent in most samples	Cluster crystals along veinlets

Mint sauce prepared with mint adulterated with hazel leaf was the subject of a court case heard in 1960.

Parsley (Carum petroselinum). Dried parsley usually has a bright green colour. Its volatile oil is readily lost, particularly during the drying. Microscopically parsley does not show either hairs or crystals. The upper epidermal cells are polygonal but those on the lower epidermis are wavy. The epidermal cells tend to be striated.

Rosemary (**Rosmarinus officinalis**). The diagnostic microscopical characters of rosemary are the thin-walled, branching hairs sometimes with globular heads and the eight-celled oil glands.

Sage (**Salvia officinalis**). Sage differs considerably in appearance according to the country of origin. Dried English sage is green, whereas that from Cyrprus is a pale bluish-grey colour, hard and leathery and the highly esteemed type from Dalmatia is grey and very fluffy. All types are pubescent (covered with hairs). Sage from the Mediterranean area usually contains rather more volatile oil (Dalmatian up to 4·5 per cent.) than the English varieties included in the table. The characteristic flavour substance in the oil appears to be the ketone thujone. *Microscopically*, the main features are the numerous long, slender, whip-like, non-warty, pointed hairs, the short hairs with globular glands and the oil glands (mostly twelve-celled). Button (*Analyst*, 1953, **78**, 679) has described and illustrated the adulteration of sage with cistus leaf, which can be detected microscopically by the star-shaped hairs. Many samples of English sage tend to give figures for the ash and acid insoluble ash in excess of the U.S. maxima (see Table). The B.P.C. (1934) prescribed maxima for foreign organic matter (3 per cent.) and ash (8 per cent.).

Savory (**Satureia**). The diagnostic microscopical characters of savory are the warty, pointed hairs (many jointed) and the very large twelve-celled oil glands. The chief odorous constituent appears to be carvacrol.

Thyme (**Thymus vulgaris**). The diagnostic microscopical characters are the one- to two-celled warty hairs, the small stomata and the numerous twelve-celled oil glands. Most commercially acceptable samples of thyme comply with the U.S. standards (*cf.* Table, p. 245). The most important constituents of the volatile oil are the phenols thymol and carvacrol. Lemon thyme (*Thymus citriodorus*) is greener than thyme proper when dried. It has a lemon odour in addition to that of thyme proper. The volatile oil content of lemon thyme is about 1 per cent.

BARKS

Cinnamon (**Cinnamomum zeylanicum**). Cinnamon bark, also referred to as Ceylon cinnamon, is imported as long, brown quills 6–9 mm. in diameter. The following table gives the typical ranges for the analytical figures together with the B.P. and U.S. standards:

CINNAMON

	Range %	*B.P. standards* %	*U.S. standards* %
Moisture	7·8–10·5	—	—
Ash	4·1– 5·7	7 (max.) (B.P. 1932)	5 (max.)
Water soluble ash . .	1·4– 2·6	—	—
Acid insoluble ash . .	0·1– 0·7	2 (max.)	2 (max.)
Alcohol extract . . .	8·7–13·6	—	—
Volatile oil (v/w) . . .	0·7– 1·4	1·0(min.)—whole 0·7(min.)—powder	—
Fixed oil	1·3– 1·7	—	—
Starch	16–22	—	—
Foreign organic matter .	—	2 (max.)	—

The B.P. requires cinnamon oil (distilled from the bark) to contan 55–70 per cent (w/w) of cinnamic aldehyde. The oil also contains 4–10 per cent. of phenols (mainly eugenol).

Microscopically, cinnamon can be identified by the long bast fibres, with thick walls, the sclerenchymatous cells (often thickened more on one side than the other), the small starch grains and the minute calcium oxalate crystals.

Cassia bark (Cinnamomum cassia), which closely resembles cinnamon, has been used as an adulterant or substitute for the latter. The main differences are (a) volatile oil of cassia contains no eugenol and more cinnamic aldehyde (at least 75 per cent), (b) area of fibres per g.—cinnamon has 90 sq. cm. per g. and cassia 13 sq. cm. per g. (Saber, *Quart. J. Pharm.*, 1940, **13**, 7) and (c) the general microscopical examination shows that cassia contains cork, has larger and wider bast fibres, and larger starch grains. Care must be taken in interpreting the results of such tests, however, as some poorer quality (" coarse ") samples of cinnamon tend to resemble cassia. The U.S. standards for cinnamon given in the table above also apply to cassia.

BULBS

Onion, Garlic. Dried onion is now available on the market in the powdered or kibbled form. The composition of the fresh and dried forms are given by Sherratt (*Analyst*, 1943, **68**, 200). The kibbled form (moisture content about 10 per cent.) has occasionally been attacked by the flour moth (*Ephestia*). Microscopically both onion and garlic powders show abundant parenchymatous cells. Garlic powder shows sclerenchymatous cells and rhombic crystals of calcium oxalate, both of which are usually absent in onion.

FLOWERS

Capers (Capparis spinosa) are cultivated in the Mediterranean area and the wilted flower buds are pickled in salt and vinegar and packed in bottles.

Cloves (Eugenia caryophyllus) are dark reddish-brown flower buds, 13–21 mm. long. The following table gives the typical ranges for the analytical figures together with the B.P. (1953) and U.S. standards:—

CLOVES

	Range %	*B.P.* (1953) *standards* %	*U.S. standards* %
Moisture	5·0–11·0	—	—
Ash	4·5– 7·0	7 (max.)	7 (max.)
Water soluble ash	2·7– 4·2	—	—
Acid insoluble ash	0 – 0·3	1·0 (max.)	0·5 (max.)
Volatile oil	14·5–20·0	15·0(min.)—whole 14·0(min.)—powder	15 (min.) (for vol. eth. extract)
Fixed oil	6·2–10·1	—	—
Alcohol extract	13·5–15·5	—	—
Crude fibre	6 –10	—	10 (max.)
Nitrogen	0·9– 1·2	—	—
Foreign organic matter (*e.g.*, fruits)	—	1·0 (max.)	—
Stalks	—	5 (max.)	5 (max.) for " stems "
Quercitannic acid	—	—	12 (min.)

The principal constituents of cloves include the volatile oil (containing 85–92 per cent. eugenol), tannin and a crystalline substance caryophyllin. The most likely adulterant is exhausted clove, which is shrunken and may be detected by the low volatile oil and low ether and alcohol extracts. Genuine whole cloves sink in water, but exhausted cloves float. *Microscopically*, cloves show rosette crystals of calcium oxalate, epidermal cells with stomata from calyx teeth, oil glands (and oil drops) and triangular pollen grains. **Clove stalks** seldom yield more than 7 per cent. of volatile oil but the crude fibre may exceed 20 per cent. Microscopically, the stalks show prisms of calcium oxalate and sclerenchymatous cells. Apart from a few stalks most samples of clove also contain a few ripe fruits (**mother cloves**) which contain starch.

FRUITS

Pepper

Black pepper consists of the dried *unripe* fruits of *Piper nigrum.* The whole peppercorns are almost spherical, about 5 mm. in diameter and the black surface is deeply reticulated.

White pepper consists of the dried *ripe* fruits of *Piper nigrum* from which the outer part of the pericarp has been removed by friction after previously fermenting or soaking in water. The whole peppercorns are spheroids and on the average are slightly larger than black peppercorns. The colour is greyish-white and there are ten to sixteen white lines running from the apex to the base like meridians.

Decorticated white pepper consists of white pepper with all the coats down to the perisperm removed. This treatment has the effect of reducing the amount of crude fibre and ash (both usually below 1 per cent.—*cf.* table on p. 251).

The pungency of pepper appears to be due to 5–9 per cent. of the

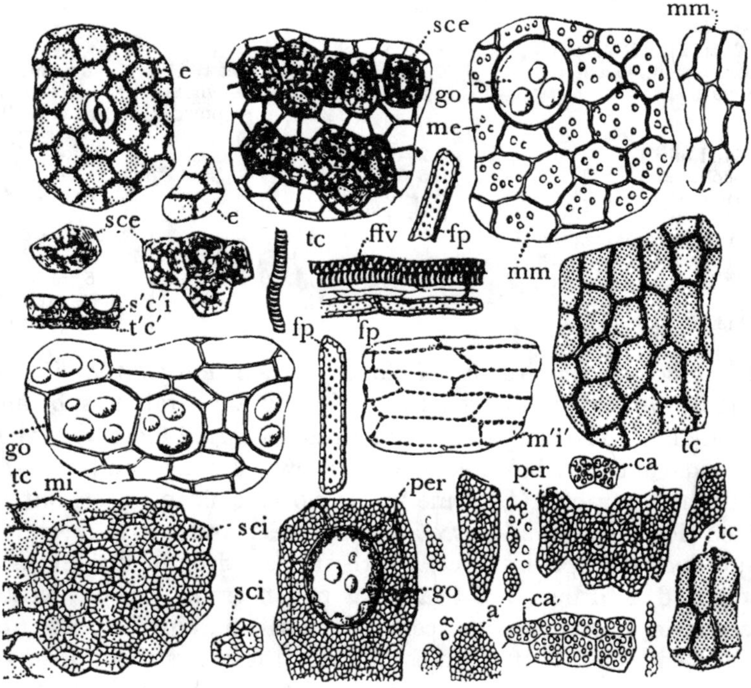

Fig. 36. Microscopical features of Black Pepper (from Greenish and Collin, *loc. cit.*).

a, Starch grains.
ca, Outer layers of perisperm containing aleurone.
e, Outer epidermis of pericarp.
ffv, Debris of fibro-vascular bundle.
fp, Pitted fibrous cells from bundle.
go, Oil-cells.
me, Outer portion of mesocarp.
mi, Inner portion of mesocarp.
m'i', The same in longitudinal section.
mm, Cells from central part of mesocarp.
per, Cells of perisperm containing starch
sce, Cells of the outer sclerenchymatous layer.
s'c'i', Cells of the inner sclerenchymatous layer, in surface view.
tc, The brown layer of seed-coat, in surface view.
t'c', The same in profile.

alkaloid piperine ($C_{17}H_{19}O_3N$) and about 6 per cent. of a green resin chavicine, which is isomeric with piperine. The aroma is derived from the volatile oil (s.g. 0·885–0·900 at 20° C.) which consists mainly of sesquiterpenes. Piperidine is probably present also due to decomposition of piperine.

Analytical Data for Black Pepper, White Pepper, Pepper Husks, Long Pepper and Exhausted Cubebs

	Black pepper %	*White pepper* %	*Pepper husks* %	*Long pepper* %	*Exhausted cubebs* %
Moisture	8·7–13·0	9·5–14·5	7·0–11·0	9·5	7·7
Total ash	3·1– 6·4	0·6– 2·5	7·0–15·0	7·1	4·5
Acid insoluble ash	0 – 1·4	0 – 0·3	0·2– 6·0	—	—
Water soluble ash	1·5– 3·0	0·2– 0·8	1·5– 4·5	3·8	2·6
Crude fibre	8·5–16·0	0·5– 5·0	20 –31	5·5	23
Fixed oil	6·1–10·7	6·2– 9·7	3·0– 9·5	6·8	6·1
Volatile oil	0·6– 2·5	0·5– 1·0	0·7– 1·1	1·5	0·4
Alcohol extract	8·5–11·9	7·2– 9·4	5·0– 7·5	8·6	—
Starch (by diastase)	22 –40	45 –64	5 –13	41	7
Total nitrogen	1·8– 2·5	1·6– 2·3	2 – 3	2	1

U.S. and B.P.C. Standards for Black and White Pepper

	Black pepper		*White pepper*	
	U.S. %	*B.P.C.* %	*U.S.* %	*B.P.C.* %
Ash (max.)	7	6	3·5	4
Acid insoluble ash (max.)	1·5	1	0·3	0·5
Crude fibre (max.)	—	15	5	6
Fixed ether extract (min.)	6·75	—	7	—
Starch (min.)	30	—	52	—
Foreign organic matter (max.)	—	2	—	2

Microscopically, black pepper shows numerous layers (Fig. 36):—

PERICARP

1. Epidermis of small polygonal cells with dark brown contents, occasional stomata and small prismatic crystals of calcium oxalate.
2. An interrupted sclerenchymatous layer of cells with dark contents.
3. Several layers corresponding to the mesocarp, the outer layers of parenchymatous tissue and the inner layers lignified. Oil cells and a little starch are also present among these layers.
4. An inner sclerenchymatous layer of lignified beaker-shaped cells.

SEED COAT

5. Three layers of collapsed cells, one brown, one yellow and one colourless.

PERISPERM

6. The outer layers consist of polygonal cells with aleurone grains. The inner layers consist of elongated cells which contain numerous aggregates of small starch grains (4–6 μ). Oil cells are also present in the perisperm.

White pepper shows the same microscopic appearance as black, except that the outer parts of the pericarp are missing (epidermis and sclerenchymatous layer). As coloured cells are present in the seed coat, most samples of genuine white pepper contain dark particles. It is necessary to consult the analytical figures (particularly the crude fibre), however, in order to obtain a more quantitative measure of excessive husk.

Long pepper (**Piper officinarum, P. longum**) takes the form of spikes up to 6·5 cm. long but the individual fruits on the spike are not unlike black pepper. Compared with black pepper, long pepper shows (under the microscope) relatively few stone cells and the larger starch grains of the perisperm (up to 10 μ) show a characteristic white appearance under polarised light.

Cubebs or **tailed pepper** (**Piper cubeba**) consists of globular fruits, up to 6 mm. diameter, dark brown and with a reticulated pericarp. Compared with pepper, cubebs shows under the microscope much larger cells in the outer layer of the seed coat and the starch grains are also larger (up to 12 μ). Cubebs yield 5–8 per cent. ash and 10–18 per cent. volatile oil (see table on p. 239 for data for exhausted cubebs).

In addition, the Australian Board of Trade Regulations (1917) prescribe minimum alcohol extract figures of 8 and 7 per cent. for black and white pepper, respectively.

Mehring (*J. Agric. Res.*, 1924, **29**, 569) has given data for the composition of the ash of black and white pepper (and also red pepper and mustard).

Quality and Adulteration of Black and White Pepper

As the husks contain much higher proportions of ash and crude fibre than the interior the results from these two determinations form the best bases for the assessment of pepper quality from chemical determinations. The routine examination should always include a microscopical examination, and the determination of total ash. If **black pepper** yields an ash in excess of about 5·2 per cent., the crude fibre should also be determined. Excessive husk is indicated if the fibre of black pepper exceeds about 15 per cent.

With **white pepper,** the colour of the powder itself (particularly if looked at after mounting on a slide preparatory to the microscopic examination) often gives a good indication of the proportion of husk present. In view of the heavy decortication which is commonly carried out, the ash of white pepper now seldom exceeds about 1·2 per cent. and the standards quoted in the table above seem especially liberal. The crude fibre figure which is correspondingly low in most samples is a better indication of the quality of white pepper. Excess husk is indicated if the crude fibre exceeds 5 per cent.:—

$$\text{Excess husk (\%)} = (\%\ \text{fibre} - 5) \times 4$$

A high crude fibre figure may also be due to the presence of other adulterants such as fruit stones and nut shells. The typical fibre figures (as percentages) for such materials are: olive stones ("pepperette" or "poivrette") 60, date stones 65, walnut shells 52, coconut shells 56 and almond shells 49. Microscopically, such adulterants contain (like pepper husks) a high proportion of stone cells. Other possible adulterants of pepper which can also be detected microscopically are: cereals (especially rice starch in view of its similarity to pepper starch), ginger, capsicum, long pepper, cubebs and turmeric. The colouring of pepper with, say, 0·2 per cent. of turmeric should not, however, be looked upon as adulteration. Turmeric may be estimated by comparison of the colour of the alcohol extract with that of dilutions of an alcoholic tincture of freshly ground turmeric. Pepper has also on occasion been adulterated with inorganic materials, *e.g.*, magnesium carbonate (*Analyst*, 1937, **62**, 121).

Capsicum, Chillies, Red Peppers—Cayenne Pepper, Paprika

Red peppers are derived from various species of Capsicum and are grown in many parts of the world. The whole pods are often referred to as chillies. The pods vary considerably in size, pungency and colour. Thus, in general, the strong cayenne types seldom exceed 3·5 cm. in length and the colour is orange red, whereas the less pungent paprika forms (mainly grown in Europe) may be as much as 13 cm. in length and have a dark red colour. Although the names used for the various types are not standardised the following represents the main types which are met:—

Cayenne pepper. The capsicums which are usually used for the preparation of cayenne pepper are grown mainly in Africa, India and America. The U.S. regulations specify that cayenne pepper is the dried ripe fruit of *C. frutescens* L., *C. baccatum* L., or some other small-fruited species of *Capsicum*.

Capsicum, B.P. (**1948 Edition**) is derived from **C. minimum,** which is indigenous to America and is cultivated mainly in Africa and South America.

Paprika is usually derived from **C. annuum** and grown in Hungary.

Pimiento is another form of paprika which is grown in Spain (Spanish paprika).

CONSTITUENTS

Capsicum fruits contain a fatty oil, a red colouring matter and traces of a liquid alkaloid (all non-pungent) Svirbely and Szent-Györgyi (*Biochem. J.*, 1933, **27**, 279) reported that Hungarian paprika is particularly rich in ascorbic acid (about 0·2 per cent.). Capsicum seeds are richer in oil than the rest of the fruit.

The pungency of capsicum is not destroyed by 2 per cent. sodium hydroxide solution (aqueous), but potassium permanganate solution destroys it. The characteristic pungent taste of red pepper is due to **0·05–0·3** per cent. capsaicin ($CH_3OC_{17}H_{24}.NO_2H$), which can be estimated by the following method:—

Shake 2 g. of powdered capsicum with 100 ml. of pure acetone for twenty-four hours, allow to stand for one hour and filter. To 5 ml. of filtrate add (drop by drop) 1 per cent. vanadyl trioxychloride solution in carbon tetrachloride until there is no further change in colour. Compare the colour obtained with that produced by mixing 20 drops of the reagent with 5 ml. of 0·016 per cent. capsaicin solution in acetone.

A Joint Committee of the Pharmaceutical Society and the S.A.C. (*Analyst*, 1959, **84**, 603) recommended a method in which capsaicin is extracted either chromatographically or by an ether-alkali partition extraction procedure and is determined by a spectrophotometric difference method. Alternatively, the extracted capsaicin can be determined by direct spectrophotometry or by a colorimetric method based on the coupling reaction between diazobenzene-sulphonic acid and capsaicin. An organoleptic test for assessing the pungency of capsicum is given in the B.P. (1948), p. 120.

Microscopically, cayenne pepper shows a large number of structures derived from the pericarp, seed, calyx and stalk. Of these the most striking features are the numerous drops of orange-coloured oil, the polygonal beaded cells of the outer epidermis of the pericarp, the somewhat sinuous cells with thickened and pitted walls of the inner epidermis of the pericarp and the very large sclerenchymatous cells of the epidermis of the seed coat with very thick sinuous walls.

CAPSICUMS—CAYENNE PEPPER, PAPRIKA, ETC.

Ranges for Analytical Data and U.S. and B.P. (1948) Standards

	Red Pepper	Cayenne Pepper		Capsicum, B.P.	Paprika				
						U.S. maxima for paprika			
							Hungarian		Spanish
	(Un-qualified) U.S. maxima	Range	*U.S. standards	B.P. (1948) standards for C. minimum	Range	(Un-qualified) from C. annum	Rosen-paprika (placenta, stalks and stems removed No added seeds)	Koenigs-paprika (whole pods)	Pimiento
Water (%)	—	3·7 – 9	—	—	7 – 9·5	—	—	—	—
Total ash (%)	8	5·1 – 6·4	8 (max.)	8 (max.)	5·6– 7·6	8·5	6	6·5	8·5
Water soluble ash (%)	—	2·8 – 4·9	—	—	4·7– 5·5	—	—	—	—
Acid insoluble ash (%)	1	0·05– 0·25	1·25 (max.)	—	0 – 0·35	1	0·4	0·5	1
Crude fibre (%)	—	17·5 –25	28 (max.)	—	14 –20	—	23	23	21
Alcohol extractive (%)	—	20 –30	—	—	14 –16	—	—	—	—
Non-volatile ether extract (%)	—	15·5 –22	15 (min.)	12 (min.)	7·5–12	—	18	18	18
Volatile ether extract (%)	—	0·7 – 2·6	—	—	0·3– 1·5	—	—	—	—
Nitrogen (%)	—	1·8 – 2·3	—	—	2·2– 2·8	—	—	—	—
Starch (by diastase) (%)	—	0·5 – 1·5	1·5 (max.)	—	—	—	—	—	—
Foreign organic matter (%)	—	—	—	1 (max.)	—	—	—	—	—
Calyces and pedicels (%)	—	—	—	3 (max.)	—	—	—	—	—
Iodine number of extracted oil	—	—	—	—	128–138	125–136	—	—	—

* Standards for *C. frutescens* L., *C. baccatum* L., or other small-fruited species of *Capsicum*.

All the capsicum-type red peppers show similar characters, although the paprikas tend to have larger epidermal cells (*cf.* Wallis, *Pharm. J.*, 1901 (4), **13**, 552; 1902 (4), **15**, 3).

Adulteration of Red Peppers

The most likely adulterant or substitute of cayenne pepper is exhausted capsicum, which can be detected by the low non-volatile ether extract, which should be at least 15 per cent. Microscopical examination should reveal the presence of the other possible adulterants, *e.g.*, farinaceous materials, ginger, mustard and turmeric.

Pimento (Pimenta officinalis); Allspice; Jamaica pepper

The fruits are dark reddish brown, brittle and rough almost globular berries about 6–9·5 cm. in diameter. The range for the analytical figures of allspice are:—

	Moisture	Ash	Water soluble ash	Fixed oil	Volatile oil	Alcohol extract	Crude fibre	Total nitrogen	Starch (by diastase)
Maximum (%)	10·5	4·8	2·7	7·7	5·2	14·3	24·0	1·0	3·8
Minimum (%)	9·4	4·1	2·3	4·3	3·0	7·4	20·0	0·8	1·8

The U.S. standards prescribe the following maxima for allspice: ash 6, acid insoluble ash 0·4, crude fibre 25 and quercitannic acid 8.

The volatile oil contains about 70 per cent. of eugenol which accounts for the close similarity of the odour to that of cloves. Microscopical examination shows variously sized and shaped sclerenchymatous cells (the most characteristic are pointed), one-celled conical thick-walled hairs, starch (many compound grains), oil glands and prismatic and cluster crystals.

Umbelliferous Fruits

The umbelliferous fruits (often called "seeds") are all rather similar in structure and are called cremocarps. They split into two one-seeded portions (mericarps). The arrangement of the ridges on the rounded dorsal surface of the mericarps is often useful for identification purposes. The following table gives ranges for some of the analytical figures obtained from the seven umbelliferous fruits which are used as spices, together with the U.S. standards and those which have been prescribed in Britain for pharmaceutical purposes (B.P. and B.P.C.):—

Ranges for the Analytical Figures of the Umbelliferous Fruits together with the British (B.P., etc.) and U.S. standards*

	Aniseed %	Caraway %	Celery %	Coriander %	Cummin %	Dill %	Fennel %
Total ash	—	4·8–7·6	about 10	—	about 8	—	—
Total ash (U.S. max.)	9	8	10	7	9·5	10	9
Water soluble ash	—	2·0–2·2	—	—	—	—	—
Acid insoluble ash (B.P., etc. *max.)	1·5	1·5	2	1·5	—	3	1·5
Acid insoluble ash (U.S. max.)	1·5	1·5	2	1·5	1·5	3	2
Crude fibre	—	17·5–22·3	—	—	—	—	—
Cold water extract	—	20–26	—	—	—	—	22–27
Fixed oil	8–20	8–20	15–30	12–20	10–14	15–18	12–20
Volatile oil	1·5–4·0	2·5–5·9	1·5–3·0	0·3–1·0	2–4	2–4	1·0–4·0
Volatile oil (B.P., etc., *min.)—whole	2	3·5	1·5	0·3	—	2·5	1·4
Volatile oil (B.P., etc., *min.)—powder	—	2·5	1·5	0·2	—	2	1
Foreign organic matter (B.P., etc., *max.)	1	2	1	2	2†	2	2
Other fruits and seeds (B.P.C. max.)	2	—	4	—	—	—	—

* Standards prescribed in present or past editions of the B.P. or B.P.C.
† U.S. maximum for harmless foreign matter 5 per cent.

Anise (Pimpinella anisum). The fruits are about 3 mm. long, greenish-grey to brown in colour, ovoid and somewhat laterally compressed. The volatile oil contains about 90 per cent. of anethole. Microscopically, the main diagnostic characters are the short, thick-walled conical hairs; the polygonal, striated epidermal cells and the aleurone grains and calcium oxalate (rosettes).

Caraway (Carum carvi). The fruits (" seeds ") are about 6 mm. long with a brown, glabrous surface and slightly curved. The volatile oil contains about 50 per cent. of carvone. Exhausted seeds can be detected from the much lower figures which are obtained for the cold water extract (below 15 per cent.), the fixed oil and the volatile oil. Microscopically, the main diagnostic characters are the different shaped sclerenchymatous cells, the striated epidermis and the aleurone grains. Hairs are absent.

The light brown Levant (or Mogador) caraway and also Indian dill have been substituted for genuine caraway (see B.P.C.).

Celery (Apium graveolens). The " seeds " are brown, subspherical and very small. Although the volatile oil consists largely of terpenes, the characteristic odour of celery appears to be due to a combination of a large number of compounds. The most outstanding feature of the microscopic appearance are the elongated " crossing " cells, which are not unlike those of fennel.

Coriander (Coriandrum sativum). The fruits are almost spherical, up to 6 mm. in diameter, brown and glabrous. Each cremocarp contains ten straight secondary ridges, which alternate with ten, primary, wavy, less conspicuous ridges. The volatile oil contains up to 70 per cent. of the alcohol *d*-linalool. The oil distilled from the unripe fruit has a disagreeable odour. Under the microscope, coriander shows layers of lignified sclerenchyma which cross one another and this results in a characteristic appearance which can be likened to a parquet floor. Hairs are absent.

Cummin (Cuminum cyminum). The cremocarps are brown, elongated oval, up to 6 mm. long and with bristly secondary ridges. Microscopically, cummin shows branched pluriseriate hairs, variously shaped sclerenchymatous cells, aleurone grains and calcium oxalate (rosettes).

Dill (Anethum graveolens). The mericarps are about 4 mm. long and 2·5 mm. broad, flat and glabrous. Like caraway, the main constituent of the volatile oil is carvone. Microscopically, the main features of dill are the striated cuticle of the outer epidermis, the fixed oil and the aleurone grains enclosing a micro crystal of calcium oxalate.

Fennel (Foeniculum vulgare). The fruits are about 12 mm. long and 3 mm. wide, greenish- (or yellowish-) brown, glabrous and oval-oblong. Each mericarp has five pale prominent primary ridges. The principal constituents of the volatile oil are anethole (about 50 per cent.) and fenchone (about 20 per cent.). Microscopically, the most diagnostic feature is the characteristic parquet-like arrangement of the narrow thin-walled cells of the inner epidermis. The polygonal tubular cells of the epidermis show occasional stomata. Aleurone grains and small rosette crystals of calcium oxalate are also present, but hairs and starch are absent.

SEEDS

Mustard as it appears on the market consists of a powdered and milled mixture of the seeds of *Brassica nigra* or *Brassica juncea* (called black or brown mustard) and *Sinapis alba* (called white or yellow mustard). If starch is present the product should be described as compound mustard or mustard condiment. When water is added in making up the condiment, the corresponding glucosides present become hydrolysed due to the enzyme myrosin in both seeds:

BLACK MUSTARD

$$\underset{\text{sinigrin}}{KC_{10}H_{18}NS_2O_{10}} \xrightarrow[\text{myrosin}]{H_2O} \underset{\substack{\text{allyl isothiocyanate}\\ \text{(volatile)}}}{C_3H_5NCS} + \underset{\text{dextrose}}{C_6H_{12}O_6} + KHSO_4$$

WHITE MUSTARD

$$\underset{\text{sinalbin}}{C_{30}H_{42}N_2S_2O_{15}} \xrightarrow[\text{myrosin}]{H_2O} \underset{\substack{\text{acrinyl isothiocyanate}\\ \text{(non-volatile)}}}{C_7H_7ONCS} + \underset{\text{dextrose}}{C_6H_{12}O_6} + \underset{\substack{\text{sinapin hydrogen}\\ \text{sulphate}}}{C_{16}H_{24}NO_5HSO_4}$$

Black mustard contains up to 2 per cent. of volatile oil of which over 90 per cent. consists of allyl isothiocyanate (rather less in *B. juncea*). Allyl isothiocyanate is the compound in mustard which is mainly responsible for the pungent odour and taste. Acrinyl isothiocyanate, which is present in white mustard, is odourless, but contributes to the pungency of the taste. Both black and white mustard seeds contain appreciable quantities of fixed oil (25–40 per cent.). Partial removal of the fixed oil should not necessarily be considered as a serious form of adulteration as it does not affect the taste to any great extent, and further, a lower proportion of oil reduces the liability of the mustard to rancidity. The following table shows typical analyses of brown and white mustard and the ranges of figures for five samples of commercial mixed mustards:—

MUSTARD

	Brown flour %	White flour %	Brown husk %	White husk %	Ranges for coml. mixed mustards %
Moisture . .	4·9	5·2	10·3	7·25	4·8 – 5·3
Starch (by diastase).	2·8	5·15	0·7	1·2	3·3 –14·3
Nitrogen . .	4·4	4·8	2·95	4·7	4·1 – 5·3
Fixed oil . .	39·8	38·1	14·5	22·8	24·0 –38·8
Allyl isothiocyanate.	1·5	*	—	—	0·5 – 0·9
Total sulphur . .	1·49	1·5	0·85	1·2	1·14– 1·57
Ash	4·1	4·1	4·6	4·25	3·7 – 4·5
Phosphorus (as P_2O_5)	1·8	1·47	0·75	1·3	1·51– 2·02
Pentosans . .	3·0	3·3	9·5	7·4	1·1 – 3·6
Crude fibre . .	3·7	3·4	15·7	13·2	1·4 – 4·2

* Volatile oil of white mustard flour is approximately 0·07 per cent.

The Food Standards (Mustard) Order requires that "**mustard, compound mustard** or **mustard condiment** shall be of such composition as to yield not less than 0·35 per cent. of allyl isothiocyanate after maceration with water for two hours at 37° centigrade and shall consist of a blend of brown and white mustard flours with or without amylaceous flours and/or spices: provided that the proportions of amylaceous flours and spices (if any) shall not together exceed 20 per cent. by weight."

The *U.S. standards* specify (a) for both **black and white mustard seed** maxima for total ash (5 per cent.) and acid insoluble ash (1·5 per cent.), (b) for **black mustard** a minimum for volatile oil (0·6 per cent. as allyl isothiocyanate) and (c) for **mustard flour** maxima for total ash (6 per cent.) and starch (1·5 per cent.) and partial removal of the fixed oil is permissible.

The routine examination of mustard should include a microscopical examination and the determination of **ash, fixed oil** and **allyl isothiocyanate.** If the microscopical examination indicates the presence of added flour, the **starch** should also be determined. Other determinations which may be necessary are **moisture, acid insoluble ash, crude fibre** (measures the amount of husk), **nitrogen, phosphorus** and **sulphur.**

In the following method the **allyl isothiocyanate** is produced by the enzyme in the presence of water and the volatile compound is then distilled over and allowed to react with silver nitrate in ammoniacal solution.

Weigh out 5 g. of mustard into a 250 ml. distillation flask (with side tube) and add 100 ml. of water. Stopper the flask (neck and side tube) and incubate at 37° C. for two hours. Remove the stoppers and add 20 ml. of alcohol.

Connect the flask to a condenser, replace the cork in the neck and distil about 65 ml. into a 100 ml. graduated flask (wide neck) containing about 10 ml. of ammonia solution (1 + 2), taking care that the end of the condenser delivery tube dips below the surface of the ammonia. Add exactly 20 ml. of N/10 silver nitrate solution to the distillate, mix and allow to stand overnight. Then heat in a boiling water-bath in order to coagulate the silver sulphide precipitate (about thirty minutes), cool and make up to 100 ml. with water. Filter. To 50 ml. of filtrate add 5 ml. of concentrated nitric acid and titrate with N/10 potassium thiocyanate using 5 ml. of 10 per cent. ferric ammonium sulphate as indicator.

1 ml. N/10 silver nitrate $\equiv$ 0·00495 g. allyl isothiocyanate.

CALCULATION

5 g. taken; 20 ml. N/10 $AgNO_3$ added to distillate and made up to 100 ml.
50 ml. filtrate ($\equiv$ 2·5 g.) required 7·25 ml. KCNS.

$$\text{Allyl isothiocyanate} = \frac{(\frac{20}{2} - 7{\cdot}25) \times 0{\cdot}00495 \times 100}{2{\cdot}5} = 0{\cdot}54\%$$

$$\text{Crude volatile oil} = 0{\cdot}54 \times \tfrac{100}{93} = 0{\cdot}58\%$$

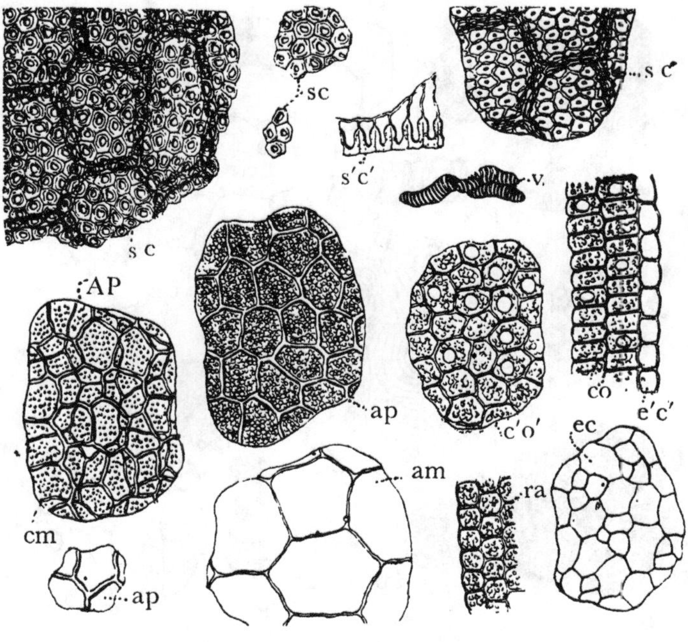

Fig. 37. Microscopical features of Black Mustard Seeds (from Greenish and Collin, *loc. cit.*).

am, Mucilaginous epidermis without distinct concentric striations.
ap, Aleurone layer, isolated.
AP, The same together with the membranous layer, cm.
co, Margin of cotyledon in transverse section, with epidermis, ec.
ec, Epidermis of cotyledon, surface view.
ra, Radicle.
sc, Sclerenchymatous layer surface view; the cells are brown and exhibit the characteristic polygonal network.
s'c', The same in profile, showing the characteristic thickenings.
v, Debris of raphe.

During storage mustard becomes moist—conditions which encourage production of allyl isothiocyanate, which tends to be lost by volatilisation.

The **microscopical examination** should preferably be carried out on the defatted material obtained after treatment with ether and then alcohol. Black and white mustard resemble one another closely in microscopical appearance. The sclerenchymatous layer in black mustard, however, is dark yellowish-brown and the upper surface shows a darker coloured polygonal network, whereas such reticulations are usually not discernible in white mustard. In addition, white mustard has an epidermis containing large polygonal cells with a colourless, striated mucilage, but that of black mustard is not striated. White mustard shows a hypodermal layer of large collen-

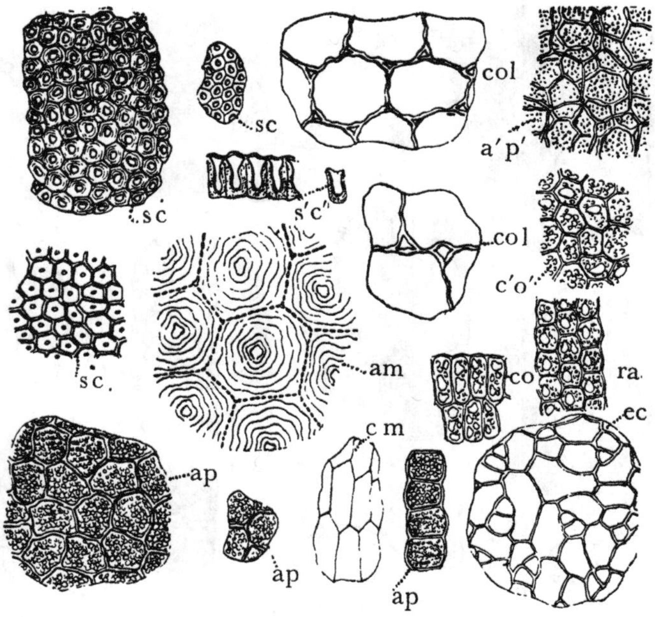

Fig. 38. Microscopical features of White Mustard Seeds (from Greenish and Collin, *loc. cit.*).

am, Large mucilaginous cells of epidermis in surface view, showing concentric striation.
ap, Isodiametric cells of proteid layer, containing aleurone grains.
a′p′, The same, with membranous layer of seed-coat adhering.
cm, Membranous layer.
co, Cells from outer part of cotyledons.
c′o′, Cells from inner part of cotyledons.
col, Collenchymatous cells.
ec, Epidermis of cotyledons.
ra, Tissue of radicle.
sc, Sclerenchymatous layer, surface view.
s′c′, The same in profile.

chymatous cells. Both types of mustard contain aleurone grains and in the unextracted material numerous round oil globules. A few minute rounded starch grains are seen occasionally due to the presence of unripe seeds. The microscopical examination should also show the presence of added starch and adulterants such as pepper, capsicum and turmeric. The presence of the latter can be confirmed by carefully running a little concentrated sulphuric acid under the coverslip so that the particles of turmeric are coloured red.

Starch. If the microscopical examination reveals the presence of added starch it should be estimated using diastase, as acid inversion gives high (incorrect) results:—

Thoroughly extract 5 g. with light petroleum, then with 95 per cent. alcohol and dry. Wash the residue into a 100 ml. graduated flask with ammonia-free water and heat on the water-bath for thirty minutes. Cool to 55° C. and add 0·1 g. of diastase dissolved in a little water and maintain at about 55° C. for three hours. Cool, add 5 ml. of N/10 sodium hydroxide solution, dilute to 100 ml. and estimate the sugar produced using Fehling's solution.

For calculation purposes assume that wheat flour (the most common in mixtures) contains 72 per cent. of starch.

The other determinations which may be necessary such as ash, fixed oil, etc., are carried out by the usual procedures. The crude fibre is useful for assessing the proportion of husk removed (see table). The proportion of black and white mustard present can be assessed approximately from the allyl isothiocyanate figure and from the amount of sulphur associated with nitrogenous constituents other than allyl isothiocyanate.

Prepared mustard is described in U.S. standards: " a paste composed of a mixture of ground mustard seed and/or mustard flour and/or mustard cake, with salt, a vinegar, and with or without sugar (sucrose), spices, or other condiments." The **fat-** and **salt-free solids** should contain a minimum of 5·6 per cent. of nitrogen and not more than 24 per cent. of carbohydrates (as starch) nor more than 12 per cent. of crude fibre.

Prepared mustards should be examined for water, ash, chloride, ether extract, allyl isothiocyanate, nitrogen, starch, acidity (calculate as acetic acid) and preservatives (*e.g.*, benzoic acid) and should also be examined under the microscope.

Nutmegs and **mace** are both derived from *Myristica fragrans*, the former being the kernel of the seed and the latter being the dried arillus (an " extra " seed coat). " True " Banda nutmegs are about

2·5 × 2 cm., ovoid and the surface is light brown, finely pitted and with reticulate markings. In comparison, Bombay nutmegs (*M. malabarica*) and Macassar (or Papua) nutmegs (*M. argentea*) are longer and narrower and tend to lack aroma.

The following table gives the ranges for the analytical figures of nutmeg and mace together with the B.P. (1953) and U.S. standards:—

NUTMEGS AND MACE

	Nutmegs			Mace	
	Range	B.P. (1953)	U.S.	Range	U.S.
	%	%	%	%	%
Moisture	4 – 8	—	—	3·5– 7	—
Ash	1·8– 4·5	3 (max.)	5 (max.)	1·6– 2·5	3 (max.)
Water soluble ash	1 – 2	—	—	0·9– 1·7	—
Acid insoluble ash	0 – 0·3	—	0·5 (max.)	—	0·5 (max.)
Fixed oil	30 –40	—	25 (min.)	24 –33	20–30
Volatile oil (v/w)	5 –15	5 (min.) whole 4 (min.) powder	—	4 –15	—
Alcohol extract	10 –16·5	—	—	21·5–25	—
Crude fibre	2 – 3·7	—	10 (max.)	4·7– 7·3	10 (max.)
Nitrogen	1·1– 1·4	—	—	0·85– 1·15	—
Starch	7·5–12	—	—	—	—

The U.S. standards permit a thin coating of lime which is sometimes met with as a protection from insect attack. On occasion, damaged nuts have been " repaired " by moulding the pieces with clay. Such factitious nutmegs have a high ash and yield little volatile oil.

The volatile oil (s.g. 0·860–0·925; o.r., 100 mm. tube + 10° to + 45°; R.I. 1·472–1·488—all at 20° C.) consists mainly of terpenes and a crystalline, toxic substance myristicin ($C_{11}H_{12}O_3$). Expressed oil of nutmeg is a yellow, aromatic fat (m.pt. 25°–43° C.). *Microscopically*, powdered nutmeg shows solid fat, simple and compound starch grains and the characteristic brown cells of the perisperm.

" True " Banda mace, when dried, consists of flattened, irregular golden-yellow pieces, about 2·5–5 cm. in length and brittle. The typical analytical ranges are given in the table. The fixed and volatile oils of mace tend to resemble those of nutmeg in composition. *Microscopically*, however, mace shows irregular grains of amylodextrin (which stain red with iodine), large, pointed thick-walled epidermal cells and the absence of starch.

Bombay mace and Macassar (or Papua) mace, which are considered inferior, can often be detected from the higher fixed ether extract (over 50 per cent.) and the higher alcohol extract (over 35 per cent.) —*cf.* Table. Further, if the spice is exhausted with light petroleum and then the extracted material is treated with ether, Bombay mace yields a much higher ether extract (30 per cent.) than genuine mace (about 2–4 per cent.). Other tests for distinguishing between spurious and genuine mace have been described by Griebel (*Analyst*, 1909, **34**, 441) and Schindler-Zwickau (*Analyst*, 1902, **27**, 327).

Cardamom (Ellettaria cardamomum). The seeds are about 4 mm. long, reddish-brown, irregularly angular and wrinkled. The fruit is up to 2 cm. long, greenish, cream or pale buff in colour, ovoid or oblong and may be longitudinally striated or nearly smooth. The B.P. prescribes the following standards for cardamom seeds: maximum ash 6 per cent. (U.S. 8 per cent.), maximum acid insoluble ash 3·5 per cent. (U.S. 3 per cent.) and minimum volatile oil 4 per cent. The seeds contain much starch and they yield about 7 per cent. extractive to 45 per cent. alcohol. *Microscopically*, the diagnostic features are the sclerenchymatous cells containing a nodule of silica and the small starch grains in the cells of the perisperm, which also contains prismatic crystals of calcium oxalate.

Fenugreek (Trigonella foenum-graecum). The seeds are up to 6·5 mm. long, 3 mm. wide and 2·5 mm. thick, hard, yellowish-brown, irregularly rhomboidal and flattened with a depression in one of the long narrow sides. The B.P.C. (1949) prescribed the following standards: maximum ash 6 per cent., maximum acid insoluble ash 2 per cent., maximum foreign organic matter 2 per cent. and minimum water soluble extractive 30 per cent. Fenugreek seeds contain about 6 per cent. of fixed oil, 28 per cent. of mucilage and also trigonelline and choline. *Microscopically*, fenugreek shows epidermal palisade cells which are about five times as long as they are wide, hourglass cells with bar-like thickenings, mucilage, aleurone grains and starch. Fenugreek is mainly of interest as the principal odorous constituent of curry powder.

ROOTS AND RHIZOMES

Ginger (Zingiber). Ginger of B.P. quality is the scraped and dried rhizome of *Zingiber officinale.* A number of different types come on to the market as being suitable for food purposes, *e.g.*, scraped gingers from Jamaica, the East Indies, India and Japan (the latter derived from *Z. mioga*) and unscraped ones from Africa. These varieties can be distinguished by the appearance of the "hands" and "fingers" (see Wallis' "Textbook of Pharmacognosy," Churchill, Lond.). Reich (*Z. Unters. Nahr. Genussm*, 1907, **14**, 549) has given the ranges for the analytical figures of various commercial varieties of ginger in the following table.

	Water %	Total ash %	Water soluble ash %	Sand %	Ether extract: Volatile %	Ether extract: Non-volatile %	Alcohol extract (after ether extraction) %
COCHIN GINGER							
Maximum	12·59	4·98	2·96	0·20	1·65	3·77	2·11
Minimum	10·40	3·33	1·64	0·05	1·05	3·02	1·49
Average	11·64	4·18	2·33	0·15	1·38	3·40	1·86
JAPAN GINGER							
Maximum	13·80	6·44	3·04	0·52	1·89	5·16	4·46
Minimum	9·20	3·24	1·52	0·14	0·90	3·72	2·69
Average	11·68	4·65	2·04	0·30	1·38	4·48	3·45
BENGAL GINGER							
Maximum	13·85	9·33	4·51	3·79	2·15	4·94	2·98
Minimum	11·69	5·48	2·56	1·21	1·17	2·84	1·12
Average	12·51	7·06	3·45	2·05	1·60	3·97	1·88
AFRICAN GINGER							
Maximum	13·65	6·22	2·56	2·47	3·08	8·08	2·42
Minimum	11·16	3·29	1·63	0·08	2·10	5·66	1·39
Average	12·74	4·37	1·97	0·84	2·54	6·50	1·70

The attractive whitish appearance of some gingers is due to treatment with calcium carbonate or sulphate, which also assist in warding off attack by insects and fungi. Limed gingers have a calcium content in excess of 1 per cent. (as CaO) whereas for natural ginger the figure is usually below 0·5 per cent. Bleaching by sulphurous acid or bleaching powder would appear to be a contravention of the Preservatives Regulations.

Analytical Figures and B.P. and U.S. Standards for Ginger

	Standards B.P. %	Standards *U.S.* %	Usual ranges for genuine ginger %	Exhausted ginger Sample A %	Exhausted ginger Sample B %
Moisture	—	—	8·4–13·9	11·5	12·7
Total ash	6 (max.)	†7 (max.)	3·2–7·6	3·8	5·75
Water soluble ash	1·7 (min.)	*2 (min.)	1·0–3·7	0·55	1·02
Acid insoluble ash	—	2 (max.)	—	—	—
Lime (CaO)	—	†1 (max.)	—	—	—
Volatile oil	—	—	1·0–3·1	0·5	1·1
Fixed oil and resin	—	—	2·8–7·5	1·2	1·4
90 per cent. alcohol extract	4·5 (min.)	—	4·5–8·1	3·1	2·5
Crude fibre	—	8 (max.)	1·7–6·5	—	—
Nitrogen	—	—	1·0–1·5	1·0	1·1
Cold water extract	10 (min.)	†12 (min.)	7·0–14·0	5·3	4·5
Starch	—	42 (min.)	48·5–53·0	—	—
Gingerol	—	—	0·9–2·5	—	—

* For cold-water soluble ash, whereas the B.P. method employs boiling water.

† Additional U.S. standards:—Jamaica ginger—cold water extract 15 per cent. (min.). Limed ginger or bleached ginger—calcium carbonate 4 per cent (max.), total ash 10 per cent. (max.).

The crude fibre, for which a maximum is prescribed in the U.S., is useful for assessing the degree to which the rhizomes have been scraped. The presence of exhausted (spent) ginger, which is a by-product of ginger beer manufacture, is indicated if the cold water and alcohol extracts are low. In addition, the water soluble ash represents only about 25 per cent. of the total ash instead of the usual 40–50 per cent. for genuine ginger.

In view of the wide variations found in natural gingers (*e.g.*, high ash in Bengal varieties), the pharmacopoeial limits should not be applied too meticulously, however (especially if the description is not

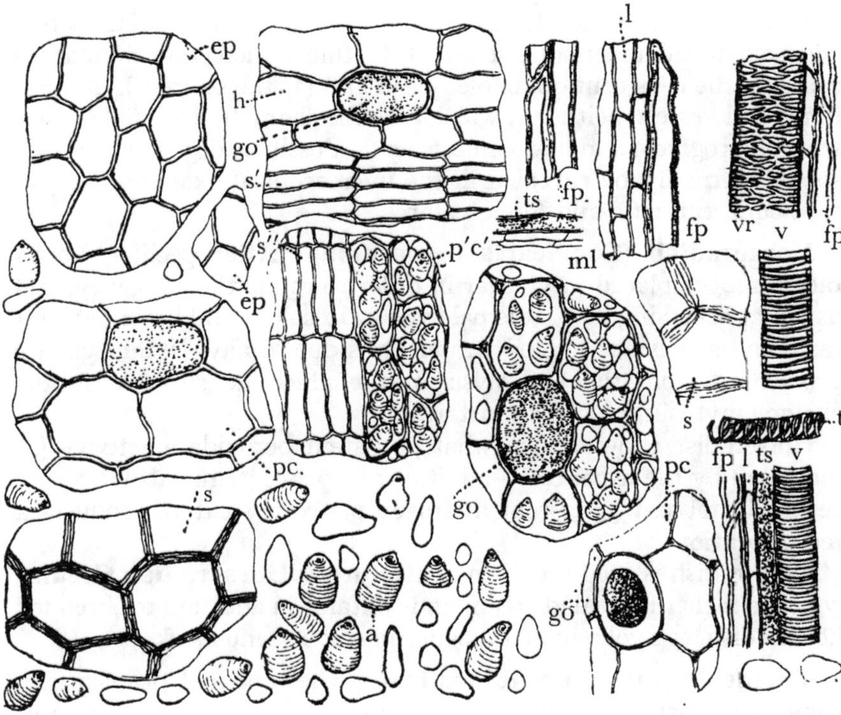

Fig. 39. Microscopical features of Ginger Rhizome (from Greenish and Collin, *loc. cit.*).

a, Starch grains.
ep, Epidermis.
fp, Sclerenchymatous fibres.
go, Oleoresin cells.
h, Hypoderma.
l, Bast.
ml, Parenchyma of stele.
pc, Cortical parenchyma, the walls somewhat swollen by treatment with potash.
p'c', The same in longitudinal section.
s, s', s'', Cork, in surface view, transverse and longitudinal section respectively.
ts, Elongated secretion cells.
v, vr**,** Vessels.

qualified by the letters " B.P."). In general the cold water extract is the most useful criterion of quality.

The aroma of ginger is mainly derived from the volatile oil which contains cineole, terpenes (d-camphene, β-phellandrene and zingiberene), citral and borneol. The pungency is due to an oleo-resin, gingerol, the chemistry of which has been discussed by Nomura (*J. Chem. Soc.*, 1917, **111**, 769). Gingerol can be separated by Garnett and Grier's method (*Yrbk, Pharm.*, 1909, 344; see also Redgrove, *Pharm. J.*, 1930, **125**, 54).

Microscopically, ginger shows (1) abundant starch grains which are either sack-shaped, ovoid or simple, are faintly striated, have an eccentric hilum (length 12–50 μ), and exhibit an asymmetric cross when examined under polarised light, (2) thin-walled parenchymatous cells, (3) the cells containing oleo-resin and (4) the sclerenchymatous fibres and vessels, neither of which give the characteristic pink colour with phloroglucin and hydrochloric acid. Possible adulterants, which should normally be revealed in the microscopical examination, are capsicum, turmeric and foreign starches.

Horseradish (**Cochlearia armoracia**). After peeling, horseradish root is placed in vinegar in order to retain the pungency, and usually packed in jars. It is pale yellow in colour and has a pungent taste and a mustard-like odour, which is due to allyl isothiocyanate production from the hydrolysis of the glucoside sinigrin by the enzyme myrosin (*cf.* black mustard).

Horseradish shows a somewhat persistent peroxidase activity (as indicated by the usual test—addition of 1 per cent. guaiacol in 50 per cent. alcohol and 1 per cent hydrogen peroxide solution produces a red coloration).

Horseradish is used for the manufacture of **Horseradish Cream,** which is rather like salad cream and contains in addition to shredded horseradish, egg yolk, milk, oil, vinegar, salt and mustard.

Turmeric (**Curcuma domestica**). The prepared primary rhizome is pear-shaped (" bulb " turmeric), whilst the secondary rhizome tends to be cylindrical ("finger" turmeric). The latter type has the greater tinctorial value. On analysis, turmeric yields 6·5–7·5 per cent. ash, 7–9 per cent. fixed oil, 3–5 per cent. volatile oil and much starch. As the rhizomes are steamed or boiled during their preparation, the yellow coloured starch appears somewhat gelatinised under the microscope. Turmeric has a bitter taste, but is used perhaps rather more for imparting colour, due to the presence of curcumin, a yellow crystalline substance. It is considered to be inferior as a spice, but is invariably present in curry powder and sometimes in ground mixed spice. It has been used as an adulterant of ginger, mustard and pepper and can be detected chemically by preparing an alcoholic tincture and using it for the usual test for borates.

A very high lead content has been reported on occasion in turmeric due to the use of lead chromate as colouring.

PRODUCTS OF HERBS

Mixed herbs usually contain three or four herbs selected from thyme, savory, sage, marjoram, parsley and rosemary.

Stuffings contain the appropriate herbs together with bread-crumbs, suet, salt and sometimes pepper and colouring. Sage and onion stuffing was required by an official order (S.R. & O. 1942 No. 1863—now revoked) to contain at least 2 per cent. of sage and 3 per cent. of dried onion. The same revoked order prescribed a minimum of 2 per cent. of thyme for parsley and thyme stuffing. The analysis of stuffings has been discussed by Sherratt (*Analyst*, 1943, **68**, 202).

PRODUCTS OF SPICES

Curry powder usually contains four or five spices selected from fenugreek, turmeric, cinnamon, cassia, coriander, fennel, ginger, capsicum, pepper, mustard and pimento. Other possible spices are cardamoms, mace, cumin, dill, the ground up leaves of an Indian plant, *Murraya koenigii* or a piece of bay leaf. Salt is invariably present. Many different mixtures are prepared, but the Food Standards (Curry Powder) Order* states an overall minimum of 85 per cent. of spices and aromatic seeds and herbs. This infers that the total of added flour (pea or bean flour were frequently present at one time) and salt shall not exceed 15 per cent. The official order (as amended)† also states a maximum of 20 p.p.m. of lead. The amount of lead present can be conveniently determined by Hamence's method (see p. 61), in which the large quantity of iron present is removed as thiocyanate.

The following data represent the typical figures or ranges obtained from samples of curry powder (all as percentages): moisture 11·2, total ash 6·0–12·0, water soluble ash 8·2, acid insoluble ash 0·1–0·7, salt 2–10, cold water extract 20 and fixed ether extract 13·1. The usual ranges for trace elements are arsenic 0·1–1·0 p.p.m. and lead 1–10 p.p.m.

* S.I. 1949 No. 1816. † S.I. 1956 No. 1166.

Ground mixed spice usually contains about five ground spices selected from cloves, ginger, coriander, fennel, cinnamon, cassia, capsicum and turmeric. Mixed spice usually yields a total ash of 4·1–4·8 per cent. and an acid insoluble ash of 0·3–1·2 per cent. The arsenic content is usually negligible but lead may be present (up to 5 p.p.m.).

Pickling spice consists of a mixture of whole spices (not ground). Most samples contain coriander, pimento, chillies, ginger (broken pieces), white mustard, cloves, black pepper and mace. Occasionally cinnamon, black mustard and white pepper are present.

Liquid and powdered seasoning extracts are found to be convenient for the incorporation of flavour into foods such as soup mixes and meat products. The liquid seasonings are virtually essences of the spice oils. The solid extracts usually contain salt or dextrose as diluent or absorbent.

Chapter 10

FERMENTATION PRODUCTS

Wines—Spirits—Beer—Cider—Vinegar

THE chemistry of wine and spirits has been much more studied in the wine-producing countries, notably France and Germany, than in England. The detailed examination of different wines does not commonly afford to the analyst so much information as is obtainable from the palate of an experienced taster. In this chapter, therefore, it is not proposed to discuss in detail the large number of estimations which have been recommended by various authorities for the examination of wine, but only to indicate broadly the commoner processes directed at the detection of adulteration and to give references to the literature where further information may be obtained.

WINES

It is difficult even to define wine. Formerly it meant simply the fermented juice of the grape, but many kinds are now fortified by the addition of more alcohol, which may or may not have been derived from the grape. Some wines are " still," and others contain much carbon dioxide. Some are red and some are white, depending upon the type and form of grape used, and there are the wide variations associated with the different classes of wine, such as port, sherry, claret, burgundy, hock, champagne, etc., indicating in many cases the country or district of origin. These may be modified by a geographical prefix such as " Australian burgundy."

By virtue of the Anglo-Portuguese Treaty Acts the term port is restricted to the produce of Portugal shipped with a certificate from the Portuguese Authority. Strictly speaking, sherry is the wine which takes its name from Xeres and is shipped only from certain parts of Spain.

Analytical data on the common types of wines are shown in the

Composition of the Commoner Wines

	Port.	Sherry.	Claret.	Burgundy.	Hock.	Champagne.
Specific gravity	0·995–1·050	0·992–1·015	0·990–1·020	0·995–1·010	0·990–1·008	1·040–1·055
Alcohol, grams per 100 ml.	13·5–20·0	13·5–20·5	7·5–12·5	7·5–12·5	7·5–12·5	10·0–14·0
Total solid matter	3·3–8·4	2·0–5·0	2·0–3·5	2·0–3·5	1·5–2·5	9·5–18·0
Free volatile acid (as acetic)	0·05–0·10	0·15–0·23	0·09–0·15	0·2–0·35	0·05–0·15	0·03–0·20
Fixed acid (as tartaric)	0·35–0·55	0·25–0·50	0·30–0·50	0·3–0·60	0·25–0·45	0·30–0·45
Ash	0·25–0·35	0·35–0·55	0·2–0·3	0·2–0·40	0·10–0·25	0·25–0·45
Sugars	2·5–6·5	2·0–5·0	0·0–0·7	0·03–0·55	0·0–0·35	8·5–16·0
Phosphoric acid (P_2O_5)	0·03–0·05	0·03–0·05	0·03–0·04	0·02–0·03	0·02–0·03	0·03–0·05
Glycerin	0·3–1·3	0·4–1·0	0·3–1·0	0·3–1·0	0·25–1·3	0·3–1·0

table opposite, information for which has been collected from various sources. The results are expressed in grams per 100 ml.

The limits found in the table must be interpreted with caution, as so much depends on the place of origin of the particular type of wine, and on the age, vintage, and other circumstances. Analyses of large numbers of wines from different localities are given by König.

As the composition of wines varies so much, no very definite conclusion can be drawn from the consideration of any particular figure, but the following generalisations are widely accepted on the Continent. The specific gravity should not be below 0·985; if the alcohol content is more than 14·5 g. per 100 ml. (in the case of port or sherry this figure is often exceeded), added alcohol is indicated; the glycerol content, which is not often more than 1 per cent., should be one-fifteenth to one-fifth of the alcohol, and the total acid should not exceed 1·5 per cent. The total solid matter or extract varies within wide limits, being largely determined by the sugar present. "Dry" wines contain about 0·1 per cent. of sugar, others contain up to 6 per cent., and champagne even up to 16 per cent. According to the Paris Municipal Laboratory, if the sum of the sugar and twice the alcohol exceeds 32·5, the wine may be considered to have received an addition of sugar or alcohol. Importance also attaches to the ratio alcohol : extract; this for red wines should not exceed 4·5, or for white wines 6·5.

Analysis of Wine

A complete examination of wine should include the estimation of **specific gravity, alcohol, extract, acidity, ash, sugar, glycerol, sulphate,** and embody a search for **sulphites,** or other preservative. **Saccharin** may also be present, and sometimes it is necessary to search for **methyl alcohol** or for **added colouring matters.**

For the estimation of **alcohol** the general method of distillation, followed by determination of specific gravity, is employed. Proteinoid matters should be precipitated by the addition of lead acetate or tannin, and volatile acid must be neutralised by alkali. Excess of the latter is to be avoided, lest any ammonia be distilled over. For wines it is convenient to take 100 ml. measured accurately at 20° C. and wash it into a distillation flask with 50 ml. of water, add a slight excess of sodium hydroxide solution, then 2·5 g. of lead acetate and then distil until about 90 ml. have passed over. This is adjusted to 100 ml. at 20°* by the addition of water, and its specific gravity is taken by means of a pyknometer. Reference to the Alcohol Tables issued by the Commissioners of Customs and Excise gives the

* Alcohol tables are also available for measurements at 15·5° C.

percentage of alcohol corresponding to the observed specific gravity. It is convenient to return the results as grams per 100 ml.

Strong spirits or alcoholic mixtures containing more than 50 per cent. of alcohol should be diluted with water before distillation. For example, 50 ml. may be diluted to 150 ml. and 100 ml. distilled.

When the material under examination contains **essential oils** or other volatile ingredients the method of Thorpe and Holmes is applied (*J. Chem. Soc.*, 1903, **83**, 314). The sample or distillate from it is diluted so that the alcohol content does not exceed about 25 per cent., then shaken in a separator with sufficient salt to saturate the liquid. It is then vigorously shaken with 50 ml. or more of petroleum spirit, the aqueous layer is run out and the alcohol distilled off; the essential oils or fatty matter will all be in the petroleum layer.

Many methods have been recommended for the detection and estimation of **methyl alcohol** in ethyl alcohol, the presence of the former being an indication of adulteration with *denatured spirit;* perhaps the best method is that of Denigès, which is as follows: The alcoholic solution is first purified, where necessary, by distillation and diluted to 10 per cent. by volume. To 5 ml. of this prepared liquid, contained in a wide test tube, are added 2·5 ml. of 2 per cent. solution of potassium permanganate and 0·2 ml. of sulphuric acid. When the reaction has proceeded for three minutes, 0·5 ml. of oxalic acid solution (9·6 per cent.) is added. On shaking, the liquid becomes clear and nearly colourless; 1 ml. of strong sulphuric acid is now run in and well mixed with the solution, which is finally treated with 5 ml. of Schiff's reagent. A violet colour is developed in the course of a few minutes unless only traces of methyl alcohol are present, when twenty or thirty minutes may be required.

It is essential that the details be strictly adhered to, and the Schiff's reagent must not be too strongly acid. *Schiff's reagent* may be prepared by dissolving 0·2 g. of basic fuchsin in 120 ml. of hot water, then decolorising the solution, after cooling, by adding 2 g. anhydrous sodium sulphite dissolved in 20 ml. of water, a little charcoal, and finally 2 ml. of hydrochloric acid. The solution should be kept in a stoppered amber-coloured bottle. Denigès' method can be applied quantitatively by comparing the colour developed by the sample with that of 10 per cent. ethyl alcohol containing small additions of methyl alcohol and treated by the same process. Alternatively, chromotropic acid (see p. 91) can be used instead of Schiff's reagent (*cf. J. Pharm. Pharmacol.*, 1949, **1**, 335, 777; 1950, **2**, 434).

The increasing industrial use of *iso*-**propyl alcohol** renders it necessary to test for it in alcoholic distillates. One method is that of Adams and Nicholls (*Analyst*, 1929, **54**, 2) which depends upon the oxidation of *iso*-propyl alcohol to acetone, which condensed with *o*-nitrobenzaldehyde forms indigo.

The solutions to be tested are diluted to a strength of approximately 10 per cent. of alcohol. To 10 ml. of each solution in test tubes 5 ml. of saturated bromine water are added. The tubes are lightly corked and allowed to stand for about three to six hours in the cold, or even overnight, in a cool dark cupboard. After the period of standing, 1 ml. of a 1 per cent. solution of *o*-nitrobenzaldehyde in 50 per cent. ethyl alcohol is added, and the solutions are mixed by gentle shaking. Finally, 2 ml. of a 30 per cent. solution of sodium hydroxide are added, and the solutions shaken once more. The tubes are then allowed to stand for about fifteen minutes in diffused daylight, and the colours developed are compared with those produced in standard *iso*-propyl alcohol solutions similarly treated. A convenient standard solution is one containing 2·5 ml. of pure *iso*-propyl alcohol in 100 ml. of 10 per cent. ethyl alcohol. The standard colours are those produced by using 1·0, 1·5, 2·0 ml., etc., of this solution, made up to 10 ml. with 10 per cent. ethyl alcohol.

Rigidly defined methods for the determination of **extract** are given by Continental workers. Perhaps the best is to evaporate 50 ml. of a dry wine or 25 ml. of a sweet wine in a flat-bottomed dish of 3-inches diameter, then dry the residue in a steam oven at 100° for two hours. If the extract exceeds 5 per cent., indirect estimation is preferable: determine the specific gravity of the wine (S) and that of the distillate (S′) when diluted to the original volume; then the extract (grams per 100 ml.) $= \frac{S - S'}{0{\cdot}00386}$; the results so derived are more accurate than those obtained by direct estimation, on account of the sugars and glycerol present.

The **ash** is obtained by ignition of the extract. Its alkalinity and solubility in water may also be determined.

The estimation of **total acidity** of wine is rendered less easy by reason of carbon dioxide and of colouring matters, which obscure the indicator; the simplest expedient is to use phenolphthalein spotted on a white tile as an outside indicator. For the separate estimation of volatile acid, which is reckoned as acetic, to 25 ml. add about 1 g. of tannin, then water, distil the liquid until the volume is reduced to one half, then introduce more water and continue the distillation until at least 200 ml. have been collected. Titrate the distillate as usual. Fixed acid is calculated by difference, but it is to be noted that the volatile acid as acetic should be calculated to its equivalent as tartaric acid before being subtracted from the total acidity, which is usually reckoned as tartaric except in France, where convention requires that acidity should be expressed as H_2SO_4 per 1,000 ml.

The **sugars** present may include sucrose as well as dextrose and lævulose, and possibly others, so that it is necessary to determine the cupric reducing power before and after inversion, the former being returned as invert sugar and the latter—after allowing for reducing sugar—as sucrose. One hundred ml. of the sample are carefully neutralised with sodium hydroxide, then evaporated to half that volume to remove alcohol; the solution is cooled, cleared with zinc ferrocyanide (p. 111), made up to 100 ml. and then filtered. The copper reducing power is determined volumetrically on the filtrate, and similarly on another 25 ml. of the sample after inversion.

Glycerol in wine may be estimated approximately by the German official process, which involves direct weighing of the glycerol. If the wine contains sugar not exceeding 2 per cent., 100 ml. are evaporated on a water bath to 10 ml. Then for each gram of extract is added 1 g. of fine sand and 2 ml. of concentrated (40 per cent.) milk of lime. The mixture is evaporated nearly to dryness; then 5 ml. of alcohol are added, and the mass worked up into a paste in the basin with the aid of a spatula, more alcohol being added if requisite. Then the paste is warmed nearly to the boiling point and is washed into a 100 ml. flask with several small quantities of warm alcohol, cooled to 15° and made up to the mark with alcohol. After shaking and filtration, 90 ml. of the clear liquid are evaporated very gently almost to dryness, and the residue taken up by a small quantity of absolute alcohol, transferred to a stoppered cylinder and rinsed in with more alcohol until the volume is 15 ml. Now three separate portions of 7·5 ml. of ether are added with shaking between each addition. The clear liquid is poured into a tared basin, rinsed in with a little alcohol-ether mixture (2 : 3) and very gently evaporated, and finally is dried in the oven for an hour and weighed.

In the case of sweet wines containing more than 2 per cent. of sugar, 50 ml. are warmed in a flask on the water bath with 2 g. of sand and an excess of milk of lime—excess is indicated by the colour changing from dark to pale—the mixture is cooled, 100 ml. of alcohol are added and the precipitate is filtered off and washed with alcohol. The filtrate is treated exactly as an unsweetened wine as above (*cf. Anal. Abstr.*, 1954, **1**, 1678; 1956, **3**, 3458).

Sulphites are commonly used in the sulphuring of casks, and may sometimes be added to wine. For the estimation the following direct process gives reasonably accurate results with white wines: Add 50 ml. of the wine to 25 ml. of N sodium hydroxide solution and, after allowing the mixture to stand for fifteen minutes, 10 ml. of dilute sulphuric acid (1 + 3) and a few drops of starch solution. Then titrate with 0·05 N iodine solution; each ml. of 0·05 N iodine is equivalent to 0·0016 g. SO_2. For coloured liquids recourse must usually be had to the distillation process described on p. 80. **Benzoic acid** and **Salicylic acid** can be detected on the alcohol-free liquid by the methods already given.

Saccharin is detected by extracting a large volume, say 250 ml., of the acidified wine from which the alcohol has been removed by evaporation and replaced by the addition of water, with three portions of 30 ml. of ether. Then, after removal of the ether, the wine is evaporated to half its bulk and again extracted with ether. The mixed ether extracts are distilled off at a low temperature and finally evaporated spontaneously in an open basin. The residue is dissolved in about 30 ml. of water and again evaporated to dryness, then dissolved in about 5 ml. of 0·2 N sodium hydroxide solution and slight excess of potassium permanganate solution is added. The pink colour should persist after a few minutes' warming and then the mixture is acidified and just decolorised by the addition of sulphurous acid. The liquid is filtered and extracted three times with ether; the ether is allowed to evaporate spontaneously and the residue taken up with water and tasted. A characteristic sweet taste is indicative of saccharin; it should be confirmed by the qualitative reaction given below.

For quantitative estimation the residue is moistened with hydrochloric acid and evaporated nearly to dryness on the water bath, then re-heated with a further ml. of hydrochloric acid and evaporated. The hydrolysed residue is dissolved in ammonia-free water and to an aliquot part is added 2 ml. of Nessler reagent. The colour is matched against standard ammonium chloride solution.

$$1{\cdot}0 \text{ mg. saccharin } (C_7H_5O_3NS) = 0{\cdot}29 \text{ mg. } NH_4Cl.$$

Dulcin is detected by extracting the alcohol-free liquid with ether after making it alkaline. The solvent is evaporated off and the sweet-tasting dulcin remains. Dulcin and saccharin may be identified by heating a few mg. with an equal quantity of resorcinol and a few drops of sulphuric acid to 180° C. for about two minutes. Then cool, add water and make alkaline with caustic soda. A red colour with green fluorescence indicates saccharin. To part of the solution add a few drops of bromine water. In the presence of dulcin a violet colour appears on standing. A quantitative method for dulcin is given by Longwell and Bass (*Analyst*, 1942, **67**, 15).

Detection of Fruit Wine in Grape Wine. See *Analyst*, 1929, **54**, 422.

Detection of Cider in Grape Wine. See *J. Sci. Food Agric.*, 1955, **6**, 387.

British Wines

Most British wines are prepared from imported grape, but on occasion raisins, currants, rhubarb and juices of other fruits have been used. Some products contain ginger or orange peel to give flavour and added colours may be present. The alcohol content varies from 10–27 per cent. of proof spirit. Samples of these products should be examined for alcohol, sugars, extract, ash, preservatives, lead, copper and tin.

Apart from the special requirements of the Labelling of Food Order given below, the M.O.F. issued a press notice in 1946, which was intended principally to give advice to the trade as to the labelling of British wines. This stated that terms such as " Ginger Wine " and " Elderberry Wine " are appropriate, but if terms such as " wine " or " sherry " are used, such words should be preceded by the word " British " in type of at least equal size. In the description of the fruit used, the word " grapes " should not be used to describe the basis of a product which, though derived from grapes, is made from imported grape juice or dried vinous fruits.

SPECIAL REQUIREMENTS OF THE LABELLING OF FOOD ORDER, 1953 (AS AMENDED), RELATING TO THE LABELLING OF WINES AND CERTAIN LIQUORS.

Article 8 of Part V states that:

" No person shall sell or have in his possession for sale any liquor described in a label attached to or printed on the wrapper or container:—

(a) in the case of intoxicating liquor, by any name or words calculated to indicate either directly or by ambiguity, omission or inference, that the liquor is, or resembles, wine obtained by the fermentation of the juice of grapes, or is a substitute for or has the flavour of such wine, unless it is derived from fruit, and, in so far as it is derived from fruit, is derived exclusively from grapes;

(b) in the case of intoxicating liquor which is not derived from fruit or which is wholly or partly derived from fruit other than grapes, by the use of the word " wine," unless that word is immediately preceded in identical lettering by a word or words accurately specifying the description of fruit or fruit product or other saccharine material used;

Provided that nothing in this sub-paragraph shall prevent the use of the description " Ginger Wine " or " Orange Wine " for products wholly or partly derived from fruit other than grapes;

(c) by any name or words calculated to indicate either directly or by ambiguity, omission or inference, that the liquor is, or resembles a sweetened liqueur or is a substitute for or has the flavour of a sweetened liqueur unless it is a suitably flavoured compounded spirit which has been rendered sweet and viscous only by the addition of sucrose, dextrose or invert sugar and not by the use of any other ingredient."

Also Article 4(d) of Part II requires the labels of wines, which are not derived exclusively from grapes, to bear a statement as to the fruit used and the alcoholic content (see S.I. 1953 No. 536 as amended by S.I. 1961 No. 440).

Claims for Medicinal Properties (*Labelling of Food Order*)

Claims of medicinal properties are not permitted for liquors by reason only of the presence of alcohol. To justify such a claim some other ingredient with such properties must be present and its amount stated. This restriction does not apply, however, to " Indian Tonic Water " or " Quinine Tonic Water " which contains not less than $\frac{1}{2}$ grain of quinine (calculated as quinine B.P.) per pint.

SPIRITS

LEGISLATION

Section 3 (4) of the Food and Drugs Act, 1955, states that in proceedings taken under S. 2 it shall be a defence, in respect of diluted whisky, brandy, rum or gin, for the defendant to prove that its strength was still not lower than 35 U.P., *i.e.*, a minimum of 65 per cent. Proof Spirit.* Part V (8f) of the Labelling of Food Order, 1953, prescribes that prepacked spirits which have an alcoholic content which is less than 65 per cent. proof spirit may not be sold as brandy, gin, rum or whisky unless such name is preceded by the word " diluted." Part II (4, (4a) (iv)) requires that the declaration on the label for such products should be in the following form:—

* See footnote on p. 280.

DILUTED WITH WATER TO NOT
LESS THAN (Y)

The declaration shall be completed by inserting at (Y) the minimum content of " PER CENT. ALCOHOL BY VOLUME " or " PER CENT. PROOF SPIRIT." This requirement shall not apply to brandy, the alcohol content of which has fallen below 65 per cent. P.S. through maturing in the cask.

Spirits differ essentially from other alcoholic liquors in being produced by distillation; they contain much less extractive matter and no glycerin, but a high percentage of alcohol together with certain esters which, though small in amount, are of much importance, as they confer the characteristic flavour of the particular spirit.

Brandy is distilled from wine and subsequently matured by age. That sold in England comes largely from the Cognac district of France, but so many other brandies are now on the market that care is needed in interpreting the results of the chemical analysis. The term Cognac can only properly be applied to brandy made in the Cognac district. Much importance attaches to the taste or characteristic " bouquet " which is produced by the appropriate conditions of manufacture and ageing. Chemically it is due to the presence of small but significant amounts of higher alcohols, acids and esters, among the more important of which are acetic, butyric, œnanthic, and valerianic esters, furfuraldehyde, and ethyl pelargonate. It is usual to express the " impurities " as milligrams per 100 ml. of absolute alcohol.

The table on p. 281 shows the usual composition of brandy. Of these numbers 1 to 6 are French brandies, probably genuine, and numbers 7, 8 and 9 are Greek, Spanish and Australian spirits respectively; in these the alcohol content is much higher. Much has been made in the past of a standard of 80 for the minimum content of esters and of 300 for the total impurities, which include volatile acids, esters, aldehydes, furfural and higher alcohols, but

* " Proof Spirit " is defined in S. 172 (2) of the Customs and Excise Act, 1952, as follows: " Spirits shall be deemed to be at proof if the volume of ethyl alcohol contained therein made up to the volume of the spirits with distilled water has a weight equal to that of twelve-thirteenths of a volume of distilled water equal to the volume of the spirits, the volume of each liquid being computed as at 51° F.". Proof spirit contains 49·3 per cent. of anhydrous alcohol by weight (≡ 57·1 per cent. by vol.). A liquid which is 35° U.P. contains 65 vols. of proof spirit and 35 vols. of water. The relationship between the s.g. and the percentage of alcohol by weight and by volume together with the corresponding percentages of proof spirit are given in the Customs and Excise Tables (H.M.S.O., 1949).

Constituents of Brandy

	1.	2.	3.	4.	5.	6.	7. Greek.	8. Spanish.	9. Australian.
Specific gravity	0·9566	0·9481	0·9477	0·9427	0·9414	0·9471	0·8896	0·8539	0·9180
Proof spirit (per cent.)	73·05	76·78	78·71	82·78	83·84	75·70	122·8	146·6	103·5
Total solids (per cent.)	1·904	0·646	1·024	0·757	0·608	0·132	0·01	0·08	0·704
Ash (per cent.)	0·012	0·012	0·006	0·005	0·004	0·004	0·001	0·004	0·006
Total acidity as acetic (per cent.)	0·046	0·039	0·072	0·054	0·041	0·004	0·022	0·034	0·027
= Milligrammes per 100 ml. alcohol	110	89	160	115	86	9	31	41	46
Total esters (ethyl acetate)	580	463	564	502	335	112	—	—	—
Volatile esters	122	57	88	101	123	44	71	48	33
Furfural	6	2	3	4	5	0	1·4	1·2	0
Aldehyde	69	41	71	76	29	24	31	6	8
Higher alcohols	146	315	782	675	611	374	464	314	316

how far these standards are really justified is debatable (*cf.* Hehner, *Analyst*, 1905, **30**, 36), and they should be applied with care and supported by tasting by an experienced palate. It appears to be the practice of certain manufacturers to import "silent spirit" (a term applied to almost pure alcohol obtained by distillation in a patent still), and add to it artificial esters to produce so-called brandy; obviously such a liquid could easily be made to pass the chemical standards suggested. The matter is further complicated by the wide variation in brandies of known purity, the "impurities" varying from 25 to 1,200 and esters from 18 to 450.

Whisky was formerly spirit distilled from malted barley only, but of late years the definition has much widened and includes spirit distilled from a variety of fermented cereals. Like brandy, therefore, the standards by which it must be judged are somewhat elastic. Much depends upon the type of still employed; the product of a simple pot-still generally has a higher content of impurities than that of a patent still in which there is considerable rectification. There is some overlapping, however, so that it is difficult to draw conclusions from the analytical figures alone. The effect of storage is important. Certain tannins and colouring matters are derived therefrom and definite changes take place in the liquid (*cf.* Crampton and Tolman, *J. Amer. Chem. Soc.*, 1908, **30**, 98), among which it may be noted that the alcohol content decreases slightly—about 1 per cent. per annum—and the furfural and esters increase during the first three or four years, afterwards remaining nearly constant.

Much information as to the different varieties of whisky is given in the Minutes of the Royal Commission on Whisky and other Potable Spirits, 1908–9, which defines whisky as the spirit obtained by distillation from a mash of cereal grains saccharified by the diastase of malt; Scotch whisky being distilled in Scotland and Irish whisky in Ireland. The following are analyses of known genuine whiskies:—

	1.	2.	3.	4
Specific gravity . .	0·9425	0·9425	0·9463	0·9400
Proof spirit (per cent.) .	80·53	80·53	76·34	82·60
Total solids (per cent.) .	0·122	0·135	0·070	0·103
Ash (per cent.) . .	0·012	0·015	0·003	0·010
Total acidity as acetic (per cent.)	0·021	0·010	0·013	0·025
= mgms. per 100 ml. alcohol	46·0	22·0	31·7	53·0
Esters	50·0	56·5	75·9	35·5
Furfural	4·8	5·5	0·7	1·7
Aldehydes . . .	65·2	52·1	126·4	87·0
Higher alcohols . .	404·3	439·1	326·4	360·5

For a large number of analyses of whiskies, see Schidrowitz (*J. Soc. Chem. Ind.*, 1902, **21**, 815; 1905, **24**, 585).

The standards by which whisky may be judged are similar to those of brandy, but care must be taken not to place too much reliance on figures alone. If the total impurities are much less than 300 there is a probability that some silent spirit has been added. The esters should always be lower than the higher alcohols. The reverse is often the case with rums.

Rum is a spirit distilled from the fermented products—often waste products—of the sugar cane. Formerly it was almost exclusively derived from Jamaica and the Indies, but imitation rum is now imported from countries in which the sugar cane is not cultivated. Some kinds are prepared from beet molasses, others from silent spirit. Rum differs essentially from whisky or brandy in the preponderance of esters over higher alcohols; ethyl acetate and butyrate predominate. Partly on this account it is particularly easy to produce fictitious rums by the addition of these esters to suitably diluted and coloured silent spirit. The proportion of higher alcohols is quite low in rum; usually it is less than 250 per 100,000, and sometimes less than 100.

Gin is a more artificial product than the other spirits, and is usually prepared by the addition of various flavouring essences to a highly rectified spirit; the essences added commonly include oils from cardamom and coriander seeds and juniper berries. In view of the process of manufacture it is clear that there can be no standard applied or conclusions drawn from the proportions of esters, higher alcohols, or other impurities as in the case of whisky, brandy or

	Rum.			Gin.	
	1	2.	3.	4.	5.
Specific gravity .	0·9436	0·9375	0·9403	0·9531	0·9504
Proof spirit (per cent.)	79·60	85·95	82· 50	69·95	72·04
Total solids (per cent.)	0·385	0·420	0·295	0·095	0·013
Ash (per cent.) . .	0·041	0·039	0·025	0·008	0·009
Total acidity as acetic (per cent.) . .	0·102	0·095	0·082	0·015	0·020
= Milligrammes per 100 ml. alcohol .	224	193	174	—	—
Esters . . .	241·3	275·0	190·3	—	—
Furfural . . .	10·5	9·6	6·8	—	—
Aldehydes . .	27·8	35·2	24·5	—	—
Higher alcohols . .	174·5	190·3	160·1	—	—

rum. The oils added from the seeds and essences used are mainly of the terpene class, and should not be included as higher alcohols.

London gin is unsweetened, whereas Old Tom gin contains some sugar. Sloe gin is usually 50° U.P. and Lemon or Orange gin 40° U.P. The table on p. 283 shows analyses of rum and gin.

Methods for the Examination of Spirits

Alcohol. The alcohol content of spirits is best determined by direct distillation of the sample. By determining the specific gravity of the distillate (made up to the original volume) the percentage of alcohol in the sample can be found by reference to the Customs and Excise Tables. If the total solid matter is small the alcohol content of spirits for routine purposes can be estimated approximately by determining the s.g. of the sample (without distillation), but it is more accurate to make a correction for the **"obscuration."** The soluble matter increases the s.g. and the alcoholic content indicated by the s.g. of the sample itself will be lower than the true figure. This difference between the apparent and real alcohol contents is referred to as the obscuration. From the analysis of a large number of spirits it would appear that each 1 per cent. w/v of extract increases the s.g. by 0·0041. Thus for whisky:—

Extract (total solids) = 0·20% w/v

s.g. of sample at 15·5° C.* = 0·9565 ≡ 64·48% v/v P.S.

Correction for extract

= 0·20 × 0·0041 = 0·0008

s.g. of alcohol in sample = 0·9557 ≡ 65·4(6)% v/v P.S.

* Temperature correction: 1° C. diff. in temp. ≡ 0·0006 diff. in s.g.

Alternatively the s.g. correction figure can be obtained by evaporating down a known volume to low bulk so that all the alcohol is removed. Then the liquid is made up to the original volume and the s.g. is taken. The figure so obtained, less 1·0000, is subtracted from the original s.g. of the sample and the percentage of proof spirit corresponding to the corrected figure is obtained by reference to the alcohol tables (see also Nelson *et al.*, *J. Ass. Off. Agric. Chem.*, 1958, **41**, 139).

Higher Alcohols

(1) *The Allen-Marquardt Process* (*Analyst*, 1891, **16**, 102, and *J. Soc. Chem. Ind.*, 1902, **21**, 815). The method is as follows (care must be given to all the details): To 200 ml. of the spirit are added 4 ml. N sodium hydroxide solution and the mixture is boiled under a

reflux condenser for an hour. The liquid is transferred to a flask for steam distillation, and distilled over a flame until the volume is reduced to about 20 ml. Then steam is introduced and a further 110 ml. are distilled over so that about 10 ml. remain in the flask and the distillate measures 300 ml. This is divided into two portions and strong brine solution is added to one portion until the specific gravity of the mixture is 1·10. This is then shaken out in a separator with 40 ml., 30 ml., 20 ml., and lastly 10 ml. of carbon tetrachloride. The mixed carbon tetrachloride solution is shaken out with 50 ml. of brine to remove any ethyl alcohol, and then with 50 ml. of saturated sodium sulphate solution to remove chlorides. It is then filtered through a dry paper and the higher alcohols oxidised to their corresponding acids in the following manner. Five grams of potassium dichromate, 2 ml. of sulphuric acid and 10 ml. of water are mixed in a flask fitted to a reflux condenser, the carbon tetrachloride is added and the mixture boiled for eight hours. The condenser is then turned to the position for ordinary distillation, 30 ml. of water are added, and the liquid is distilled to a volume of 20 ml., when steam is admitted and the heating continued until 300 ml. have passed over and only 5–10 ml. remain in the flask. The distillate is now neutralised with 0·1 N barium hydroxide solution, methyl orange being used as indicator. When neutral, phenolphthalein is added, and the titration is continued to the neutral point of this indicator. One ml. of 0·1 N barium hydroxide = 0·0088 g. of higher alcohols expressed as amyl alcohol. The original distillate having been divided into two portions, it is convenient to make the estimation in duplicate. It is essential that pure carbon tetrachloride be used and it is advisable to use apparatus with ground glass joints. The methyl orange acidity of the distillate should not exceed 2 ml. 0·1 N.

(2) *The Colorimetric Method* (Royal Commission on Whisky and Potable Spirits: Report, 1909, Appendix).—This method depends upon the colour produced by furfural and sulphuric acid in contact with higher alcohols, but not with ethyl alcohol. It is much more expeditious than the Allen-Marquardt process. A standard solution of higher alcohols is prepared by dissolving 1 g. of a mixture of propyl alcohol 1 part, *iso*butyl alcohol 2 parts, amyl alcohol 3 parts, and capryl alcohol 1 part in 100 ml. of pure 50 per cent. ethyl alcohol. This solution is further diluted 1 : 10 in 50 per cent. alcohol for use, then each ml. = 0·001 g. of higher alcohols. The spirit to be examined is diluted to contain 50 per cent. of alcohol. Then 200 ml. are distilled until 190 ml. have passed over and the distillate is made up to 200 ml. Ten ml. are mixed in a small flask with 0·5 ml. of

1 per cent. furfural solution and 10 ml. of sulphuric acid. The acid should be run in to form a layer on the bottom of the flask. Then the mixture is shaken for thirty seconds while cooled in iced water. In similar flasks mixtures are made containing 0·5 ml., 1·0 ml., 2·0 ml., and upwards of the standard higher alcohol solution in 10 ml. of 50 per cent. alcohol. In each case the mixture is made by shaking gently for thirty seconds in iced water. The reddish-violet colours produced are matched after standing for an hour at ordinary temperature and the higher alcohols calculated in milligrams per 100 ml. of absolute alcohol (*cf.* Osborn and Mott, *Analyst*, 1952, **77**, 260).

Esters are estimated by hydrolysis with alkali and returned as ethyl acetate. One hundred ml. of the sample are distilled into a flask which has been well steamed out, until there remains about 10 ml. Steam is then passed in and distillation continued till the distillate measures 150 ml. and the residue 5 ml. The distillate and the residue are titrated separately until exactly neutral to phenolphthalein and the total acidity calculated in terms of acetic acid. If desired, the volatile acid may be returned separately as acetic and the non-volatile acid as tartaric acid. To the neutralised distillate are added 10 ml. of 0·1 N sodium hydroxide solution and the mixture is boiled under a reflux condenser for half an hour, then cooled and titrated back with 0·1 N acid. The alkali used is calculated in terms of ethyl acetate, although other esters are probably present in the spirit.

The total acidity may also be estimated by the direct titration of 50 ml. with 0·1 N alkali; barium hydroxide is better than soda for this purpose.

For the estimation of **furfural,** Hewitt's method (*J. Soc. Chem. Ind.*, 1902, **21**, 98) is perhaps the best, although Schidrowitz (*ibid.*, 816) criticises it and points out that there is the possibility of the formation of some furfural during distillation. Fifty ml. of the sample are distilled to a low bulk; to the residue is added pure alcohol, and the distillation repeated; this operation is repeated several times, then the mixed distillate is diluted to contain 50 per cent. of alcohol and the volume noted. To 10 ml. of the distillate is added 1 ml. of a mixture of equal volumes of aniline and glacial acetic acid, and the colour compared after twenty minutes with that of 10 ml. of standard 50 per cent. alcohol to which has been added 10 mg. per litre of furfural. The comparison should be made in a colorimeter, and the depths of liquids which show an equivalent colour are noted and the proportion of furfural is calculated therefrom. If a colorimeter is not available, the process of "nesslerising" may be applied.

For the determination of **aldehydes** advantage is taken of their property of restoring the colour of magenta which has been decolorised with sulphurous acid—Schiff's reagent, the formula of which is given on p. 274. A standard aldehyde solution is required and is prepared by dissolving 1·386 g. of pure dry aldehyde-ammonia in 50 ml. of pure 95 per cent. alcohol, then adding 22·7 ml. N sulphuric acid solution and making the volume up to 100·8 ml. with alcohol (the 0·8 ml. is to compensate for the volume of ammonium sulphate). The liquid is allowed to stand for twenty-four hours and filtered; each 1 ml. = 0·01 g. aldehyde. To carry out the estimation 20 ml. of the spirit (which must be decolorised with lead acetate and de-leaded with sodium sulphate if it is highly coloured) are treated with 5 ml. of the Schiff reagent and the colour compared after twenty minutes with that of 20 ml. of alcohol of similar strength to which known volumes of aldehyde solution have been added. The method is approximate, not absolute, as the colour is to some extent influenced by the particular aldehydes present and in a small degree by the furfural, so is not quite proportional to the amount of aldehyde present.

The estimation of **sulphates, total solids, ash,** and, if necessary, of **tannins,** may be carried out by the ordinary methods.

BEER

Section 307 of the Customs and Excise Act, 1952, states that: "beer" includes ale, porter, stout, black beer and any other description of beer, and any liquor which is made or sold as a description of beer or as a substitute for beer and which on analysis of a sample thereof at any time is found to be of a strength exceeding two degrees of proof, but does not include liquor made elsewhere than upon the licensed premises of a brewer for sale which on analysis of a sample at any time is found to be of an original gravity not exceeding one thousand and sixteen degrees and to be of a strength not exceeding two degrees of proof.

Beer consists essentially of a fermented liquor prepared from malt and other cereals, flavoured with hops. Gentian or quassia or other non-deleterious bitter may be used as a flavouring agent, but such are not employed to any great extent in England. Although the products of the fermentation are very numerous, and it is possible to make detailed analyses showing, in addition to alcohol, acidity, specific gravity and extract, such minor constituents as albuminoids, sugar, dextrin, glycerin, tannin, and the composition of the ash, detailed analyses are not often required.

Analysis of Beer

The estimations commonly required on a beer include that of **specific gravity, alcohol, original gravity, acidity, total solids,** and an examination for **preservatives, lead** and **arsenic.** Analyses of various types of beer are given below:—

	India pale ale	Mild ale	Bock	Lager beer	Irish stout
Specific gravity . .	1012·5	1010·1	1020·5	1014·5	1016·0
Alcohol, per cent. by weight . . .	4·30	3·15	4·50	3·20	4·30
Total solids % . .	4·95	3·55	6·80	5·38	5·70
Acidity as acetic % .	0·15	0·08	0·12	0·17	0·17
Ash %	0·26	0·20	0·29	0·20	0·23
Original gravity . .	1053·0	1042·1	1063·5	1046·5	1057·5

The estimation of **alcohol** in beer is made during the course of the determination of the original gravity, *i.e.*, the indicated specific gravity of the wort before attenuation. The investigations of Thorpe, Brown, and others, have made it possible to calculate with accuracy the quantity of saccharine matter destroyed during the fermentation, thus arriving at the loss of density, and hence the original gravity of the unfermented wort.

If the beer contains yeast and much carbon dioxide, it is filtered through paper, the funnel being kept covered to obviate loss of alcohol, or, if there is no appreciable turbidity, the carbon dioxide may be removed by gently pouring the beer from one beaker to another a few times. One hundred ml. of the beer measured at 15·5° C are transferred to a distillation flask and rinsed in with about 40 ml. of water, then distilled into a 100 ml. flask until 80–90 ml. have passed over. The distillate is made up to 100 ml. at 15·5° C; the residue is cooled and diluted also to 100 ml. at 15·5° C. The specific gravity of the distillate and of the residue is taken with an accurate pyknometer. The specific gravity of the distillate may be used for determining the alcohol by reference to the usual tables; the difference between this specific gravity and 1,000·0 gives the so-called *Spirit Indication.* Reference to the table on p. 290 shows the degrees of gravity lost. (This table is reproduced from that adopted in the Finance Act of 1914.)

Allowance must be made for the acid present, if it exceeds 0·1 per cent. For this purpose 10 ml. of the beer are titrated with 0·1 N sodium hydroxide, litmus being used as an outside indicator. The result is calculated as acetic acid and the allowance reckoned for the acid in excess of 0·10 from the following table:—

CORRECTION FOR EXCESS ACID.

Acetic acid	Corresponding degress of spirit indication.									
	0·00	0·01	0·02	0·03	0·04	0·05	0·06	0·07	0·08	0·09
0·0	—	0·02	0·04	0·06	0·07	0·08	0·09	0·11	0·12	0·13
0·1	0·14	0·15	0·17	0·18	0·19	0·21	0·22	0·23	0·24	0·26
0·2	0·27	0·28	0·29	0·31	0·32	0·33	0·34	0·35	0·36	0·37
0·3	0·39	0·40	0·42	0·43	0·44	0·46	0·47	0·48	0·49	0·51
0·4	0·52	0·53	0·55	0·56	0·57	0·59	0·60	0·61	0·62	0·64
0·5	0·65	0·66	0·67	0·69	0·70	0·71	0·72	0·73	0·75	0·76
0·6	0·77	0·78	0·80	0·81	0·82	0·84	0·85	0·86	0·87	0·89
0·7	0·90	0·91	0·93	0·94	0·95	0·97	0·98	0·99	1·00	1·02
0·8	1·03	1·04	1·06	1·07	1·08	1·09	1·10	1·11	1·13	1·14
0·9	1·15	1·16	1·18	1·19	1·21	1·22	1·23	1·25	1·26	1·28
1·0	1·29	1·31	1·33	1·35	1·36	1·37	1·38	1·40	1·41	1·42

The gravity lost equivalent to the corrected spirit indication is now added to the specific gravity of the extract, *i.e.*, residue in the flask after being made up to the original volume; the sum gives the original gravity. An example will make this plain:—

Specific gravity of distillate	0·9935
Specific gravity of extract	1·0204
Acidity as acetic acid (per cent.) . . .	0·31
Specific gravity of water	1,000·0
Specific gravity of distillate	993·5
Spirit indication	6·5
Allowance for 0·21 excess acid . . .	0·28
Corrected spirit indication	6·78
Equivalent to gravity lost	30·0
Add specific gravity of extract	1,020·4
Original gravity	1,050·4

Essery (*J. Inst. Brewing*, 1956, **62**, 153) has shown that the original gravity of beer can be calculated from equations derived from the refractive index and present gravity values (*cf.* Hudson's "Development of Brewing Analysis" (Institute of Brewing, Lond.) 1960).

FINANCE ACT, 1914 (Sess. 2).

Table of Gravity Lost for Determining the Original Gravity of Worts of Beer.

Degrees of Spirit Indication.	0·0	0·1	0·2	0·3	0·4	0·5	0·6	0·7	0·8	0·9
0	0·00	0·42	0·85	1·27	1·70	2·12	2·55	2·97	3·40	3·82
1	4·25	4·67	5·10	5·52	5·95	6·37	6·80	7·22	7·65	8·07
2	8·50	8·94	9·38	9·82	10·26	10·70	11·14	11·58	12·02	12·46
3	12·90	13·34	13·78	14·22	14·66	15·10	15·54	15·98	16·42	16·86
4	17·30	17·75	18·21	18·66	19·12	19·57	20·03	20·48	20·94	21·39
5	21·85	22·30	22·76	23·21	23·67	24·12	24·58	25·03	25·49	25·94
6	26·40	26·86	27·32	27·78	28·24	28·70	29·16	29·62	30·08	30·54
7	31·00	31·46	31·93	32·39	32·86	33·32	33·79	34·25	34·72	35·18
8	35·65	36·11	36·58	37·04	37·51	37·97	38·44	38·90	39·37	39·83
9	40·30	40·77	41·24	41·71	42·18	42·65	43·12	43·59	44·06	44·53
10	45·00	45·48	45·97	46·45	46·94	47·42	47·91	48·39	48·88	49·36
11	49·85	50·35	50·85	51·35	51·85	52·35	52·85	53·35	53·85	54·35
12	54·85	55·36	55·87	56·38	56·89	57·40	57·91	58·42	58·93	59·44
13	59·95	60·46	60·97	61·48	61·99	62·51	63·01	63·52	64·03	64·54
14	65·10	65·62	66·14	66·66	67·18	67·70	68·22	68·74	69·26	69·78
15	70·30	70·83	71·36	71·89	72·42	72·95	73·48	74·01	74·54	75·07
16	75·60	—	—	—	—	—	—	—	—	—

Arsenic. In the early part of the century numerous instances were reported of beer contaminated with arsenic due to the use of (a) arsenical fuel for drying the malt or hops or (b) arsenical pyrites in the sulphuric acid used in the manufacture of the glucose. The Arsenic in Food Regulations, 1959 (see p 53) prescribe a maximum for arsenic in alcoholic beverages of 0·2 p.p.m. (as As). For routine purposes arsenic can be determined " direct " by the Gutzeit method, *i.e.*, without preliminary destruction of organic matter. The addition of one drop of amyl alcohol sometimes assists in reducing frothing.

Lead. Traces of lead are liable to be present in beer if it has been in contact with the metal and the F.S.C. Report on Lead (1954) recommends that the use of lead pipes and lead containers should be discontinued in all breweries, bottling plants and licensed premises. A maximum of one p.p.m. is recommended for lead in beer (see p. 64). For the determination of lead it is preferable to employ the S.A.C. method described on p. 66, but the sulphide coloration method (p. 69) is also suitable provided a fairly large quantity of sample (*e.g.*, 200 g.) is evaporated down prior to dry ashing.

Preservatives. Samples of beer should be examined for sulphur dioxide. The possibility of finding other preservatives such as benzoic acid, salicylic acid and fluoride now seems somewhat remote.

CIDER

Although there is in this country no legal definition of cider or perry, such exists in France, where " no drink is to be sold under the name of cider unless it is derived exclusively from the fermentation of the juice of fresh apples, or a mixture of fresh apples and pears extracted with or without the addition of water. . . . " The term cider or perry is reserved for cider or perry containing at least 2·5 per cent. of alcohol, 12 g. per litre of extract (sugar being deducted), and 1·2 g. of mineral matter per litre." The following standard has been suggested in the U.S.A. (all figures as percentages): Alcohol (max.) 8·0, extract (min.) 1·8, ash (min.) 0·2. Such standards, although affording a rough guide by which to judge the authenticity of a cider, must be accepted with reserve, for it would be easy for an unscrupulous person to prepare a fictitious liquid which would pass such standards. For a discussion of the composition of cider, see Barker and Russell (*Analyst*, 1909, **34**, 125), from whose paper the following analyses of genuine cider are taken:—

CIDER

	1.	2.	3.	4.	5.	6.
Specific gravity	1·000	1·022	1·0125	1·009	1·007	1·0075
Acid (as malic acid), per cent.	0·48	0·59	0·32	0·33	0·34	0·45
Tannin, per cent.	0·26	0·26	0·21	0·25	0·18	0·27
Alcohol, per cent.	5·83	3·55	4·20	4·76	5·07	4·82
Solids, per cent.	2·41	6·88	5·74	3·92	3·88	3·61
Ash, per cent.	0·32	0·34	0·31	0·31	0·32	0·35
Alkalinity as K_2CO_3, per cent.	0·116	0·127	0·095	0·113	0·108	0·026
P_2O_5, per cent.	0·015	0·023	0·019	0·017	0·019	0·013

These figures represent blends such as are common on the market. From other data it appears that the acidity varies from 0·22 to 0·59 per cent., and the tannins from 0·04 to 0·37 per cent. Some light may be thrown on the question of whether or not a cider contains added water by the calculation of the original solids. This may be done by adding the sum of 2·1 × the alcohol and 1·5 × the malic acid to the total solids found. The original solids thus calculated should not be below 12 per cent. in cider from undiluted apple juice. As it may happen that an artificial product containing no apple juice may appear as cider, the following test is given by Barker and Russell for its detection, based on the colour reaction of the tannins extracted by ethyl acetate, with lime water. One hundred ml. of the sample are evaporated to 10 ml. and shaken in a test tube for five minutes with 10 ml. of ethyl acetate; when the mixture has separated the upper layer is drawn off and poured carefully on the surface of a few millilitres of lime water in a test tube. At the junction of the liquids a clear band of yellow colour appears in the presence of apple or pear juice. The colour soon disappears, but the reaction is sufficiently sensitive to detect as little as 1 part of apple juice in 1,000.

Many methods have been proposed for the quantitative differentiation of small quantities of tartaric, malic and citric acids, but it is doubtful whether the results afford information comparable with the labour involved.

Search should be made for **preservatives,** particularly benzoic and salicylic acids or sulphites. In this connection it is important to note that apples usually contain small amounts of boric acid, so that this is also present in cider; also that cider casks are not infrequently treated with sulphur dioxide, hence traces of sulphites may be present in the liquid from this cause. The quantity found will generally indicate whether boric or sulphurous acid has been added intentionally or not. The estimation of **boric acid** in cider and similar products is complicated by the presence of phosphates in comparatively large amounts. These must be removed by calcium chloride, which may be effected in the following way. To 100 ml. of the liquid add 2 ml. N sodium hydroxide more than is required to neutralise the free acid present, evaporate to dryness and ignite over an Argand burner to a white ash. Dissolve the ash in a slight excess of dilute hydrochloric acid, making the volume up to about 80 ml., and warm for half an hour on the water bath to expel carbon dioxide. Then cool and add 1 ml. of 5 per cent. calcium chloride solution and 0·5 ml. of phenolphthalein solution. Titrate slowly with N sodium hydroxide until there is a faint permanent pink colour, not more, and make up to 100 ml. and filter off 50 ml. for the final titration. Make the 50 ml. quantity just neutral to methyl orange, add 20 ml. of neutralised glycerol and titrate with 0·1 N sodium hydroxide in the usual way.

1 ml. N/10 sodium hydroxide = 0·0062 g. boric acid.

Labelling of Food Order, 1953 (Part V, 8e)

Cider and *Perry* which have been artificially aerated, must not be described on the label by any name or words calculated to indicate either directly or by ambiguity, omission or inference, that the liquor resembles or is a substitute for, or has the character of champagne.

VINEGAR

The following definitions for various types of vinegar were agreed upon between the S.P.A. and the Malt Vinegar Brewers' Federation (*Analyst*, 1935, **60**, 2):—

(1) Vinegar is a product of the alcoholic and acetous fermentation of a saccharine solution without any intermediate distillation, except in the case of spirit vinegar as defined in (4).

(2) Malt vinegar should be derived, without intermediate distillation, wholly from malted barley, with or without the addition of entire cereal grain malted or otherwise, the starch of which has been saccharified by the diastase of malt.

(3) When vinegar is demanded, a purchaser should be supplied with malt vinegar, unless due notification is given to the purchaser of the article supplied.

(4) The name " Vinegar " may be applied to other products which comply with the definition of vinegar (No. (1) above), provided a prefix is used to denote the origin of the product; thus, " Distilled Vinegar," " Spirit Vinegar."

Distilled vinegar is the product of the distillation of vinegar as defined in No. (1) above, and its source should be denoted; such as, distilled malt vinegar.

ANALYTICAL FIGURES OF VINEGARS AND NON-BREWED CONDIMENT

	Malt Vinegar	Non-Brewed Condiment	Cider Vinegar	Wine Vinegar	Spirit Vinegar (concentrated)
Specific gravity	1·013–1·022	1·007–1·022	1·013–1·024	1·013–1·021 (1)	0·015–1·020
Total acidity as acetic acid (% w/v)	4·3 –5·9	4·1 –5·3	3·9 –9·0	4·4 –7·4 (1)	11·5 –12·2
Fixed acidity as acetic acid (%)	0·2 –0·4	Negligible	0·1 –0·2	—	—
Malic acid (%)	—	—	0·07 –0·16	—	—
Total solids (%)	1·4 –3·5	0·30 –0·45	1·9 –3·5	1·4 –3·2 (1)	0·15 –0·60
Total ash (%)	0·18 –0·45	0·02 –0·05	0·20 –0·45	0·16 –0·69 (1)	0·02 –0·05
Alkalinity of ash (ml. N/100 ml.)	—	—	2·2 –4·0	—	—
Total nitrogen (%)	0·04 –0·14	0 –0·02	—	—	0·003–0·03
Phosphoric acid as P_2O_5 (%)	0·05 –0·12	0 –0·02	0·04 –0·30	—	0·005
SO_3 (%)	0·05 –0·12	0 –0·01	—	—	—
Sodium Chloride (%)	0·15 –0·25	0·01 –0·12	—	—	—
Total sugars (%)	—	—	0·15 –0·7	0·22 –0·56 (1)	—
Alcohol (% v/v)	—	—	0·03	—	0·15
Oxidation value	500–1800	0–20	Up to 3500	600–2000	90–650 (2)
Alkaline oxidation value	70–180	0–10	—	60–180	3–20 (2)
Iodine value	380–1500	0–25	—	380–1000	5–30 (2)
Ester value	30–140	0–15	—	50–220	0–20 (2)

(1) Analyses by the Paris Municipal Laboratory, quoted by Mitchell. (2) Calculated to 5 per cent. acetic acid.

Spirit vinegar is the product of the acetous fermentation of a distilled alcoholic fluid.

(5) " Imitation " or " Artificial " vinegar* should in every case be sold specifically marked " Imitation " or " Artificial " vinegar. It is any substitute for vinegar containing acetic acid which is not wholly the product of alcoholic and subsequent acetous fermentation, and shall not contain any acid other than acetic acid.

(6) All vinegars and imitation or artificial vinegars shall contain not less than 4 per cent. w/v of acetic acid. They shall not contain any acid other than acetic acid or those acids produced by normal fermentative processes.

(7) Caramel may be used as a colouring matter in all vinegar and in " Imitation " or " Artificial " vinegars.

N.B.—The minimum of 4 per cent. for the total acidity as acetic acid is usually accepted by the courts for all types of vinegar. Possible standards for vinegar have been suggested by McLachlan (*Food Manuf.*, 1949, **24**, 58).

Malt vinegar should be prepared from malted barley, but other cereals such as rice and maize may be used. The wort is made by steeping the malt (sometimes mixed with unmalted grain) in water at 140° F. and at a controlled pH so that the starch is converted by diastatic action to maltose, dextrose and dextrins (" mashing "). Yeast is then added in order to bring about alcoholic fermentation (at 85° F.). The " gyle " so produced is then inoculated with *Mycoderma aceti*, which oxidises the alcohol so that acetic acid is produced.

The best vinegar is said to be made by the old slow Orleans Process. The Quick Vinegar Process, however, enables the time of acetification to be reduced from a matter of weeks to a few hours. This involves passing air continuously over the liquid in the acetifier (wood shavings being used as a " support "), the process being maintained at about 95° F. For Fring's improved process, Mackin's continuous atomising method of acetification, etc., the reader should refer to Prescott and Dunn, " Industrial Microbiology," 1949, pp. 377–383 (McGraw-Hill, N.Y.).

Distilled malt vinegar is a colourless liquid prepared by distilling malt vinegar from a still, often constructed from silver, which is not attacked by acetic acid. Obviously the product only contains the volatile constituents of the vinegar from which it is derived. It is commonly used in the manufacture of pickled white onions.

Spirit vinegar is sometimes derived from molasses and when manufactured usually has a total acidity of about 12 per cent. as acetic acid. It is diluted to about 4–5 per cent. for use in pickling, etc. The composition of spirit vinegars has been discussed by Manley (*Analyst*, 1940, **65**, 594).

* *cf.* Section on " Non-Brewed Condiment " (p. 296).

Spiced vinegars such as tarragon, garlic, elder, etc., are prepared by steeping the leaves or spices in an ordinary vinegar. The presence of the flavouring material does not appear to affect the analytical figures so that figures in the composition table (p. 294) can be used to give some indication of the type of vinegar used.

Wine vinegar is mainly produced on the Continent. It is prepared by the acetous fermentation of grape juice or of white or red wines and has a characteristic aroma. Wine vinegars usually contain 0·1–0·4 per cent. of potassium acid tartrate and small amounts of alcohol and reducing sugars.

Cider vinegar is not as common in Britain as in the U.S.A. It contains a considerable amount of malic acid, is lævo-rotatory, yields a strongly alkaline ash rich in potassium and often gives a very high oxidation value.

The analytical figures usually found with wine and cider vinegars are given in the general table.

Non-brewed condiment. Artificial non-brewed "vinegars" usually contain about 4·5 per cent. (w/v) of synthetic acetic acid and are coloured with caramel. In 1950 the High Court of Justice ruled in an appeal case (Kat *v.* Diment) that the description "non-brewed vinegar" as applied to a solution of synthetic acetic acid was a false trade description and an offence against the Merchandise Marks Act. The trade subsequently decided to describe such artificial products as non-brewed condiment (see also Circular M.F. 15/50). The minimum of 4 per cent. acetic acid applies to non-brewed condiment as well as to brewed vinegar.

Concentrated solutions of acetic acid. Non-brewed condiments are usually prepared from 80 per cent. acetic acid or from "vinegar essence" or a concentrated non-brewed condiment (usually with added caramel). In this connection the following requirements of the Labelling of Food Order are of interest:—

"No person other than a pharmacist or an authorised seller of poisons shall sell or have in his possession for sale any pre-packed concentrated acetic acid or solution of acetic acid, whether coloured or not, containing more than 15 per cent. weight in volume of acetic acid unless—

(a) it is described as 'Concentrated solution of acetic acid (X) per cent. the description to be completed by inserting at (X) the appropriate figure; and
(b) there appears on the main label in red type not less than 1/8th inch in heignt upon a light coloured ground the words 'Dangerous—not to be used unless diluted'."

Analysis of Vinegars and Non-Brewed Condiments

The " complete " analysis of a vinegar (or a non-brewed condiment) involves: (a) the organoleptic examination, (b) specific gravity, (c) total acidity, (d) volatile acidity, (e) total solids, (f) ash, (g) nitrogen, (h) phosphoric acid and tests for (i) metals, (j) preservatives, (k) free mineral acid, (l) ferrocyanides, (m) caramel and (n) special determinations for elucidating the type of vinegar. The routine examination in " official " laboratories should involve (c) the total acidity, (i) arsenic and heavy metals, (j) sulphur dioxide, (k) mineral acid and (n) elucidation of the type of vinegar present. Industrially the colour may be determined in the Lovibond Tintometer using the half-inch cell.

(a) **Organoleptic examination.** The odour and taste assist in giving information as to the type of vinegar (see also *A.O.A.C.*). The sample should be examined for vinegar eels (nematode worms—*Anguillula aceti*), possibly with the aid of the microscope.

(b) **Specific gravity.** This should be about 1·019 for a good malt vinegar.

(c) **Total acidity.** Dilute 10 ml. of sample with water in a porcelain basin (about 5 in. diameter), add phenolphthalein and titrate carefully with N/2 sodium hydroxide.

Total acidity = Titration (ml. N/2) × 0·3 = % w/v as acetic acid. The total acidity should not be less than 4 per cent. as acetic acid. Good quality products usually contain over 5 per cent.

(d) **Volatile acidity.** This can be determined by titrating the steam distillate from 10 ml. of sample. Alternatively the *fixed acidity* can be obtained by evaporating down 10 ml. at least 5 times with water and titrating and obtaining the *volatile acidity* by difference. Express the volatile acidity as acetic acid or lactic acid.

(e) **Total solids and** (f) **Ash.** Evaporate and dry (at 100° C.) 25 ml. of sample in a weighed platinum (preferable) or silica dish. The acid which tends to remain in the total solids can be removed by 3 evaporations with water. Weigh the residue and then ignite at 550° C. to give the ash. Determine the salt in the ash. For a good malt vinegar the t.s. should be at least 2 per cent. (w/v).

(g) **Nitrogen.** Determine by the Kjeldahl method (p. 34) on 25 ml.

(h) **Phosphoric acid.** Determine on the ash or after wet oxidation. The volumetric phosphomolybdate method is convenient for the estimation.

(i) **Metals.** The sample should be examined for *arsenic, copper* and *lead* (see Chapter 3).

The presence of traces of **mercury** are strongly indicative of the presence of commercial strong (80 per cent.) acetic acid. The small amounts involved can be detected by electrolysis. To 200 ml. of the sample add 10 ml. of nitric acid and electrolyse (using 6 volts and 3 amps.) for one hour using a platinum anode and a gold cathode (area 1 sq. cm.) to which a piece of gold wire is attached. The latter may be attached to the copper leads in a glass rod covered with a short piece of black rubber tubing. A silvering effect on the gold cathode indicates that a considerable quantity of mercury is present. If none can be seen, wash the gold foil with alcohol and ether and allow to air-dry. Then fold it and place it in a glass tube —drawn out and sealed at one end, the sealed end of which contains a minute fragment of iodine. Close the open end with a cork and heat the tube gently from the wide end towards the narrow part, so that any mercury present sublimes into the constricted part of the tube. Here it meets the vapour of iodine and on cooling a ring of red mercuric iodide (not unlike a Marsh arsenic mirror) is formed in the narrow part if about 0·02 mg. of the element is present.

Alternatively the small amounts of mercury present may be detected using other techniques such as spectrophotometry.

(j) **Preservatives.** The possibility of 30 p.p.m. of sulphur dioxide being present in the vinegar has been discussed by the author (*Analyst*, 1927, **52**, 397). Other preservatives, *e.g.*, borates, have on occasion been detected.

(k) **Mineral acid.** Traces of sulphate are usually present in vinegar but if the amount exceeds 0·03 per cent. (as H_2SO_4) it gives rise to some suspicion. Added mineral acid can be detected by the following simple test:—

Mix 2 ml. of sample with 2 ml. of alcohol and add 2 drops of methyl orange —a red colour indicates a low pH due to added mineral acid. Other indicators which may be employed are methyl violet and metanil yellow. Jenkins (*Analyst*, 1941, **66**, 328) employs thymol blue, which is added to the sample after diluting to 4 per cent. acetic acid and decolorising with charcoal. The pH is then determined in a comparator.

The pH of an artificial product will fall below 2·5 in the presence of mineral acid. The pH of malt vinegar (normally about 2·9 at 4 per cent. acidity) is also reduced if mineral acid is present, but the effect is not quite so marked as with non-brewed products due to the presence of buffering salts.

Added sulphuric acid can be estimated by the method of Hehner

(*Analyst*, 1877, **1**, 105), which is based on the fact that the ash of genuine vinegar is alkaline in reaction. A neutral or acid ash is indicative of the presence of added mineral acid.

Evaporate 50 ml. of sample with 25 ml. of N/10 sodium hydroxide in a platinum dish and ignite at as low a temperature as possible. Boil the ash with 25 ml. N/10 sulphuric acid, filter and wash with hot water. Titrate the filtrate + washings with N/10 sodium hydroxide using phenolphthalein as indicator (1 ml. N/10 = 0·0049 g. H_2SO_4). A blank should be performed on the reagents at the same time.

(l) **Ferrocyanides** have on occasion been used for clarifying vinegar (a process which is considered to be objectionable). They can be detected by the blue colour formed on the addition of a trace of ferric chloride.

(**m**) **Caramel** usually gives a positive reaction with Fiehe's test: Shake 100 ml. of sample with 50 ml. ether. Transfer the ether layer to a porcelain dish and allow the solvent to evaporate off spontaneously. To the residue add 3 drops of resorcinol solution (1 per cent. in HCl)—in the presence of caramel a rose colour is produced. Care must be taken in interpreting the results from Fiehe's test, however, as any liquid containing furfural (*e.g.*, cider vinegar) will give a positive reaction.

Methods for Differentiating between Different Types of Vinegar

The figures for nitrogen, phosphoric acid and ash give some indication of the type of vinegar. A reasonably good quality malt vinegar derived from malted or unmalted barley gives figures for nitrogen and phosphoric acid (as P_2O_5) of not less than 0·05 per cent. each. Where rice has been used in conjunction with malt the nitrogen figure tends to be considerably higher than the phosphoric acid. The total ash figure also gives some indication as to the starting material, *e.g.*, molasses produces a low ash and grain converted by acid hydrolysis leads to a high ash figure. Non-brewed products may contain traces of mercury, which is used as catalyst in the preparation of acetic acid synthetically from acetylene (see (i) above).

Useful methods for differentiating between the various types of vinegar are based on the values obtained on the volatile substances present, *i.e.*, the oxidation value, alkaline oxidation value, iodine value and the ester value. Brewed vinegars give comparatively high values, but those for artificial products are low as the latter are almost devoid of volatile reducing substances.

Cider vinegars can be distinguished by the tests for malic acid given on p. 303

In addition to the substances mentioned above, brewed vinegars contain significant proportions of vitamin B_1, riboflavin, nicotinic acid, pantothenic acid and pyridoxin, all of which are, of course, absent in artificial products.

Formol titration. In general brewed vinegars give a definite formol titration, but artificial products do not. The method is, therefore, useful as a rapid sorting test, but it is not specific for differentiating between brewed and non-brewed products as distilled vinegars do not, of course, give a formol titration. The method is simple to carry out:—

Add phenolphthalein to 10 ml. sample, *nearly* neutralise the acid with N/2 sodium hydroxide and then make *exactly* neutral with N/20 alkali. Then add 5 ml. of formalin (previously made *exactly* neutral to phenolphthalein), mix well and after standing for five minutes titrate the acidity produced with N/20 or N/10 sodium hydroxide. Most malt vinegars give a formol titration equivalent to 1·3–4·0 ml. of N/10 alkali (using 10 ml. sample). Artificial products and distilled vinegars give a negligible formol titration.

Oxidation value, alkaline oxidation value, iodine value and ester value. Methods for differentiating between various types of vinegar based on the oxidation, iodine and ester values have been usefully developed by Edwards and Nanji (*Analyst*, 1938, **63**, 410). The modified distillation technique described below is that due to Illing and Whittle (*Analyst*, 1939, **64**, 329). The alkaline oxidation value method of Lyne and McLachlan (*Analyst*, 1946, **71**, 203), which may have certain advantages over the other values, can be determined on the same distillate as the other three.

Whitmarsh (*Analyst*, 1942, **67**, 188) studied the individual volatile reducing substances which affect the oxidation and iodine values and found that alcohol and acetyl methyl carbinol appear to be mainly responsible for the distinctive values obtained with malt vinegar (*cf.* Lyne and McLachlan, *ibid.*).

DISTILLATION

Distil 60 ml. of the sample from a 350 ml. flask fitted with a small tap funnel. When 45 ml. of distillate has come over, add 15 ml. of water to the flask via the tap funnel and distil a further 15 ml. to give a total volume of distillate of 60 ml.

Oxidation Value is the number of ml. of N/100 potassium permanganate used by 100 ml. of sample in thirty minutes under standard conditions.

To a 250 ml. glass-stoppered bottle add 5 ml. of distillate from a malt or wine vinegar (or 10 ml. from a dilute spirit vinegar or artificial product), 10 ml. dilute sulphuric acid (1 + 3) and exactly 15 ml. of N/10 potassium permanganate. Allow to stand at about 18° C. for thirty minutes and then add 5 ml. of potassium iodide solution (10 per cent.). Titrate the liberated iodine with N/50 thiosulphate (a ml.) using starch near the end point. Carry out a blank at the same time (titration b ml.).

If 10 ml. distillate used, Oxidation Value = 20 (b−a).
If 5 ml. distillate used, Oxidation Value = 40 (b−a).

Alkaline Oxidation Value is the number of parts by weight of oxygen required to oxidise 100,000 parts of sample under standard conditions (Lyne and McLachlan, *ibid.*).

To a 250 ml. glass-stoppered bottle add 2 ml. of distillate, 100 ml. water, 10 ml. of sodium hydroxide solution (10 per cent.) and exactly 10 ml. of N/10 potassium permanganate. Allow to stand for thirty minutes and then acidify with 10 ml. dilute sulphuric acid (25 per cent.). Add 0·5 g. of potassium iodide and titrate the liberated iodine with N/50 thiosulphate (a ml.) using starch near the end point. Carry out a blank at the same time (titration b ml.).

Alkaline Oxidation Value = 8 (b−a).

About a minute after adding the permanganate a deep green colour develops with malt and wine vinegars, but with spirit vinegars and artificial products the liquid remains a pink or purple colour. Lyne and McLachlan use this as a rapid visual sorting test.

Iodine Value is the number of ml. of N/100 iodine absorbed by 100 ml. of sample under standard conditions.

To a 250 ml. glass-stoppered bottle add 5 ml. of distillate from a malt vinegar (or 10 ml. from a spirit vinegar or artificial product) and make just neutral to litmus with 10 N potassium hydroxide. Add 10 ml. N/1 potassium hydroxide and exactly 10 ml. N/10 iodine. Allow to stand in the dark for fifteen minutes and then add 10 ml. dilute sulphuric acid (1 + 3) and titrate the liberated iodine with N/50 thiosulphate (a ml.) using starch near the end point. Carry out a blank at the same time (titration b ml.).

If 10 ml. distillate used, Iodine Value = 20 (b−a).
If 5 ml. distillate used, Iodine Value = 40 (b−a).

Ester Value is the number of ml. of N/100 potassium hydroxide required to saponify the esters contained in 100 ml. of sample under standard conditions.

To a flask (suitable for fitment to a reflux condenser) add 25 ml. of distillate and make just alkaline with N/1 potassium hydroxide to phenolphthalein. Just remove the pink colour by the dropwise addition of N/50 hydrochloric acid. Then add exactly 10 ml. N/10 potassium hydroxide and heat under reflux for two hours on a boiling water bath to saponify the esters. Cool and titrate with N/50 hydrochloric acid (a ml.) after adding additional phenolphthalein and, if necessary, 100 ml. water (*e.g.*, with malt vinegar which tends to produce a reddish-brown colour which obscures the end point unless the liquid is diluted). Carry out a blank at the same time commencing with about 25 ml. water instead of the distillate (blank titration = b ml.).

$$\text{Ester Value} = 8\,(b-a).$$

Most vinegars and artificial products fall within the ranges given on the main composition table (p. 282).

Albuminoid ammonia value. Mitra (*Analyst*, 1953, **78**, 499) has found that the "albuminoid ammonia value" (commonly determined in water analysis) can be used for differentiating between malt vinegars and artificial products.

Special Reagents

Alkaline potassium permanganate solution. Mix together one litre of potassium permanganate solution (0·8 per cent.) and 500 ml. of sodium hydroxide solution (40 per cent.). Boil the mixture until the volume is reduced to one litre. Unless it is freshly prepared the reagent should be diluted with an equal volume of water and boiled down to the original volume before use.

Ammonia-free water. To a distillation flask add a large volume of tap water and a little dilute sulphuric acid. Distil and collect the distillate after it no longer gives a yellow colour with Nessler's reagent.

Method

Pipette 5 ml. of sample into a 100 ml. graduated flask, neutralise with 0·2 N sodium hydroxide and dilute to the mark with water.* Pipette 20 ml. into a distilling flask, add 230 ml. water,* a pinch of ignited sodium carbonate and ignited pumice. Distil 100 ml. and reject the distillate. Then to the liquid remaining in the distilling flask add 50 ml. of alkaline permanganate and 100 ml. of water* and distil again. Collect exactly 100 ml. of distillate and determine the ammonia by Nessler's colorimetric method (a Nessleriser is convenient). If the ammonia has to be matched against standard ammonium chloride solutions:—

To an aliquot of the distillate (10 ml. is usually suitable for malt vinegars) dilute to 50 ml.* in a Nessler glass, add 2 ml. of Nessler's reagent and match the colour after five minutes against dilutions* of standard ammonium chloride solution (containing 0·0382 g. NH_4Cl/litre. 1 ml. ≡ 0·01 mg. N).

$$\text{Albuminoid ammonia} = \frac{1000\,a}{b}\ \text{p.p.m. (as N).}$$

where a = no. of ml. standard NH_4Cl required for matching, and
b = no. of ml. distillate taken in sample Nessler glass.

* Use *ammonia-free* water throughout.

Mitra found that there is a broad correlation between the *total* nitrogen and the albuminoid ammonia and that the latter figure is very high when compared with those obtained with artificial products:—

	Albuminoid ammonia p.p.m. (as N, w/v)	Total nitrogen p.p.m. (w/v)
Malt vinegars . . .	208–400	440–840
Artificial condiments . .	0–4·0	11–34

Malic acid. Cider vinegar can be detected by the malic acid content. For a *qualitative* test, add 1 ml. of lead acetate solution to a few ml. of sample. If no precipitate settles in a few minutes (leaving a clear supernatant liquor), the sample is not cider vinegar. Also add 1 ml. of calcium chloride solution (10 per cent.) to 5 ml. of the vinegar and make alkaline with ammonia. Then filter off the precipitate. Add 3 volumes of alcohol to the filtrate and heat just to boiling. In the presence of malic acid a flocculent precipitate settles.

For the *determination* of malic acid, add 10 ml. of the calcium chloride solution to 100 ml. of sample, make alkaline with ammonia and filter after standing for one hour. Evaporate the filtrate to 25 ml. and add 75 ml. of alcohol. Heat to boiling and filter. Wash the calcium malate precipitate with alcohol (75 per cent.), dry and ignite. Boil the ash with excess of N/10 hydrochloric acid and back titrate with N/10 sodium hydroxide. The volume of N/10 acid used × 0·0067 gives the percentage of malic acid.

Chapter 11

FLESH FOODS : TABLE JELLIES

Meat—Meat Products (including Sausages)—Meat Extract—Fish—Fish Products—Gelatine—Table Jellies—Agar—Isinglass

MEAT

Composition of meat. The composition figures for various cuts of meat have been tabulated by Jackson and Jones (*Analyst*, 1932, **57**, 562) and Winton and Winton (*The Structure and Composition of Foods*, Vol. 3, Wiley, N.Y.). The water, fat, protein and ash contents vary considerably, not only between beef, mutton and pork but also between different cuts from the same animal, *e.g.*, the fat/protein ratio may vary from 0·3/1 (for veal) to over 7/1 (for pork). If the figures are calculated on the fat-free basis, however, the composition figures are reasonably constant for meats with normal water contents (*i.e.*, not dried up):—

Mean Nitrogen, Water and Ash Contents of Fat-free Beef and Pork

	Nitrogen	Protein $N \times 6{\cdot}25$	Water	Ash
	per cent.	per cent.	per cent.	per cent.
Beef . .	3·4	21·3	77·7	1·0
Pork . .	3·6*	22·5	76·5	1·0

Although these average figures have sometimes been questioned* (*e.g.*, *J. Sci. Food Agric.*, 1955, **6**, 317) those quoted for the fat-free nitrogen were accepted officially by the S.P.A. for the calculation of the meat content of meat products (*Analyst*, 1940, **65**, 257; 1952, **77**, 543).

Preparation of samples from flesh and flesh products. Raw meat consists essentially of muscle and fatty tissue held together by

* See also S.A.C. Report (*Analyst*, 1961, **86**, 557) p. 299.

a skeleton of tough connective tissue. It is somewhat difficult therefore to obtain a homogeneous sample from such a mixture which is suitable for analytical work. It is necessary to pass meat, sausages, etc., through a mincer. Then the minced sample should be mixed thoroughly in a pestle and mortar. The whole process should be repeated at least once more. The amount of mincing and mixing, however, which is necessary varies according to the coarseness of the original material. A single process may be sufficient for instance with such products as meat and fish pastes. Benne *et al.* (*J. Ass. Off. Agric. Chem.*, 1956, **39**, 937) have described a process which employs a motor driven grinder fitted with a fine mesh cutting plate and divider for preparing samples from fresh meat. Whatever method of preparation is employed the sample should be transferred to a screw-capped jar and stored under refrigeration. In view of the lack of homogeneity of the sample and the possibility of drying out during weighings, micro weighings should be avoided in the analysis of flesh foods and their products.

Assessment of Meat Spoilage

A convenient method for assessing general spoilage of meat is the determination of the *Dissolved Oxygen Absorption*:—

Two 5 g. quantities of the minced lean meat are mashed with tap water at 22°–23° and transferred to 250 ml. stoppered bottles, which are completely filled with water at this temperature (all air bubbles being carefully excluded) and incubated at 22° for two and four hours respectively. At the end of these periods there is added to each one 1 ml. of sulphuric acid, then excess of 0·1 N potassium permanganate. After five minutes the colour is discharged by 1 ml. of solution of oxalic acid, then is added 1 ml. of 33 per cent. solution of manganous chloride and 1 ml. of a mixture containing 33 per cent. of sodium hydroxide and 10 per cent. of potassium iodide. The precipitate will be practically white if all oxygen has been absorbed, or brown if there is oxygen still present in solution. If there is any putrefaction, the dissolved oxygen will all have disappeared within two hours, and if it is in the incipient stage, all will have disappeared within four hours.

A more exact assessment of the degree of spoilage of the lean meat can be obtained by determining the *Total Volatile Bases* (*T.V.B.*) using the macro method described under fish. A T.V.B. value below 20 mg. N per 100 g. indicates fresh meat and 30 mg. or over may be synonymous with inedibility. The bases which distil over consist almost entirely of the free ammonia (*cf.* fish).

The state of spoilage of meat can also be assessed by studying the *rancidity values* of the fat. With " whole " meat (unminced) the fat can be easily separated by cutting the outside fat into small pieces and rendering it in a beaker placed in the 100° C. oven for half to one hour. With minced samples mix 20 g. of sample with an excess of anhydrous sodium sulphate and then add to the mixture fresh diethyl ether. After stirring, filter the fat solution into a flask and remove the solvent. Alternatively if the fat content has been estimated on the sample the weighed fat can be used provided only solvent (*i.e.*, no acid or alkali) has been employed for the extraction. An F.F.A. figure in excess of 1·5 per cent. (as oleic acid) or a peroxide value much above 5 ml. of N/500 thiosulphate per g. is often an indication of incipient spoilage (the figures suggested are calculated on the *extracted fat*). As the proportion of fat in the original sample is small, such degrees of rancidity, however, may be insufficient to affect the taste.

Broumand *et al.* (*Food Tech.*, 1958, **12**, 159) have discussed the use of chemical tests for meat which are associated with organoleptic quality factors.

Preservation of meat. Meat is normally preserved by placing in a cold store (34–40° F.), by chilling (29° F.) or freezing (approx. 14° F.), by canning or by dehydration (maximum moisture 7·5 per cent.). Apart from the use of certain substances for curing, the Preservatives Regulations do not allow meat to be preserved with artificial preservatives (except that up to 450 p.p.m. of sulphur dioxide is permitted in minced meat *in Scotland* from June to September). For the *estimation of sulphur dioxide* in meat the iodimetric method is suitable. Formalin, salicylates, benzoates and borates are the principal non-permitted preservatives which were formerly found on occasion in meat.

Cured Meat

The term curing refers to treatment by salting or salting and smoking. Meat is salted either by rubbing in the salt mixture (dry curing), or, by immersion in a brine containing 15–25 per cent. of salt (wet curing), or, by pumping a weaker brine into the meat so that the salt is distributed by the arteries (arterial method). Salt used alone produces a dull appearance and small amounts of other ingredients (*e.g.* nitrate, nitrite) are incorporated also into most cures. The nitrite (either in the form of added nitrite or produced from the nitrate by reducing micro-organisms) reacts with the hæmoglobin pigments in the meat to produce the familiar pink colour. Hornsey and Mallows (*J. Sci. Fd. Agric.*, 1954, **5**, 573; 1955, **6**, 705) have made a most useful study of the chemical and bacteriological changes which occur in meat brines. They have shown that the nitrate reduction is associated with an increase in the numbers of halophilic gram-negative rods, which are present in the brine in very large numbers.

The *analysis of cured meats* should include the determination of **salt, nitrite** and **nitrate** and **added colour.** If the **meat content** is required it is determined by the methods used for sausages. **Canned products** should also be examined for **arsenic, lead** and **tin.**

Salt may be determined after careful ashing at 550° C. Alternatively, prepare an approximately 10 per cent. extract by blending (preferably mechanically) 25 g. of cured meat with 225 ml. cold water and filter. Then titrate 10 or 20 ml. of the filtrate with silver nitrate. The **nitrite** can also be determined on the filtrate. **If cereal is present,** however (*e.g.*, with Continental sausages, luncheon meat), a clearing agent must be used as in the A.O.A.C. method:—

Mix 5 g. cured meat thoroughly with 40 ml. water heated to 80° C. and transfer to a 500 ml. graduated flask using a further 260 ml. of hot water. Place on the water bath with occasional shaking, for two hours. Then add 5 ml. saturated mercuric chloride solution, cool, dilute to the mark and mix. Filter and determine the nitrite on a suitable aliquot of the filtrate using the Griess-Ilosvay reagents.* Measure the pink colour visually (a Nessleriser disc is available) or absorptiometrically.* Calculate the result as $NaNO_2$ in p.p.m.

Cooked pickled meat in general may contain up to 200 p.p.m. of nitrite (as $NaNO_2$), but no limit is recommended for bacon and ham (see also pp. 439, 443).

Nitrates. The A.O.A.C. describes the following method for nitrates in meats (see also *J. Ass. Off. Agric. Chem.*, 1935, **18**, 459; 1939, **22**, 597):—

SPECIAL REAGENTS

m-Xylenol.—1-hydroxy-2.4-dimethylbenzene.

Silver-ammonium hydroxide.—Dissolve 5 g. of silver sulphate in 60 ml. ammonia. Boil, concentrate to 30 ml. and dilute to 100 ml.

Standard nitrate solution.—0·1804 g. KNO_3 per litre.

METHOD

Weigh 10 g. of mixed sample into a beaker, add 80 ml. of warm water, break up any lumps and heat on a water bath for one hour, with occasional stirring. Transfer to a 100 ml. graduated flask, cool and dilute to the mark. Filter. Pipette 40 ml. of filtrate into a 50 ml. graduated flask, add three drops of bromocresol green and then dilute sulphuric acid (1 + 10) until the colour changes to yellow. Add N/5 potassium permanganate solution dropwise to oxidise any nitrite present to nitrate until a faint pink colour persists for about a minute. Add 1 ml. of dilute sulphuric acid (1 + 10), 1 ml. of 20 per cent. phosphotungstic acid solution and make up to the mark. Filter. Pipette into a 500 ml. conical flask a suitable aliquot of the filtrate (containing 0·025–0·25 mg. nitrate N and total volume not exceeding 20 ml.) and sufficient

* For details of the Griess-Ilosvay method see pp. 88–89.

silver-ammonium hydroxide to precipitate chlorides and also most of the excess of phosphotungstic acid. To the liquid in the flask add three times its volume of sulphuric acid (3 + 1). Stopper the flask, mix, cool to 35° C., add 0·05 ml. of *m*-xylenol, stopper, shake and hold at 30°–40° C. for thirty minutes. The presence of nitrates is indicated by a yellowish-brown colour appearing (a bright red precipitate due to incomplete removal of phosphotungstic acid may also appear—large excess may cause interference). After completion of the nitration, add 150 ml. of water and distil 40–50 ml. into 5 ml. of 1 per cent. sodium hydroxide solution. Transfer the distillate to a 100 ml. graduated flask, dilute to the mark and determine the nitrate N by comparing the colour produced by a suitable aliquot with that of a standard.

Prepare the colour standard by taking 10 ml. of the standard nitrate solution (10 ml. ≡ 0·25 mg. nitrate N) and add 0·05 ml. of *m*-xylenol and 30 ml. sulphuric acid (3 + 1). Cool and then carry out the nitration at 30°–40° C., distil, etc., as for the sample, but make the volume of distillate up to 500 ml. The result obtained represents the nitrate + nitrite calculated as nitrogen. The proportion of nitrite N is usually comparatively small.

Barnes (*Analyst*, 1950, **75**, 388) replaces the distillation by a double extraction technique, which also reduces interference due to organic substances. The " incubated " solution is shaken with toluene, the separated organic layer is shaken with 1·6 per cent. sodium hydroxide solution and the optical density of the alkaline aqueous layer measured against water using a violet filter (*cf.* Landmann *et al.*, *J. Ass. Off. Agric. Chem.*, 1960, **43**, 531).

The composition of various types of bacon and ham has been discussed by Baker (*J. Sci. Food Agric.*, 1956, **7**, 179).

Sausages

Fresh sausage meat consists of a mixture of various cuts of pork and/or beef, cereal filler (usually wheat rusk), water, seasoning (salt, spices, etc.) and usually sulphite preservative. Some products contain milk powder and added colour is sometimes present. Both animal and artificial casings are used. The Offals in Meat Products Order, 1953 (S.I. 1953, No. 246) prohibits the use of certain offals in *uncooked open meat products*, viz. brains, feet, fries, gut (including chitterlings), manifolds, paunches, udder, sweetbreads, tripe, melts or lites, spinal cord, uteri, pigs' maws and calves' vells. The meat content of sausages was controlled under Defence Regulations during World War II and standards were retained until the revocation of The Meat Products Orders in 1953. In 1956, the Food Standards Committee Report on Sausages (S.O. Code No. 24–247) was published and its main *recommendations* for *uncooked* sausages were as follows:

(1) Sausages made wholly or mainly with pork—minimum 65 per cent. meat.

(2) Other meat sausages (including beef)—minimum 50 per cent. meat.

(3) Proportion of fat not to exceed 50 per cent. of the total meat content.

(4) The majority of the Committee considered that the terms "pork sausages" and "beef sausages" should apply where at least four fifths of the meat content consists of the named meat.

In addition the Preservatives Regulations prescribe a maximum for sulphur dioxide of 450 p.p.m. for "Sausages and sausage meat containing *raw meat*, cereals, and condiments."

Analysis of sausage meat

After preparing the sample by double mincing and mixing the examination for **sulphur dioxide** should be commenced as soon as possible as the amount of preservative present falls during storage. The analytical methods and the calculations to be employed for assessing the **meat content** were originally proposed by Stubbs and More (*Analyst,* 1919, **44**, 125). The S.P.A. (*ibid.*, 1952, **77**, 544) recommended that the same type of calculation be used but incorporated the revised factors suggested by Jackson and Jones (*ibid.*, 1932, **57**, 562):—

(a) **Water.** Dry 5 g. at 100° C. on 20 g. sand with the assistance of a glass rod.

(b) **Fat.** (A) Extract the fat from the dried residue obtained in (a) by macerating with ether. Filter the extracts into a weighed flask, remove the solvent, etc.
(B) *Alternatively* carry out a Werner-Schmid type extraction by digesting 5 g. with 40 ml. of 4 N hydrochloric acid for thirty minutes at 60° C. and then on the water bath. Extract the fat from the cooled digest three times with ether, wash the combined extracts with water, remove the solvent,* etc. A rapid method for determining fat in sausages has been described by Talbot (*Analyst,* 1949, **74**, 462).

(c) **Nitrogen.** Use 2·5–5·0 g. for the macro-Kjeldahl method. Carry out in duplicate and return the protein as N × 6·25.

(d) **Ash.** Incinerate 5 g. at 500° C. Salt can be determined on the ash by adding water and potassium chromate and titrating with N/10 silver nitrate solution. By determining the calcium in the ash it is possible to get some idea as to whether the sample contains milk (see (h) p. 299).

(e) **Starch** (*S.P.A.*). Digest 20 g. with 300 ml. of 5 per cent. alcoholic potash on the water bath. Filter off the insoluble matter and wash with alcoholic potash. Wash the insoluble matter into a beaker with 200 ml. of warm water, add 40 ml. of N/1 aqueous potassium hydroxide and warm to dissolve the starch. Cool and make up to 250 ml. with

* A simple method for removing the solvent is illustrated on p. 39.

water in a graduated flask. Add a 500 ml. aliquot to a beaker containing 300 ml. of 90 per cent. v/v alcohol acidified with acetic acid (equivalent to 8 ml. of N/1 potassium hydroxide), stir thoroughly and allow to stand overnight. Filter off the starch, wash with alcohol, then with ether and dry at 100° C. After weighing, incinerate and deduct the ash from the weight of the residue to give the starch in 4 g. of original sample.

Illing and Whittle (*Food*, 1944, **13**, 32–) proposed the following method for determining starch in sausages:—

Make 5 g. of the sausage into a paste with 5 ml. of 5 per cent. alcoholic potash solution. Add 50 ml. more of the alcoholic potash solution and heat on the water bath for fifteen minutes. Allow to settle and decant the liquid through a filter paper in a Buchner funnel. Extract the residue twice more with 25 ml. of alcoholic potash and then boil with 2 × 25 ml. lots of alcohol (80 per cent.), washing all the extracts through the filter. Add the filter paper and residue to the main residue, add 75 ml. of 0·7 per cent. aqueous potassium hydroxide and gelatinise the starch by placing in boiling water for thirty minutes. Transfer the hot liquid to a 200 ml. graduated flask, cool, make up to volume and filter.

Pipette 20 ml. of the filtrate into a centrifuge tube and neutralise with 5 per cent. acetic acid to phenolphthalein. Then add 4 ml. of N/10 iodine solution and 4 ml. of 10 per cent. potassium acetate solution. Allow to stand until the precipitate settles (at least one hour), centrifuge and pour off the supernatant liquid. To the residue add 12 ml. of a mixture containing 10 ml. of 95 per cent. alcohol and 2 ml. of N/10 sodium thiosulphate, adding in several small quantities and breaking up the residue thoroughly with a rod. Add 25 ml. of 80 per cent. alcohol, filter through a Gooch crucible, wash with 95 per cent. alcohol, dry and weigh as starch.

(f) **Calculation of the meat content** (S.P.A., 1952)

(1) % (Dry carbohydrate + crude cellulose)
$= C = 100 - (\%\ \text{Water} + \%\ \text{Fat} + \%\ \text{Protein} + \%\ \text{Ash})$

(2) If a wheat filler is present:

$$\%\ \text{Filler nitrogen} = N_c = \frac{2C^*}{100}$$

For fillers with a negligible nitrogen content such as potato starch ("farina") or cornflour, $N_c = 0$.

(3) $$\%\ \text{Total Meat} = M_T = \frac{\%\ \text{Total N} - N_c}{N_m} \times 100 + \%\ \text{Fat}$$

where Nm (nitrogen in the fat free) = 3·4 (beef), 3·6* (pork), 3·5 (mixed meats).

(4) Assuming that the filler is bread containing 40 per cent. of water—

$$\%\ \text{Cereal filler} = 2C$$

N.B. If wheat rusk is used the percentage figure for cereal filler is about 1·3 C.

(5) $\%\ \text{Added Water} = W_A = 100 - (M_T + 2C + \%\ \text{Salt})$

(6) If the determined starch differs from the figure obtained by difference by more than 2 per cent., the presence of materials such as milk-powder or soya flour is indicated (see also (h) below).

* See also page 311.

The factors for the average nitrogen in the fat-free (N_m) of 3·6 and 3·4 for pork and beef respectively have been subjected to some criticism (*see* Marshall, *Analyst*, 1955, **80**, 776 and Reith *et al.*, *J. Sci. Food Agric.*, 1955, **6**, 317). The A.M.C.'s Meat Products Sub-Committee (*Analyst*, 1961, **86**, 557) have collected data for various cuts of meat and make the following recommendation in relation to pork: " After due consideration of the types of meat normally used in the manufacture of comminuted products, and of the use of pigs of varying live weight, the Sub-Committee recommends an average nitrogen factor of 3·45 as the best compromise for general use." In a further report, the Sub-Committee (*ibid.*, p. 560) state that the values obtained from analyses of the wheat rusk now used indicate nitrogen contents ranging from 1·9 to 2·6 per cent. (calculated on dry carbohydrate). They therefore recommend as the best compromise a figure of 2·3 per cent. as the correction for nitrogen in the rusk filler (*see also* Fraser and Holmes, *Analyst*, 1958, **83**, 371). Compared with the method outlined on the previous page, the revised nitrogen factor for pork has the effect of increasing the calculated meat content, but the higher subtraction figure recommended for the rusk influences the result in the opposite direction.

The following are not included in the S.P.A. report:—

(g) **Soya bean meal.** Methods for the detection of soya meal by means of the urease reaction and the microscope are given on p. 183 . No completely satisfactory method exists for determining soya quantitatively in meat products. Some usedata, however, is given by Illing and Whittle (*Food*, 1944, **13**, 32) who rely on the fact that dry oil-free soya contains 11·4 per cent. sucrose (expressed as dextrose), 6·8 per cent. of salt-free ash and 39 p.p.m. of manganese. " Low fat " soya such as was used in sausages during World War II has a protein content equal to three times its weight of lean meat.

Hendry (*J. Ind. Eng. Chem., Anal. Edn.*, 1939, **11**, 611) assesses the soya content from a determination of the insoluble non-fermentable sugars.

(h) **Dried milk.** Some brands of sausage contain from about 2–6 per cent. of dried milk, the presence of which can sometimes be deduced if the ash less salt figure is in excess of about 0·7 per cent. This difference would, however, also be increased if inorganic emulsifying agents such as sodium phosphates were present and it is always advisable to search for lactose in a clarified filtrate by paper chromatography

The amount of dried milk present in sausage can be calculated from the calcium and lactose figures. Dried skim milk contains 80 times more calcium than meat or the other ingredients in sausages. McVey and McMillin (*J.A.O.A.C.*, 1940, **23**, 811) have described a suitable procedure for the determination of calcium in sausages. The ash from 25 g. is dissolved in dilute hydrochloric acid, ammonium oxalate solution added and the liquid boiled. Then the liquid is neutralised to methyl red with ammonia and the calcium oxalate is filtered off, dissolved in dilute sulphuric acid and titrated with N/20 permanganate at 90° C. (*cf.* p. 30).

1 ml. N/20 potassium permanganate ≡ 0·0014 g. CaO. In order to arrive at the composition of the sausage, assume that dried skim milk powder contains **1·84** per cent. of calcium (as CaO), **5·81** per cent. of nitrogen and 50 per cent. of lactose. Then the percentage of dried milk(S) equals $\frac{\%\ \text{CaO} - 0{\cdot}024}{0{\cdot}0184}$

The total meat (M_T) in the sausage is then given by the following modified formula which allows for the nitrogen and lactose due to the milk:—

$$M_T = \frac{\%\ \text{Total N} - 0{\cdot}06\ S - 0{\cdot}02^*\ (C - 0{\cdot}5\ S)}{N_m} \times 100 + \%\ \text{Fat}$$

where C = Total Carbohydrate, etc. by difference (which includes lactose).

N.B. From the point of view of the calculation of meat contents, 6 per cent. of milk powder is equivalent to 10 per cent. of lean meat. This method, based on the calcium, can be applied to samples which have badly decomposed.

McVey and McMillin (*loc. cit.*) also describe a selective fermentation method for determining lactose in sausage. The lactose is determined after the other common reducing sugars have been destroyed by fermentation with yeast (see also *A.O.A.C.*, and *J.A.O.A.C.*, **1958, 40,** 57). The " direct " determination of lactose (*i.e.*, without fermentation) is probably only suitable for very fresh sausage as there often appears to be a marked incrėase in the *total* reducing sugars during storage presumably due to enzymic production of maltose.

(i) **Added colour**—see p.100.

Tests for Horseflesh

Many reactions have been proposed for the detection of horseflesh in sausages and meat products generally. These mostly depend on the recognition or estimation of glycogen, which is a carbohydrate present in the liver of most animals, but only to a small extent in other parts of their flesh, except in the case of the horse. Further investigation by various chemists has shown that it is not safe to conclude the presence of horseflesh unless the glycogen content is

* See also p. 311.

upwards of 2 per cent. on the dry matter, and as this amount is not always present in raw horse-meat itself, the test is of limited value.

Glycogen forms a characteristic wine red colour with iodine solution. The reaction may be applied by boiling 50 g. of the meat with 200 ml. of water for an hour, then cooling and adding enough dilute nitric acid to reduce the colour, and then filtering. To the filtrate is

	Water.	Glycogen.	Glycogen on dried substance.
	Per cent.	Per cent.	Per cent.
Horseflesh	61·83–72·90	0·174–1·366	0·64–4·62
Beef	73·62	0·206	0·74
Beef	75·55	0·018	0·07
Veal	76·12	0·346	1·44
Veal	74·47	0·066	0·25
Pork	54·0–66·3	Traces	Traces
Horse sausages :			
Red sausage	70·04	0·504	1·68
Liver sausage	67·00	1·762	5·34
Salami sausage	33·00	0·034	0·05

added a layer of iodine solution, which gives a red ring at the point of contact in the presence of glycogen. It has been suggested that the test so carried out will reveal the presence of 5 per cent. of horseflesh, but this is improbable. For the estimation of glycogen, Piettre's method is to boil 25 g. of the minced meat with 80 ml. of 10 per cent. alcoholic potassium hydroxide solution until all the tissue is dissolved. Then the insoluble residue is filtered off and washed with acidified alcohol until free from alkali. It is then dissolved by heat in water made just alkaline. An equal volume of strong alcohol is added to precipitate the starch, which is filtered off and washed with 50 per cent. alcohol. The filtrate is evaporated to low bulk, and the glycogen is precipitated by adding strong alcohol, collected, dried and weighed. Obviously, the process is not susceptible of great accuracy. The amounts shown in the table above have been reported by Bujard.

Another test of limited value, but useful for confirmation, is that of Ehrlich, who shows that when horse-meat is treated with formalin, it develops an intense and characteristic smell, suggestive of roast goose, within forty-eight hours. This smell is not given by any other meat. The only specific test is a biological one, which, unfortunately, is not always readily performed by the chemist, owing to the difficulty of obtaining a supply of the appropriate serum. The serum is obtained by injecting fresh filtered horse serum into a rabbit. After some days blood is withdrawn from the rabbit and a dilute solution of its serum in physiological salt solution is prepared. This has the property of precipitating an extract from meat containing horse-flesh, but not extracts from other flesh (Fiehe, *Z. Unters. Nahr. Genussm.*, 1907, **13**, 744).

The precipitin test is a sensitive one, but is only applicable to uncooked meat. The technique employed is described in Hawk, Oser and Summerson's " Practical Physiological Chemistry," (Churchill, London, p. 485), by Roche Lynch (*Analyst*, 1928, **53**, 5) and by Bonnand and Thiery (*Anal. Abstr.*, 1956, **3**, No. 2284), who have put the test on a semi-quantitative basis.

Corroboration of the presence of horseflesh may sometimes be obtained by examination of the fat, which may be extracted by solvents in the ordinary way. Horse-fat crystallises from ether in tufts resembling beef-stearin, also it has a higher iodine value (80–95) than other fats such as those of pork, veal, beef and mutton, but as the iodine values of these are so variable and are sometimes rather high, deductions from the iodine value are uncertain, especially in cases of admixture.

Dalley (*Analyst*, 1950, **75**, 336) has reported favourably on the method proposed by Paschke (*Analyst*, 1939, **64**, 47), which depends on the presence in horse-fat of up to 2 per cent. of linolenic acid (when determined as the hexabromide) and its relative scarcity in most other fats. Dugan and Petheram (*J. Ass. Off. Agric. Chem.*, 1952, **35**, 767) state, however, that in the presence of beef or pork fat the hexabromide value gives variable results and suggest that it is preferable to employ a spectrophotometric method for determining the trienoic fatty acids (expressed as linolenic acid).

Meat Paste

The Food Standards (Meat Paste) Order, 1951 (S.I. 1951, Nos. 1457, 2242) requires meat paste (or spread) to contain a minimum of 55 per cent. meat (calculated as *raw meat* after waste has been eliminated). In addition a code of practice has been published which relates mainly to labelling (*Brit. Food J.*, 1955, **57**, 51):—

(a) The term " Potted Meat " must only apply to a product which consists solely of meat and contains no binder.

(b) If the term " paste " or " spread " is preceded by the name of a specific meat (or meats), the meat content should normally be derived exclusively from the named meat (or meats), *e.g.*, the meat in " Chicken and Ham paste " should consist entirely of chicken and ham. The code also adds, however, that other meats may be present provided that there is present at least 55 per cent. (the statutory minimum) of the named meats.

(c) Where the minimum statutory meat content includes meat(s) other than those named, the product may be described as " Meat Paste " (or " Spread ") followed by the name of the variety provided certain other stated minima are complied with, *e.g.*, the label

Meat Paste CHICKEN AND HAM

is permissible provided that there is present at least 25 per cent. chicken and 25 per cent. ham (calculated as *a percentage of the total meat content*). The code states various minima of this type for different varieties of meat paste. If these minima are not complied with, the term " Meat Paste " (or " Spread ") is permissible provided the product complies with the statutory minimum of 55 per cent. meat.

Analysis of meat pastes. The analysis of meat paste is similar to that applied to sausages, *i.e.*, the determination of **water, fat, protein** (**N** $\times$ **6·25**), **ash, salt, starch** and **trace metals** and an examination for **added colour.** The meat content of meat pastes can be calculated by means of the Stubbs and More method for sausages using the appropriate factors. For this purpose the following average percentage figures for the nitrogen in the fat-free have been given more or less official recognition (at the time of writing; *cf.* p. 299): pork 3·6; beef, mutton 3·4; mixed meat or unspecified meat (*e.g.*, meat roll) 3·5; chicken, turkey 3·5. Other figures for the nitrogen/f.f. obtained from various sources are: liver 3·2, kidney 2·9, tongue 4·0, pheasant 3·8, wild duck 3·6.

Minced Chicken, Chopped Chicken in Chicken Stock and Similar Products

Agreement has been reached between the Food Manufacturers' Federation and the Association of Public Analysts relating to the meat content and labelling of minced chicken, etc. (*Brit. Food. J.*, 1958, **60**, 117):—

(a) Products sold under an unqualified description, *e.g.*, "Minced Chicken," " Chopped Ham," etc., shall contain a minimum of 95 per cent. of the designated meat(s).

(b) Products sold under a qualified description, *e.g.*, " Minced Turkey in Jelly," " Chopped Chicken in Chicken Stock," etc., shall contain a minimum of 70 per cent. of the designated meat(s).

(c) The lean meat is calculated from the formula:—

$$\% \text{ Lean Meat} = \frac{\% \text{ Nitrogen}}{\text{Conversion factor}} \times 100$$

Conversion factors: Chicken 3·5, Turkey 3·5, Pork 3·6, Beef 3·4 (*cf.* p. 311).

Luncheon Meat

In 1959 a standard for luncheon meat was agreed upon between the F.M.F. and the A.P.A. This provided that " Products sold under any description which includes the words " luncheon meat " shall have a meat content of not less than 80 per cent. The meat content is calculated as raw meat." Meat includes pork, bacon, ham, beef, mutton and veal but does not include " prohibited offals " as defined in the Offals in Meat Products Order 1953 (see p. 308). This standard does not necessarily apply to imported articles (see *Brit. Food J.*, 1959, **61**, 96; *cf. ibid.*, 1958, **60**, 102).

MEAT EXTRACT

Much of the concentrated meat extract used in Britain comes from S. America, where it is made as a supplement to corned beef manufacture. For the process, several batches of boned meat are successively heated in the same tank of water, which dissolves out the soluble extractives. The meat which is removed is hot pickled for the preparation of corned beef. When the dissolved solids in the cooking water reaches about 5 per cent., the fat is skimmed off and the liquid is concentrated to a solids content of about 80 per cent. The product is then sealed whilst hot in 56 lb. containers. During the process some of the creatine gets converted to its anhydride creatinine. During storage, however, there is a tendency for some of the creatinine to revert back to creatine and, as the latter is less soluble, it sometimes separates out as crystals. Also a clear fluid layer exudes from some extracts during prolonged storage.

For commercial purposes the quality of meat extract is often assessed from the creatine + creatinine figure, which should preferably not fall much below 10 per cent. when calculated on the dry matter. The amount of gelatin present should preferably be below 1 per cent. The following standards for meat extract have been suggested in the U.S.A.:—

" Meat extract is the product obtained by extracting fresh meat with boiling water and concentrating the liquid by evaporation after removal of the fat. It contains at least 75 per cent. of total solid matter, of which not more than 27 per cent. is ash and not over 12 per cent. sodium chloride. The fat should not exceed 0·6 per cent. and the nitrogen be not less than 8 per cent. The nitrogenous compounds contain not less than 40 per cent. of meat bases and not less than 10 per cent. of creatine and creatinine."

S.P.A. recommended methods of analysis

Suitable methods for the partial analysis of meat extract and similar products have been proposed by The Analytical Methods

Committee of the S.P.A. (*Analyst,* 1951, **76**, 329) who suggest that the determination of **water, ash, chloride, total nitrogen** and **total creatine and creatinine** are sufficient for general quality assessment. If a more detailed analysis is required, particularly for the detection of adulterants and non-meat ingredients, further determinations may be necessary, viz., **phosphate, fat, soluble and insoluble nitrogen, amino nitrogen, gelatin, tannic acid precipitate, nicotinic acid, starch and trace elements.** Subsequently a method for determining gelatin was described in an interim report (*Analyst,* 1953, **78**, 134).

The A.M.C. methods are described below:—

Sampling. Blend pasty samples prior to weighing out by warming so that sediment is incorporated. Prepare a 10 per cent. (w/v) stock solution using hot water to ensure solution of all soluble materials (solution A).

Water. Dry 10 ml. of stock solution A in a nickel dish fitted with a lid (see *Analyst,* 1927, **52**, 402) firstly on the water-bath and then for eight hours in the 100° C. oven.

Ash. Evaporate 10 ml. of stock solution A in a platinum dish on the steam-bath, char on a low flame and complete the ashing at 550° C.

Chloride. Evaporate 20 ml. of stock solution A with 10 ml. of 5 per cent. sodium carbonate solution and ignite at a temperature not greater than dull redness. Extract with hot water, filter and wash. Retain the filtrate. Return the filter paper and residue to the dish, moisten with a little carbonate solution and ignite to a white ash. Dissolve the ash in dilute nitric acid, filter and wash the paper thoroughly with water. Combine the filtrates and determine the chloride by the Volhard method.

Total Nitrogen. Determine on 5 ml. of stock solution A by the macro-Kjeldahl method. The digestion time should proceed for 3 hours after clearing.

Total Creatine + Creatinine

(a) Prepare a 0·1603 per cent. (w/v) stock solution of creatinine zinc chloride in N/10 hydrochloric acid. *Dilute ten times* prior to use so that 1 ml. ≡ 0·1 mg. creatinine (solution B).

(b) Hydrolyse 10 ml. of stock meat extract solution A with 10 ml. of 2N hydrochloric acid by refluxing in a boiling water bath for two hours (minimum) or alternatively autoclave for twenty minutes at 117°–120° C. Cool, add 10 ml. of 2N sodium hydroxide solution and dilute to 500 ml. (C) or, if no absorptiometer is available, to 250 ml. (D).

(c) To a series of dry 100 ml. graduated flasks add 5 ml. (or more) of the hydrolysed sample solution C and for the standard curve 0 (reagent blank), 2, 4, 6, 8 and 10 ml. of the standard solution B (equivalent to 0–1·0 mg. creatinine). To each flask add water to bring the volume to 20 ml. and then add 20 ml. of 1 per cent. picric acid solution (standardised by titration to phenol red) and 2·5 ml. of 2N sodium hydroxide solution. Allow the flasks to stand at 20° C. for fifteen minutes and then dilute each to the mark and filter. Reject the first runnings and measure the orange colour on the absorptiometer using an Ilford green filter No. 604 (520 mμ) and a 1 cm. cell.

(d) *If no absorptiometer is available,* dilute the hydrolysed solution to 250 ml. (D) instead of 500 ml. Then to three 100 ml. graduated flasks add 20 ml. of standard solution B (≡ 2 mg. creatinine) and two trial aliquots of 7 and 10 ml. of the hydrolysed solution of the sample (D). Make the volume in each flask up to 20 ml. and then add 20 ml. of 1 per cent. picric acid solution, 2·5 ml. of 2N sodium hydroxide solution, etc., as in the absorptiometer method above and compare the colours visually by Nesslerising or by using a Duboscq colorimeter. If necessary calculate a more suitable aliquot to employ for the colour comparison and repeat the picric acid procedure.

Whichever method is used, calculate the percentage of total creatine and creatinine (as creatinine).

Gelatin (*Analyst,* 1953, **78**, 134)

Weigh out into a 250 ml. beaker 10 g. of meat extract and add 125 ml. water. Bring to the boil, stirring constantly, and add 0·5 ml. glacial acetic acid. Coagulate the insoluble material by digesting the mixture on the steam-bath for 15–30 minutes. Filter through a No. 4 Whatman into a 250 ml. graduated flask washing through well with hot water. Cool the filtrate and make up to 250 ml. Pipette 25 ml. into a porcelain basin (capacity approx. 200 ml.), add 0·25 ml. of formalin and mix thoroughly. Concentrate to a thick consistency and then add a further 0·25 ml. of formalin and mix thoroughly again with the glass rod so that the mixture is spread evenly over the basin to within one inch of the rim. " Bake " hard on a *boiling* steam-bath for two hours and extract the contents of the dish twice with 100 ml. of diluted formalin (≡ 1 per cent. formaldehyde) at 40° C. Allow one hour for each extraction maintaining throughout at 40° C. Filter each washing through a No. 54 Whatman, breaking up the complex during the final extraction. Loosen the complex, transfer it to the filter paper and wash it with a further 100 ml. of the diluted formalin solution at 40° C. Finally determine the nitrogen in the gelatin-formaldehyde complex by the Kjeldahl method (Gelatin = N × 5·55).

Wood (*J. Sci. Food Agric.*, 1956, **7**, 196) has described paper chromatographic methods for the detection of various classes of compounds present in meat extracts (*cf. Chem. & Ind.*, 1947, p. 302).

A semi-fluid product sold on the retail market in Britain contains meat extract, stock, hydrolysed protein, yeast extract, dried beef, caramel, starch, salt and other flavourings. This product contains about 1·3 per cent. of creatine + creatinine. A brand of compressed cubes on the market contains meat and yeast extract, hydrolysed protein, stock, dried beef, beef fat, flour, caramel, salt and other flavourings. A series of cubes examined by Cox gave the following ranges for the composition figures (as percentages): water 6·5–14·7, organic matter 34·4–66·2, mineral matter 25·7–59·1, salt 18·8–50·3, fat 0·9–2·4, carbohydrates 3·4–40·4, total nitrogen 1·7–8·2 (≡protein 10·9–51·4), creatine + creatinine 0·7–2·0. Cox stated that such cubes appear to contain up to 15 per cent. of added gelatin.

A sample of **yeast extract** examined in the same way as for meat extract gave the following percentage figures: Water 12·4, organic matter (soluble) 34·7, organic matter (insoluble) 12·2, mineral matter 40·7, salt 38·5, total nitrogen 7·3, creatine + creatinine—nil.

FISH AND FISH PRODUCTS

Composition of fish. For the purposes of the analysis of fish products it is convenient to divide fish into three groups (i) White (non fatty) fish, (ii) Fatty fish and (iii) Shellfish. The following table shows the main composition figures from various sources for many of the fish which are incorporated into fish products:—

Percentage composition of the edible portion of fish

	Water	Fat	Protein N × 6·25	Ash	Nitrogen on fat-free basis
WHITE FISH (RAW)					
Cod	80–83	0·1–0·9	15–19	1·2	2·7
Haddock	79–84	0·1–0·6	15–20	1·2	2·8
Whiting	81	0·2–0·6	16–19	1·1	2·8
FATTY FISH (RAW)					
Herring	73 (mean)	0·4–28*	16–20	1·5	3·4
Mackerel	73 (mean)	2–20*	17–23	1·2	3·2
Salmon	61–79	2–15	17–24	1·4	3·0–4·0 (mean 3·5)
Sprat	60–77	1–17	14–19	1·8	3·2
SHELLFISH (RAW)					
Crab	74	1–7	22·4	2·7	3·7
Lobster	71–75	0·5–3	19–21	—	3·3
Prawn, Shrimp	69	2	21	—	3·4
FISH ROES (RAW)					
Cod roe, Haddock roe	73	1–2	21·5	—	3·4
Herring roe	73	3·3	22	—	3·6
PROCESSED FISH					
Bloater	64	14	19	—	—
Kipper	47–63	8–28	20	3–5	—
Salmon (canned)	57–67	5–22	19–24	2–4	—
Sardine (canned) drained	50–65	13	23	3–6	—
Caviar	40–60	7–17	23–28	5–12	—
Fish Paste (most types)	60–70	5–13	15–23	2·5–7	—

* Occasionally even higher figures for fat have been reported.

A study of the composition tables in the literature shows that the sum of the percentages of water, fat, protein (N × 6·25) and ash in fish amounts usually to between 100·2 and 100·6. This discrepancy is caused by the use of the conventional factor for the conversion of nitrogen to protein of 6·25, whereas the factor for the major protein in fish, myosin, is close to 6·0. With the fatty fishes there are marked differences in the fat content according to various factors such as the age (and size) of the fish and the season of the year in which the fish was caught.

Extensive tables relating to the composition of fish are given in the following works:

"The elementary chemical composition of marine organisms," by Vinogradov (transl. Efron and Setlow), Yale, U.S.

"The chemical composition of Fish tissues," D.S.I.R., Food Investigation Special Report No. 69, H.M.S.O., 1959.

Fish

Inspection of fish. Fish is inspected for the presence of parasites and diseases and also for quality assessment, a term which usually refers to the degree of spoilage. Parasites such as worms and lice have occasionally been reported in fish, but as they tend to reside in the stomach and intestine they are normally removed during gutting (see *A Guide to Marketable Fish*, by Daniel, Livpl. Univ.). Diseases causing tumours, ulcers and "boils" have also been reported occasionally in fish (see "*Fish and Fish Inspection*," by Syme, Lewis, Lond.), but poisoning in humans due to the ingestion of fresh (unprocessed) fish is very rare (see Shewan and Liston, *J. Appl. Bact.*, 1955, **18**, 522). Shellfish have been the cause of several outbreaks of poisoning, however, and a limit of 200 E. coli per mussel or oyster seems useful as a general criterion of "infection."

As fish spoils, various physical changes become apparent, *viz.* the odour deteriorates, the surface becomes dull and the slime on it becomes opaque, the flesh softens, the eye pupil becomes turbid and a reddish coloration develops along the backbone (see *D.S.I.R. Food Inv. Leaflet No.* 3, *H.M.S.O.* and Keywood, *Roy. Soc. Hlth. J.*, 1957, **77**, 520). In spite of the extensive search by the world's fish research stations for a rapid scientific method which would be suitable for the routine assessment of the degree of spoilage in fish, most inspectors still rely largely on appearance, odour, etc. In borderline cases, however, chemical determinations may be of considerable value.

Assessment of fish spoilage

Fish spoils somewhat rapidly and the analyst is sometimes called upon to assess its edibility. The relevant biochemical changes and their application to scientific methods of spoilage assessment have been the subject of numerous reviews, *e.g.*, Adv. in Food Res., Vol. 2, p. 344; Vol. 7, p. 41, Academic Press, N.Y.

The spoilage of fish stored in ice is due to bacterial and enzymic action which results in the production of various volatile compounds, in particular *trimethylamine* (T.M.A.), dimethylamine (D.M.A.), ammonia and volatile acids. Trimethylamine oxide is reduced during spoilage to T.M.A. and the ammonia formed is mainly a product of protein breakdown. The amounts of T.M.A. and the *total volatile bases* (T.V.B., *i.e.*, volatile amines + ammonia) are most commonly used for assessing the degree of spoilage in white fish. The T.M.A. and T.V.B. contents are often determined by the Conway microdiffusion technique (Beatty and Gibbons, *J. Biol. Bd. Can.*, 1937, **3**, 77) or by semi-micro steam distillation (Ronold and Jakobsen, *J.S.C.I.* 1947, **66**, 160). The T.V.B. content can be readily determined, however, without recourse to special equipment by employing the following modification of the macro distillation method proposed by Lucke and Geidel (*Z. Untersuch. Lebensm.* 1935, **70**, 441):—

DETERMINATION OF TOTAL VOLATILE BASES IN FLESH FOODS

Set up a macro-Kjeldahl distillation apparatus. To the distilling flask add 10 g. minced fish, 2 g. magnesium oxide, 300 ml. tap water and, if necessary, porous pot and an anti-foaming agent (an "Antifoam" silicone preparation or octyl alcohol). To the 500–700 ml. receiving flask add 25 ml. of 2 per cent. boric acid solution and a few drops of screened methyl red indicator as used in the determination of protein, as described in Chapter 2, p. 22. Connect up the apparatus with the receiver tube dipping below the boric acid solution. Heat the distilling flask so that the liquid boils in exactly ten minutes and, *using the same rate of heating*, distil for exactly twenty-five minutes. Wash down the condenser, etc., with distilled water and titrate the distillate with N/10 sulphuric acid. Multiply the titration (less blank) by 14 to obtain the T.V.B. as mg. N per 100 g. flesh.

If the amount of T.V.B. is less than 20 mg. N/100 g. fish can be considered as "fresh." Most fishery research stations maintain that fish is "stale" when the rate of T.M.A. production begins to be appreciable and this corresponds with T.V.B. in excess of 30 mg. N/100 g. Most "conditioned" consumers, however, would be prepared to eat fish showing rather higher values.

With **fatty fish**, the rancidity of the fat is a more useful criterion than the T.V.B. The rancidity values are determined on the oil which is extracted with ether after mixing the fish with anhydrous sodium sulphate (see also p. 26). Fish oils being highly unsaturated are somewhat prone to oxidation and peroxide values of the order of 10–20 ml. N/500 thiosulphate per g. of extracted oil correspond to incipient spoilage. Fat hydrolysis is also considerable and the increase in the acid value of the fat represents another possible measure of the degree of spoilage (Sigurdsson, *Ind. Eng. Chem. Anal. Ed.*, 1947, **19**, 892). A maximum *acid value* of 2·75 (calculated in the oil) has been proposed for the grading of canned herring (Charnley and Davies, *Analyst*, 1944, **69**, 302).

Freshwater fish quality can also be assessed from fat rancidity values. Freshwater fish (like meat) contain no trimethylamine oxide so that the volatile bases formed consist almost entirely of ammonia. The most serious form of quality deterioration in **frozen fish** is protein denaturation which affects the texture of the thawed, cooked fish (see *D.S.I.R. Food Inv. Leaflet No.* 11, H.M.S.O.). Determination of the T.V.B. is usually of little use for frozen fish, but fat rancidity values are sometimes of value in assessing the quality of frozen fatty ash.

Determinations of the type mentioned above bear out (for unfrozen fish) the following comparisons which are well-known in the trade:—

Flat white fish (*e.g.*, plaice, halibut) keep best.

Round white fish (*e.g.*, cod, haddock) keep reasonably well (except whiting).

Fatty fish (*e.g.*, herring, mackerel) are poor keepers.

Fish Paste

The Food Standards (Fish Paste) Order, 1951 (S.I. 1951 No. 1456 as amended by S.I. 1951 No. 2241) requires fish paste to contain a minimum of 70 per cent. fish (calculated as *raw fish* after waste has been eliminated even if the fish used has been smoked or otherwise processed). In addition a code of practice has been published which relates (as does the code for meat paste quoted previously) mainly to labelling (*Brit. Food J.*, 1955, **57**, 51):—

(a) The term " Potted Fish " must only apply to a product which consists solely of fish and contains no binder.

(b) If the term " paste " is preceded by the name of specific varieties of fish, the fish content should normally be derived exclusively from the named fish, *e.g.*, the fish in " Salmon & Shrimp Paste " should consist entirely of salmon and shrimp. The code also adds, however, that other varieties of fish may be present provided that there is present at least 70 per cent. (the statutory minimum) of the named fish.

(c) Where the minimum statutory fish content includes fish other than those named, the product may be described as " Fish Paste " followed by the name of the variety provided certain other stated minima are complied with, *e.g.*, the label:—

Fish Paste SALMON AND SHRIMP

is permissible provided that there is present at least 25 per cent.

salmon and 2·5 per cent. shrimp (each calculated as *a percentage of the named fish in the finished product, cf.* meat paste). The code states various minima of this type for 26 different varieties of fish paste. If these minima are not complied with, the term " Fish Paste " is permissible provided the product complies with the statutory minimum of 70 per cent. of fish.

Analysis of fish pastes. The analysis of fish paste is very similar to that applied to sausages, *i.e.*, the determination of **water, fat, protein** (N × **6·25**), **ash, salt, starch, arsenic** and **lead** and an examination for **added colour.** By using appropriate factors the total fish content (as raw fish) can be assessed with a reasonable degree of accuracy. It is still apparent, however, that although specialists are now able to differentiate between different species of the raw fish by the examination of electrophoretic patterns of the proteins and from fractions separated by paper (or column) chromatography, there does not appear to be any really satisfactory method of identifying the fish present in a cooked product. For routine purposes useful information can, however, be obtained from the chemical composition and from a search for fish tissues (*e.g.*, scales, pieces of shell) by means of the microscope. The examination of the extracted fat for oil values should give some indication of the presence of butter or margarine-type fat. Salmon fat shows the following values:—

Salmon Fat

Specific gravity (15·5° C./15·5° C.) . .	0·924–0·926
Refractive index at 40° C.	1·469–1·472
Iodine value	135–165
Saponification value . .	185–190

Heavy traces of arsenic (up to 3 p.p.m.) have been found in fresh fish (see *Analyst*, 1925, **50**, 3) such as plaice and haddock and relatively higher quantities in shellfish.* It is always advisable therefore to examine fish pastes for the element.

Fish Cakes

Fish cakes are usually made from pre-cooked fish, a filler (often mashed potato), seasoning (salt, spices, etc.). The moulded cakes are either fried in batter (the commonest type), baked or may be uncooked and they are often coated with breadcrumbs. The Food

* Monier-Williams (" Trace Elements in Food ") quotes the following figures: oysters 3–10 p.p.m.; up to 120 p.p.m. in other shellfish (especially mussels); very high amounts in crustacea (up to 170 p.p.m. in prawns).

Standards (Fish Cakes) Order, 1950 (S.I. 1950 No. 589) prescribes a minimum fish content of 35 per cent. Subsequently, the Food Standards Committee *recommended* that "fish cakes should contain not less than 40 per cent. by weight of fish expressed in terms of raw fish in the mix excluding any batter coating" (see *Brit. Food J.*, 1951, **53**, 53).

The analysis of fish cakes is very similar to that applied to sausages, *i.e.*, the determination of **water, fat, protein, ash** and **salt.**

The amount of lean fish can be calculated from the following formula, which assumes that the filler present is mashed potato:—

$$\text{Lean Fish } (\%) = \frac{\%\ \text{Total N} - 0{\cdot}015 \times \%\ \text{Carbohydrate}}{0{\cdot}028} \times 100$$

If the only type of fish present is white fish, 0·4 per cent. should be added to give the total fish content. With doubtful samples the oil values should be determined on the extracted fat in order to assess the proportion of fish fat present.

Canned Fish

As far as the official analyst is concerned the main analyses which should be carried out on canned fish are determinations of trace elements, particularly of tin, lead and arsenic. Cans containing crustacea are especially liable to contain relatively large amounts of arsenic.

Cans of fish (particularly of crustacea) are occasionally found which contain small harmless crystals of magnesium ammonium phosphate ("struvite"), which are liable to be mistaken by customers for glass (see Purcell and Hickey, *Analyst*, 1922, **47**, 16; McFee and Swaine, *Food Eng.* 1953, **25**, pp. 67, 190, 192). Simple chemical tests (solubility in water, molybdate test for phosphate, etc.) should be sufficient to allay doubts.

Occasionally a general assessment of the edibility of the fish is required and the determinations of the T.V.B. and the acid value of the extracted fat (see p. 321) should be useful for this purpose. Many fish are canned in a salad-type oil, *e.g.*, olive oil or arachis oil. The oil should be separated and identified from the oil values and examined also for possible rancidity.

GELATINE

Mainly because of its jelly-forming properties, gelatine is present as an ingredient in a number of foods. In the manufacture the collagen present in "waste" material such as hides, skin and bones is converted to gelatin by simmering with water. Two processes are employed for the manufacture, viz., acid extraction and lime extraction.

Lime gelatine usually has a higher ash, higher calcium content and its solution has a higher pH than an acid gelatine. The higher calcium content of lime gelatine tends to form insoluble salts in some foods in which it is used, *e.g.*, calcium tartrate crystals may be formed if tartaric acid is present and calcium phosphate is sometimes deposited in jars of lamb's tongues in jelly. The two types are best distinguished from the iso-electric point (I.E.P.), which is at a pH of about 8 for an acid gelatine and about pH 5 for a lime gelatine (Ward, *Food,* 1951, **20**, 255).

The *ultimate composition* of gelatin is approximately: carbon 50·5, hydrogen 6·7, nitrogen 17·9, sulphur 0·6 and oxygen 24·3 (all as percentages).

In determining gelatin (dry protein) the appropriate Kjeldahl factor is Total N $\times$ 5·55. Manufactured gelatine contains 12–17 per cent. moisture and up to 2·5 per cent. ash. Tannin solution and platinum salts precipitate gelatin, but iron, aluminium, lead, copper and gold salts do not. It is soluble in acetic acid but insoluble in alcohol or ether.

LEGAL REQUIREMENTS FOR GELATINE

The Food Standards (Edible Gelatine) Order, 1951, states that "Edible gelatine shall be clean wholesome protein which:—

(a) is obtained by extraction from collagenous material;
(b) is free from objectionable taste and offensive odour;
(c) dissolves completely in warm water to give a clear or translucent colloidal solution" and a 3 per cent. solution should set to a jelly when cooled to and maintained at 60° F. (lower setting gelatines are allowed to be sold provided an appropriate statement appears on the label).

The order also prescribes maxima for ash (3·25 per cent.) arsenic (2 p.p.m. as As), copper (30 p.p.m.), lead (7 p.p.m.) and zinc (100 p.p.m.) and the Preservatives Regulations prescribe a maximum for sulphur dioxide (1000 p.p.m.).

Analysis of gelatine

The routine examination of gelatine in official laboratories normally includes the simple **setting test** prescribed in the Order together with the determination of **ash, trace elements** (**arsenic, copper, lead, zinc**), **sulphur dioxide** and the **acidity.** Industrially other common determinations are those for **calcium, iron** and **moisture** and the assessment of various physical properties are also of considerable importance, e.g. **jelly strength** (often measured quantitatively), **colour, clarity** and the **pH value.** Methods for the chemical analysis of gelatin are described in B.S. 757: 1959. A **bacteriological examination** is also of importance industrially, particularly if the gelatine is to be incorporated in preserved flesh products.

Moisture. Weigh out 1 g. into a metal dish, add 10 ml. water and allow to soak. Then place on the water-bath to dissolve the gelatin and evaporate off the water. Complete the drying in the 100° C oven for 18 $\pm$ 1 hours.

The moisture content is not as useful a measure for assessing quality as with other materials. In fact a low moisture content may infer poor quality as excessive heating during the drying process causes protein degradation with with consequent loss of jelly strength.

Ash. Incinerate carefully 5 g. at 500–550° C.

Trace elements. Determine arsenic, copper, lead and zinc by the methods described in Chapter 3. Maxima for each of these metals have been prescribed (see above). A useful routine sorting test for metals in gelatine has been described by Scott-Dodd (*Analyst,* 1937, **62,** 18; *cf.* B.S. 757: 1959).

Calcium and Iron. Ash 10 g. Boil the ash with 10 ml. of 5 N. hydrochloric acid and filter the extract into a 100 ml. graduated flask. Make the filtrate up to the 100 ml. mark and use aliquots for calcium and iron.

Calcium. To 50 ml. of the filtrate add methyl red, make alkaline with ammonia and then *just acid* with dilute hydrochloric acid. Boil gently, add ammonium oxalate solution in slight excess and then make the liquid *just alkaline* with dilute ammonia. Allow to stand and filter off the calcium oxalate precipitate. Then determine the calcium in the precipitate either by ignition or by dissolving in dilute sulphuric acid and titrating with standard permanganate. Full details of the methods for determining calcium are given in the standard textbooks on quantitative inorganic analysis. A calcium figure in excess of about 0·3 per cent. (as CaO) usually indicates the properties of a lime gelatine.

Iron. Boil gently a suitable aliquot of the filtrate (5–25 ml.) with 2 ml. concentrated hydrochloric acid and five drops of 10 vol. hydrogen peroxide. Transfer the cooled, oxidised extract to a 100 ml. graduated flask, add 20 ml. of 10 per cent. potassium thiocyanate solution and make up to the mark. Determine the optical density of the solution (red) at 480 mμ (blue-green filter; Ilford No. 603). Prepare a standard graph using a solution of ferric ammonium sulphate (0·173 g./litre; 1 ml. = 0·02 mg. Fe), using dilutions over the range 0–0·1 mg. Fe. Boil each dilution with acid, peroxide, etc., as for the sample (*cf.* p. 151).

Ferric iron imparts a reddish-brown tinge to the gelatine solution.

Sulphur dioxide. Determine by the iodine method.

Acidity. Dissolve 2 g. in 50 ml. water on the water bath and titrate with N/10 sodium hydroxide using phenolphthalein. *Formerly* the B.P. required that gelatine for pharmaceutical purposes should not require more than 5 ml. N/10 alkali.

The pH value is determined on a 1–2 per cent. solution. The pH value may vary from about 4·0–6·3. Higher values in this range suggest the presence of lime gelatine. For the determination of the *iso-electric point,* prepare a series of tubes containing 2 per cent. solutions of the gelatine and including varying proportions of N/50 acid and alkali so that successive tubes differ by 0·2 per cent. acidity (as H_2SO_4). The I.E.P. is then obtained by measuring the pH of the tube with the maximum turbidity (Ames, *J.S.C.I.*, 1944, **63**, 200). Janus *et al.* (*Research*, 1951, **4**, 247) determine the I.E.P. by mixed bed deionisation.

The *Jelly Strength* can be measured in absolute units on the Bloom Gelometer (B.S. **757** : 1959) or more easily using the Boucher Jelly Tester (Koprowski, *Analyst*, 1951, **76**, 732). Industrially a comparative measure against an agreed standard is often sufficient (see B.S. **757** : 1959, p. 12). *Colour* (degree of " brownness ") and *clarity* (turbidity) are also important industrially and can be approximately assessed in the solution prepared for the jelly strength (see *J. Sci. Food Agric.*, 1953, **4**, 523). Solutions for jelly strength must be prepared by soaking in cold water for 1–3 hours and then dissolving at 60° C. (higher temperatures tend to cause loss of strength).

Gelatine which is used for food purposes usually comes within the following ranges of composition:—

		Legal maxima
Moisture	12–17%	16% (B.P.)
Ash	0·9–2·5%	3·25%
Calcium (as CaO)	0·1–0·6%	—
Iron (as Fe)	30–150 p.p.m.	—
Arsenic (as As)	⩽1 p.p.m.	2 p.p.m.
Copper	2–10 p.p.m.	30 p.p.m.
Lead	⩽5 p.p.m.	7 p.p.m.
Zinc	⩽20 p.p.m.	100 p.p.m.
Sulphur dioxide	⩽700 p.p.m.	1000 p.p.m.
pH	4·0–6·3	—

TABLE JELLIES

The **legal requirements** for table jelly tablets, crystals and compounds are stated in The Food Standards (Table Jellies) Order, 1949 (S.I. 1949 No. 1656).

Table Jelly Tablets are usually manufactured from sugar (sucrose syrup and invert syrup), acid gelatine, citric or tartaric acid, water, flavour and colouring. Less frequently liquid glucose and natural fruit juice are present. The main requirements for table jelly tablets in the Order are:—

(i) Compliance with a setting test (see below).

(ii) Minimum of soluble solids of 72 per cent. by weight as determined at 20° C. by refractometer on the sucrose scale (uncorrected).

(iii) Minimum sugar content of 63 per cent. by weight (as sucrose + reducing sugars as dextrose).

(iv) Jelly forming ingredients other than gelatine may be present, but farinaceous products are not permitted.

Analysis of Table Jelly Tablets

The analysis of table jelly tablets normally includes the determination of the **soluble solids, sugars, sulphur dioxide, arsenic, copper, lead** and **zinc** in addition to the official setting test and the examination for the **added colour** present. Other possible determinations are **total solids, ash, total acidity, pH** and **gelatine.** The sample should preferably consist of at least two tablets so that one can be used for the setting test and the second for the other determinations:—

I. Determinations on the tablet itself

(a) **Nett weight.**

(b) The **Soluble solids** at 20° C. can be determined on a portion of the tablet. The refractometer used should preferably be fitted with back illumination (see also (n) on p. 329).

(c) **Ash.** Incinerate 5 g.

(d) **Gelatine.** Carry out the macro-Kjeldahl procedure using 2 g. (Gelatin = N × 5·55).

(e) **Sulphur dioxide.** Carry out the hydrogen peroxide method, using 25 g. (see p. 81).

(f) **Trace metals.** Determine arsenic, copper, lead and zinc.

(g) **Setting test.** (S.I. 1949 No. 1656).

Mark a litre beaker at a level indicating one pint. Then completely dissolve the tablet (broken into sections) by adding water at 88° C. and make up to the one pint mark. Stir well. Then:—

"85 mls. of the solution shall be immediately introduced into each of six beakers of approximately 5 cm. internal diameter (100 ml. squat form) and the beakers cooled for eighteen hours in a water-bath maintained at 16° C. ± 1°C. At the end of this period the contents of the beakers shall be turned out on to a plate or dish by the following method. Each beaker shall be immersed in a water-bath at approximately 50° C. for eight seconds. Upon removal each beaker shall be immediately dried and the contents transferred to a plate or dish by inversion of the beaker. If not less than four out of the six jelly table sweets shall retain for thirty minutes the general shape of the beaker and shall not at the end of such period have collapsed or split so as to alter their shape, then the setting test shall be deemed to have been satisfied."

II. Determinations on a 10 per cent. solution of the tablet

(h) **Stock solution.** Prepare a 10 per cent. (w/v) solution by dissolving 20 g. in *warm* water and make up to 200 ml.

(i) **Water.** Evaporate 10 ml. of the stock solution on the water-bath and complete the drying in the 100° C. oven.

(j) **Sugars** (before inversion). Pipette 20 ml. of the stock solution into a 200 ml. graduated flask and add 5 ml. of each of the ferrocyanide clearing agents (as used for determining sucrose in condensed milk), mixing between each addition. Make the solution up to the 200 ml. mark, mix and allow the precipitate to settle. Filter. Determine the reducing sugars on the filtrate (1 per cent. w/v) using Lane and Eynon's volumetric method. If 10 ml. of Fehling's solution is used the titration usually falls within the desired 15–50 ml. range. Calculate as invert sugar (A per cent.) and dextrose (B per cent.). Use the remainder of the 1 per cent. filtrate for (k) below.

(k) **Sugars** (after inversion). Pipette 25 ml. of the 1 per cent. filtrate obtained in (j) into a 100 ml. graduated flask, add 5 ml.* concentrated hydrochloric acid and about 25 ml. distilled water. Place the flask in a water-bath maintained at 70° C.* for ten minutes. Cool, add phenolphthalein solution, neutralise with sodium hydroxide solution (approx. 5 N) and make up to the 100 ml. mark. Determine the sugars in the 0·25 per cent. solution using Lane and Eynon's method (10 ml. of Fehling's usually gives a suitable titration). Calculate the total sugars as invert sugar (C per cent.).

For the purposes of The Food Standards Order, the percentage (w/w) of sugar in the tablet is calculated as follows:—

$$\% \text{ Sugar} = \% \text{ Sucrose} + \% \text{ Reducing Sugars (as dextrose)}$$
$$= 0{\cdot}95\,(C - A) + B$$

(l) **Acidity.** Titrate 20 ml. of the stock solution with N/10 sodium hydroxide solution using phenolphthalein as indicator. Calculate the total acidity as citric or tartaric acid according to which acid is present.

(m) **pH value.** Determine the pH of the stock solution using a pH meter.

(n) **The Soluble solids** at 20° C. can be determined by refractometer on the 20 per cent. stock solution (see also (b) above).

Alternative methods for determining some of the above constituents have been described by Chatt (*Analyst,* 1952, **77**, 335).

The following table shows the usual range of figures obtained:—

Table Jelly Tablets

	Usual range	Legal requirements
% Soluble solids (refractometer)	73–79	72 (min.)
% Reducing sugars (as dextrose)	15–40	—
% Sucrose	30–50	—
% Total sugars	63–70	63 (min.)
% Total acid (as citric or tartaric acid)	0·7–1·5	—
% Gelatine (N × 5·55)	6–12	—
% Ash	0·1–0·3	—
Setting test	Satisfactory	At least 4 out of 6 should set
Sulphur dioxide (p.p.m.)	0–20	Depends on gelatine content

N.B. The amounts of arsenic, copper, lead and zinc present are usually negligible.

Separation of white crystals (sometimes mistaken for "mould" by customers) may occur during storage of table jelly tablets due to excessive reducing sugars or excessive sucrose being present. Crystallisation is most likely in tablets with less than about 20 or more than 40 per cent. of reducing sugars. **Mineral oil** is sometimes used as lubricating agent for the moulds during manufacture and a light smear on the surface of the tablet is common. Checking for compliance with The Mineral Oil in Food Order (maximum 0·2 per cent.), however, is difficult as waxed wrapping paper is commonly used.

Coal-tar dyes can usually be extracted by simply boiling a solution of the sample with wool.

Table Jelly Crystals usually consist of a mixture of sugar (invariably sucrose), acid gelatine, citric or tartaric acid, flavour and

* *Cf.* p. 112.

colouring. The main requirements for table jelly crystals in the order are:—

(i) Compliance with the setting test described under tablets.

(ii) Minimum sugar content of 84 per cent.

The amending order (S.I. 1955 No. 828) provides for the use of a low setting gelatine by prescribing an alternative minimum of 2½ oz. of sugar in each pint of table jelly sweet when the latter is made up in the manner specified in the official jelly test.

(iii) Jelly forming ingredients other than gelatine may be present but farinaceous products are not permitted.

Analysis of Table Jelly Crystals

The analytical methods for table jelly crystals are similar to those employed for tablets:—

I. Determinations on the solid jelly crystals

Determine (a) **Nett weight,** (c) **Ash,** (d) **Gelatine,** (e) **Sulphur dioxide,** (f) **Trace Metals** and carry out (g) the official **Setting test** as described for the analysis of table jelly tablets (pp. 328–9) and identify the **added colour** present (p. 100).

II. Determination on a 10 per cent. solution of the crystals

Prepare a 10 per cent. (w/v) stock solution of the crystals (h) and determine the moisture (i) as for tablets.

(j) **Sugars** (before inversion). Prepare a 1 per cent. cleared filtrate as for tablets. It is usually sufficient to confirm by a qualitative test that the 10 per cent. solution only gives a trace of reduction with Fehling's solution.

(k) **Sugars** (after inversion). Prepare a 1 per cent. cleared filtrate as for tablets (see (j) for tablets). Invert 50 ml. of this 1 per cent. cleared solution with 5 ml. concentrated hydrochloric acid at 70° C. in a 100 ml. graduated flask, cool, neutralise and make up to the mark as usual. Then titrate the 0·5 per cent. solution against 25 ml. Fehling's solution using the Lane & Eynon method. Calculate the result as invert sugar and multiply by 0·95 to give the concentration of sucrose in the crystals (w/w).

(l) and (m) Determine the *Acidity* and *pH value* as for tablets.

Table Jelly Crystals

	Usual range	Legal requirements
% Moisture	1·5–3·5	—
% Sugar (sucrose)	84–88	84 (min.)
% Total acid (as citric or tartaric acid)	1·0–2·1	—
% Gelatine (N × 5·55)	8–13	—
% Ash	0·3–0·8	—
Setting test	Satisfactory	At least 4 out of 6 shall set
Sulphur dioxide (p.p.m.)	0–50	Depends on gelatine content (see p. 439).

N.B. The amounts of arsenic, copper, lead and zinc present are usually negligible.

Table jelly compounds usually consist of a mixture of sugar (sucrose), lime gelatine, cornflour, flavour and colouring. Milk is used for making table jelly compounds into a jelly. No acid is present as it would curdle the milk. The main requirements for table jelly compounds in the order are:—

(i) Compliance with the setting test described under tablets except that the table jelly sweet is made up with milk instead of water.
(ii) Minimum sugar content of 50 per cent.
(iii) Maximum starch content of 20 per cent.

Analysis of Table Jelly Compounds

Nett weight.

Moisture. Weigh out 5 g. into a prepared dish containing a rod. Add a little water, mix in well with the rod and evaporate on the water-bath. Complete the drying in the 100° C. oven.

Sugar. Macerate 25 g. of sample with *cold* water in a mechanical blender and transfer to a 500 ml. graduated flask. Add 10 ml. each of the ferrocyanide clearing agents and make up to 500 ml. with water. Filter. Invert 50 ml. of the filtrate with acid in a 100 ml. graduated flask and make up to the mark after neutralisation. Titrate the inverted solution (after suitable dilution) by Lane & Eynon's method. Sucrose is usually the only sugar present, but the " direct " filtrate should be examined qualitatively for reducing sugars and the method modified accordingly (see methods for crystals).

The **ash, gelatine, sulphur dioxide** and **trace metals** can be determined by the methods used for tablets.

Setting test. Carry out in exactly the same way as for tablets, except that instead of making up the pint of solution in water, jelly compounds must be dissolved in " milk at 88° C."

Added colour. Stir the sample with dilute ammonia, filter and boil the acidified filtrate with the wool in the usual manner.

AGAR

Agar is the dried, gelatinous substance obtained by concentrating a decoction of seaweed. The moisture is removed by successive freezing, thawing and drying at about 35° C. Agar is composed chiefly of a calcium salt of a sulphuric ester of a carbohydrate complex R $(O.SO_2.O)_2$ Ca. Agar for food purposes usually contains 15–20 per cent. of moisture up to 4 per cent. of ash, 1–2 per cent. of protein and traces of elements such as iron, arsenic and boron. It produces a gel from a very dilute solution and as a small amount enhances the setting powers of gelatine considerably it may be found in meat pie jellies in hot weather. Agar is also used in the confectionery industry. Agar yields galactose on hydrolysis with consequent loss of gelatinising power. Its gelling powers are reduced much more by heating or the presence of acid than is gelatine. The fact that agar shows diatoms on microscopical examination is made use of for its detection.

ISINGLASS

Isinglass consists of the dried prepared swimming bladder of the sturgeon and other fish. It contains 66–88 per cent. of gelatin, 13–20 per cent. of moisture and up to 1·50 per cent. of fat. It should not yield more than 3 per cent. of matter insoluble in boiling water and the ash should not exceed one per cent. It differs from gelatine in giving a reddish ash and from the fibrous structure which is seen under the microscope after soaking in cold water. Isinglass has often been employed for clarifying wines.

Chapter 12

DAIRY PRODUCTS (I)

Milk—Cream—Condensed Milk—Dried Milk—Infant Foods—Casein—Malted Milk

MILK

Constituents of Milk

Milk contains the following constituents: water; fat, which consists of the mixed glycerides of palmitic, stearic, lauric, oleic and lower fatty acids, principally butyric, caproic and capric acids; proteins, principally casein, albumin and globulin; carbohydrate (lactose); organic acids, *e.g.*, lactic and citric acids and traces of acetic acid. Other constituents present include urea, lecithin, creatinine, alcohol and mineral matter. The following enzymes have been stated to be present in fresh milk: peroxidase, which gives a blue colour on the addition of guaiacol and hydrogen peroxide solutions; catalase, which decomposes hydrogen peroxide; lipase, which splits fats; phosphatase, which hydrolyses phosphoric esters; reductase, which is capable of reducing various organic substances (as in the methylene blue and resazurin tests); and amylase, which is capable of decomposing starch. Milk contains protease which decomposes proteins, and also lactase which converts lactose into dextrose and galactose.

PHYSICAL PROPERTIES OF MILK

	Mean value
Specific gravity at 15·5° C.	1·032
Refractive index at 20° C.	1·348
Freezing point (Hortvet)	−0·544° C.
Specific heat	0·93
Electrical conductivity (K_{15})	0·0041*
pH	6·6
Diameter of fat globules	0·003 mm.

* Coste and Shelbourne, *Analyst*, 1919, **44**, 159.

Composition of Milk

Average Percentage Composition of Milk of the cow and other mammals

	Cow	*Goat*	*Human*	*Sheep*	*Mare*	*Ass*
Water . . .	87·70	86·0	88·2	81·3	89·8	90·1
Fat . . .	3·61	4·6	3·3	6·9	1·2	1·3
Lactose . .	4·65	4·2	6·8	5·2	6·9	6·5
Protein (N × 6·38)	3·29	4·4	1·5	5·6	1·8	1·6
Ash . . .	0·75	0·8	0·2	1·0	0·3	0·5

Vieth's ratio for cows milk. For milks containing a normal amount of S.N.F. (8·4–9·0 per cent.) the ratio of lactose : protein : ash is 13 : 9 : 2. This ratio is, of course, unaffected by the addition of water.

Colostrum, the thick liquid secreted immediately after parturition, differs markedly in composition from normal milk. Colostrum is richer in fat, protein (especially albumin) and ash and gives a higher F.Pt. depression than ordinary milk (see Elsdon, *Analyst*, 1934, **59**, 665 and Moody *et al.*, *J. Dairy Sci.*, 1951, **34**, 106). The change in composition to normal milk takes place in about 4 days. Colostrum exhibits large compound granular cells under the microscope.

Factors which Affect the Composition of Milk

Apart from pathological conditions which give rise to the production of abnormal milk, there are a number of factors which affect the composition. Large variations, especially in the fat content, are given by the different breeds of cows, as can be seen from the following table due to Vieth and others:—

Breed	Average total solids %	Fat %	Solids-not-fat %
Dairy Shorthorn .	12·90	4·03	8·87
Jersey . . .	14·89	5·66	9·23
Kerry . . .	13·70	4·72	8·98
Red Polled . .	13·22	4·34	8·88
Sussex . . .	14·18	4·87	9·31
Welsh . . .	14·15	4·91	9·24
Ayrshire . . .	13·46	4·24	9·22
Friesian . . .	11·50	3·00	8·50
Devon . . .	13·77	5·07	8·70

The composition is also influenced by the time of year, *i.e.*, it tends to be poorest in summer and richest in autumn. This can be seen from the following table which shows the mean composition of milk delivered from a number of farms to a manufacturer of milk products (*cf.* Davis, *Analyst*, 1952, **77**, 499):—

MEAN MONTHLY VARIATIONS

	Fat. Per cent.	Solids-not-fat. Per cent.
January	3·63	8·84
February	3·65	8·84
March	3·56	8·80
April	3·56	8·80
May	3·60	8·82
June	3·54	8·80
July	3·62	8·75
August	3·68	8·74
September	3·75	8·75
October	3·84	8·85
November	3·88	8·90
December	3·78	8·91

The period of lactation has an effect on the composition of the milk. The total solids, fat and S.N.F. fall during the second month and then rise until the ninth or tenth month, after which there is a general diminution as the lactation ceases. The milk yield progressively decreases as lactation continues. Also with cows which are far advanced in lactation, the fat tends to contain a smaller proportion of volatile fatty acids and this results in changes in the Reichert and other values of the butter prepared therefrom.

The composition is also influenced by the time interval which elapses between milkings. In general the evening milk is richer in fat than that drawn in the morning from the same animal. If the period between milkings were 12 hours in each case, the milk would be of similar composition, but the period from evening to morning is usually considerably greater than that from morning to evening. The percentage of fat is approximately in inverse proportion to the interval between milkings. Collins's formula is:—

$$E - M = 0{\cdot}25\ (e - m) - 0{\cdot}2$$

where E and M are the fat percentages in the evening and morning milk and e and m are the times in hours between the evening and morning, and morning and evening milkings respectively. In addition, the milk from individual cows may fluctuate in quality and quantity from day to day. Such variations may be influenced by the mental and physical condition of the cow. Excitement, worry or discomfort is liable to have an adverse effect on the quality of the milk as well as on the quantity produced. It should also be borne in mind that, during milkings, the first runnings or "fore milk" are low in fat, whereas the last runnings or "strippings" are rich in fat (*cf.* "*Richmond's Dairy Chemistry*", Vth Edn., pp. 60–87).

TRENDS IN MILK COMPOSITION

The Report of the Interdepartmental Committee on Milk Composition (see p. 337) summarises the average values for the fat and s.n.f. contents of milk (collected from various sources) for various periods:—

England and Wales			Scotland		
Period	Fat (%)	S.N.F. (%)	Period	Fat (%)	S.N.F. (%)
1897–1916	3·75	8·89	1910–1915	3·14	8·65
1900–1915	3·58	8·70	1916–1930	3·45	8·76
1910–1915	3·59	8·76	1921–1922	3·95	8·78
1916–1930	3·67	8·86	1932–1940	3·68	8·74
1916–1930	3·65	8·83	1940–1950	3·72	8·75
1923–1931	3·61	8·91	1950–1957	3·76	8·72
1930–1940	3·74	8·78			
1931–1940	3·66	8·80			
1931–1934	3·64	8·80			
1934–1935	3·65	8·76			
1941–1950	3·56	8·65			
1941–1950	3·63	8·75			
1941–1950	3·60	8·69			
1951–1958	3·53	8·65			
1951–1959	3·61	8·69			

Difficulties due to sampling and possible changes in analytical methods mean that data of this type must be interpreted with caution. From the available evidence, however, it would appear that the changes in composition which have taken place in England and Wales over the years are relatively small, although there has been a slight fall in the fat content. In Scotland, however, the s.n.f. content has remained fairly steady between 1910 and 1957, but there has been an appreciable rise in the fat content (see also Davis, *Analyst*, 1952, **77**, 499).

Reaction. Cow's milk is acid to phenolphthalein, alkaline to methyl orange, but amphoteric to litmus due to the presence of phosphates. The total acidity of freshly drawn milk is usually about 0·15 per cent. (as lactic acid). On storage, the acidity increases due to the action of micro-organisms, and a sour taste is perceptible at an acidity of about 0·35 per cent. When the acidity reaches 0·7 per cent., milk curdles at ordinary temperatures.

Mineral matter. The total ash of genuine milk varies from about 0·72 to 0·82 per cent. The amounts of the principal constituents in the ash are as follows (per cent. w/w of the ash):—

Potassium (as K_2O)	22 (23–30)
Calcium (as CaO)	22·4 (20–28)
Sodium (as Na_2O)	8·9 (5–12)
Phosphorus (as P_2O_5)	25·8 (21–30)
Chloride (as Cl)	14·3 (13·5–16·5)

Traces of magnesium, iron, sulphate, etc., are also present. Milk has been reported to contain about 0·2 p.p.m. of lead and 1–3 p.p.m. of zinc.

Recommendations of Interdepartmental Committee on Milk Composition in the U.K.

An Interdepartmental Committee was set up in May, 1958, " to consider the composition of milk sold off farms in the United Kingdom from the standpoint both of human nutrition and of animal husbandry and to recommend any legislative or other changes that may be desirable." The Committee's Report entitled " Milk Composition in the U.K." (H.M.S.O., 1960, Cmnd. 1147) included the following recommendations:—

1. The definition of " milk " in the Food and Drugs Act, 1955, is inadequate and the Committee suggest the following (para. 86):—

   " ' Cows' milk ' means the secretion, excluding colostrum, which can be gained by normal milking methods from the lactating mammary gland of the healthy, normally fed cow."

2. *Paragraph* 88
   (a) consideration should be given to the adoption within ten years of a fixed minimum standard for whole milk of 8·5 per cent. s.n.f.;
   (b) in the meantime, the presumptive minimum standard for s.n.f. should be abolished when Ministers are satisfied that suitable marketing standards are in operation within the dairying industry to maintain and improve the s.n.f. content of milk;
   (c) the presumptive minimum standard for fat should be continued for the present but replaced within five years by a fixed minimum standard for whole milk of 3 per cent.;
   (d) the fixed minimum standards should apply to milk only at the point of sale to the consumer;
   (e) the fixed minimum standards should not apply to special grades of milk for which other standards may be necessary;
   (f) the evidence of the Hortvet test should be accepted in legal proceedings (with the provisos stated in Panel Report 3) as proof of the presence or absence of added water in a milk sample;
   (g) abstraction of fat from milk intended for sale as whole milk should be a specific legal offence;
   (h) a general statutory definition of milk should be introduced.

Milk standards. The presumptive standards for milk were prescribed in The Sale of Milk Regulations, 1939 (S.R. & O. 1939

No. 1417), which have been perpetuated by the Food and Drugs Act, 1955 (in the Twelfth Schedule and Sections 33 and 136 (2)). These regulations prescribed that where a sample of milk (not being sold as skimmed, separated or condensed milk) contains less than 3 per cent. of milk fat it shall be presumed, until the contrary is proved, that the milk is not genuine by reason of the abstraction therefrom of milk fat, or the addition thereto of water. Also, it is prescribed that where a sample contains less than 8·5 per cent. of milk solids other than milk fat it shall be presumed, until the contrary is proved, that the milk is not genuine by reason of the abstraction therefrom of milk solids other than fat or the addition thereto of water. For skimmed or separated milk the regulations prescribe a standard of not less than 8·7 per cent. of milk solids other than milk fat. Below this figure the skimmed or separated milk is presumed not to be genuine until the contrary is proved.

Further, The Milk and Dairies (Channel Islands and South Devon Milk) Regulations, 1956 (S.I. 1956 No. 919) require that all milk sold as Channel Islands, Jersey, Guernsey or South Devon milk shall contain a minimum of 4 per cent. of milk fat.

Section 32 of the 1955 Act prohibits the addition of colouring matter, water, dried or condensed milk, or any fluid reconstituted therefrom, to milk or skimmed milk. The addition of preservatives is prohibited by the Preservatives Regulations.

The Milk (Special Designation) Regulations, 1960. The Milk (Special Designation) Regulations (S.I. 1960 No. 1542) prescribe the conditions under which producers' licences and dealers' licences may be granted in relation to milk which is sold as " Pasteurised," " Sterilised " and " Tuberculin Tested."

" *Tuberculin Tested Milk* " is required to comply with the methylene blue test (see p. 356). Also a Clot-on-Boiling Test is to be applied to samples of milk taken from producers (see p. 354). " T.T. (Pasteurised) " Milk must also comply with the phosphatase test (p. 359).

Pasteurised Milk. The Regulations require that " Pasteurised Milk " shall be either retained at 145°–150° F. for at least thirty minutes or be retained at not less than 161° F. for at least fifteen seconds. The milk is then to be cooled immediately to not more than 50° F. following either of these two methods of pasteurisation. Other conditions of pasteurisation which may be specified by the licensing authority require the approval of the Minister. Pasteurised milk is required to comply with the methylene blue test and the phosphatase test (see pp. 356, 361).

Sterilised Milk. The Regulations require that " Sterilised Milk " shall be filtered or clarified, homogenised and thereafter heated to and maintained at not less than 212° F. for such a period that it will comply with the turbidity test.

Analysis of Milk

Prior to *each determination* the milk should be thoroughly mixed by *continuous slow inversions* of the sample bottle or by *slowly* pouring it into a beaker and back into another beaker and repeating the process many times. Very sour samples should be mixed with a wire whisk.

In the *routine examination* of milk it is convenient to take the specific gravity by a lactometer and, after determining the **fat** by the rapid Gerber method, it is possible to calculate the **total solids** and hence the **non-fatty solids** (N.F.S.) by means of Richmond's formula. The results obtained are then confirmed by gravimetric methods on any doubtful samples and the determination of the **freezing point** is then also advisable while the milk remains reasonably fresh. **Sourness** can be assessed by determining the **acidity** of the milk. The assessment of the **degree of pasteurisation** from the results of the **phosphatase test** is also a common routine procedure for milks which are sold as " Pasteurised." Other determinations which may be required are those of **proteins, lactose, ash** (and **chloride**) and **citric acid** and examination for **dirt, added dyes, preservatives** and **detergents.** The **methylene blue test** and the **bacteriological examination** must be carried out on samples taken under aseptic precautions and these tests should be " put up " first.

Specific gravity. The specific gravity of milk varies according to the proportions of fat (s.g. 0·93), N.F.S. (s.g. 1·614) and water (s.g. 1·0). The s.g. of milk can be determined by a specific gravity bottle, Sprengel tube, Westphal balance, etc., but is more conveniently measured by means of the lactometer, which is a special hydrometer calibrated over the range 1·025–1·035 (25·0°–35·0° as lactometer degrees). For the determination, the milk should be as fresh as possible and be thoroughly (but gently) mixed, avoiding incorporation of air. If a lactometer is used, a small allowance should be made for the rise up the stem due to the meniscus. The temperature of the sample being measured must always be taken. For lactometers which are calibrated at 60° F. a correction of +0·11 is applied for each 1° F. above 60° F. and vice versa. As large corrections tend to produce errors it is preferable to bring the milk as near 60° F. as possible before taking the reading.

If the percentage of fat is determined, *e.g.*, by the rapid Gerber method (see below), the total solids figure can be calculated from *Richmond's formula*:—

$$\text{Total solids (\%)} = T = 0{\cdot}25G + 1{\cdot}2F + 0{\cdot}14 \quad \ldots\ldots(1)$$

where G = lactometer reading (degrees) at 60° F. and F = fat (per cent.). Non-fatty solids (N.F.S.) = Solids-not-fat (S.N.F.) = T−F. For routine purposes the figures are usually reported to the nearest 0·05 per cent.

B.S.734 has modified the method and formula to some extent. The specific gravity hydrometer is replaced by a density hydrometer which is calibrated at 20° C. in relation to milks with a normal surface tension. The following formula then applies:—

$$T = 0{\cdot}25D + 1{\cdot}22F + 0{\cdot}72 \quad \ldots\ldots(2)$$

where D = the B.S. density hydrometer reading at 20° C. The correction for densities measured at temperatures other than 20° C. is approximately +0·24 for each 1° C. above 20° C. (*cf.* B.S.734).

The specific gravity of milk taken an hour or more after milking (when all air bubbles have disappeared) is lower than that observed subsequently (**Recknagel's phenomenon**). During the first twelve hours after milking the s.g. (observed) of the milk may rise as much as 0·0013. The change has been attributed to alterations in (a) the casein and (b) the physical condition of the fat. Richmond's formula applies to aged, cooled milks with the Recknagel effect virtually completed. The B.S. method, however, requires that the density is measured after the milk has been warmed to 40° C., maintained at that temperature for five minutes (with gentle mixing) and cooled to 20° C. This ensures that all milks are examined under the same conditions. *If, however, the milk contains fat in the solid state*, the B.S. requires that the density is measured after first cooling to 4·5° C. (or lower) for at least six hours (or to 7·2° C. for at least sixteen hours) and then raising the temperature of the sample to 20° C. After such treatment the following further modified formula is applicable:—

$$T = 0{\cdot}25D + 1{\cdot}22F + 0{\cdot}55 \quad \ldots\ldots(3)$$

Slide rules and other mechanical devices based on formulæ (1) and (2) above have been devised (*Analyst*, 1920, **46**, 218; *Dairy Ind.*, 1957, **22**, 1030; 1958, **23**, 649). Also B.S.734 includes extensive tables which relate the fat content and the N.F.S. Corrections for temperature to be applied to the hydrometer reading are also contained in such rules and in the B.S. tables.

Total solids. The following is an outline of the gravimetric method for determining the total solids of milk (S.P.A., *Analyst*, 1945, **70**, 105 and B.S.1741: 1951):—

Weigh out about 5 g. of mixed milk into a round flat-bottomed metal dish (about 7 cm. diam.) and provided with a closely fitting lid which is weighed with the dish. Place the uncovered dish on a boiling water-bath for thirty minutes. Then wipe the bottom of the dish and transfer it to a well-ventilated oven at 98°–100° C. Place the lid next to the dish in the oven. Dry for three hours in the oven and then cover the dish with the lid, cool for thirty minutes in a desiccator, and weigh. Then heat the dish and lid as before for one hour periods in the oven, cool, weigh, etc., until the loss in weight between successive weighings does not exceed 0·5 mg. If the acidity of milk exceeds 0·20 per cent. (as lactic acid) some of the acid present may volatilise on drying and neutralisation of the milk with N/10 strontium hydroxide is advisable. If this is done a deduction must be made from the weight of the residue obtained equivalent to 0·00428 g. for each ml. of N/10 strontia used.

Fat. Rapid **volumetric methods** are often used for routine purposes for determining fat in milk. In Britain, the **Gerber method** is commonly used:—

Part 1 of B.S.696: 1955 describes the necessary apparatus and Part 2 covers the method of determination and gives specifications for the reagents. The test is performed in butyrometer tubes which are closed with special rubber stoppers. For the test, pipette the following liquids into a milk butyrometer, ensuring that they do not mix with one another: 10 ml. of sulphuric acid (density 1·813–1·817 at 20° C.≡90 per cent. approx.), 10·94 ml. of mixed milk and 1 ml. of fat-free amyl alcohol (density 0·810–0·812 at 20° C.). Close the tube with a stopper, mix the contents throughly and centrifuge the hot tube at 1100 r.p.m. for four minutes (if a number of determinations are being made the tubes can be conveniently placed in water at 65° C. until it is convenient to spin all the tubes together). After centrifuging, transfer the tube (stopper downwards) to a litre beaker filled with water at 65° C. for at least three minutes and read off the percentage of fat directly from the scale. Readings to an accuracy of 0·05 per cent. are usually adequte for routine purposes. As it is difficult to separate the small fat globules in homogenised (*e.g.*, "sterilised") milk, it is advisable to re-centrifuge after warming in the 65° C. bath until the reading reaches a maximum. Other modifications are necessary with skim milk, separated milk and buttermilk (see B.S.696: 1955, Part 2).

By adding a trace of ferric chloride to the sulphuric acid used, a violet colour is produced if the milk contains *formaldehyde* (*cf.* p. 91). Numerous reagents have been proposed for replacing the sulphuric acid used in the Gerber method by less corrosive liquids. Those which incorporate synthetic detergents appear to be the most promising (*e.g.*, Macdonald, *Analyst*, 1959, **84**, 287, 747 and Houston, *J. Soc. Dairy Tech.*, 1955, **8**, 47).

Gravimetric methods for determining fat in milk are suitable for reference purposes.

The Rose-Gottlieb process (see also p. 39) gives very accurate results provided that attention is paid to detail. The Gottlieb extraction tubes are fitted with tubes for blowing off the fat solution (B.S.1741: 1951):—

Weigh out 10 g. of milk into the tube, add 1 ml. of ammonia solution (s.g. 0·88) and mix. Then add 10 ml. of alcohol (95 per cent.) and again mix well. Add 25 ml. of diethyl ether ("peroxide-free"—see *Analyst*, 1949, **74**, 85), cork the tube and shake vigorously for one minute. Add 25 ml. of light petroleum (b.pt. 40–60° C.) and shake vigorously for thirty seconds. After separation is complete, blow the fat solution using wash-bottle tubes into a suitable flask (dried at 100° C., cooled, weighed, etc.). To the tube add two successive lots of 5 ml. of mixed " ethers " and transfer (without shaking) to the flask in order to wash the siphon tubes. Then repeat the extraction (with 15 ml. of ether and 15 ml. of light petroleum) and the subsequent operations twice more. Distil off the solvents from the flask (see p. 27), dry the fat at 100° C., cool and weigh. If any non-fatty matter appears to be present, wash out the fat from the flask with light petroleum, dry, re-weigh and correct the result accordingly.

The Werner-Schmid process (see also p. 38). The extraction can be carried out in similar tubes to those used for the Gottlieb process (see B.S.1743: 1951, p. 7):—

Weigh out 10 g. of the milk into the tube, add 10 ml. of concentrated hydrochloric acid and immerse in boiling water until all the casein has dissolved. At this stage the mixture should be brown (or violet) in colour and the fat will be seen to collect on the surface. Cool the tube rapidly. Extract the fat by shaking with 30 ml. of diethyl ether and blow the extract, after allowing the layers to separate into a weighed flask. Separation is often assisted by the addition of a little alcohol. Repeat the extraction three times more and distil off the solvent. Then dry the fat at 100° C., cool and weigh. If any non-fatty matter appears to be present, dissolve the fat in light petroleum, etc., as in the Gottlieb method above.

The Werner-Schmid method is applicable to sour milk. If the special tubes are not available the milk can be heated with the acid in a beaker over a gauze and the ether extraction carried out using a separating funnel.

Adams method (*Analyst*, 1885, **10**, 46; 1886, **11**, 71) was adopted by the S.P.A. in 1886 as the official method for determining fat in milk, but is now seldom used. Briefly, the method involves weighing out 5 ml. of milk on to a piece of filter paper 22 × 2½ inches. After air-drying the paper is rolled into a coil and the fat extracted with dry ether for at least four hours in a Soxhlet apparatus. For further details and comments on Adams method the reader should refer to "*Richmond's Dairy Chemistry*," Vth Edn., pp. 342–5, Griffin, Lond.

Analysis of Sour Milk—The Maceration Method

On occasion *sour or altered milk* has to be examined and its composition when fresh deduced. In such cases the method worked out by Thorpe

(*J. Chem. Soc.*, 1905, **87**, 206) and criticised by Richmond and Miller (*Analyst*, 1906, **31**, 317) is adopted. The fat and solids-not-fat are determined in one operation known as the maceration method.

Before proceeding with the analysis of a sample of sour milk, the contents of the bottle are transferred to a suitable vessel and thoroughly mixed with a wire whisk. Portions of about 10 g. of the sample are weighed into flat-bottomed platinum basins which have been tared with a short glass rod having a flattened end. The acidity is neutralised with 0·1 N strontia solution, phenolphthalein being used as indicator. The milk is evaporated on the water-bath with constant stirring towards the end, until it attains the consistency of dry cheese. About 20 ml. of dry ether are poured over the solids, which are then carefully triturated with the glass rod. The ethereal solution of the fat is passed through a dried and tared filter and the maceration of milk is continued with at least eight successive quantities of ether. At the conclusion of this process the solids should be in a fine state of division like precipitated chalk. Before becoming quite dry they are transferred as far as possible to the filter paper, washed free from fat and dried in the oven in a weighing bottle to constant weight, as is also the platinum basin with the remaining solids. From the total weight of non-fatty solids is deducted 0·00428 g. for each ml. of 0·1 N strontium hydroxide added. The weight of fat is obtained by evaporating the ethereal solution and drying in the usual way.

The following corrections, which are additive, must be applied to the non-fatty solids. The *alcohol correction* is obtained by distilling 50–100 g. of the milk, then neutralising the distillate, using litmus paper as indicator, and re-distilling it. The final distillate is made up to the original volume and its specific gravity is determined by means of a pyknometer. The corresponding percentage by weight of alcohol multiplied by 90/46 gives the percentages of lactose which has disappeared in the production of the alcohol. Next is determined the *volatile acid correction.* Ten g. of the milk are neutralised to the extent of one-half of the total acidity with the use of 0·1 N sodium hydroxide and phenolphthalein. The mixture is evaporated to dryness on the water-bath with frequent stirring, and after treatment with 20 ml. of boiling water so as thoroughly to break up and detach the solids from the basin, a further addition of alkali is made until complete neutrality is reached. The difference between the original acidity of the milk and that of the evaporated portion is regarded as the acetic acid (60 parts of this acid correspond to a loss of 62 parts of lactose, so an addition must be made of 62/60 times the acetic acid found).

The other corrections may be made—for butyric acid and for ammonia. These are, however, very small, and may be omitted when the sum of the other two corrections is less than 0·2 per cent., as they are then not likely to exceed 0·05 per cent. For the estimation of butyric acid the volatile acids are separated from the quantity of milk which has been taken for the determination of the alcohol. A portion of the mixed aqueous acids is neutralised with barium hydroxide, evaporated and dried to constant weight. From the percentage of barium contained in the mixed salts the proportions of the two solids, acetic and butyric, are calculated. Each 88 parts of butyric acid corresponds to a loss of 92 parts of lactose. The loss due to acetic acid has been calculated previously.

To estimate the small quantity of *ammonia* formed, 2 g. of the milk are made up to a volume of 100 ml. with ammonia-free water and filtered through a washed paper. The ammonia in 10 ml. of the filtrate diluted to 50 ml. is estimated by Nessler's reagent in the usual way, a solution of ammonium chloride equivalent to 0·01 mg. per ml. being used as standard.

Acidity. The titratable acidity is determined by the following method (*cf.* B.S.1741: 1951, p. 13; *Analyst,* 1951, **76**, 504):—

To a small porcelain basin add 10 ml. of milk and 1·0 ml. of phenolphthalein indicator (0·5 per cent. in 50 per cent. alcohol) and rapidly titrate with N/10 sodium hydroxide solution. The colour at the faint pink end-point should be equivalent to a mixture of 10 ml. of the milk and 1 ml. of 0·00024 per cent. rosaniline acetate* solution (in 96 per cent. alcohol) contained in a similar basin. Titration with N/10 strontia instead of soda tends to give a better end-point. The titratable acidity is usually calculated as lactic acid (per cent. w/v).

$$1 \text{ ml. N/10} \equiv 0{\cdot}009 \text{ g. lactic acid.}$$

Other modes of expression of the acidity are as lactic acid on a w/w basis and as *degrees of acidity,* which is the number of mls. of *normal* alkali required per litre of milk.

Fresh milk usually gives a titration of 1·7–2·0 ml. of N/10 alkali (for 10 ml. sample taken). The acidity of freshly drawn milk is mainly due to casein, phosphates and citrates and it appears doubtful whether any lactic acid is in fact present. Lactic acid is, however, formed during souring, the lactobacilli eventually causing the pH to drop to about 4·3. For further information relating to the acidity and pH of milk reference should be made to Caulfield and Riddell (*J. Dairy Sci.*, 1936, **19**, 235), Ling (*J. Dairy Res.*, 1937, **8**, 173) and Davis (*Food,* 1950, **19**, 44, 84).

The *true* lactic acid content can be determined by the method of Davidson (*J. Dairy Res.*, 1949, **16**, 209) in which interfering substances are precipitated with copper sulphate and calcium hydroxide at 46° C. The lactic acid is then oxidised to acetaldehyde with sulphuric acid in the presence of copper sulphate. Acetaldehyde develops a purple colour with p-hydroxydiphenyl and its intensity is measured at 570 mμ (yellow-green filter) (see also Ling *J. Sci. Food Agric.*, 1951, **2**, 279).

Protein. The protein can be determined on 10 g. by the macro-Kjeldahl method using the factor Total N × 6·38.

The *formol titration* is useful for the rapid determination of protein in fresh milk. The method depends on the fact that when formaldehyde is added to neutralised milk, free acid (which can be titrated with alkali) is produced in proportion to the amount of proteins present. The protein content is then obtained by multiplying the titration by an empirical factor, which depends on the

* *Rosaniline acetate solution. Stock solution. Solution A.* Dissolve 0·12 g. of rosaniline acetate in approximately 50 ml. of ethyl alcohol (95–96 per cent. v/v) containing 0·5 ml. of glacial acetic acid. Make up to 100 ml. with ethyl alcohol.

Bench solution. Solution B. Dilute 1 ml. of Solution A to 500 ml. with a mixture of ethyl alcohol (95–96 per cent. v/v) and distilled water in equal proportions by volume.

Note.—Solutions A and B should be stored in the dark in bottles securely stoppered with rubber bungs.

ratio of casein to albumin and also the particular technique employed. The method proposed by Pyne (*Biochem. J.*, 1932, **26**, 1006), in which interference due to calcium is prevented by the addition of oxalate, gives reasonably accurate results:—

To 10 ml. of milk add 0·5 ml. of 0·5 per cent. phenolphthalein indicator and 0·4 ml. of neutral saturated potassium oxalate. Mix, allow to stand for a few minutes and neutralise with N/10 sodium hydroxide to the standard pink colour described under "Acidity." (*N.B.*—The colour standard employed by Pyne consisted of 10 ml. milk, 0·4 ml. oxalate and 1 ml. of alcoholic magenta* solution.) Add exactly 2 ml. of formalin, mix, allow to stand for a few minutes and titrate the "new" acidity produced with N/10 sodium hydroxide to the same pink colour (titration a mls.). Then titrate separately 2 ml. of the formalin + 10 ml. of water with the N/10 alkali (b mls.) as blank. Then the protein content of the milk (equivalent to N × 6·38 from the Kjeldahl method) is 1·7 (a − b) per cent.

If the oxalate is omitted the first titration gives the acidity and a higher formol factor is usually appropriate, *i.e.*, Protein = 1·95 (a − b) per cent. For further information on formol titration methods and factors employed the reader should refer to *Analyst*, 1906, **31**, 224; 1906, **33**, 115; 1911, **36**, 9; 1924, **49**, 559; 1957, **82**, 488.

Casein can be determined by means of the formol titration (*Analyst*, 1936, **61**, 824). Further methods for the determination of casein and other proteins in milk have been described by Moir (*Analyst*, 1931, **56**, 147, 228).

Ash is determined by heating 10 g. at a temperature not exceeding 500° C. so that the chlorides are not volatilised. Any black particles which remain can usually be removed by mixing the ash with distilled water, then evaporating off the water on the water-bath and re-igniting. An ash figure below 0·75 per cent. may be due to the presence of added water. The ash of normal milk is composed of (all figures as percentages of the total ash w/w): K_2O 25·5, Na_2O 9·0, CaO 22·5, MgO 2·6, Fe_2O_3 0·2, P_2O_5 25·7, SO_3 3·0, Cl 14·3, Carbonate trace. The **chloride content** of normal milk seldom exceeds 0·13 per cent. when calculated on the whole milk. Milk derived from cows suffering from mastitis (**Mastitis milk**), however, when compared with normal milk, shows increased chloride and soluble nitrogen figures, together with a reduction in the amounts of lactose and casein (and consequently the N.F.S.). The *casein number* $\left(\frac{\text{Casein N\%}}{\text{Total N\%}} \times 100\right)$ of mastitis milk falls below the normal value and the *chlorine-sugar number* $\left(\frac{\text{Chlorine \%}}{\text{Lactose \%}} \times 100\right)$

* Rosaniline acetate (See Footnotes on p. 344).

tends to be correspondingly high (*Analyst Abs.*, 1921, **46**, 460). The average figures for normal milk are a casein number of 80 and a chlorine-sugar number of about 2·3. For a discussion of the effect of mastitis on the composition of milk reference should be made to Rowland *et al.* (*J. Dairy Res.*, 1938, **9**, 47, 182; 1939, **10**, 108, 267). The *chloride* can be determined on the ash or by direct titration methods as proposed by Davies (*Analyst*, 1932, **57**, 79; *J. Dairy Res.*, 1938, **9**, 327) and Sanders (*J. Dairy Sci.*, 1939, **22**, 841), who described the following procedure:—

To 10 ml. of milk add excess (say 15 ml.) of 0·0291 N silver nitrate solution (4·9438 g. silver nitrate, 200 ml. concentrated nitric acid and 300 ml. of saturated ferric alum solution per litre). Then back titrate with 0·0291 N potassium thiocyanate solution. Allowing for the average specific gravity of milk (1·032):—

% Chlorine (w/w) = 0·01 (ml. AgNO3 added − ml. thiocyanate used).

Lactose may be estimated either gravimetrically, volumetrically or by the polarimeter.

To 20–30 g. of the milk add about 50 ml. of water, then 5 ml. of zinc acetate solution (p. 99) and 5 ml. of potassium ferrocyanide, mixing gently after each addition. Dilute to 250 ml. and filter. The lactose in the filtrate is conveniently determined by titration with Fehling solution, using the Lane and Eynon process as described on p. 107. Calculate the lactose as anhydrous lactose.

For the polarimetric estimation, the proteins must be removed, for which purpose mercuric nitrate is suitable, although Richmond (*Analyst*, 1910, **35**, 516) shows that this does not completely precipitate all proteins, and that phosphotungstic acid should also be added; the error so introduced is negligible except in dried milk. To 100 ml. of milk are added 3 ml. of mercuric nitrate (mercury dissolved in twice its weight of concentrated nitric acid), the mixture is shaken, filtered and polarised in a 200 mm. tube at 20°; then the reading in angular degrees $\times \dfrac{0{\cdot}95 \times (100 - 1{\cdot}057 \times \text{fat } \%)}{100 \times 105 \times \text{sp. gr.}}$ gives the percentage by weight of anhydrous milk sugar. For a discussion of the errors involved by the volume of fat and proteins, see Richmond and Boseley (*Analyst*, 1897, **22**, 98).

Citric acid can be determined in milk by the pentabromacetone method recommended by Lampitt and Rooke (*Analyst*, 1936, **61**, 654):—

To 150 g. of the milk heated to 50°–60° in a 250 ml. flask add 25 ml. of 2 per cent. potassium oxalate solution, then after shaking, 20 ml. of diluted sulphuric acid (1 : 1). After cooling add 10 ml. of phosphotungstic acid solution (15 per cent.), dilute to 250 ml. and allow to stand for the serum to separate. To 50 ml. of the serum add 5 ml. of 37·5 per cent. potassium bromide solution. Then add potassium permanganate solution (5 per cent.) from a burette until a slight brown precipitate persists after standing for an hour. Sufficient ferrous sulphate solution* is added to produce a pale yellow solution with a white precipitate; then the mixture is set aside in an ice chest overnight. The precipitate is collected on a sintered glass crucible, washed with 10, 10, and 5 ml. of cold water and dried in a vacuum desiccator to constant weight, W. The weight of anhydrous citric acid is $0{\cdot}424\left(W+\frac{0{\cdot}005V}{100}\right)$ where V is the volume of the filtrate.

* Twenty g. $FeSO_4.7H_2O$ in 100 ml. of 1 per cent. sulphuric acid.

Adulteration of Milk

In addition to the more obvious forms of adulteration by adding water or removing fat, it is necessary to consider the possible presence of preservatives, dyes, thickening substances, and condensed milk, and whether the milk has been heated or pasteurised, though this is not strictly adulteration. Tests may also be required for the amount of dirt and bacteriological contamination. In connection with the application of the official minimum standards of 3·0 per cent. of fat and 8·5 per cent. of non-fatty milk solids, it has to be borne in mind that they raise only a presumptive proof of the addition of water or abstraction of fat, which may be rebutted by evidence either internal or external. There is no doubt that sometimes the milk from an individual cow may fall below these minimum figures, but that of a mixed herd of cows such as is sold in most dairies seldom falls materially below these limits if unsophisticated. Information of value is given sometimes by a more detailed examination, for, when milk is watered, the proportion of non-fatty solids is reduced, but the ratio of the various constituents is unaltered. When cows yield abnormal milk it is almost always the lactose which is deficient, but the ash tends to be slightly higher. Vieth, as the result of the examination of some thousands of samples, draws attention to the ratio of anhydrous lactose, proteins and ash, which is 13 : 9 : 2 (*Vieth's Ratio*). A material divergence from this ratio would therefore suggest abnormality rather than adulteration. Vieth has also shown that the ash is almost invariably 8 per cent. of the solids-not-fat. A higher proportion of albumin than 0·6 per cent. in cows' milk is suggestive of the presence of colostrum or of disease in the animal.

Nitrates do not occur in genuine milk even if the cow has been dosed therewith, but most natural waters contain small quantities of nitrate; so the detection of nitrate affords confirmatory evidence of the addition of water. To detect nitrate add 6 drops of mercury reagent to 5 ml. of the milk, shake, filter through a filter paper well washed with distilled water into a test tube containing 2 ml. of diphenyl-benzidine reagent*, so that the filtrate runs down the side of the tube and forms a layer on the diphenyl-benzidine solution. A distinct blue colour develops in the presence of nitrate. A genuine milk should be tested as a " blank " to avoid errors due to traces of nitrate from the laboratory or apparatus.

Freezing point of milk. It is well-known that cows produce milk (particularly in April and August) which contains less than 8·5 per cent. (the presumptive minimum) of N.F.S. It is inadvisable, therefore, to certify a milk as containing extraneous (or added) water merely on the basis of a low N.F.S. figure alone but confirmation should be sought from the freezing point. It has been definitely established that the circumstances, which affect the N.F.S. markedly, do not alter the osmotic pressure of the milk and hence the freezing point is approximately a constant. Using the semi-empirical Hortvet technique the freezing point of genuine milk usually falls within the range —0·530° to —0·550° C. Macdonald (*Analyst*, 1948, **73**, 423) states that the average freezing point of bulk milk is —0·544° C. (equivalent to a depression Δ_1 of 0·544°). Depressions below 0·530 have been reported under exceptional circumstances (see *Chem: and Ind.*, 1950, 333, 636; *Analyst*, 1955, 80, 623), *e.g.* for the evening milk from cows which had been deprived of water during the previous night for sixteen hours and then after being let out drank copious amounts of water. If the depression for such evening milk is low, that for the morning milk is correspondingly high (extreme limits 0·518–0·572). In spite of such possibilities, a minimum depression of 0·530 is usually assumed for legal purposes.

The observed depression increases as the milk gets sour and therefore the determination should be made when the milk is as fresh as possible. The Report of a Sub-Committee on the Freezing Point (Hortvet) Test of Milk (Dept. of Health for Scotland, 1946; H.M.S.O. Code 49–279) states that where the acidity exceeds 0·18 per cent., the best estimate of the original freezing point of the fresh milk is given by a correction equivalent to 0·0034° C. per 0·01 per cent. of lactic acid (calculated on a wt/wt basis), which should be subtracted from the apparent freezing point (Hortvet) depression, the relevant formula being:—

$$\text{Original F.Pt.(°C.)} = \text{Observed F.Pt.(°C.)} - 0{\cdot}34\ (\text{Acidity \% lactic acid w/w} - 0{\cdot}18).$$

The calculation is of doubtful accuracy for milks with acidities in excess of 0·30 per cent. Evans (*Analyst*, 1936, **61**, 666) has,

* The reagents are (1) 20 g. mercuric chloride, 5 g. ammonium chloride, and 20 ml. hydrochloric acid made up to 100 ml. with water. (2) Diphenylbenzidine 0·085 g., water 50 ml. and sulphuric acid 450 ml.

however, proposed a method for calculating the original freezing point of a very sour milk (*cf. ibid.*, **1945**, **70**, 323).

The presence of preservative also affects the freezing point and Macdonald (*Analyst*, **1943**, **68**, 171) has described a method which corrects for any formalin which may be present (*e.g.*, in a sample which has been sent by post). For every 100 p.p.m. of formaldehyde, he subtracts 0·009° from the observed F.Pt. depression.

The Hortvet Apparatus and Method

The apparatus and technique described in the A.O.A.C. has been commonly employed in most laboratories in Britain for many years. The essential parts of Hortvet's cryoscope are as follows: A cylindrical Dewar flask of capacity 1 litre fitted with a cork through which passes a thin-walled metal tube 250 × 33 mm. external measurements. At one side of the cork is a metal inlet tube which passes into a perforated loop near the bottom of the flask and serves for the blowing in of air. On the other side is an outlet T piece which serves for the escape of the ether vapour. At the back of the cork is a control thermometer of which the bulb reaches nearly to the bottom of the Dewar flask. Inside the thin-walled tube is the glass freezing tube, which should be 240 × 29 mm. in size. In the centre of the rubber stopper of this tube is the Hortvet thermometer; at one side is a small looped stirrer and at the other a metal tube with a small hole to carry a fragment of ice which serves as a starter for the freezing. At one side of the apparatus is a Folin drying tube containing sulphuric acid (conc.) and at the other a drainage tube for carrying any vapours away from the operator. A small telescope is provided for reading the thermometer more accurately, and it is convenient to have a mechanical stirrer and blower. It is also convenient to connect the outlet to a pipe leading out of the window so as to remove all inflammable vapour from the laboratory.

The Hortvet thermometer has a solid stem 580–600 mm. in length, and a bulb 6·5–7 mm. × 55–65 mm. Its total scale is from +1° to −2°, each degree being 99–100 mm. in length and divided into tenths and hundredths. This thermometer should be standardised with water and against pure sucrose solutions, using the same procedure as when determining the freezing point of milks. Solutions of pure sucrose 7 g. in water made up to 100 ml. at 20°, and of 10 g. made up to 100 ml. have freezing point −0·422° and −0·621° respectively. The depression Δ is the difference between the observed freezing point of water and that of the sucrose solution. The zero point of the thermometer is likely to vary from day to day, or even a little during the same day. If the mean of two or more determinations of Δ obtained for the sucrose solutions differs materially from the figures given another thermometer should be sought; if the differences are but small a correction may be applied.

The procedure in making a freezing-point determination is as follows: Pour 400 ml. of cold ether through the T piece into the Dewar flask—note the depth in the flask—close the inlet with a cork and start the blower so that air bubbles through the drying tube and through the ether at such a rate that the thermometer in the ether falls to 0° in about ten minutes, and continue till the temperature is about −3°. Then gauge, by means of a glass tube through the T piece, the depth of the ether and add enough to restore the volume to 400 ml. Now pour into the freezing tube enough boiled cold water to cover the bulb of the Hortvet thermometer; this should be about 35 ml. Also pour a few ml. of alcohol into the space between the two test tubes to serve as conducting medium. Start the stirrer at a steady motion of about 30 strokes per minute and continue the blowing until the temperature in the cooling bath is $-2·5°$. Continue the slow passage of air through the ether until the temperature on the Hortvet thermometer indicates a super-cooling of 1°–1·2°. At about this temperature the water begins to freeze and the temperature rises. Stop the stirrer and move it up and down three or four times as the mercury in the thermometer becomes stationary. It should remain stationary for about a minute. The thermometer is gently tapped and the temperature carefully read with a small telescope, taking precautions to avoid parallax, to the nearest 0·001°. Pour out the water, rinse the tube with the milk and put in 35 ml. of the cooled milk. Bring the level of the ether up to 400 ml. again and maintain the cooling bath 2·5° below the freezing point of the milk by blowing air through it as before. When the milk has reached about 1° super-cooling as shown on the Hortvet thermometer, with a penknife place a tiny fragment of ice in the hole of the starter rod and insert it quickly into the super-cooled milk. A rapid rise of the thermometer should occur; stop the stirrer and manipulate it three or four times as before, tap the thermometer and observe the maximum reading, which is the freezing point. The depression is given by the difference between the zero with water and the observed temperature. At least two determinations should be made, and these should not differ by more than 0·002° or 0·003° with the same worker and the same thermometer.

A suitable formula for calculating the amount of added water present is given under (f) below.

Temple (*Analyst*, 1937, **62**, 709) maintains the temperature of the cooling bath by means of an automatic refrigeration unit instead of using ether evaporation. The Temple system is commonly used where large numbers of determinations are performed.

The Hortvet apparatus which is specified in B.S.3095: 1959 shows a number of differences from that described in the A.O.A.C. The more important differences are:—

(a) Dimensions of the various parts of the apparatus are specified in greater detail than in the A.O.A.C.

(b) The B.S. prescribes that the range of the control thermometer should be $-5°$ C. to $+5°$ C. and graduated in 0·5 degrees.

(c) Five hundred ml. of ether are used in the B.S. technique.

(d) If sucrose is used for the calibration of the Hortvet thermometer the following five solutions (per cent. w/v at 20° C.) should be used (corresponding depressions in the brackets): 7·0 (0·422° C.), 7·5 (0·454), 8·0 (0·487), 8·5 (0·520) and 8·75 (0·537).

(e) For the purpose of calculating the true freezing point the formula proposed in the Report to the Dept. of Health for Scotland (see p. 336) is employed for slightly sour milks having acidities between 0·18–0·30 as lactic acid (on a w/v basis).

(f) For the purpose of calculating the proportion of added water present in milk the modified calculation due to Elsdon and Stubbs (*Analyst*, 1936, **61**, 383) is used:—

$$\text{Added Water (per cent. w/w)} = \frac{\Delta_1 - \Delta}{\Delta_1} \times (100 - T)$$

where T = total solids (per cent. w/w) and Δ = F.Pt. depression of the milk samples and Δ_1 = the F.Pt. depression of the genuine milk.

The following values of Δ_1 are stated in B.S.3095: 1959:—

F.Pt. Depression (Δ_1° C.)	*Type of milk*
0·545	Average for cows' milk in bulk
0·540	Minimum for quantities of 200 gallons or more
0·530	Minimum for milk from a single cow.

The B.S. adds that the proportion of added water present can be calculated by substituting an adopted figure for Δ_1, in the formula and should be reported as minimum percentage of added water, stating the arbitrary figure chosen and the formula used.

For legal purposes* a similar formula is used and Δ_1 is the F.Pt. depression of the appeal-to-the-cow sample or more likely the usually accepted minimum, *i.e.*, 0·530. For "industrial disputes" the figure 0·540 may be more expedient (see also *Analyst*, 1933, **58**, 318).

Dirt in Milk

The determination of dirt in milk has been thoroughly studied by a committee of the Society of Public Analysts (for details, see *Analyst*, 1937, **62**, 287). This committee considered that a limit of 2 parts of moist dirt per 100,000 is a proper maximum, though legal action is not recommended in respect of an isolated sample showing 3 parts. The method for the determination of dirt consists, briefly, of adding formalin to the milk and allowing it to stand in a covered sedimentation vessel connected to a standardised centrifuge tube, also filled with the milk, for seventy-two hours. Then the lower end of the vessel is stoppered by a glass rod with a rubber cork, the centrifuge tube is removed and spun at 2000 r.p.m. for five minutes. After pouring off the milk the separated dirt is washed by filling the tube with water and centrifuging again for three minutes, then washing it similarly with N ammonia solution and N hydrochloric acid. The volume of moist dirt is read off in the calibrated tube. For details of the calibration and size of the tubes, see the report above quoted. A microscopical examination of the dirt can be made before the ammonia wash, and the part removed returned to the tube before the final washes and measurement.

Preservatives in Milk

Preservatives have seldom been reported as being present in milk in recent years. In view of their effectiveness as preserving agents there is always the possibility that boric acid or formaldehyde may be used on occasion. Other preservatives which were formerly used are hydrogen peroxide, benzoic acid, fluoride, salicylate, β-naphthol, sodium carbonate and glycerine.

Boric acid. For the *detection* of boric acid or borax, make about 5 ml. of the milk just alkaline with lime water, evaporate and ignite the residue, acidify the ash with hydrochloric acid, and dip into the liquid a piece of turmeric paper, then dry it on the water-bath. In the presence of boric acid the paper turns pink or red according to the amount present.

For the *estimation* of boric acid Richardson and Walton's process is quick and accurate. To 25 ml. of the milk are added 2·5 ml. of 5 per cent. solution of copper sulphate, the mixture is well stirred, then heated to the boiling point and filtered. The precipitate is washed four or five times with small quantities of boiling water. To the cold filtrate is added 1 ml. of 1 per cent. phenolphthalein solution and 0·1 N sodium hydroxide until a blue shade appears.

* See also the recommendations (Paragraph 88, f) of the Interdepartmental Committee on Milk Composition in the U.K. (p. 325).

Then there is added 15 ml. of neutralised glycerine and the liquid is titrated with 0·1 N alkali until the characteristic blue shade again appears. Under these conditions 1 ml. 0·1 N = 0·0071 g. H_3BO_3.

An elaboration of Thomson's method in use at the Government Laboratory (*cf. Analyst*, 1923, **48**, 416), though exact, is tedious.

Formaldehyde can be detected and determined by the methods described on p. 91. It can be detected when the routine Gerber method for the determination of fat is carried out using sulphuric acid which contains a trace of iron (a violet colour is produced as in Hehner's test when formalin is present).

A preservative for milk known as "Mystin," which contained about 0·3 per cent. formaldehyde and 10 per cent. sodium nitrite, appeared on the market many years ago. Milk which contains mystin does not usually give Hehner's test for formalin, but gives a positive reaction with the Griess-Ilosvay reagents for the detection of nitrites. Monier-Williams found, however, that if 5 ml. of milk containing mystin is heated in boiling water for two minutes with 0·05 g. urea and 1 ml. N/1 sulphuric acid and cooled, the nitrite is decomposed and the resultant liquid gives a positive reaction with Hehner's test.

Hypochlorites may be present in milk due to its being used for disinfecting utensils. Hypochlorites may be detected by the test originally proposed by Wright and Anderson (*Analyst*, 1938, **63**, 252; see also B.S.1741: 1951).

To 3 ml. milk (cooled to 0°–5° C.) in a test tube add 3 ml. of 73·5 per cent. sulphuric acid* containing 0·025 per cent. stannous chloride (also cooled to 0°–5° C.). Shake in a freezing mixture of ice and salt, allow to stand in the mixture for three minutes and then centrifuge at 2500 r.p.m. for three minutes. When examined under u.v. light, the presence of a yellow fluorescence is indicative of the presence of hypochlorite. The test detects the presence of chlorate not hypochlorite (stock solutions of hypochlorite used in the dairy industry must contain at least 0·7 per cent. of sodium chlorate).

Quaternary ammonium compounds may be present in milk due to some residual detergent solution remaining after bottle washing. The following test due to Miller and Elliker (*J. Dairy Sci.*, 1951, **34**, 273) detects about 5 p.p.m. in milk:—

Indicator Solution. Prepare a stock solution by dissolving 0·05 g. eosin in 100 ml. acetone. Shake 10 ml. of stock solution with 90 ml. of tetrachloroethane and 1 g. of citric acid and filter before use.

* Mix the requisite amount of stannous chloride with 1 vol. water and add carefully 3 vols. conc. sulphuric acid.

Buffer. Dissolve 25 g. citric acid in 100 ml. water and adjust to pH 3·5 with 50 per cent. sodium hydroxide solution (approximately 12 ml. required).

Method

To a centrifuge tube add 1 ml. milk, 5 ml. water, 1 ml. indicator solution and 0·2 ml. buffer solution and shake hard for ten seconds. Centrifuge for five minutes at 3200 r.p.m. If Q.A.C. is present the bottom layer assumes a red colour (*cf. J. Dairy Sci.*, 1953, **36**, 225).

Artificial colouring matters are prohibited in milk. Formerly annatto, caramel and coal-tar dyes were added to milk. *Annatto* can be detected by pouring a few ml. of milk into a flat dish and adding sodium bicarbonate solution. Then a strip of filter paper or a piece of cotton wool is immersed therein. After some hours the paper or wool is stained brown in the presence of annatto and turns pink on the addition of a drop of stannous chloride solution (*cf.* Gardiner, *Analyst,* 1925, **50**, 550). *Caramel* can be detected by coagulating 10 ml. of the milk by warming with a drop of acetic acid, separating the curd and covering it in a basin with concentrated hydrochloric acid. In the presence of caramel a blue colour develops. *Coal-tar dyes* can be detected by boiling with wool (see p100). Milk containing an azo dye usually gives a pink colour on the addition of hydrochloric acid to the milk, or better, to the coloured curd. For further tests for the detection of colouring matters in milk, the reader should refer to Richmond's " Dairy Chemistry " (5th edition, pp. 410–413).

Assessment of Cleanliness and Efficiency of Heat-treatment of Milk

The Milk (Special Designation) Regulations prescribe that the methylene blue test be applied to T.T. and pasteurised milk. Industrially the unofficial resazurin test is frequently employed for routine control work. The clot-on-boiling test instead of the methylene blue test, is now to be applied to samples of T.T. milk taken from producers to assess its keeping quality. The phosphatase test is prescribed for assessing the efficiency of pasteurisation and the turbidity test is applied to sterilised milk. Formerly T.T. milk and pasteurised milk were required to conform to certain bacteriological standards.

Sampling. " Provisions as to Sampling " are stated in Part I of the Third Schedule to the Regulations :—

* * * * *

2. When the milk is in containers not exceeding one quart in capacity, the sample shall consist of one such container which shall be delivered intact to the testing laboratory.

3. When the milk is in containers exceeding one quart in capacity, the sample shall consist of not less than two fluid ounces of the milk. The milk shall be thoroughly stirred before sampling and the sample shall be taken from well below the surface of the milk. The instruments used for stirring and sampling shall be sterile and the sample shall be poured into a sterile bottle which shall thereupon be immediately stoppered. The part of the stopper which may come into contact with the milk shall be sterile. Where the person taking the sample breaks the seal on a container he shall, after taking the sample, re-seal the container and attach to it a label certifying that it has been opened and re-sealed by him.

Identification of sample

4. For the purpose of identification in the testing laboratory, the person taking the sample shall mark the container of the sample with a number or other suitable identification mark at the time of sampling and shall enter in a book or on a paper, which shall accompany the sample, the following particulars:—

(a) the identification number or mark;

(b) the name and address of the holder of the licence by whom the milk was consigned, or by whom it was being delivered, or on whose premises the sample was taken;

(c) in the case of a sample to which the clot-on-boiling test is to be applied, the date of milking and whether the sample comprises milk of the morning or evening milking or mixed milk, if known; and for this purpose milk produced before 12 noon on any day shall be treated as morning milk and milk produced at or after 12 noon on any day shall be treated as evening milk.

Transport of sample

5. The bottle or container containing any sample of milk shall be transferred forthwith to an insulated container, which shall not be artificially cooled, for transport to the testing laboratory. The sample shall be transported to the testing laboratory with the least possible delay. Any sample which does not arrive at the testing laboratory on the day on which it is taken shall be discarded.

The Clot-on-Boiling Test. The Clot-on-Boiling Test has been introduced for T.T. milk to which a producer's licence relates. This test is used for assessing the keeping quality (or commercial " life ") of milk. After incubation at 22° C. the sample is heated at 100° C. for five minutes and examined for clots (*cf.* Davis, " *Milk Testing,*" Dairy Ind., 1959).

" The Clot-on-Boiling Test for Tuberculin Tested Milk " is described in Part II of the Third Schedule to the Regulations:—

Treatment of sample

1. On arrival at the testing laboratory the sample of milk shall be treated as follows:—

(a) Evening milk or mixed milk from two or more milkings on one and the same day sampled on the day of production.

The sample shall be removed from the insulated container and placed in the cold store or refrigerator at 3–5° C. until 9.0 a.m. on the following morning when it shall be removed and stored in the atmospheric shade box until immediately before the beginning of the incubation period which shall be at 4.0 p.m. on the day following sampling.

(b) Evening milk sampled on the day following production and morning milk sampled on the day of production.

The sample shall be removed from the insulated container and stored in the atmospheric shade box until immediately before the beginning of the incubation period which shall be at 4.0 p.m. on the day of sampling.

(c) Mixed milk from a morning milking and the evening milking of the previous day and milk insufficiently labelled to enable the milking to be identified.

The sample shall be removed from the insulated container and stored in the atmospheric shade box until immediately before the beginning of the incubation period which shall be at 4.0 p.m. on the day of sampling.

2. Where any sample referred to in sub-paragraph (b) or (c) of the last preceding paragraph does not arrive at the testing laboratory before 4.0 p.m. on the day on which it is taken, it shall be discarded.

Apparatus

3.—(1) The bottle used for the incubation of the sample at 22° C. shall conform to British Standard 809: 1949.

(2) Test tubes shall conform to British Standard 625:1959, nominal size 150/16, and shall be accurately marked at 5 ml. They shall be plugged with cotton wool or covered with closely fitting aluminium caps or stored in such a way as to prevent contamination.

(3) Glassware and rubber stoppers shall be sterile immediately before use.

(4) The water bath shall be fitted with a reliable automatic thermo-regulator and shall be capable of maintaining the water at a temperature of 22° C. ± 0·5° C. and returning it to that temperature as soon as possible after the addition of a sample.

(5) The maximum thermometer referred to in paragraph 5 of this Part of this Schedule shall be of the clinical type.

Incubation of sample

4.—(1) Immediately before the time specified in paragraph 1 of this Part of this Schedule as the beginning of the incubation period, the sample shall be removed from the atmospheric shade box and shall be mixed thoroughly by inverting and shaking. Where the sample is in a 3 oz. bottle, sufficient milk shall be poured out to leave approximately 2 oz. in the bottle for testing. Where the sample is in a container of larger capacity than 3 oz., approximately 2 oz. shall be transferred to a sterile 3 oz. bottle which shall immediately be stoppered. Aseptic precautions shall be taken in carrying out all these operations.

(2) At 4.0 p.m. the 3 oz. sample bottle containing the 2 oz. test portion shall be transferred to the water bath. The water in the bath shall be kept at least ½ in. above the level of the milk in the bottle and at least ½ in. below the rim of the bottle. The temperature in the water bath shall be restored as soon as possible to, and thereafter maintained at, 22° C. ± 0·5° C. by means of the automatic thermo-regulator. The duration of incubation of the sample at 22° C. ± 0·5° C. shall be determined by the mean atmospheric shade temperature between 9.0 a.m. and 4.0 p.m. on the day on which the incubation of the sample was begun.

5. If during the period of incubation the temperature as indicated by a maximum thermometer placed in a 3 oz. sample bottle containing 2 oz. of water stored in the bath with the sample exceeds 22·5° C., the test shall be void.

Mean atmospheric shade temperature

6. The mean atmospheric shade temperature shall be the arithmetic mean of the maximum and minimum temperatures in the atmospheric shade box between 9.0 a.m. and 4.0 p.m. on the day incubation is begun. The maximum thermometer and minimum thermometer shall be set at 9.0 a.m. each day and the thermometers shall remain in the atmospheric shade box until 4.0 p.m. on the same day when the maximum and minimum temperatures shall be read to the nearest degree and the arithmetic mean calculated. A permanent record of the daily maximum and minimum temperatures and of the arithmetic mean of the two readings shall be kept in the laboratory.

Time of applying the test

7. The time of applying the test and the duration of incubation of the sample at 22° C. ± 0·5° C. shall be determined by the mean atmospheric shade temperature as follows:—

Mean atmospheric shade temperature between 9.0 a.m. and 4.0 p.m. on day incubation of test samples is begun	Time of applying the Clot-on-Boiling test on following day
67·6° F. and above	9.30 a.m.
64·6° F. to 67·5° F.	10·30 a.m.
60·6° F. to 64·5° F.	11.30 a.m.
55·6° F. to 60·5° F.	12.30 p.m.
50·6° F. to 55·5° F.	1.30 p.m.
43·6° F. to 50·5° F.	2.30 p.m.
Under 43·6° F.	3.30 p.m.

Method of carrying out the test

8. The sample shall be removed from the water bath just before the time of test, allowing only sufficient time to mix and pour it. If the sample is found to have clotted, the result of the test shall be recorded as positive (+). If the sample has not obviously clotted, it shall be mixed thoroughly by inverting and shaking the sample bottle, and 5 ml. of the milk to be tested shall be poured into a clean test tube marked at 5 ml. The tube shall be placed in a bath of boiling water for 5 minutes, care being taken that the level of the water is at least ½ in. above the level of the milk in the test tube and that the water is kept boiling.

After being in the boiling water for 5 minutes the tube shall be removed and tilted almost horizontally to expose a thin film of milk for observation. When the whole of the milk clots or when discrete particles are seen to be adhering to the side of the tube when it is returned to the vertical, the result shall be recorded as positive (+). If clots or discrete particles are absent or if there is any doubt as to their presence, the result shall be recorded as negative (−).

Interpretation

9. A sample shall be regarded as satisfying the Clot-on-Boiling test when a negative result is obtained after boiling for 5 minutes in accordance with the last preceding paragraph.

The Methylene Blue Reduction Test is officially prescribed by The Milk (Special Designation) Regulations for T.T. milk and pasteurised milk. The methylene blue test virtually assesses the bacterial quality of the milk and hence the keeping qualities. The test depends on the fact that the reducing activities of the micro-organisms and any reducing substances in the milk cause a fall in the redox potential and the change is assessed visually by means of methylene blue. Most of the reducing activity of the bacteria is used for removing the dissolved oxygen in the milk (for respiration) and the dye is not in fact reduced until all the oxygen is removed. The methylene blue test is more rapid than plate count methods and the results are more reproducible.

"The Methylene Blue Test for Tuberculin Tested Milk and Pasteurised Milk" is described in Part III of the Third Schedule to the Regulations:—

Treatment of sample

1.—(1) On arrival at the testing laboratory the sample of milk shall at once be removed from the insulated container. Thereafter it shall be stored as follows:—

(a) a sample taken at any time during the period from 1st May to 31st October, inclusive, in any year shall be kept at atmospheric shade temperature until 9.30 a.m. on the following day.

(b) A sample taken at any time during the period from 1st November to 30th April, inclusive, in any year shall be kept in its original container or in a sterile 3 oz. sample bottle at atmospheric shade temperature until 5.0 p.m. on the day of sampling and thereafter at a constant temperature at 65° F. ± 2° F. until 9.30 a.m. on the following day.

(2) If during the period of storage at atmospheric shade temperature to which a sample is subjected this temperature at any time exceeds 70° F., the test shall not be applied.

(3) The test shall be begun between 9.30 and 10.0 a.m. on the day after the sample is taken.

Reagent—Methylene Blue

2.—(1) Tablets manufactured under arrangements made by the Minister shall be used for the test. A solution shall be prepared aseptically by adding one tablet to 200 ml. of cold, sterile, glass-distilled water in a sterile flask, shaking until the tablet is completely dissolved, and making up the solution to 800 ml. with cold, sterile, glass-distilled water. The resultant solution shall be stored in a stoppered sterile flask in a cool, dark place, and shall not be used if—

(a) it has been exposed to sunlight, or

(b) a period of two months has elapsed since the date of preparation.

(2) The amount of methylene blue required for a day's work shall be poured off from the stock bottle into a suitable glass container. The pipette used for transferring the methylene blue solution to the tubes of milk shall not be introduced into the stock bottle.

Apparatus

3.—(1) Test tubes shall conform to British Standard 625:1959, nominal size 150/16, and shall be accurately marked at 10 ml. They shall be plugged with cotton wool or covered with closely fitting aluminium caps or stored in such a way as to prevent contamination.

(2) Pipettes shall be 1·0 ml. straight-sided blow out delivery pipettes, and shall be plugged with cotton wool at the upper end.

(3) Glassware and rubber stoppers shall be sterile immediately before use.

(4) The water bath shall be fitted with a reliable automatic thermo-regulator capable of maintaining the water at a temperature of 37·5° C. ± 0·5° C.

Method of carrying out the test

4.—(1) The sample shall be mixed thoroughly by inverting and shaking and the milk shall be transferred to a test tube up to the 10 ml. mark in such a manner that one side of the interior of the test tube is not wetted with milk. 1 ml. of methylene blue solution shall be added without letting the pipette come into contact with the milk in the tube or with the wetted side of the interior of the tube. After a lapse of 3 seconds, the solution remaining in the tip of the pipette shall be blown out. The test tube shall be closed with a rubber stopper, aseptic precautions being taken, and shall then be inverted twice slowly, so that the whole column of contained air rises above the level of the milk. Within a period of 5 minutes the test tube shall be placed in a water bath. The water in the bath shall be kept above the level of the milk in the test tube, and its temperature, which shall be between 37° C. and 38° C., shall be maintained as nearly uniform as possible by means of a reliable automatic thermo-regulator. The interior of the bath shall be kept completely dark.

(2) A control tube shall be used for comparison with each batch of experimental tubes to indicate when decolourisation is complete. The control tube shall be prepared by immersing in boiling water for 3 minutes a stoppered test tube containing 1 ml. of tap water and 10 ml. of mixed milk having a fat content and colour similar to that of the milk being tested.

(3) The milk shall be regarded as decolourised when the whole column of milk is completely decolourised or is decolourised up to within 5 mm. of the surface. A trace of colour at the bottom of the tube may be ignored provided that it does not extend upwards for more than 5 mm.

Interpretation

5. The test shall be deemed to be satisfied by milk which fails to decolourise methylene blue in 30 minutes.

The Resazurin Test also assesses the reducing activities of the micro-organisms present in the milk. The resazurin dye used changes from blue in the oxidised state through various shades of pink to colourless. As changes in the colour (or tint) are seen in a shorter time than with methylene blue, the unofficial resazurin test is frequently employed industrially as a rapid platform test and for routine grading. The resazurin test appears to be more sensitive than methylene blue for revealing " faults " in *certain* milks, *e.g.*, for milks rich in weakly reducing organisms, for aged milks with a large number of cells. The test, which must be performed with the usual aseptic precautions, is now briefly described:—

Shake the milk to incorporate air during the mixing and pour into a sterile test tube up to the 10 ml. mark. Add 1 ml. of 0·005 per cent. resazurin solution, close the tube with a sterile rubber bung and place in a water bath at 37° C. Assess the colour after one hour at 37° C. Reduction to a lilac or pink colour is indicative of an unsatisfactory bacterial condition. Complete reduction of the dye to colourless takes a little longer than with methylene blue. A special disc is available for assessing the colour produced. Where the factor is particularly important the reading is taken at ten minutes (*cf. Milk Testing*, Davis, 1951, p. 185–; *Dictionary of Dairying*, Davis, 1955, p. 904–; *Dairy Ind.*, 1940, **5**, 244; 1943, **8**, 215; 1945, **10**, 177).

The Phosphatase Test for Pasteurised Milk. The processing times and temperatures to be employed for the pasteurisation of milk are stated in The Milk (Special Designation) Regulations (see p. 328). These Regulations also prescribe the Phosphatase Test for the assessment of the adequacy of pasteurisation. The principle on which the test is based is that although phosphatase enzymes are invariably present in raw milk, they are inactivated during pasteurisation. It has also been shown that this enzyme is more difficult to destroy than the most heat-resistant pathogenic organisms which are likely to be present, *viz.*, tubercle bacilli.

The prescribed phosphatase test was devised by Aschaffenburg and Mullen (*J. Dairy Res.*, 1949, **16**, 58). The test involves incubation of the milk with disodium p-nitrophenyl phosphate under alkaline conditions. If the milk contains phosphatase a yellow colour is produced due to the formation of p-nitrophenol. The degree of destruction of the phosphatase in the milk during pasteurisation is then assessed by comparing the colour produced with the standard colours on an APTW comparator disc (see Tramer and Wight, *J. Dairy Res.*, 1950, **17**, 194; *Dairy Ind.*, 1952, **17**, 54). A positive reaction in the phosphatase test may, on occasion, be due to contamination of pasteurised milk with raw milk.

"The Phosphatase Test for Pasteurised Milk" is described in Part IV of the Third Schedule to the Regulations:—

Examination of sample

1. The sample of milk shall be examined as soon as possible after arrival at the testing laboratory. If it is not examined immediately on arrival at the testing laboratory, it shall be kept at a temperature of between 3° C. and 5° C. until examined. The sample shall be raised to room temperature immediately before being tested.

Precautions

2. The following precautions shall be taken:—
   (a) A sample which shows evidence of taint or souring shall not be tested.
   (b) All glassware shall be clean immediately before use.
   (c) A fresh pipette shall be used for each sample of milk. Pipettes shall not be contaminated with saliva.
   (d) The test shall not be carried out in direct sunlight.
   (e) Distilled water shall be used throughout.

Reagents

3.—(1) Whenever possible, reagents of analytical quality shall be used.

(2) The buffer-substrate solution shall be prepared as follows:—

(a) Buffer solution: 3·5 g. of anhydrous sodium carbonate and 1·5 g. of sodium bicarbonate shall be dissolved in distilled water, and made up to one litre.

(b) Substrate: Disodium p-nitrophenyl phosphate. The solid substrate shall be kept in a refrigerator.

(c) Buffer-substrate solution: 0·15 g. of the substrate shall be placed in a 100 ml. measuring cylinder, and made up to 100 ml. with the buffer solution. The solution shall be stored in a refrigerater and protected from light. It shall give a reading of less than the standard marked 10 on the comparator disc A.P.T.W. or A.P.T.W.7 when viewed in transmitted light through a 25 mm. cell in the " all purposes " comparator, distilled water being used for comparison. The solution shall not be used for more than one week.

Apparatus

4. The following apparatus shall be used:—

(a) A Lovibond " all purposes " comparator complete with stand for work in reflected light.

(b) A Lovibond comparator disc A.P.T.W. or A.P.T.W.7.

(c) Two fused glass cells, 25 mm. depth.

(d) A water bath or incubator capable of being maintained at 37·5° C. ± 0·5° C.

(e) A pipette to deliver 5·0 ml.

(f) A supply of 1·0 ml. straight-sided pipettes of an accuracy equal to that of N.P.L. grade B.

(g) A 1000 ml. graduated flask.

(h) A 100 ml. measuring cylinder.

(i) A supply of test tubes conforming to British Standard 625: 1959, nominal size 150/16, with rubber stoppers to fit.

Care of apparatus

5.—(1) After use, each test tube shall be emptied, rinsed in water, well washed in hot water containing soda, rinsed in warm water, rinsed in distilled water and finally dried.

(2) If after treatment in accordance with sub-paragraph (1) hereof a test tube does not appear to be clean, the treatment shall be repeated with the addition that after being rinsed in warm water it shall be soaked in 50 per cent. commercial hydrochloric acid and then rinsed again in warm water before being rinsed in distilled water and finally dried.

(3) New glassware shall be cleaned by soaking in chromic acid solution prepared as follows:—

5 volumes of 8 per cent. W/V potassium bichromate

4 volumes of concentrated sulphuric acid. This should be added slowly and carefully to the mixture of bichromate and water.

The solution shall be kept covered and shall be discarded when it becomes green. After cleaning in chromic acid solution, new glassware shall be rinsed in warm water, rinsed in distilled water and finally dried.

(4) Pipettes shall be well rinsed in cold water and then cleaned by soaking for 24 hours in chromic acid solution in a 250 ml. glass cylinder or other suitable container. The pipettes shall then be well rinsed in warm water, rinsed in distilled water and finally dried.

(5) Glassware used for the test shall not be used for any other purpose and shall be kept apart from all other apparatus in the laboratory.

Method of carrying out the test

6. 5 ml. of the buffer-substrate solution shall be transferred to a test tube using a pipette and the test tube shall be stoppered and brought to a temperature of 37° C. 1 ml. of the milk to be tested shall be added, the test tube stopper replaced and the contents well mixed by shaking. The test tube shall then be incubated for exactly 2 hours at 37° C. One blank prepared from boiled milk of the same type as those undergoing the test shall be incubated with each series of samples. (Where the sample consists of highly coloured milk, such as homogenised milk or milk from Channel Island cows, a separate blank of such milk shall be prepared.) After incubation the test tube shall be removed from the water bath and its contents shall be well mixed. The blank shall be placed on the left hand ramp of the stand and the test sample on the right. Readings shall be taken in reflected light by looking down on to the two apertures with the comparator facing a good source of daylight (preferably north light). If artificial light is needed for matching, a "daylight" type of illumination must be used. The disc shall be revolved until the test sample is matched. Readings falling between two standards shall be recorded by affixing a plus or minus sign to the figure for the nearest standard.

Interpretation

7. The test shall be deemed to be satisfied by milk which gives a reading of 10 μg. or less of p-nitrophenol/ml. of milk.

Prior to the 1960 Regulations, the prescribed phosphatase test was that proposed by Kay and Graham (*J. Dairy Res.*, 1935, **6**, 191) as modified by Neave (*ibid.*, 1939, **10**, 475). This involved incubation of the milk with disodium phenyl phosphate, so that free phenol was liberated by the phosphatase. Then Folin and Ciocalteu's reagent was added, so that the phenol produced gave a blue colour in alkaline solution.

The Turbidity Test for Sterilised Milk. When milk is heated to 80° C. or above, all the albumin becomes denatured and if solutions of inorganic salts or acids are added the albumin separates with the casein. In Aschaffenburg's test for sterilised milk, the sample is shaken with ammonium sulphate, filtered and the filtrate is then heated. Any albumin which is still present in solution due to insufficient heat treatment during manufacture, will reveal itself as a turbidity in the heated filtrate. If the milk has been sterilised adequately, however, all the albumin will have been precipitated and no turbidity will be produced. The officially prescribed Turbidity Test for Sterilised Milk is described in Part V of the Third Schedule to the Regulations (*cf. J. Soc. Dairy Technol.*, 1950, **3**, 236):—

Examination of sample

1. The sample of milk may be examined at any time after delivery to the testing laboratory but shall be at room temperature when the test is begun.

Reagent

2. Ammonium sulphate A.R. shall be used.

Apparatus

3. The following apparatus shall be used:—

(a) Conical flasks of 50 ml. capacity.

(b) Graduated cylinders of 25 ml. capacity.

(c) Test tubes conforming to British Standard 625: 1959, nominal size 150/16.

(d) Filter funnels of 6 cm. diameter.

(e) Beakers of 400 ml. capacity.

(f) 12·5 cm. No. 12 Whatman folded filter papers.

Method of carrying out the test

4. 4 ± 0·1 g. of ammonium sulphate shall be weighed into a 50 ml. conical flask. 20 ± 0·5 ml. of the milk sample shall be measured out and poured into the conical flask, the flask being shaken for 1 minute to ensure that the ammonium sulphate dissolves. The mixture shall be left for not less than 5 minutes and then filtered through a folded filter paper into a test tube. When not less than 5 ml. of a clear filtrate have collected, the tube shall be placed in a beaker of water, which has been kept boiling, and kept therein for 5 minutes. The tube shall be transferred to a beaker of cold water, and when the tube is cool, the contents shall be examined for turbidity by moving the tube in front of an electric light shaded from the eyes of the observer.

Interpretation

5. The test shall be deemed to be satisfied when a sample of milk treated as in paragraph 4 hereof gives a filtrate showing no sign of turbidity.

A test has been proposed by Fellows (*Dairy Ind.*, 1953, **18**, 309) which is perhaps more suitable for industrial purposes as it assesses the extent to which sterilised milk has been heated. This measures the " reducing capacity " of sterilised milk.

Bacteriological examination. Memo. 139/Foods (Jan. 1937) described tests and prescribed certain bacteriological standards for graded milk, including:—

T.T. Milk—No coliforms in 0·01 ml. (37° C.).

T.T. Milk (*Pasteurised*)—Maximum count on agar at 37° C. of 30,000 per ml.

Pasteurised Milk—Maximum count on agar at 37° C. of 100,000 per ml.

Although not included in The Milk (Special Designation) Regulations, these bacteriological standards are useful as confirmation for doubtful samples. The tests which were prescribed in Memo. 139 are summarised on the next page.

SAMPLING

Morning milk is to be kept until 6 p.m., and evening milk until 10 a.m. if unbottled, or may be kept for two hours if bottled, before testing. Thereafter if it is not convenient to begin the test at once the samples may be kept at 32°–40° F., but not for more than eighteen hours.

Bacteriological Count

The medium for plates contains: yeastrel, 3 g., peptone (B.D.H. or equivalent), 5 g., agar, 15 g., milk, 10 ml., water, 1000 ml. and has pH 7·2. The dilutions required are $\frac{1}{10}$, $\frac{1}{100}$ and $\frac{1}{1000}$ ml., made from (a) 90 ml. water (or $\frac{1}{4}$ strength Ringer* solution) + 10 ml. milk, (b) 90 ml. of water + 10 ml. of (a), (c) 90 ml. of water + 10 ml. of (b).

Three straight-sided, not bulbed, pipettes are required for each sample, one for each dilution. In making the dilutions, the sample and each dilution bottle should be shaken 25 times with an up-and-down motion of about 1 foot. To the required quantity (1 ml.) of the diluted milk in a sterile tube are added 10 ml. of melted agar cooled to 45° C., the mixture is poured into a petri dish and incubated for forty-eight hours at 37° C.

If among the dilutions there are plates containing from 30 to 500 colonies, these should all be counted and the number, multiplied by the dilution, reported as the final count. No plate containing less than 30 colonies is to be counted unless it was made with a $\frac{1}{10}$ dilution.

Tests for Coliforms

Three tubes each containing 5 ml. of MacConkey broth and a Durham's fermentation tube are inoculated with 1 ml. of $\frac{1}{100}$ dilution of the sample, and incubated at 37° C. The milk is satisfactory if two of the three tubes are free from acid and gas at the end of forty-eight hours. MacConkey broth is prepared by dissolving 10 g. of lactose, 5 g. each of sodium chloride and sodium taurocholate, and 20 g. of peptone in 1 litre of water, by heating. The liquid is filtered, adjusted to pH 7·4, and 10 ml. of 1 per cent. solution of neutral red added.

Skimmed Milk with Non-Milk Fat

The Skimmed Milk with Non-Milk Fat Regulations (S.I. 1960 No. 2331) impose requirements as to the labelling and advertising of certain " specified foods " which have the appearance of milk, condensed milk or dried milk, and which contain skimmed milk and non-milk fat (" filled milk "). The regulations require labels and advertisements for these foods to bear the words " Skimmed milk with non-milk fat." The regulations also prohibit (subject to certain savings) the labelling or advertising of the specified foods, and beverages containing skimmed milk, in a manner suggestive of milk or anything connected with a dairy interest.

The labels for such products are required to bear the words " Unfit for babies " or " Not to be used for babies " except that

* Ringer's solution contains in 1 litre NaCl 4·25 g., KCl 0·11 g., $CaCl_2$ 0·12 g., $NaHCO_3$ 0·05 g.

where the kind and amount of fat used is specified, the words " Should not be used for babies except under medical advice " may be substituted.

None of these declarations about baby feeding is required in the case of five products which are named in the Order, *viz.*, Alfonal, Alfonal Evaporated, White Dove Brand, Efalac and S.M.A. These latter products are, however, required to conform to the following standards:—

(a) shall contain poly-unsaturated fatty acids of the *cis-cis* form to the extent of not less than 12 per cent. of the total fatty acids present in such food;

(b) shall not contain any protein other than protein derived from milk; and

(c) shall not contain any ingredient of no nutritional value.

	Protein derived from milk (min.) % w/w	Fat (min.) % w/w	Vitamin A (min.) I.U./fl. oz.	Vitamin D I.U.	Ascorbic acid (min.) mg./oz.
Alfonal . .	3·6	3·5	44	13±4/fl. oz.	—
Alfonal Evaporated, White Dove Brand	9·6	9·3	110	33±10/fl. oz.	—
Efalac . . .	23·1	28·6	357	100±31/oz.	—
S.M.A. . .	11·6	27·0	321	92±28/oz.	11·5

Where any claim is made on the label as to the effect of the specified food in the container in relation to coronary diseases, the label shall also bear a statement specifying the quantity or proportion and the appropriate designation of each fat or oil present in the food in the container and whether or not each such fat or oil is hydrogenated.

" Specified foods " (as defined in the above regulations) do not have to comply with the requirements of the regulations which relate to condensed milk and dried milk (see also Circular FSH 13/60).

CREAM

Legal standards for cream are prescribed in The Food Standards (Cream) Order, 1951 (S.I. 1951 No. 668). This order states the following minimum milk fat contents (percentage w/w): "double cream," "thick cream," "clotted cream"—48; "sterilised cream" —23; "single cream," "cream," "pouring cream," "coffee cream," "fruit cream"—18. The addition of preservatives and thickening agents to cream is prohibited by the Preservatives Regulations.

The Food and Drugs Act, 1955 (Section 135) defines " cream " as " that part of milk rich in fat which has been separated by skimming or otherwise." Nowadays cream is usually prepared in mechanical separators, which can produce a product containing up to about 65 per cent. fat. Although the proportion of fat present in cream varies greatly, the relative proportions of the other ingredients remain the same as in the milk from which the cream was derived. This applies to the ratio of S.N.F. to water unless water has been lost, *e.g.*, by heating as in the preparation of Devonshire or pasteurised cream. It is useful to compare the S.N.F. (as determined analytically) with that calculated from: S.N.F. = 0·102 × Water. If the S.N.F. is substantially lower than that calculated, watering is indicated. As Vieth's ratio still applies it is also useful to check the lactose, protein and ash are $\frac{13}{24}$ths, $\frac{9}{24}$ths and $\frac{2}{24}$ths of the S.N.F. respectively.

The following figures due to Richmond give the relationship between the total solids, S.N.F., fat and ash of genuine samples of cream:—

Total solids. Per cent.	Solids-not-fat. Per cent.	Fat. Per cent.	Ash. Per cent
32·50	6·83	25·67	0·57
37·59	6·14	31·45	0·52
50·92	5·02	45·90	0·42
55·05	4·65	50·40	0·38
55·18	4·77	50·41	0·39
55·97	4·47	51·50	0·38
56·37	4·40	51·97	0·38
57·99	4·17	53·82	0·41
68·18	3·30	64·88	0·28

Richmond has also suggested that the following formulae may be used:—

$$S = \frac{10 - 0{\cdot}1\,F}{1{\cdot}08};\quad S = 10{\cdot}1 - 0{\cdot}1\,T;\quad F = 1{\cdot}102\,T - 10{\cdot}2$$

where S = S.N.F. (%), F = Fat (%) and T = Total solids (%).

Analysis of Cream

Especially in view of the requirements of The Food Standards (Cream) Order it is most important to determine the **fat content.** Other possible determinations are of the **total solids, protein,**

lactose, ash and the **acidity** and on occasion it may be necessary to carry out tests for added **preservatives, thickening agents, colouring matters** and **foreign fats.**

Fat can be determined accurately by the **Rose-Gottlieb process** using 2 g. of cream and 8 ml. of water. The **Werner-Schmid** method is also suitable. For routine purposes the rapid **Gerber method** is, however, more convenient:—

If the special cream butyrometers are available, weigh out 5 g. of cream into a stoppered weighing funnel (specifications for the tubes, funnel, etc., are given in B.S.696: Part 1: 1955). Then wash all the cream from the funnel into the butyrometer with 6 ml. of hot water (90° C.), add 10 ml. Gerber sulphuric acid and 1 ml. of amyl alcohol. Add more hot water to bring the level of the contents to 5 mm. below the shoulder of the tube. Stopper, mix, place in a water-bath at 65° C. for three to ten minutes, centrifuge at 1100 r.p.m. for five minutes and return to the 65° C. bath for three minutes. Then read off the percentage of fat directly on the scale. The method is given in detail in Part 2 of B.S.696: 1955, which also describes the use of ordinary milk tubes for the determination. The quantities to be employed using ordinary milk butyrometers are 0·9–1·1 g. cream, 6 ml. hot water, 10 ml. Gerber acid, 1 ml. amyl alcohol and sufficient water to bring the contents up to the correct level. The procedure is the same as for the special tubes. The fat content can be obtained by reference to the tables in B.S.696. Davis and Macdonald (*Richmond's Dairy Chemistry*, 1953, p. 350, Griffin, Lond.) employ 1·1 g. of cream and then obtain the fat content from the formula:—

$$\text{Fat (\%)} = (10 \times \text{Butyrometer reading}) - 1$$

For the investigation of the **type of fat** present carry out a Werner-Schmid type separation (with hydrochloric acid) on a quantity of the cream, extract the fat with ether and wash the ether extracts well with water. After removing the solvent the fat can be examined by the methods given under butter fat.

Other determinations on cream which may be necessary are summarised below:—

TOTAL SOLIDS

Dry 2 g. in sand for three hours at 100° C. with the assistance of a glass rod and sand.

PROTEIN

Determine by the macro-Kjeldahl method (p. 22).

LACTOSE

Determine by the Lane and Eynon process or polarimetrically.

ASH

Carefully ignite 5 g. at 550° C.

ACIDITY

Carry out as for milk, except that thick creams cannot be pipetted and must be weighed out instead.

PRESERVATIVES

Are not normally found in cream nowadays. Of those which were formerly used, borates and formalin were the most common. These can be detected in the same way as for milk.

THICKENERS

Starch may be detected by the addition of iodine.

Gelatin may be detected by this test due to Stokes (*Analyst*, 1897, **22**, 320): Mix together 10 ml. of cream, 20 ml. of water and 20 ml. of Stokes reagent (dissolve mercury in twice its weight of concentrated nitric acid and dilute to 25 times the volume with water). Shake, allow to stand for five minutes and filter. To the filtrate add an equal volume of saturated picric acid solution. If gelatin is present, a yellow precipitate is produced (*cf.* J. Ass. Off. Agric. Chem., 1934, **17**, 314).

Sucrose can be detected by Cayaux's test:—

Mix together 15 ml. of cream, 0·1 g. of resorcinol and 1 ml. of concentrated hydrochloric acid and heat the mixture on the water-bath. The presence of sucrose is indicated if a red colour is produced. A positive reaction for sucrose may be due to calcium saccharate.

Calcium saccharate (*viscogen*) can be detected by determining the calcium content. The percentage of lime present in the ash of milk (and cream) does not normally exceed 27 per cent. of CaO (*cf. Analyst*, 1908, **33**, 401; 1930, **55**, 747).

Glycerin—see *Analyst*, 1928, **53**, 335.

COLOURING MATTERS

Have occasionally been added to cream. Annatto can be detected by the method described under milk.

Reconstituted Cream and Imitation Cream

Section 47 of The Food and Drugs Act, 1955, places restrictions on the use of the word " cream " and defines " *reconstituted cream* " as a substance which resembles cream in appearance and contains no ingredient not derived from milk except water. This definition is virtually the same as that used previously for " artificial cream," which is considered to be a rather misleading title (Food Stds. Comm., " Report on Artificial Cream," May 1951). Section 47 also defines " *imitation cream* " as a substance which resembles cream in appearance and is produced by emulsifying edible oils or fats with water. The F.S.C. " Report on Synthetic Cream " (May 1951) *recommended* that imitation cream should contain at least 25 per cent. of fat and/or oil in the form of triglycerides.

No really satisfactory method exists whereby a well-emulsified reconstituted cream can be definitely differentiated from genuine cream. Determination of the lactose, protein, ash and fat may disclose an abnormality in the relative amounts present if the reconstitution has been made in incorrect proportions. The freezing point of the serum obtained by centrifuging is also of value. The method proposed by Richardson (*Analyst*, 1933, **58**, 686), based on miscibility, is useful provided the cream is fresh and has not been submitted to too thorough emulsification:—

Shake about 5 ml. of cream with 5 ml. of acetone in a centrifuge tube and centrifuge. Natural cream does not separate but reconstituted cream usually separates into layers. The test should be performed alongside a fresh genuine sample of cream.

The type of fat present may be recognised by the Reichert value, iodine value, etc. of the extracted fat (obtained by heating with hydrochloric acid, extracting with ether, etc.). The most common emulsifying agents used are methyl cellulose, sodium alginate and glyceryl monostearate.

CONDENSED MILK

Manufacture, Properties and Composition of Evaporated and Condensed Milk

It is now clearly recognised industrially that the milk used for the manufacture of condensed milk should be bacteriologically good and have a normal chemical composition and a low acidity. The following processes are involved in the production of *evaporated* (*unsweetened*) *milk:* pre-warming, vacuum evaporation, homogenising, stabilising (heat stability is often increased by addition of sodium phosphate or citrate), filling, sealing, sterilising (at 240° F. for fifteen to twenty minutes) and cooling. For the manufacture of *sweetened condensed whole milk*, the milk used is standardised by the addition of cream so that the ratio of fat to N.F.S. is 1 : 2·44. The subsequent operations in the process are pre-heating (at about 170–180° F.), addition of sugar (to give a final concentration of 42–45 per cent.), vacuum evaporation (at 130–140° F.), rapid cooling to about 86° F., seeding with lactose (to produce rapid crystallisation of small crystals of lactose so that a smooth texture is obtained), filling and sealing. It will be noted that sweetened condensed milk (unlike unsweetened evaporated milk) is not heat sterilised after filling into the can but keeps well, even after opening the can, because of the comparatively high sugar concentration. Blown cans due to contamination with yeasts have been met with on occasion. For further information on this subject the reader should refer to Davis's "Dictionary of Dairying." If whole milk containing 3·6 per cent. of fat and 8·8 per cent. of N.F.S. is used as the raw material, it must be reduced during manufacture to 1/2·5 of its original bulk to produce a product containing the minimum percentages of 9 and 22 respectively. The following table shows the results of the analysis of various samples of condensed milk.

	Sweetened milk.				Unsweetened milk.		
	Full cream.		Skimmed.		Full cream.		Skimmed.
	1.	2.	3.	4.	5.	6.	7.
	Per cent.	Per cent.	Per cent.	Per cent.	Per cent.	Per cent.	Per cent.
Water	25·05	25·14	26·62	27·03	66·18	67·47	76·62
Milk solids	33·39	32·94	27·21	26·38	33·82	32·53	23·38
Fat	10·64	9·60	0·22	0·29	9·28	9·10	0·75
Lactose[1]	12·25	13·07	14·93	14·25	13·33	12·74	12·50
Protein	8·57	8·41	9·61	9·53	9·16	8·75	8·35
Sucrose	41·56	41·92	46·17	46·59	—	—	—
Ash	1·93	1·86	2·45	2·31	2·05	1·94	1·78
Acidity as lactic acid, %	0·32	0·28	0·35	0·30	0·35	0·40	0·41
Specific gravity	1·305	—	—	—	1·089	1·085	-

[1] Hydrated lactose.

The legal quality standards and labelling requirements for condensed milk are now stated in The Condensed Milk Regulations, 1959 (S.I. 1959 No. 1098). These regulations amend and consolidate The Public Health (Condensed Milk) Regulations, 1923–1953:—

The regulations state that " *condensed milk* " means milk, partly skimmed milk or skimmed milk which has been concentrated by the removal of part of its water, whether with or without the addition of sugar, and includes the article commonly known as evaporated milk but does not include the article commonly known as dried milk or milk powder.

Legal minima for condensed milk are stated in the Second Schedule to the 1959 regulations:—

Description of Condensed Milk	Percentage of milk fat	Percentage of milk solids including fat
1. Full cream, unsweetened	9·0	31·0
2. Full cream, sweetened	9·0	31·0
3. Half cream, unsweetened	4·5	26·5
4. Half cream, sweetened	4·5	26·5
5. Skimmed, unsweetened	—	20·0
6. Skimmed, sweetened	—	26·0

The Condensed Milk Order, 1951 (now revoked) defined " Special full cream condensed milk " as condensed milk which contains not less than 10 per cent. of butter fat. This applied to both sweetened and unsweetened products.

Labelling. The First Schedule of the 1959 regulations requires cans of condensed milk to be clearly labelled in one of the following ways:—

(1) *Unsweetened full cream milk:*—

> CONDENSED FULL CREAM MILK, UNSWEETENED.
>
> THIS TIN CONTAINS THE EQUIVALENT OF
> (X) PINTS OF MILK

(2) *Sweetened full cream milk:*—

> CONDENSED FULL CREAM MILK, SWEETENED.
>
> THIS TIN CONTAINS THE EQUIVALENT OF
> (X) PINTS OF MILK, WITH SUGAR ADDED.

(3) *Unsweetened half cream milk:*—

> CONDENSED PARTLY SKIMMED MILK, UNSWEETENED
> (HALF CREAM).
>
> SHOULD NOT BE USED FOR BABIES EXCEPT
> UNDER MEDICAL ADVICE.
>
> THIS TIN CONTAINS THE EQUIVALENT OF
> (X) PINTS OF HALF CREAM MILK.

(4) *Sweetened half cream milk:*—

> CONDENSED PARTLY SKIMMED MILK, SWEETENED
> (HALF CREAM).
>
> SHOULD NOT BE USED FOR BABIES EXCEPT
> UNDER MEDICAL ADVICE.
>
> THIS TIN CONTAINS THE EQUIVALENT OF
> (X) PINTS OF HALF CREAM MILK
> WITH SUGAR ADDED.

(5) *Unsweetened skimmed milk* :—

CONDENSED, MACHINE-SKIMMED MILK (OR CONDENSED SKIMMED MILK), UNSWEETENED.

UNFIT FOR BABIES
(OR
NOT TO BE USED FOR BABIES)

THIS TIN CONTAINS THE EQUIVALENT OF (X) PINTS OF SKIMMED MILK.

(6) *Sweetened skimmed milk* :—

CONDENSED MACHINE-SKIMMED MILK (OR CONDENSED SKIMMED MILK), SWEETENED.

UNFIT FOR BABIES
(OR
NOT TO BE USED FOR BABIES)

THIS TIN CONTAINS THE EQUIVALENT OF (X) PINTS OF SKIMMED MILK, WITH SUGAR ADDED.

The insertion (X) is the appropriate number of pints with fractions expressed as eighths, quarters or half.

For the purposes of the First Schedule the terms " full cream milk " and " half cream milk " mean milks containing at least 12·4 and 10·8 per cent. (respectively) of total milk solids (including not less than 3·6 and 1·8 per cent. milk fat respectively). Also " skimmed milk " means milk which contains not less than 9 per cent. of milk solids other than milk fat (all percentages w/w).

Analysis of Condensed Milk

The analysis of canned condensed milk should include the determination of the **net weight, total solids, fat, protein** and **sugars** and the calculation of the **equivalent pints.** Other determinations which may be required are those of **ash, acidity, citric acid** and **tin.** Methods for the analysis of condensed milk are reviewed in the reports of the S.P.A.'s Milk Products Sub-Committee to the A.M.C., *Analyst,* 1927, **52,** 402; 1930, **55,** 111; 1932, **57,** 630 (see also B.S.1742: 1951).

Sampling

After determining the gross weight of the canned condensed milk, pour the contents into a glass sample jar and mix thoroughly with a spoon so that any crystals present are incorporated and also to ensure that no air bubbles are formed. In order to calculate the net weight, clean out the can, dry and weigh.

Total Solids (S.P.A.)

(1) *Sweetened Condensed Milk.* Place about 25 g. of prepared sand (passes 30 but retained by 90 mesh sieve, acid washed and ignited) and a glass stirring rod in a metal dish (with lid) and dry to constant weight in an oven at 90°–100° C., the lid being removed whilst drying and replaced before removing the dish from the oven. Allow the dish to remain for forty-five minutes in the desiccator before weighing.

Tilt the sand to one side of the dish; place on the clear space about 1·5 g. of the well-mixed sample and weigh rapidly. Add 5 ml. of water to the milk and mix these; then mix the diluted milk thoroughly with the sand by means of the rod.

Place the dish on a rapidly boiling water-bath for twenty minutes, carefully stirring during the earlier period. Transfer the dish, with rod and cover, to a well-ventilated oven at 90°–100° C., as recorded by a thermometer in the air immediately above the dish. After one and a half hours, cover the dish and place in the desiccator for forty-five minutes; weigh; return the dish to the oven, and heat for one hour with lid removed; remove and weigh as before; repeat this process until the loss of weight between successive weighings does not exceed 0·0005 g.

(2) *Unsweetened Condensed Milk.* Weigh out 3 g. of condensed milk and use 3 ml. of water; otherwise proceed as in (1).

Fat (S.P.A.). Transfer to a suitable apparatus from 2 to 2·5 g., accurately of the well-mixed sample; add 8 ml. of warm water and mix well; cool; add 1 ml. of concentrated ammonia solution, mix, add 10 ml. of alcohol and again mix. Add 25 ml. of ether and shake vigorously for one minute; add 25 ml. of petroleum spirit and again shake vigorously for thirty seconds. Allow the liquids to stand for not less than half an hour, until the ethereal layer is perfectly clear, or centrifuge at a low speed. Transfer the ethereal layer to a suitable flask. To the milk residue add 5 ml. of ether, and transfer without further shaking; repeat this operation in the same manner with a further 5 ml. of ether. Add 0·5 ml. of alcohol, and repeat the extraction with 25 ml. of ether and 25 ml. of petroleum spirit, as before, shaking vigorously for one minute after the addition of the ether and for thirty seconds after the addition of the petroleum spirit. As before, allow the ethereal layer to separate completely and transfer to the flask. Repeat the extraction once more with alcohol, ether and petroleum spirit in the same manner. Cautiously distill the solvents from the flask and dry the residual fat at 98°–100° C. to constant weight, taking the ordinary precautions to remove all traces of volatile solvent. Completely extract the fat from the flask by repeated washings with petroleum spirit, allowing any sediment to settle before each decantation. Finally dry the flask at 98°–100° C. The difference in weights before and after the petroleum spirit extractions is the weight of fat contained in the quantity of condensed milk taken. Make a blank determination, using the specified quantities of reagents and distilled water in place of the milk, and deduct the figure found, if any, from the weight of fat obtained.

Fat may also be estimated by the rapid Gerber process using ordinary milk butyrometers:

Use 11 ml. of a 20 per cent. solution (w/v) of the sample and carry out the method as described for milk (p. 329):—

$$\text{Fat }(\%) = \frac{\textit{Milk butyrometer reading} \times 100}{20}$$

The tube should be heated to 65° C. before centrifuging. At least two spins are necessary.

Added Sucrose (S.P.A.)

SPECIAL REAGENTS

Zinc acetate. 21·9 g. crystallised zinc acetate, $Zn(C_2H_3O_2)_2,2H_2O$ and 3 ml. glacial acetic acid in water and make up to 100 ml.

Potassium ferrocyanide solution. A 10·6 per cent. aqueous solution.

Dilute ammonia. 10 ml. of 0·88 ammonia diluted to 100 ml.

Dilute acetic acid. Equivalent to the dilute ammonia solution.

METHOD

Transfer to a 100 ml. beaker an accurately weighed quantity, approximately 40 g. of the well-mixed sample; add 50 ml. of hot distilled water (80°–90° C.), mix, transfer to a 200 ml. measuring flask, washing in with successive quantities of distilled water at 60° C., until the total volume is from 120 to 150 ml. Mix, cool to air temperature, and then add 5 ml. of the dilute ammonia solution. Again mix, and allow to stand for fifteen minutes. Add a sufficient quantity of the dilute acetic acid solution to neutralise the ammonia added (the exact equivalent is determined beforehand by titration), and again mix. Add, with gentle mixing, 12·5 ml. of zinc acetate solution and mix, followed in the same manner by 12·5 ml. of potassium ferrocyanide solution. Bring the contents of the flask to 20° C. and add distilled water at 20° C. up to the 200 ml. mark.

Up to this stage all additions of water or reagents should be made in such a manner as to avoid formation of air bubbles.

Close the flask with a dry stopper and mix thoroughly by shaking. Allow to stand for a few minutes and then filter through a dry filter paper, rejecting the first 25 ml. of filtrate.

Determine the direct rotation of the filtrate at 20° C.

Pipette 40 ml. of the filtrate obtained as above into a 50 ml. flask; add 6 ml. of 6·34 normal hydrochloric acid. Immerse for twelve minutes the entire bulb of the flask in a water-bath maintained at 60° C., mixing by rotatory movement during the first three minutes, in which time the contents of the flask should have attained the temperature of the bath. Cool, dilute to 50 ml. at 20° C. with distilled water, mix and allow to stand for one hour.

Determine the rotation at 20° C.

Calculation.—W = Weight of sample taken, in grms.
F = Percentage of fat in the sample.
P = Percentage of protein (N × 6·38) in the sample.
V = Volume to which the sample is diluted before filtration.
v = correction in ml. for volume of precipitate produced during clarification.
D = observed direct polarimeter reading.
I = observed invert polarimeter reading.
l = length in dm. of polarimeter tube.
Q = Inversion divisor factor (see pp. 103–4).

Then
$$v = \frac{W}{100}[(F \times 1{\cdot}08) + (P \times 1{\cdot}55)]$$

and the percentage of sucrose in the sample

$$= \frac{D - (\frac{5}{4} \times I)}{Q} \times \frac{V - v}{V} \times \frac{V}{l \times W}$$

(*cf.* Wood, *Analyst*, 1954, **79**, 779; Murphy, *ibid.*, 1960, **85**, 720).

Total Milk Solids (in Sweetened Condensed Milk) = Total Solids − Sucrose.

For methods which are applicable to condensed milks in which hydrolysis or other changes in the sugars has taken place during storage S.P.A. Report

No. 3 (*Analyst*, 1932, **57**, 630) should be consulted (see also B.S.1942: 1951, pp. 14, 18).

Lactose. Weigh 10–12 g. of sample into a 250 ml. graduated flask, dilute with 200 ml. hot water and allow to stand for at least thirty minutes. Cool, add 4 ml. of zinc acetate solution, mix and add 4 ml. of potassium ferrocyanide solution (reagents of same concentration as for determination of sucrose above). Dilute to the mark, filter and titrate the filtrate by Lane and Eynon's method (p. 107) using 25 ml. of mixed Fehling's solution. The result may be calculated as hydrated lactose. The correction for the volume of precipitate which should be deducted from the total 250 ml. is similar to that given for condensed milk above (v). Vieth's ratio (see p. 334) applies to condensed milk as for liquid milk.

Ash. Determine by igniting 5 g. of sample.

Proteins. Determine on about 2 g. of sample by the macro-Kjeldahl method and calculate the result as N × 6·38.

Acidity. The difference between 100 and the sum of the constituents is usually about 0·5–0·6 per cent. and this discrepancy is largely due to citric and lactic acids. *Citric acid* can be determined by diluting 30 g. of condensed milk and continuing by means of the process described on p. 346.

The *Total Acidity* obtained by direct titration and using phenolphthalein as indicator is usually between 0·25 and 0·45 per cent. (as lactic acid).

Tin. If the container shows internal corrosion, the condensed milk should be examined for tin (p. 71). Unsweetened condensed milk is rather more liable to attack the tinplate than sweetened products.

Thickeners. The presence of thickeners in condensed milk would now appear to be a relic of the past. If suspected, samples should be examined for gelatin and starch by the methods described under " Cream " (p. 367).

Calculation of the Equivalent Pints

The Condensed Milk Regulations require that the equivalent in pints shall be declared on the label. In order to calculate the equivalent pints (E.P.) it is necessary to take into account the specific gravity of the standard milks as defined in the regulations.

For full cream condensed milk the standard whole milk containing 3·6 per cent. of fat (F) and 12·4 per cent. of total milk solids (T.M.S.) has a specific gravity at 60° F. of 1·032 (such specific gravities can be calculated from Richmond's formula).

1 pint of the standard milk = 20 fl. oz. = 20·64 oz. (weight).

Total milk solids in one pint of the milk = 12·4 per cent. of 20·64 oz. = 2·559 oz. = 2·559 × 28·35 = 72·55 g.

In a full cream condensed milk the total milk solids in the can = $\frac{\text{T.M.S.} \times \text{W}}{100}$, where W = the weight of the contents of the can in grams.

$$\text{Equivalent pints (E.P.)} = \frac{\text{T.M.S.} \times \text{W}/100}{72{\cdot}55} = \frac{\text{T.M.S.} \times \text{W}}{7255}$$

The equivalent pints can be similarly calculated for the fat and solids not fat present in the condensed milk (see table on p. 375). The lowest figure obtained from the three calculations for the equivalent pints should be taken and compared with the statement on the label (*cf. Analyst*, 1923, **48**, 596).

DATA FOR CALCULATING THE EQUIVALENT PINTS OF CONDENSED MILK

Type of condensed milk	Data for standard milk				Formulae for calculating the equivalent pints of condensed milk		
	Fat (minimum) per cent.	Total milk solids (minimum) per cent.	Solids not fat (minimum) per cent.	*Specific gravity at 60° F.	Based on T.M.S. content E.P. =	Based on fat (F) content E.P. =	Based on S.N.F. content E.P. =
Full cream . .	3·6	12·4	—	1·032	$\frac{\text{T.M.S.} \times \text{W}}{7255}$	$\frac{\text{F} \times \text{W}}{2106}$	$\frac{\text{S.N.F.} \times \text{W}}{5149}$
Half cream . .	1·8	10·8	—	1·034	$\frac{\text{T.M.S.} \times \text{W}}{6335}$	$\frac{\text{F} \times \text{W}}{1056}$	$\frac{\text{S.N.F.} \times \text{W}}{5279}$
Skimmed. . .	—	—	9·0	1·0355	—	—	$\frac{\text{S.N.F.} \times \text{W}}{5284}$

W = weight of contents of can.

* Not stated in the regulations.

DRIED MILK

Manufacture

Dried milk is manufactured by a number of different processes. Most of the processes, however, can be considered to come under the heading of roller drying or spray drying. **Roller-dried milk** is produced by running the milk on to revolving metal rollers which are internally heated to about 300° F. by means of superheated steam. The milk dries almost instantaneously on the roller and the powder is automatically removed by metal scraper blades set at an angle to the surface.

Spray-dried milk is produced by atomising a spray of hot pre-condensed milk (of about 40 per cent. T.S.) inside a large chamber where it meets hot air entering at about 330° F. The milk dries instantaneously and the powder falls into the bottom of the chamber. For further information on this subject the reader should refer to " Davis' Dictionary of Dairying."

Particularly in view of its higher solubility, spray-dried milk is usually considered to be superior to the roller-dried product. After manufacture the moisture content of spray-dried milk is often less than 1 per cent. and that of roller-dried milk up to about 3 per cent. Due to its hygroscopic nature, however, powdered milk as retailed may have a moisture content of over 6 per cent. It is possible to differentiate between the two types of dried milk by microscopical examination. When mounted in water, spray powder appears as regular approximately spherical particles varying in size (about 10–130 mμ). The particles contain pockets of entrapped air. Roller powder, however, appears as irregularly shaped particles which vary in size.

The legal quality standards and labelling requirements for dried milk are stated in The Public Health (Dried Milk) Regulations, 1923–1948. The following minimum percentages of milk fat are stated in the regulations:—

Dried milk described as:	Percentage of milk-fat
Dried full cream milk	26
Dried three quarter cream milk . . .	20
Dried half cream milk	14
Dried quarter cream milk	8

The regulations do *not*, however, apply to dried milk in large containers (gross weight exceeding 10 lbs.) nor to articles containing less than 70 per cent. of dried milk.

Labelling. The Schedule of the regulations requires containers containing dried milk to be clearly labelled in one of the following ways:—

(1) *Full cream Dried Milk* (*minimum* 26 *per cent. milk-fat*)

DRIED FULL CREAM MILK THIS TIN CONTAINS THE EQUIVALENT OF (a) PINTS OF MILK

(2) *Partly Skimmed Dried Milk* (*not less than* 8 *per cent. but less than* 26 *per cent. milk-fat*)

DRIED PARTLY SKIMMED MILK (b) CREAM SHOULD NOT BE USED FOR BABIES EXCEPT UNDER MEDICAL ADVICE THIS TIN CONTAINS THE EQUIVALENT OF (a) PINTS OF (b) CREAM MILK

(3) *Skimmed Dried Milk* (*less than* 8 *per cent. milk-fat*)

DRIED MACHINE-SKIMMED MILK (OR DRIED SKIMMED MILK) UNFIT FOR BABIES THIS TIN CONTAINS THE EQUIVALENT OF (a) PINTS OF SKIMMED MILK

The declaration shall be completed as follows:—

(i) There shall be inserted at (a) the appropriate number in words and figures, *e.g.*, " one-and-a-half ($1\frac{1}{2}$)," any fraction being expressed as eighths, quarters or a half.

(ii) There shall be inserted at (b) the word " Three-quarter " if the percentage of milk fat is not less than 20; " Half " if such percentage is less than 20 but not less than 14; and " Quarter " if such percentage is less than 14 but not less than 8.

Further the heading shall be modified if other substances are added: " Sweetened "—if the only added substance is sugar; " Modified " if the only added substance is a milk constituent; " Compounded "—in any other case. In each case the words " with (c) added " shall be added to the last sentence where (c) specifies the addition.

The *Equivalent Pints* can be calculated from the formulae given in the table on p. 366, which also gives the data for the standard milks defined in the regulations (*Analyst*, 1924, **49**, 471).

DATA FOR CALCULATING THE EQUIVALENT PINTS OF DRIED MILK

Type of dried milk	Data for standard milk				Formulae for calculating the equivalent pints of dried milk (W = weight of contents in grams)		
	Fat (minimum) per cent.	Total milk solids (minimum) per cent.	Solids not fat (minimum) per cent.	*Specific gravity at 60° F.	Based on T.M.S. content E.P. =	Based on fat (F) content E.P. =	Based on S.N.F. content E.P. =
Full Cream . .	3·6	12·4	—	1·032	$\frac{\text{T.M.S.} \times \text{W}}{7258}$	$\frac{\text{F} \times \text{W}}{2107}$	—
Three quarter cream	2·7	11·6	—	1·033	$\frac{\text{T.M.S.} \times \text{W}}{6792}$	$\frac{\text{F} \times \text{W}}{1582}$	—
Half cream . .	1·8	10·8	—	1·034	$\frac{\text{T.M.S.} \times \text{W}}{6335}$	$\frac{\text{F} \times \text{W}}{1056}$	—
Quarter cream . .	0·9	9·9	—	1·034	$\frac{\text{T.M.S.} \times \text{W}}{5811}$	$\frac{\text{F} \times \text{W}}{528}$	—
Skimmed . .	—	—	9·0	1·0355	—	—	$\frac{\text{S.N.F.} \times \text{W}}{5285}$

* Not stated in the regulations.

Analysis of Dried Milk

The analysis of dried milk is similar to that of fresh milk, but in view of the higher concentration small errors tend to become of greater importance. For example, the citric and lactic acids which would be ignored in the ordinary methods of analysis of fresh milk may amount to over 1 per cent. Vieth's ratio, however, applies to dried or fresh milk.* The main determinations to be carried out on dried milk are detailed in the Report of the Milk Products Sub-Committee to the S.P.A. (*Analyst*, 1936, **61**, 105) and in B.S.1743: 1951. In view of the hygroscopic nature of dried milk the sample should be stored in a tightly stoppered bottle.

Moisture is determined by drying 1 g. at 102°–103° C. for two to three hours. The metal dish should have a close fitting lid.

Fat can best be determined by the Werner-Schmid process. The following procedure is suitable for both sweetened and unsweetened dried milk:—

Weigh out 1 g. of sample into the extraction tube and add 8 ml. of water and 2 drops of 0·88 ammonia. Boil gently to disintegrate any lumps. Then add 10 ml. of concentrated hydrochloric acid and boil gently for three minutes. Cool, add 10 ml. of alcohol, mix and extract the fat with mixed ethers as for fresh milk.

A more rapid result can be obtained by employing the **Gerber process** (B.S.696: Part 2: 1955):—

To an ordinary milk butyrometer add 10 ml. of Gerber sulphuric acid and cold water so that a layer is formed on top of the acid about 6 mm. deep. Add 1·69 ± 0·01 g. of dried milk by means of the dry stemless funnel, followed by 1 ml. of amyl alcohol. Then add sufficient hot water (about 80° C.) to bring the level of the liquid up to the shoulder below the neck, allowing all air entrained in the neck to escape. Stopper the tube, mix, place in a water-bath at 65° C. for three to ten minutes and centrifuge at 1100 r.p.m. for five minutes. It is advisable to replace the tube in the 65° C. bath and centrifuge again.

$$\text{Percentage of fat} = 6{\cdot}66 \times \text{butyrometer reading.}$$

Direct extraction of the fat gives low results (*Analyst*, 1923, **48**, 529).

As dried milk deteriorates, fat-peroxides are formed. The peroxide value of the fat therefore rises during storage.

Nitrogen is determined on 1 g. of sample using the macro-Kjeldahl method (Protein = N × 6·38).

Lactose can be determined on the cleared solution by the volumetric, polarimetric or gravimetric method.

* *cf.* Halliday *et al.*, *Analyst*, 1960, **85**, 839.

Ash is determined by igniting 2 g. of sample at 550° C. If the *alkalinity of the ash* is also determined it should be expressed as Na_2CO_3.

Acidity is determined by the method described in B.S.1743: 1959:—

Weigh out 1 g. of sample into a porcelain basin, add 10 ml. of boiling water and stir thoroughly until a smooth liquid is obtained. Cool, add 1 ml. of of 0·5 per cent. phenolphthalein solution and titrate with N/9 (or N/10) sodium hydroxide solution. For the colour standard for the end-point, stir 10 ml. of boiling water into 1 g. of the sample in another basin, cool and add 1 ml. of 0·00024 per cent. rosaniline acetate solution (see p. 344). The acidity should not exceed about 1·4 per cent. (w/w) as lactic acid (1 ml. N/10 ≡ 0·009 g. lactic acid).

Solubility of dried milk is of some considerable commercial importance. As milk is an emulsion rather than a solution the figure obtained is not true solubility but represents how far the powder will yield on mixing a liquid similar to natural milk. The solubility figure depends on a number of factors including the method of drying, the drying temperature, the acidity and the method of carrying out the test. The tests used are somewhat empirical and the results obtained comparative. The results show that spray-dried powders are almost 100 per cent. soluble, whilst the solubility of roller-dried powders is usually 80–95 per cent. For the determination of solubility Lampitt and Hughes (*Analyst*, 1931, **56**, 781) recommend that 5 g. of the milk powder should be shaken for three minutes with 38 ml. of water, or 45 ml. in case of skim milk powder, at 20° C. The mixture is then centrifuged and the fat layer poured off. Five ml. of the aqueous layer are evaporated to dryness and weighed. The rest of the liquid is decanted and any fat carefully removed from the side of the tube, and the weight of residue plus the associated liquid is noted. Then the residue is dried and weighed. Calculate the weight of soluble solids associated with the water lost on drying the residue. Subtract this weight of the soluble solids from the weight of the dried residue to obtain the weight of insoluble residue on the weight taken.

Industrially, rapid methods of the volume sediment type are often used. Crossley has proposed a method of this type in which 10 g. of sample is shaken 80 times with 100 ml. of cold water. The 50 ml. of reconstituted milk is centrifuged for fifteen minutes at 1000 r.p.m. in a graduated tube and the upper 30 ml. is decanted off and water added to restore the volume to 50 ml. After centrifuging again for fifteen minutes the volume of sediment is read off and expressed as the **solubility index.** Tubes are available which can be used in the conventional Gerber centrifuge. Volume sediment methods yield good comparative results and are sensitive enough to show slight differences in the solubility of highly soluble spray-dried powders (*cf. J. Dairy Res.*, 1943, **13**, 180; 1945, **14**, 163; 1949, **16**, 377).

Methods for the determination of **bulk density** and **sediment** are described in B.S.1743: 1951.

COMPOSITION OF DRIED MILKS

The following data represent typical average figures for the two main types:—

	Full Cream Milk Powder per cent.	Skimmed Milk Powder per cent.
Moisture	3·5	3·9
Fat	26·5	1·3
Protein (N× 6·38)	28·8	38·6
Ash	5·7	7·6
Lactose (by difference)	35·5	48·6
Calcium (as CaO)	1·37	1·84

The moisture content varies according to the method of drying, the average figures for roller and spray-dried powders being 5 and 2·5 per cent. respectively.

Infant Foods

There are three distinct classes of infant foods on the market: pure milk preparations, those prepared from cows' milk with various additions or alterations and those prepared from farinaceous materials rendered soluble by the action of heat or enzymes, and which are to be used with milk. Doubtless all are, to some extent, modelled on human milk, but they usually differ widely therefrom in composition in respect of the ratio of fat and carbohydrate. The tables which follow show the composition of some British brands.

Of the samples shown, *a* is dried full-cream milk with an addition of cane sugar; *b* is mainly casein precipitated from skimmed milk; *c* is dried milk plus added lactose and cane sugar; *d* and *e* are milk casein with fat, cane sugar and maltose; *f*, *g* and *h* are baked flours.

1. MILK PRODUCTS

—	*a*	*b*
	Per cent.	Per cent.
Water	4·01	12·80
Fat	19·20	1·05
Casein	24·40	76·50
Lactose	35·50	Trace
Cane sugar	10·58	—
Ash	5·74	8·70

2. Milk Products with Additions

—	*c*	*d*	*e*
	Per cent.	Per cent.	Per cent.
Water	5·60	4·02	3·80
Fat	13·10	5·31	8·50
Casein	10·20	11·85	14·10
Lactose	32·90	Trace	Trace
Cane sugar	33·55	36·10	30·50
Ash	3·80	2·01	3·15
Maltose and dextrin	—	21·50	42·40
Starch	Absent	Present	Present

3. Cooked Cereal Products

—	*f*	*g*	*h*
	Per cent.	Per cent.	Per cent.
Water	6·10	9·03	4·35
Fat	1·32	0·62	1·20
Protein	12·50	11·75	10·54
Lactose	? 1·50	—	—
Other carbohydrates, mainly starch, maltose and dextrin	77·58	77·29	80·12
Ash	1·00	1·31	3·79

It is impracticable to describe methods of analysis which are applicable to all infant foods. The determination of water, fat, protein, ash, salts and fibre follow the well-known lines. A qualitative and microscopical examination will usually indicate what substances are present. If the constituents are those of dried milk plus sugar, lactose or sucrose, the methods applicable to condensed milks may be used, though it may be convenient to determine the sucrose volumetrically by inverting the solution containing the lactose and sucrose after precipitation with zinc and ferrocyanide. The inversion is conveniently done by heating 50 ml. with 5 ml. of hydrochloric acid at 70° C. for five minutes, then cool, neutralise and titrate the inverted sugar by the Lane and Eynon method. If starch is present it may be determined approximately by the method of Thorne and Jeffers. Five g. of the sample are triturated in a mortar with 15 ml. of water and 15 ml. of hydrochloric acid added in small quantities. The mass is allowed to stand for half an hour, then diluted; 10 ml. of 4 per cent. phosphotungstic acid solution and 20 ml. of hydrochloric acid are added and the volume made up to 200 ml. The mixture is allowed to stand for a further hour then filtered and the rotation (R) is observed in a 200 mm. tube at 20° C. The percentage of starch is $\frac{40\ \text{R.}}{11{\cdot}6}$

This figure is based on the average specific rotation of starch which is taken as +200°. If other sugars are present the rotation is corrected by adding 0·028 S and deducting 0·07 (D + L), where S, D and L are the percentages of sucrose, dextrose and lactose respectively.

When hydrolysed starch products such as maltose and maltodextrins are present the analysis becomes complicated. The maltose and dextrin can be calculated from the total soluble matter, cupric reducing power and optical activity in the manner indicated in Chapter 4. For the determination of lactose in such mixtures advantage may be taken of the fact established by Baker and Hulton that ordinary brewer's yeast ferments dextrose, maltose and sucrose, but not lactose. So if a neutral inverted solution of the material is incubated with washed brewers' yeast for about 72 hours at 25°–27° C. all the sugars except the lactose will have been decomposed, and the lactose may be titrated after clarifying the liquid in the usual way. Some other useful methods are quoted by Davis, *J. Soc. Chem. Ind.*, 1916, **35**, 201.

Casein

Reference has already been made to casein as the principal nitrogenous constituent of milk. Much casein is now prepared, both for food purposes and for various industrial processes.

The usual method is by curdling separated milk, either by acid or by rennet, then washing and drying the precipitated casein. The details of the process vary considerably, so that there are produced two distinct qualities of casein, one for food purposes and one for industrial use. The following analyses show the difference between the two grades:—

	Industrial		Edible	
	1	2	3	4
	Per cent.	Per cent.	Per cent.	Per cent.
Water	10·16	8·40	2·45	3·70
Ash	0·65	4·73	0·36	0·54
Fat	3·95	2·60	0·40	0·62
Nitrogen	13·25	12·90	14·99	14·81

An edible casein should not contain more than 10 per cent. of moisture, nor more than 1 per cent. of ash and be almost free from fat (not more than 1 per cent. on dry basis).

Lactic or sulphuric acid are commonly used for the precipitation, but should be completely removed by washing. To test for these, 10 g. of the sample are extracted with 100 ml. of water, then filtered; the filtrate should not be acid in reaction, and evaporation should leave no residue, indicating the absence of lactose. For the determination of fat the Werner-Schmid process should be used (direct extraction with ether gives quite erroneous results). When estimating the ash, it is expedient to extract the charred residue with water, and ignite the carbonaceous matter separately, then return the extract to the basin, evaporate and ignite.

Casein has the property of being soluble in alkalies or alkaline salts, such as sodium phosphate, borate or carbonate. This property is utilised in the manufacture of certain well-known foodstuffs, such as Plasmon, Sanatogen and Nutrose, which contain casein with alkaline salts and phosphates.

Malted Milk

Malted milk is made by combining whole milk with a mash of ground barley malt and meal with or without the addition of salt and bicarbonate of soda. Sometimes it is prepared from liquid milk and malt extract, sometimes from the mixture of dried milk and malt powders. Additions of skim milk, lactose, sucrose, wheat or casein are sometimes present. Malted milk should not contain more than 3·5 per cent. moisture. A Code of Practice was published in 1949 which stated " It has been agreed, as a temporary measure, that a product labelled ' malted milk ' should be required to contain a minimum butter-fat content of 7 per cent., and that no exception would be taken to the presence of additional milk solids derived from skim milk." The following table shows the analyses of various samples of malted milk:—

	1	2	3	4	5	6
	Per cent.	Per cent.	Per cent.	Per cent.	Per cent.	Per cent.
Moisture . . .	3·48	3·25	3·41	3·64	3·29	5·83
Fat . . .	9·31	7·91	7·42	7·50	8·43	8·17
Ash . . .	5·28	3·49	3·43	4·39	4·04	3·30
Proteins . . .	20·07	14·63	10·23	19·95	15·06	15·06
Lactose . . .	12·72	10·95	11·89	16·93	24·71	12·29
Other reducing sugars as maltose . .	37·88	45·45	45·00	46·82	34·75	37·16
Sucrose . . .	9·52	—	—	—	—	—
Non-reducing sugars .	—	14·3	18·60	0·80	9·7	19·8

Of these No. 1 contains added sucrose and skim milk powder, No. 2 contains wheat products, Nos. 4 and 5 contain added lactose. The methods of analysis follow the same lines as dried milk or infant foods and a microscopical examination should not be overlooked.

There are on the market several milk products containing added vitamins. Of these aneurine and ascorbic acid can be determined by appropriate modification of the methods given on pp. 169 and 201. Methods for the determination of vitamin A, which is fat-soluble, are given on p. 438. The estimation of vitamin D except by bio-assay is only possible under certain favourable circumstances. The spectrophotometric method of Nield, Russell and Zimmerlie (*J. Biol. Chem.*, 1940, **136**, 73) works reasonably satisfactorily if the pink colour produced by the acetyl chloride-antimony chloride reagent is not overborne by the persistent blue colour due to preponderance of vitamin A.

Chapter 13

DAIRY PRODUCTS (II)

Butter—Margarine—Cheese—Ice-Cream

BUTTER

" BUTTER " is defined in The Food Standards (Butter and Margarine) Regulations, 1955 (S.I. 1955 No. 1899) as " the substance usually known as butter made exclusively from milk with or without salt and with or without the addition of colouring matter." These regulations also prescribe a maximum water content of 16 per cent.

Butter is usually manufactured by churning cream which has been mechanically separated from the milk. The butter so produced is then " worked," *i.e.*, excess water is removed and the mass is rendered homogeneous. Salt is often added. Methods of butter manufacture are described in " Dictionary of Dairying," by Davis and in " The Butter Industry," by Hunziker.

Composition of Butter

Most butters contain over 80 per cent. of fat. The non-fatty constituents include water and the curd, which consists of protein (casein), lactose and mineral matter. The following table shows the composition of six samples of butter:—

	1	2	3	4	5	6
	Fresh	Fresh	Salt	Salt	Salt	Salt
	Per cent.	Per cent.	Per cent.	Per cent.	Per cent.	Per cent.
Water	12·15	14·26	15·03	11·74	14·15	14·00
Fat	86·82	84·94	79·97	82·22	83·28	81·79
Casein	0·73	0·53	0·47	0·73	0·61	0·76
Lactose	0·20	0·18	0·18	0·22	0·30	0·31
Ash*	0·10	0·09	2·35	5·09	1·06	3·14
(* Including salt)	0·03	0·05	2·30	5·03	0·61	2·90

Analysis of Butter

In " official " laboratories the routine analysis of butter should include the determination of **water** and **salt** and an examination for preservatives. Also the filtered fat should be examined by the **Reichert-Polenske process** (to ascertain that the fat consists of genuine butter-fat) and for rancidity values (*e.g.*, acid value or F.F.A., peroxide value). " Methods for the Chemical Analysis of Butter " are described in B.S. 769 : 1952.

Sampling. Warm the sample, contained in a screw-capped jar, in an oven at 32°–35° C. and shake vigorously from time to time in order to produce a homogeneous fluid emulsion.

Water. Weigh out 5 g. of the sample into a metal dish containing a rod with a flattened end and dry in an oven at 100° C. to constant weight. To obtain an approximate result, rapidly heat the sample over a low Argand burner with constant stirring. Stop the heating when the curd just begins to darken in colour.

Curd + salt. Melt the residue from the moisture determination by gentle heat, add light petroleum, stir and filter through a tared filter paper or a weighed Gooch crucible. Wash the dish and filter with more light petroleum until the filtrate is free from fat. Dry the paper or crucible in an oven at 100° C. and weigh the curd + salt.

Salt. (a) Transfer the curd + salt to a flask, add hot water, cool and titrate with N/20 silver nitrate solution using potassium chromate as indicator.

(b) *Rapid direct method.* Weigh out 3 g. of butter into a small porcelain dish, add 20 ml. of water and place on a steam bath to melt the fat. Stir, *cool*, add potassium chromate and titrate with N/10 silver nitrate (*cf.* Arup, *Analyst*, 1929, **54**, 658). 1 ml. N/10 silver nitrate ≡ 0·00585 g. NaCl.

Salt in butter acts as a preservative by retarding the growth of most bacteria and moulds. Evidence exists, however, that butters which contain over 2 per cent. of salt do not, in general, keep as well as ones which contain only a small amount of salt. Butters sold in most areas in recent years have usually contained between 0·8 and 2·5 per cent. of salt.

Curd is usually obtained by subtracting the percentage of salt from the percentage of curd + salt (see also estimation of protein below).

Fat is usually obtained by difference:—

$$\text{Fat (\%)} = 100 - (\%\ \text{water} + \%\ \text{salt} + \%\ \text{curd})$$

Alternatively the fat can be extracted with light petroleum and weighed after removing the solvent, or the Gerber method may be used with 2·5 g. in a cream tube.

Protein can be determined by estimating the total nitrogen in the curd + salt residue by the Kjeldahl method, using the factor N×6·38. The curd consists mainly of casein.

Lactose can be determined in the curd by means of Lane and Eynon's method.

Ash. Cream which has an excessive acidity is sometimes partially neutralised with sodium carbonate or bicarbonate before being used for butter making. Such treatment produces higher than average figures for the ash, ash less salt and the alkalinity of the ash.

Titratable acidity. Shake 18 g. of butter with 90 ml. of hot, previously boiled, water and titrate hot with N/50 sodium hydroxide using phenolphthalein as indicator.

$$\begin{array}{l}\text{Titratable acidity} \\ \text{(as \% lactic acid)}\end{array} = \frac{\text{mls. N/50 alkali required}}{100}$$

Boric acid can be detected by the turmeric test: Melt 5–10 g. of butter and pour away the upper fatty layer. Then pour the aqueous liquid remaining into a flat porcelain dish, add three drops of saturated oxalic acid solution, two drops of 5 N hydrochloric acid and alcoholic tincture of turmeric and evaporate on the water bath. In the presence of boric acid, a reddish residue is produced.

For the **quantitative estimation,** Richmond's process is adequate: Melt 25 g. of butter in a beaker in the water oven. Then add 25 ml. of hot water and stir well. Allow the aqueous portion to settle, mix again and allow to settle. Withdraw 20 ml. of the lower layer by pipette, add 1 ml. of phenolphthalein solution (0·5 per cent.), bring the liquid just to the boil and neutralise with N/10 sodium hydroxide. Add 10 ml. of glycerol and titrate the boric acid (see also p. 90). If V = the no. of mls. of N/10 alkali required in this final titration:—

$$\text{Boric acid (\%) in sample} = V \times 0{\cdot}0062 \times \left(\frac{100 + W}{20}\right)$$

where W = the percentage of water in the sample.

Other preservatives such as benzoates, sulphites, nitrates and fluorides have on occasion been reported as being present in butter.

Diacetyl is one of the substances produced during cream ripening. It is produced by oxidation of its precursor, acetyl-methyl-carbonol and has a pronounced buttery odour and taste. It has been shown that in general the greater the acidity of the cream used, the greater is the amount of diacetyl in the butter (up to about 5 p.p.m.) and the stronger is its flavour (see Davies, *Dairy Ind.*, 1936, **1**, 165; *Agric. Progress.*, 1938, **15**, 30). Some butters appear to contain added diacetyl. Diacetyl may be detected by steam distilling about 500 g. of butter just acidified with acetic acid. The condenser should dip into a mixture of 4 ml. each of 20 per cent. sodium acetate and 20 per cent. hydroxylamine hydrochloride solution and 2 ml. of 5 per cent. nickel chloride solution. The presence of diacetyl is shown by a red colour or precipitate of nickel glyoxime (see also Barnicoat *Analyst*, 1935, **60**, 653).

Trace metals in butter. It has been shown that the presence of 1–2 p.p.m. of copper in butter encourages fat rancidity with the production of a fishy or tallowy taint in a few days. Rancidity may also be encouraged by the presence of other metals such as chromium, iron, manganese and nickel, but the action is less marked than with copper. The rancidity formation appears to increase with increased acidity. Copper and iron can be determined by the methods described by McDowell (*J. Dairy Res.*, 1947, **15**, 70; B.S. 769 : 1952; see also p. 51).

Added colouring matters such as annatto, saffron and coal-tar dyes may be present in butter. Most added colours can be extracted with alcohol (see also *Analyst*, 1905, **30**, 244; 1908, **33**, 421).

Examination of Butter Fat

The following table shows the usual range of constants for genuine butter fat:—

	Normal maximum.	Normal minimum.	Average.	Abnormal.
Specific gravity at $\frac{37 \cdot 8^\circ \text{ C.}}{37 \cdot 8^\circ}$	0·913	0·910	0·9118	0·909–0·914
Refractive index at 40°	1·4561	1·4524	1·4548.	1·4517–1·4566
= Zeiss refractometer number	45·5	40·0	43·5	39–46·0
Saponification value	232	222	228·5	218–233
Iodine value	40·0	26·0	—	22–52
Reichert value	32·8	24·5	27–28	20·5–24·5
Polenske value	3·7	1·5	2·3–2·8	1·4– 4·0
Kirschner value	27	21	24	19·0–20·0
Baryta value	— 0·7	— 23·8	— 9·6	—
Mean molecular weight of fatty acids	267	258	260–261	—

The rather wide variation in the analytical figures obtained with genuine butter fat makes it somewhat difficult to assess accurately the proportion of another fat which may be present. Butter fat consists mainly of the glycerides of the higher fatty acids, including varying percentages of butyric, caproic, caprylic, capric, lauric, myristic, stearic, palmitic and oleic acids, together with small quantities of other acids. The most reliable methods for determining butter fat are based on the presence of a comparatively high proportion of glycerides containing steam volatile water soluble acids, particularly butyric acid, which is mainly responsible for the high Reichert and Kirschner values obtained.

Various methods for the analysis of butter fat are given below:—

Specific gravity. The determination of the specific gravity at 37·8° C. (≡ 100° F.) as compared with water at the same temperature is now seldom performed. Due to the presence of glycerides of low molecular weight, butter fat has a higher s.g. than most of the fats used for its adulteration. The difference between the s.g. of butter fat and other fats is almost at a maximum at 37·8 C.

Refractive index. Formerly the refractive index of butter fat at 40° C. gave a valuable indication of its purity. Its value is now considerably reduced, however, due to the use of fats for margarine which give somewhat similar figures to those obtained with butter fat. A table showing the conversion of refractive index readings to those on the Zeiss Butyro Scale is given by Williams (Bolton's *"Oils, Fats and Fatty Foods"*, 1950, p. 490–2, Churchill, Lond.).

Iodine and saponification values are of little help in detecting adulteration in butter fat as the presence of other fats would not alter the figures obtained to any great extent.

The Reichert-Polenske-Kirschner Processes

These processes, which are most commonly used for assessing the proportion of other fats in butter fat, are the outcome of the classic work of Hehner and Angell, who, in 1872, first devised a method which aimed at the estimation of butyric acid. Later work showed that better results could be obtained by determining the soluble and insoluble volatile acids (corresponding to what we now term the Reichert and Polenske values respectively). The historical development of these processes is surveyed in Richmond's *Dairy Chemistry*.

For consistent results it is essential to adhere strictly to a standard procedure. The method described below is the procedure adopted by the A.M.C. of the S.P.A. (*Analyst*, 1936, **61**, 404) and subsequently by the B.S.I. (B.S. 769: 1952, p. 8). A semi-micro modification of the process is given on p. 392.

Preparation of the Sample for Analysis of Butter Fat

Heat a portion of the sample of butter in a beaker at 50–60° C. until the fat separates from the water and the curd. Filter the warm fat layer through a dry paper and re-filter, if necessary, until it is clear and free from water.

Reagents

Concentrated sodium hydroxide solution (50 per cent. by weight).—Sodium hydroxide is dissolved in an equal weight of water and the solution is stored in a bottle protected from carbon dioxide. The clear portion free from deposit is used.

Glycerol.

Dilute sulphuric acid solution.—Approximately 25 ml. of concentrated sulphuric acid are diluted to 1 l., and adjusted until 40 ml. neutralise 2 ml. of the sodium hydroxide (50 per cent.) solution.

Pumice powder.—Pumice, ground, passing through a sieve B.S. No. 50,* and remaining on a sieve B.S. No. 90.*

Phenolphthalein solution.—0.5 g. of phenolphthalein dissolved in 100 ml. of industrial methylated spirit.

Alcohol.—Industrial methylated spirit neutralised to phenolphthalein immediately before use.

Sodium hydroxide solution.—Approximately N/10 solution of sodium hydroxide, of accurately determined strength.

Barium hydroxide solution.—Approximately N/10 solution of barium hydroxide, of accurately determined strength.

Silver sulphate, powdered.

The assembly of the apparatus for the distillation is shown below (Fig. 40). For full details of the method the original report should be consulted.

* See British Standard Specification for Test Sieves, No. 410, 1931.

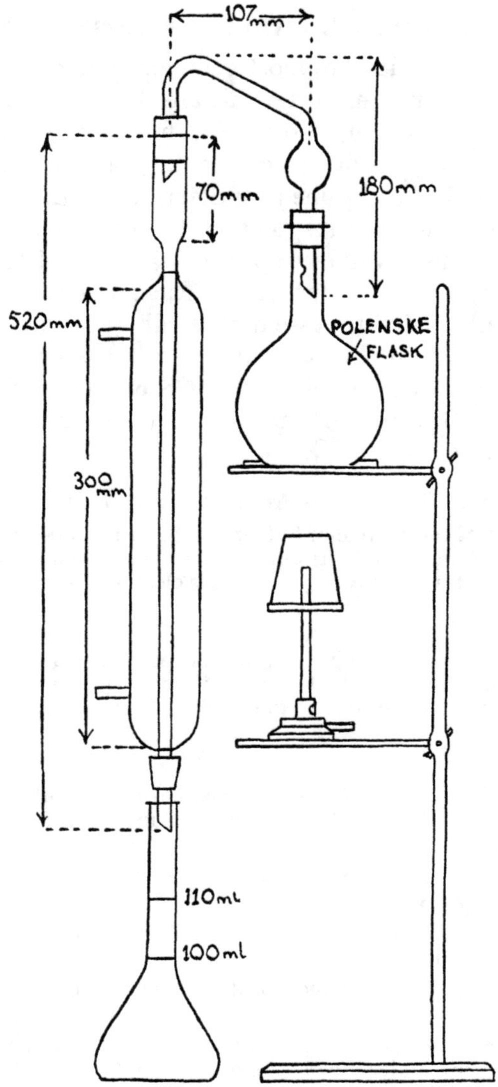

Fig. 40. Standard Reichert-Polenske Kirschner Apparatus.

PROCEDURE

Weigh 5 g. (tolerance not exceeding 0·01 g.) of the fat into a Polenske flask. Add 20 g. of glycerol and 2 ml. of concentrated sodium hydroxide solution. (The burette containing the soda solution must be protected from carbon dioxide, and, before withdrawal of the solution for tests, the nozzle must be wiped clean from carbonate, and the first few drops of solution rejected.) Heat over a naked flame, with continuous mixing, until the fat, including drops adhering to the upper parts of the flask, is saponified, and the liquid becomes perfectly clear. Cover the mouth of the flask with a watch-glass.

[Make a *blank* test without fat, but using the same quantities of reagents and following the same procedure, and avoiding over-heating during the heating with soda (indicated by darkening of the solution).]

Measure 93 ml. of boiling distilled water, which has been vigorously boiled for fifteen minutes, into a 100 ml. graduated measuring cylinder. When the soap is sufficiently cool to permit addition of the water without loss, but before the soap has solidified, add the water to the flask, draining the cylinder for five seconds, and dissolve the soap.

If the solution is not clear (indicating incomplete saponification), or is darker than light yellow (indicating over-heating), the saponification must be repeated on a fresh sample of the fat.

Add 0·1 g. of powdered pumice, followed by 50 ml. of the dilute sulphuric acid solution, and connect the flask at once with the distilling apparatus, shown in Fig. 40. Heat the flask, without boiling, until the insoluble acids are completely melted, then increase the flame and distil 110 ml. in nineteen to twenty-one minutes. The flow of water in the condenser must be sufficient to keep the temperature of the issuing distillate between 18° and 23° C. When the distillate reaches the 110-ml. mark, remove the flame and replace the 110-ml. flask by a cylinder of about 25 ml. capacity to catch drainings. Close the 110 ml. flask with a stopper, and, without mixing, place it in water at 15° C. for ten minutes so as to immerse the 110 ml. mark. Remove the flask from the water, dry the outside, and invert the flask carefully, avoiding wetting of the stopper with the insoluble acids. Mix the distillate by four or five double inversions without violent shaking. Filter through a dry 9 cm. No. 4 Whatman filter paper, which fits snugly into the funnel. Reject the first runnings and collect 100 ml. in a dry flask; cork the flask, and retain the filtrate for titration as at *R* below. The filtrate must be free from insoluble fatty acids.

Detach the still-head and wash the condenser with three successive 15 ml. portions of cold distilled water, passing each washing separately through the cylinder, the 110 ml. flask and the filter, nearly filling the paper each time, and draining each washing before filtering the next. Discard the washings.

Dissolve the insoluble acids by three similar washings of the condenser, the cylinder and the filter with 15 ml. of neutralised alcohol, collecting the solution in the 110 ml. flask and draining the alcohol after each washing. Cork the flask and retain the solution for titration as at *P* (p. 392).

REICHERT (OR SOLUBLE VOLATILE ACID) VALUE (*R*)

Pour 100 ml. of the filtrate containing the soluble volatile acids into a titration flask, add 0·1 ml. of solution of phenolphthalein, and titrate with N/10 barium hydroxide solution* until the liquid becomes pink, rinsing the 100 ml. flask with the nearly neutralised liquid towards the end of the titration.

[If the Kirschner value is to be obtained, the titration flask must be dry before use; note the actual volume of barium hydroxide solution used; drain the 100 ml. flask into the titration flask, close with a cork, and continue as at *K*, p. 392.]

* N/10 sodium hydroxide solution may be used for the titration if the Kirschner value is not required.

If the amounts of barium hydroxide solution used for the titration of the sample and the blank are equivalent to x ml. and x_b ml. N/10, respectively, the

$$\text{Reichert value} = (x - x_b) \times \frac{110}{100}$$

Polenske (or Insoluble Volatile Acid) Value (*P*)

Titrate the alcoholic solution of the insoluble volatile acids after addition of 0·25 ml. of phenolphthalein solution with N/10 barium (or sodium) hydroxide solution, until the solution becomes pink.

If the amounts of barium (or sodium) hydroxide used for the titration of the sample and the blank are equivalent to y and y_b ml. N/10, respectively, the Polenske value $= P = (y - y_b)$.

Kirschner Value (*K*)

Add 0·5 g. of finely powdered silver sulphate to the neutralised solution from (*R*), above. Allow the flask to stand in the dark for one hour, with occasional shaking, and filter the contents through a dry filter. Transfer 100 ml. of the filtrate to a dry Polenske flask, add 35 ml. of cold distilled water (recently boiled for fifteen minutes), 10 ml. of dilute sulphuric acid solution and a loosely-wound 5 mm. coil of 30 cm. of aluminium wire (about 1 mm. thick or S.W.G. about 18 to 20), or 0·1 g. of pumice powder. Connect the flask with the standard apparatus and repeat the process as described above, *i.e.*, the distillation of 110 ml. in nineteen to twenty-one minutes, the mixing (but without cooling for ten minutes), the filtration, and the titration of 100 ml. of the filtrate with N/10 barium hydroxide solution.

If the amounts of barium hydroxide solution used for the titration of the fat and the blank are equivalent to z ml. and z_b ml. of N/10, respectively,

$$\text{Kirschner value } (K) = (z - z_b) \times \frac{(100 + a) \times 121}{10{,}000}$$

where a represents the actual volume in ml. of barium hydroxide solution used in the titration for determination of the Reichert value (p. 391).

Note.—Polenske values, and, to a much slighter extent, Reichert values, have been found to be low when determined at low barometric pressures, such as may occur at high altitudes. The following factors may be applied to values determined at a barometric pressure of p mm. of mercury, to convert them to the values determined under normal pressure (Kirkham).

Corrected Reichert value

$$= \left(\frac{(\text{Observed value} - 10) \log 760}{\log p}\right) \times 10$$

Corrected Polenske value

$$= \text{Observed value} \times \left(\frac{760 - 45}{p - 45}\right)$$

Semi-micro Method for Determining Reichert, Polenske and Kirschner Values

Dyer *et al.* (*Analyst*, 1941, **66**, 355) have described a semi-micro modification of the standard Reichert process which is carried out on a 1 g. sample. Apart from the saving in materials the semi-micro method can be completed in a shorter time than the standard process. The method is particularly suitable where the amount of fat available for analysis is small, *e.g.*, the fat extracted from " butter " confectionery.

Apparatus

The shape of the apparatus used is similar to that used for the standard Reichert process, but the still-head and condenser are scaled down to about one-third the size. The distilling flask has a capacity of 60–100 ml. and the receiving flask is graduated at 22 ml. (*cf.* Dyer *et al.*, *loc. cit.*)

Special Reagents

Sodium hydroxide solution. Fifty per cent. by weight.

Dilute sulphuric acid. Dilute 25 ml. of concentrated sulphuric acid to one litre and adjust so that 40 ml. neutralises 2 ml. of the 50 per cent. sodium hydroxide solution.

The following procedure is modified slightly from that described by Dyer *et al.* (*loc. cit.*):—

Method

Weigh 1 g. of the sample into the distilling flask, add 0·4 ml. of the sodium hydroxide solution and 4 ml. glycerol and saponify the fat by gently heating over a small flame, with continuous mixing, until the liquid clears (as in the standard process). Partially cool and dissolve the soap by the careful addition of 20 ml. of hot water. Add 10 ml. of the dilute sulphuric acid, swirl round and confirm that the liquid is acid by the addition of a " speck " of litmus paper. Add pumice powder and connect the flask to the distillation apparatus. Distil so that 22 ml. of distillate are collected in 10–12 minutes. Remove the 22 ml. flask at the end of the distillation and place a small beaker under the condenser. Immerse the 22 ml. flask in cold water, filter through a small filter paper and titrate 20 ml. of filtrate with N/50 alkali (use baryta if the Kirschner value is also to be determined) using phenolphthalein.

$$\text{Reichert Value} = 1{\cdot}1\ (\text{Titration} - \text{Blank})$$

The blank is usually about 0·5 ml.

For the *Polenske value* wash the condenser twice with 3 ml. portions of cold water and pass each wash separately through the beaker, the 22 ml. flask and the filter paper which was used for filtering the Reichert portion. Then dissolve the washed water-insoluble acids by washing three times with 3 ml. portions of neutral alcohol, passing each wash through the condenser, the beaker and finally through the filter paper into a small flask. Titrate the combined alcoholic washes with N/50 baryta solution using phenolphthalein.

$$\text{Polenske value} = (\text{Titration} - \text{Blank})$$

For the *Kirschner value*, add 0·1 g. of silver sulphate to the titrated solution from the Reichert determination and place in the dark for one hour with occasional shaking. Filter, pipette 20 ml. of the filtrate into the distilling flask and add 7 ml. of water, 2 ml. of the dilute sulphuric acid and two small pieces of aluminium wire. Distil and collect 22 ml. of distillate. Stopper the 22 ml. flask, mix the contents by inversion, filter and titrate 20 ml. of distillate with N/50 baryta using phenolphthalein (titration = a ml.). If the blank titration is b ml. and r was the number of mls. of N/50 alkali added for the Reichert titration:—

$$\text{Kirschner value} = \frac{(a - b) \times 1{\cdot}21 \times (20 + r)}{20}$$

Baryta Value of Avé-Lallemant

The following modification of the Avé-Lallemant procedure is that proposed by Bolton and Revis (*Analyst*, 1907, **32**, 382):—

Saponify 5 g. of the fat with a measured volume of 0·5 N alcoholic sodium hydroxide solution by boiling under a reflux condenser for thirty minutes, then titrate back with 0·2 N hydrochloric acid and so estimate the saponification value; this value × 1·367 gives the barium oxide value (*a*) for 1 g. of fat. Remove all the alcohol by heating on a water bath and blowing in air, then dissolve the soap in a little hot water, dilute to 250 ml. at 38°, pipette out 100 ml. of this solution into a 250 ml. flask, add about 75 ml. of hot water and exactly 50 ml. of 2·5 per cent. barium chloride solution. Heat the mixture on a water bath for a few minutes to make the barium soap coalesce, then cool, dilute to 250 ml., filter off 200 ml., heat to boiling, add hydrochloric acid and 10 ml. of dilute sulphuric acid. Filter off, ignite and weigh the barium sulphate, add to this one-quarter of its weight and convert to BaO by multiplying by 0·6571. Treat 50 ml. of the barium chloride solution in a similar manner to determine its equivalent of BaO, subtract from this the BaO equivalent already found and so obtain the barium oxide equivalent of the acids forming insoluble barium salts. This, calculated on 1 g. of fat, is the insoluble barium oxide value (b); $a - b$ is the soluble barium oxide value (*c*). Calculate $b - (200 + c)$.

Interpretation of data. The analysis of butter fat and mixtures containing butter fat can provide information on several important points. In the routine examination of butter the estimation of the Reichert and Polenske values generally suffices. With margarine, when the question of the presence or amount of butter fat arises, the Kirschner value is of next importance, and the Avé-Lallemant process is of special merit in the examinations of borderline samples of butter.

Certain practical points may be noticed. Genuine butter fat usually melts clear; margarine fat very seldom does so. During the distillation it is useful to watch the appearance of the insoluble acids (Polenske). In the presence of much coconut oil these are generally in the form of oily drops, whereas, if there is much palm-kernel oil they take the form of white flakes. A genuine butter usually gives not less than 24·5 for the Reichert value. Siberian, South Russian butters and those from badly fed cows often fall below this figure, occasionally even down to 20, but a sample giving Reichert value 24 or less should always be the subject of further investigation. The majority of genuine butters give a value of 26 or more, so that, if the value found be 23, there may be quite an appreciable addition of other fat or margarine. Consideration of the Polenske value is often a help at this stage. Coconut and palm-kernel oils are almost the only common fats giving a considerable Reichert or Polenske value (the figures are given in the table on p. 427), and these are perhaps the commonest ingredients of a margarine mixture. There exists in genuine butter an approximate proportionality between the Reichert and Polenske and Kirschner values; thus the average results are:—

Reichert	Polenske	Kirschner
32	3·5	26·4
31	3·2	25·8
30	3·0	25·0
29	2·9	24·8
28	2·6	24·0
27	2·4	23·3
26	2·0	21·8
25	1·8	20·9
24	1·7	20·5
23	1·6	20·0

These values do not often fluctuate more than, say, 0·5 in the Polenske, and 0·6 in the Kirschner processes, hence a low Reichert value with a high Polenske value is suspicious and suggests the presence of margarine, whereas a low Reichert value with a correspondingly low Polenske value may be simply due to the origin of the butter. It has been pointed out (see, for example, Cranfield, *Analyst*, 1915, **40**, 439; 1916, **41**, 240, 336) that the diet affects these figures and that certain substances may lead to high Polenske values. Since the Kirschner value is approximately a measure of butyric acid, it is often a great help in deciding whether a low Reichert value is due to abnormality or to adulteration. As an example, an unknown butter gave a Reichert value 23, and Polenske value 3·6; there is the possibility that it is abnormal, or that it is a mixture of 80 per cent. of butter having Reichert 26 and Polenske 2 with 20 per cent. of palm-kernel oil which has a Reichert value of 6, and Polenske 10. The Kirschner value was found to be 17·6, which is too low for a genuine butter; the normal Kirschner corresponding to Reichert 23 is 20. Thus the Reichert and Polenske figures cast suspicion on the sample, and the Kirschner value showed definitely that it was adulterated. If the Reichert—Polenske—Kirschner values are in the correct ratio, it is not safe to assume adulteration just because they are somewhat low. Positive proof of the adulterant should be sought by the phytosterol acetate test, the determination of *iso*-oleic acid or the various colour reactions of particular oils.

The Kirschner value is also a most useful indicator of butter in margarine. Bolton, Richmond and Revis (*Analyst*, 1912, **37**, 183) have shown that the proportion of butter fat in a margarine composed of fats other than coconut or palm-kernel is in agreement with the expression

$$K = 0{\cdot}244\,B + 0{\cdot}28, \text{ or } B = \frac{K - 0{\cdot}28}{0{\cdot}244}$$

When palm-kernel or coconut oil is present, the proportion of butter fat in margarine may be deduced from the Reichert, Polenske and Kirschner values by the equation

$$\text{Butter fat per cent.} = \frac{K - 0{\cdot}1\,P - 0{\cdot}24}{0{\cdot}244}$$

Alternatively the proportion of butter fat in admixture with coconut oil and palm kernel oil can be deduced from graphs based on formulæ of this type (see Williams, *Analyst,* 1949, **74**, 508).

The above expressions are not strictly applicable when the amount of butter fat is much over 10 per cent., but as this is the maximum permissible in margarine it is unusual to find samples containing more than this quantity. Elsdon and Smith (*Analyst,* 1925, **50**, 53) prefer the equation

$$\text{Butter fat per cent.} = \frac{K - 0{\cdot}2 - 0{\cdot}1\,P}{0{\cdot}235}$$

and give a series of corrections according to the exact value of P. The most satisfactory results are obtained if the analyst works out his own data, using known mixtures; in this way the personal factor may be eliminated. In the case of unknown mixtures the percentage of butter fat cannot be returned with certainty nearer than say 1 per cent. When the presence of butter is claimed in margarine the amount must be stated within 2 per cent.

The method of Avé Lallemant is also useful in the examination of butter which gives an abnormally low Reichert value, although it is rather laborious and does not possess any great advantage over the more expeditious Kirschner determination. A genuine butter, even with a low Reichert value, almost always gives a negative value for $b - (200 + c)$, whereas the addition of any other fat tends to make this positive. Thus a butter giving Reichert value 23·5 gave $b - (200 + c) = -1{\cdot}2$, and the same butter mixed with 5 per cent. of margarine gave $+ 1{\cdot}5$.

A further method for the estimation of butter and coconut oil in margarine is that described by Elsdon and Smith (*Analyst,* **1927**, **52**, **317**), which depends on the solubility of the volatile acids in concentrated sodium chloride solution (*cf.* Gilmour, *ibid.*, 1920, **45**, 2).

The Valenta test for the detection of the adulteration of butter depends on the solubility of the fat in acetic acid (see Fryer and Weston, *Analyst*, 1918, **43**, 3).

Other tests which are sometimes of value for the examination of margarine and doubtful samples of butter are the determination of unsaponifiable matter and the detection of cottonseed oil (Halphen's test, p. 436) and sesame oil (Baudouin's test, p. 436). The addition of 10 per cent. of sesame oil is compulsory in some countries to facilitate detection.

Hydrogenated oils can be detected from the determination of the solid unsaturated fatty acids as described by Cocks, Christian and Harding (see p. 426). Solid unsaturated acids do not occur in butter, but constitute from 25–50 per cent. of the acids in hardened oils. If *nickel* is present as a residue from the hydrogenation process it can be detected by the following method: Warm the ash from 50 g. of sample with 1 ml. of concentrated hydrochloric acid, make alkaline with 0·88 ammonia, boil and filter. Then add 2 ml. of α-benzildioxime reagent (0·05 g. in 95 ml. alcohol and add 5 ml. 0·88 ammonia). A rose-red colour forms slowly in the presence of nickel (*cf. Analyst*, 1913, **38**, 316). It should be borne in mind that nickel is not necessarily present in any particular hydrogenated oil, even if the metal has been used as catalyst.

The phytosterol acetate test, although somewhat tedious, is of great value as a definite indication of vegetable oil when admixed with animal fats. The latter contain cholesterol but no phytosterol, and even though a butter may give a slight Halphen reaction, due to the cow having been fed on cotton-cake, such butter never gives any phytosterol. The test also detects hydrogenated vegetable fats in animal fats. Of the clear fat 15 g. are boiled under a reflux condenser with 10 ml. of potassium hydroxide solution (1000 g. in 1400 ml. of water) and 20 ml. of alcohol until all the fat is dissolved and then for a further half hour. Larger quantities of the fat and alkali are preferable if available. The mixture is cooled, 60 ml. of water and 108 ml. of alcohol are added, and 20 ml. of 1 per cent. alcoholic solution of digitonin. After standing over night the sterol digitonin compound is filtered off on a small Buchner funnel, washed with a few ml. of alcohol, then acetylated by heating in a boiling water bath with 3 ml. of acetic anhydride. The acetic anhydride is allowed to evaporate and the sterol acetates dissolved in a minimum amount of hot absolute alcohol and allowed to crystallise. The crystals are separated by withdrawing the alcohol through a glass tube plugged with a piece of cotton wool, redissolved in more hot alcohol and recrystallised. This recrystallisation process is repeated at least three times, and in doubtful cases five times; it may be carried out on quite small quantities of the acetate. Then the melting point of the pure acetate is determined, correcting the observed temperature for the emergent stem of the thermometer. Phytosterol acetate has a melting point of 127° C. or higher, and cholesterol acetate 113°–115° C. If, therefore, the melting point observed is above 116° C. the acetate should be recrystallised again, and if it is higher, or is still above 116° C., the presence of phytosterol, hence of vegetable oil, is certain. In the converse way animal fat in vegetable fat may be detected by the melting point being substantially lower than that of pure phytosterol acetate, 127° C. (*cf. Analyst*, 1923, **48**, 155).

Microscopic examination of butter and margarine sometimes yields valuable information. Butter which has not been melted is devoid of crystalline structure, whereas margarine fat or renovated butter often shows crystalline forms. A fresh piece of the fat is pressed out under the cover slip on a micro slide and examined with a 1-inch objective, a Nicol's prism and analyser being used. If extraneous light is carefully excluded, with pure butter all is dark, but with margarine or renovated butter bright portions and indefinite crystalline forms may usually be seen. Care should be taken to ascertain that the crystals are not those of salt or boric acid, by examining the slide without the polarising apparatus.

Effect of Storage on Butter Fat

The F.F.A. of the fat of fresh butter is usually about 0·2–0·3 per cent. (calculated as oleic acid) and the peroxide value less than 5 (calculated as mls. N/500 thiosulphate per g.). For commercial purposes an F.F.A. figure in excess of about 0·7 per cent. would be considered to be undesirable. The modified form of the Kreis test described on page 411 was proposed by Walters *et al.* (*J.S.C.I.*, 1938, **57**, 53) for indicating incipient rancidity in butter fat. Changes in the Reichert-Polenske-Kirschner values which occur during storage are discussed in " Richmond's Dairy Chemistry " (5th Edn., 1953, p. 243).

MARGARINE

Definition and Legislation

" Margarine " is defined in The Food Standards (Butter and Margarine) Regulations, 1955 as " the food usually known as margarine being an emulsion of edible oils and fats with water or skimmed milk or other substances, with or without the addition of colouring matter, which is capable of being used for the same purpose as butter."

The above regulations also prescribe that margarine shall not contain more than 16 per cent. of water and that *margarine fat* shall not contain more than 10 per cent. of *butter fat.*

The Food Standards (Margarine) Order, 1954, (S.I. 1954 No. 613) prescribes the amounts of vitamins A and D which must be present in imported and home-produced margarine sold by retail (see below).

The labelling of margarine is controlled by The Labelling of Food (Amendment) Regulations, 1955 (S.I. 1955 No. 1900), which prescribe that the wrapper or container must bear the word " MARGARINE " in block letters, legibly printed and conspicuously visible. The size of any brand or descriptive name used (and the word " butter," " cream " or " milk ") is also restricted in relation to the size of the word " MARGARINE." If the presence of butter in margarine is claimed, the amount must be stated within 2 per cent.

The manufacturing processes involved in the manufacture of margarine are described in Andersen's " Margarine " (Pergamon, Lond., 1954).

Analysis of Margarine

Apart from the fat, most samples of margarine contain 13–15·5 per cent. water, 1–2 per cent. salt and about 1 per cent. N.F.S. excluding salt. As with butter, the analysis of margarine should include the determination of *water* and *salt* and an examination for *boric acid.* The fat should give a low Reichert value, but the latter will be raised slightly if butter fat is present. A number of brands on the market contain butter fat (see standard above) and the amount present can be calculated from the Reichert, Polenske and Kirschner values (see method and discussion on pages 394–396). It should be borne in mind, however, that a large number of different oils and fats (often hydrogenated to a greater or lesser extent) are used in margarine manufacture and the interpretation of oil values may be difficult (see p. 426). Rancidity values should also be determined on the fat (when manufactured the fat of margarine has an acid value of 0·3–0·35). The methods described earlier in this Chapter for butter are also applicable to margarine. In view of the requirements of the Food Standards Order, it is necessary to determine the amount of vitamin A present in margarine (sold by retail) by the prescribed method.

The Food Standards (Margarine) Order, 1954

FIRST SCHEDULE

The stardard for margarine shall be as follows:—
Each ounce of margarine shall contain—

(a) not less than 760 international units and not more than 940 international units of vitamin A determined in accordance with the method set forth in the Second Schedule to this Order. The vitamin A content shall be calculated as the sum of the vitamin A present as such or as its esters plus 0·8 times the beta-carotene equivalent of any carotenes present; any alpha-carotene being considered as equivalent in potency to half its weight of beta-carotene; and when red palm oil is used as a source of carotenes, the beta-carotene equivalent shall be taken as 53·5 per cent. of the total carotenes;

(b) not less than 80 international units and not more than 100 international units of vitamin D.

SECOND SCHEDULE

Method for Determination of Vitamin A Content

Principle

The margarine is saponified, the unsaponifiable matter is extracted with diethyl ether and passed through a process of double column chromatography, which separates carotene from vitamin A. The optical density of the carotene solution is measured over the spectral range 440–450 mμ and the vitamin A fractions, after being identified by the Carr-Price reaction, are combined in light petroleum. The optical density of the solution is measured at the wavelength of maximum absorption, normally at 324 mμ. All operations should be carried out under subdued lighting conditions.

Apparatus (Fig. 41)

1. Chromatographic apparatus (see sketch of assembly and of components). It consists of an upper and lower column. The lower column is fitted with

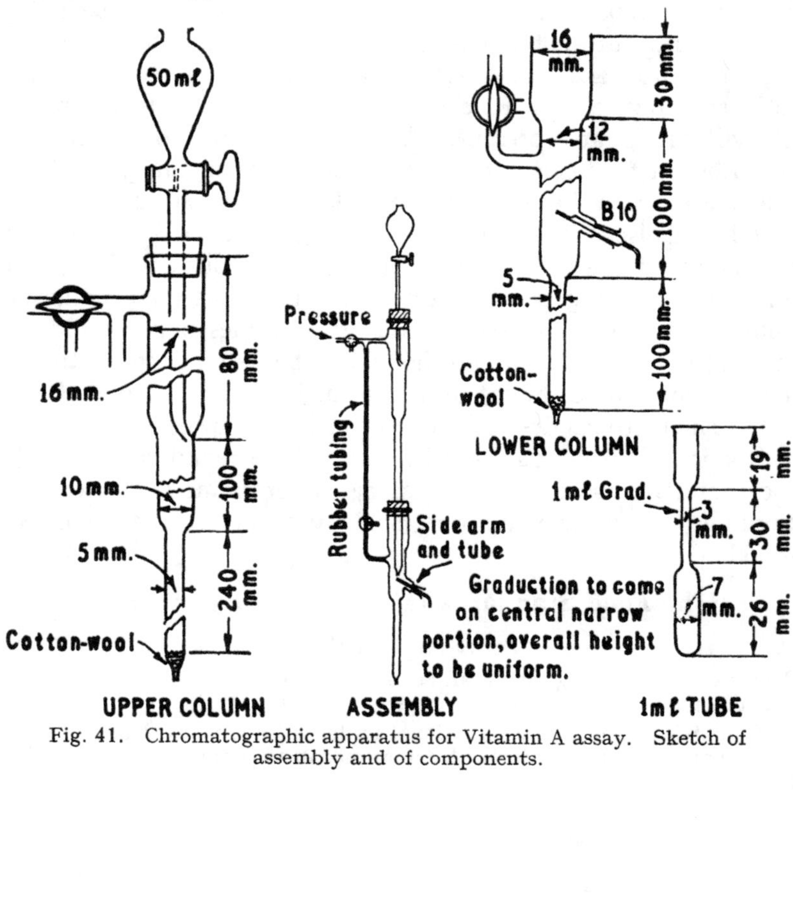

Fig. 41. Chromatographic apparatus for Vitamin A assay. Sketch of assembly and of components.

a side arm which can be adjusted so that the eluate from the upper column can either by-pass, or pass through, the lower column. The assembly is fitted with three-way taps to enable air pressure to be applied either to the top column alone or to both columns.

2. Graduated tubes—1 ml. (see sketch).

3. Spectrophotometer—For measuring ultra-violet absorption of vitamin A solution.

4. Quartz cuvettes—Two; each of 1 cm.

5. Saponification flasks—300 ml. resistance glass.

6. Separators—500 ml. capacity.

7. Graduated flasks (" A " grade)—10 ml., 50 ml.

8. Pipette (" A " grade)—1 ml., the tip drawn out to a capillary to fit graduated tubes.

Reagents

All reagents must be of the quality required for quantitative chemical analysis.

1. Alumina (Type 1).—Prepare from alumina trihydrate.

Activate the fraction passing through a 150 mesh British Standard sieve by heating at 800° C. for 7 hours. After cooling, add distilled water, 2 g. per 98 g. activated alumina, and mix well. Keep in glass stoppered bottle.

2. Alumina (Type 2).—Place 10 g. of activated alumina (Type 1) in a 50 ml. round bottom flask, add 10 ml. of sodium hydroxide solution, and mix to a thin paste. Attach the flask to a vacuum pump and evacuate for 30 minutes at room temperature at 20 mm. pressure or lower. While still under this reduced pressure, place the flask in an oil bath at 135° C. for 30 minutes. Cool under reduced pressure and shake out the alumina-sodium hydroxide into a small mortar, ignoring any which adheres to flask. Add 1·2 ml. water slowly while stirring gently with a pestle. Immediately transfer the smooth powder to a small well stoppered bottle.

3. Antimony trichloride.—A saturated solution in chloroform (Carr-Price reagent).

4. Diethyl ether.—Freshly distilled over sodium hydroxide pellets.

5. Ethyl alcohol.—Absolute.

6. Light petroleum (40° to 60° C.).—To obtain an optically pure fraction it may be necessary to distil and use only that portion boiling below 45°C.

7. Potassium hydroxide solution.—Dissolve 60 g. of potassium hydroxide pellets in distilled water and make up to 100 ml.

8. Sodium hydroxide solution.—Dissolve 10 g. of sodium hydroxide pellets in distilled water and make up to 100 ml.

9. Carbon dioxide or hydrogen supply.

10. Hydroquinone.

Saponification and Extraction of Unsaponifiable Matter

Weigh 10·0 g. margarine into a 300 ml. flat bottom flask, add 20 mg. hydroquinone, 40 ml. absolute alcohol, 10 ml. potassium hydroxide solution and boil under reflux for twenty minutes. Transfer quantitatively to a 500 ml. separator using two 40 ml. quantities of distilled water and extract once with 100 ml. and then three times with 50 ml. quantities of diethyl ether. Combine the ether extracts and wash with four successive portions, each of 50 ml. of distilled water. Transfer the ethereal solution in suitable aliquots quantitatively to a 300 ml. wide-mouth flat-bottom flask. All subsequent operations are conducted in an inert atmosphere, *e.g.*, carbon dioxide. Distil off the ether, add to the residue about 5 ml. absolute alcohol and reheat to remove the last traces of water. Make sure that no trace of alcohol remains, as otherwise the chromatogram will not develop properly. Dissolve the residue in 5 to 10 ml. of light petroleum.

Chromatography and Measurement of Optical Density

Place a small wad of cotton wool in the lower tip of each of the two columns of the chromatographic apparatus. Fill the upper column to the middle of the tube of 10 mm. diameter with light petroleum and pour in sufficient alumina (Type 1) to fill the tube of 5 mm. diameter. Prepare the lower column in a similar manner with alumina (Type 2), half filling the 5 mm. diameter section. Close the tip of the lower column with rubber tubing plugged with a short piece of glass rod. Connect the lower column to the upper column so that the outflow passes through the side arm.

Apply air pressure to the top column. When the excess of light petroleum has been forced through the column, release the pressure and transfer quantitatively the light petroleum solution of unsaponifiable matter to the column. Develop under pressure, first with 5 ml. of light petroleum, and then successively with 5 ml. quantities of light petroleum and diethyl ether mixtures containing respectively 4, 8, 12, 16, 20, 24 and 36 per cent. of diethyl ether. During these operations the level of liquid should not fall below the upper surface of the alumina.

Carotene

Carotene passes quickly down the column and is eluted before the 16 per cent. diethyl ether solvent mixture is introduced, the eluate being by-passed through the side tube into a receiver. Reduce the volume of the carotene solution, if necessary, by evaporation in an inert atmosphere, transfer to a 50 ml. graduated flask and make up to volume with light petroleum.

Measure the optical density of the solution in a 1 cm. quartz cuvette, against a light petroleum blank over the spectral range 440–450 mμ at 2 mμ intervals. From the optical density at λ max.* calculate the E (1 per cent., 1 cm.) value and multiply this by 358 to obtain the equivalent β-carotene potency in International Units per g. This value multiplied again by 0·8 will give the equivalent vitamin A potency of the carotenoids in International Units per g.

Vitamin A

Immediately prior to adding the 16 per cent. diethyl ether developing solvent, remove the plug from the tip of the lower column and the tube from the side arm of the lower column and plug the side arm orifice. Adjust the flow into and out of the lower column by means of the three-way cock and collect the eluate in calibrated 1 ml. tubes. Mix the solution in each tube by blowing through the liquid a few air bubbles with the drawn out pipette. Withdraw, from each tube, approximately 0·3 ml. of liquid and test for vitamin A by adding about five drops of antimony trichloride solution. The development of a blue colour indicates the presence of the vitamin. From each tube in which a positive test is obtained, take exactly 0·5 ml. of solution, combine together in a 10 ml. graduated flask and adjust to volume with light petroleum. Measure the optical density of the solution in a 1 cm. quartz cuvette, against a light petroleum blank, over the spectral region 290 to 340 mμ at the following wavelengths 290, 307, 309, 311, 322, 324, 326, 328, 332, 334, 336 and 340 mμ. Transfer the test solution to the blank cuvette, and the blank solution to the test cuvette, re-measure the optical density at λ max. and from the mean of the two measurements calculate the value of E (1 per cent., 1 cm.) at λ max. Multiply this value by 1830 to obtain the vitamin A potency in International Units per g. The total vitamin A potency is equal to the sum

* λ max. means the wavelength of light at which the optical density reaches its maximum value.

of the potency of the vitamin A and the equivalent vitamin A potency of the carotenoids.

The validity of the test is assessed from the optical density measurements at (λ max. minus 15 mμ) and (λ max. plus 10 mμ). The ratio of the optical density at each of these wavelengths to the optical density at λ max. (normally 324 mμ) should fall within the range of 0·86 $\pm$ 0·02. If this is not so the whole determination must be repeated.

Lord and Bradley (*Analyst,* 1955, **80**, 429) employ de-fatted bone meal (as recommended by Nelson, *Analyst,* 1947, **72**, 200) as absorbent instead of the specially activated alumina used in the official procedure for determining vitamin A. A more rapid result can be obtained by measuring the absorption at 620 mμ of the colour produced on adding the antimony chloride reagent to a solution of the unsaponifiable matter (see also Edisbury, *Analyst,* 1940, **65**, 484).

Vitamins A and D are usually incorporated into margarine as a " master-mix " concentrate consisting of synthetic vitamin A acetate or palmitate dissolved in vegetable oil together with ergo-calciferol in the correct ratio. No reliable chemical methods for the determination of vitamin D in margarine appear to be available and it is necessary to employ biological assay (see Gridgeman *et al., Analyst,* 1940, **65**, 493).

CHEESE

Cheese is the solid or semi-solid substance produced from the curd of milk. According to Richmond, the distribution of the various milk constituents when made into curd and whey is:—

	Milk.	Curd.	Whey.
	Per cent.	Per cent.	Per cent.
Water	87·30	6·50	80·80
Fat	3·75	3·45	0·30
Lactose	4·70	0·30	4·40
Casein	3·00	2·60	0·40
Albumin	0·40	Trace	0·40
Ash	0·75	0·15	0·60

For the production of curd, pasteurised milk is coagulated at 80–95° F. by means of rennet, which is obtained from the inner lining of the fourth stomach of the calf and contains various enzymes including **rennin.** The acidity is an important factor in cheese manufacture and this is nowadays controlled by using " starters," *i.e.*, organisms which produce lactic acid. The resultant curd is cut and some of the whey is removed by means of heat and acid development. After salting, the product is pressed into moulds which determine the shapes of the cheese. Mild treatment results in the production of a moist, soft cheese, whilst drier, harder varieties with much better keeping qualities are produced by applying more heat after cutting. Differences also arise according to the acidity development.

The ratio of the constituents, apart from water, of a whole milk cheese should be similar to that of the curd. From the analyst's point of view, most types may be classified as (1) **Soft cheeses** made from (a) milk and cream or (b) whole milk; (2) **Hard cheeses** made from (a) whole milk or (b) partly skimmed milk; and (3) **Processed cheese.** It should also be noted that a few cheeses are prepared from goat's or sheep's milk, *e.g.*, Roquefort. The composition of various types of cheese is given in the following table. The figures quoted are those of Richmond and Cox:—

1 (*a*). SOFT CHEESES MADE FROM CREAM

	Water.	Fat.	Protein.	Acidity. lactic.	Ash.
	Per cent.	Per cent.	Per cent.	Per cent.	Per cent
Lactic . .	25·4	65·30	6·95	0·75	0·40
Lactic . .	30·3	60·52	5·34	0·56	0·35
Gervais . .	42·32	49·18	7·75	0·27	0·49
Pommel. .	44·9	45·5	7·2	1·16	0·50

1 (*b*). SOFT CHEESES MADE FROM WHOLE MILK

	Water.	Fat.	Protein.	Acidity.	Ash.
	Per cent.	Per cent.	Per cent.	Per cent.	Per cent.
Camembert .	45·24	30·31	19·75	—	4·70
Camembert .	49·31	22·50	20·10	0·44	4·40
Brie . .	50·04	27·50	18·32	—	4·12
Bondon. .	59·5	24·4	9·4	—	0·7
Bondon. .	55·2	20·8	15·4	1·65	6·98
Neufchâtel .	37·90	41·3	23·6	—	3·4
Caerphilly .	57·5	18·4	19·5	—	3·2

2 (*a*). HARD CHEESES MADE FROM WHOLE MILK

	Water.	Fat.	Protein.	Acidity.	Ash.
	Per cent.	Per cent.	Per cent.	Per cent.	Per cent.
Stilton . .	23·57	30·13	32·55	1·24	3·51
Stilton . .	28·60	30·70	35·60	1·08	4·02
Gorgonzola .	31·85	34·34	27·88	1·35	4·58
Gorgonzola .	27·80	35·46	28·00	1·07	4·02
Wensleydale .	28·3	38·3	27·43	—	3·7
Gruyère. .	33·16	30·69	30·67	0·27	4·71
Cheddar. .	30·50	30·70	30·20	0·85	4·10
Cheddar. .	28·54	31·20	30·70	1·25	4·30
Cheshire .	37·11	30·68	26·93	0·86	4·42
Roquefort .	36·85	30·61	25·25	1·90	5·39

2 (*b*). CHEESES MADE FROM PARTLY SKIMMED MILK

	Water.	Fat.	Protein.	Acidity.	Ash.
	Per cent.	Per cent.	Per cent.	Per cent.	Per cent.
Dutch . .	41·30	22·78	28·25	0·57	7·10
Dutch . .	45·24	17·20	27·35	0·84	5·60
Gloucester .	25·25	25·80	30·05	—	4·80
Parmesan .	32·5	17·1	43·6	—	6·2

3. PROCESSED CHEESES

	Water.	Fat.	Protein.	Acidity.	Ash.	Stabiliser.
	Per cent.	Per cent.	Per cent.	Per cent.	Per cent.	
1.	35·92	28·65	26·91	0·80	5·43	Citrate present.
2.	39·61	30·52	23·35	1·10	4·16	Citrate and Tartrate.
3.	42·64	26·32	25·66	1·05	5·06	Tartrate.

Processed cheese is prepared by heating a mild cheese such as Cheddar with water and an emulsifying agent such as disodium phosphate, sodium citrate or sodium potassium tartrate. Leather (*Analyst*, 1949, **74**, 51) has reported on the presence of calcium tartrate crystals, which had been described as "bits of glass." The presence of tin derived from the tinfoil has also been reported. Standards of composition for processed cheese which have been recommended by the F.S.C. are given below.

Recommended Standards for Processed Cheese and Cheese Spread

The Food Standards Committee have recommended the following standards (Second Report on Processed Cheese and Cheese Spread, H.M.S.O., 1956):—

	Minimum butter fat content	Maximum moisture content
Processed Cheddar or Cheshire Cheese	48% in dry matter	42%
Processed Cheese . . .	45% in dry matter	45%
Cheese spread . . .	20% on the product as sold	60%

In addition, a maximum of 3 per cent. of added anhydrous emulsifying salts is recommended.

Cream cheese should presumably be prepared from cream. Davis (*Dairy Ind.*, 1960, **25**, 196) has suggested that *single cream cheese* and *double cream cheese* should contain at least 48 and 58 per cent. of fat respectively.

Quality considerations. In general, the proportion of fat in the dry matter is important when considering the quality of cheese. If this is much below 50 per cent. it suggests that the product has been prepared from partly skimmed milk. Another criterion of genuineness is that the fat/protein ratio should not be less than 1/1 (it is often as high as 1·25/1). Richmond has devised the following formula for calculating from the composition of the cheese the probable composition of the milk from which it was made:—

$$\text{Fat in milk} = \frac{100\ \text{F}}{35{\cdot}4\ \text{P} + \text{F}} + 0{\cdot}25$$

where F and P are the percentages of fat and protein respectively in the cheese. Owing to the natural variations in the ratio of fat and protein in cheese it is not wise to certify a cheese to be made from skim milk unless the calculation indicates less than 2·75 per cent. of fat, or to condemn a cream cheese unless the indicated fat percentage is less than about 10. For a study of the water-fat ratios in about 2000 samples of English cheese, see Nicholls (*Analyst*, 1941, **66**, 265).

Surprise has been expressed in some quarters that, although milk and its curd are white, many consumers seem to prefer a coloured cheese. Although preferences vary from one area to the next there would appear to be some association of flavour and colour. Annatto is the colour most commonly employed for cheese.

Outbreaks of food poisoning due to cheese have been reviewed by Vergue (*Chem. Abstr.*, 1948, 8988).

Analysis of Cheese

The routine analysis of cheese usually involves the determination of **moisture*** and **fat*** and an examination for the **type of fat present.** Other determinations which may be necessary are those of **ash, salt,* protein,* acidity** and the **pH value.*** Davis and Macdonald (*Richmond's Dairy Chemistry*, 5th Edn., 1953, p. 517–) describe methods for the determination of a number of the **products of ripening** (mainly nitrogen substances). Processed cheese should be examined for **emulsifying salts** (phosphate, citrate, etc.) and **tin** (if wrapped in tin foil).

Sampling

As cheese tends to dry out quickly on the outside care must be taken to ensure that the sample used for analysis is representative of the whole material. A grater should be used for hard cheeses prior to grinding in a mortar. Soft cheese should be ground in a mortar and all samples should be kept in tightly stoppered jars.

Moisture can be determined by drying 5 g. on sand at 100° C. to constant weight. The B.S. method, however, stipulates that 3 g. should be dried with about 20 g. of sand at 102° C., using water in the early stages so that the sample can be spread out evenly with a rod. Drying at 100° C. is included as an " alternative procedure for routine purposes."

Fat can be most accurately determined by the Werner-Schmid process. Weigh 1–2 g. into the extraction tube, add 8 ml. water and 10 ml. of concentrated hydrochloric acid and boil gently (with processed cheeses particularly, the particles often break up more easily if the sample is treated with a few drops of ammonia prior to the addition of the dilute acid). When all the particles have dissolved, add 10 ml. of alcohol, mix and cool. Then extract the fat four times using mixed ethers (25 ml. ether followed by 25 ml. light petroleum).

The **Gerber process** can also be employed. Using the special cheese butyrometer, add to it 10 ml. of Gerber sulphuric acid followed by a 6 mm. layer of warm water (30–40° C.) and 3 g. of sample. Then add 1 ml. of amyl alcohol and sufficient of the warm water so that the tube is filled to the shoulder below the neck. Stopper, shake and invert the tube and place in a water bath at 65° C. for three to ten minutes. Make sure that no solid particles remain and centrifuge at

* These determinations are described in B.S.770: 1952.

1100 r.p.m. for five minutes. Return to the 65° C. bath and read off the percentage of fat directly on the scale. The method is given in detail in Part 2 of B.S. 696 : 1955, which also describes the use of ordinary milk tubes (on 0·9–1·1 g. of sample). Butyrometers designed for cream may also be employed using 5 g. of sample (see Davis and Macdonald, *Richmond's Dairy Chemistry,* 1953, p. 352, Griffin, Lond.).

Apart from the percentage of fat in the sample the fat in the dry matter should also be calculated.

Ash is determined by igniting the dried sample at as low a temperature as possible.

Salt may be determined by titrating the ash with silver nitrate solution (*cf. Analyst,* 1932, **57**, 79).

Protein. Determine the total nitrogen on 1–2 g. of sample using the macro-Kjeldahl method. The protein is usually calculated as N × 6·38, but the figure obtained is only an approximation as numerous non-protein nitrogenous substances are present in cheese. Methods which have been proposed for the determination of proteins and products of ripening are described by Davis and Macdonald (*loc. cit.*, p. 517).

Acidity. Shake 10 g. of the cheese vigorously with 85 ml. of warm water (40° C.), add water to produce a total volume of 105 ml. and filter. Then titrate 25 ml. (≡ 2·5 g. sample) with N/10 sodium hydroxide using phenolphthalein as indicator. Express the result as lactic acid. The *pH value* is preferably determined by pressing the electrodes directly into the sample. Hard cheeses must, however, be softened first by mixing with the minimum quantity of water (see B.S. 770 : 1952).

Emulsifying salts. Citrate and **tartrate** can be detected after shaking the cheese thoroughly with warm water (40°–50° C.). **Phosphate** can be detected and determined on the ash. In considering the figure obtained from processed cheese as a measure of added emulsifying agent allowance should be made for the natural presence of phophate in cheese. In this connection, hard-pressed cheeses contain about 0·5 per cent. of phosphorus. The Food Standards Committee recommended a maximum of 3 per cent. emulsifying salts in processed cheese or cheese spread.

Tin may present in processed cheese which is wrapped in tin foil. The metal can be estimated by the usual methods.

Type of fat present. Margarine-type fats or other non-butter fats can be detected by warming 30 g. of the grated cheese (no traces of rind should be present) in a funnel plugged with fat-free cotton wool, keeping it in the water-oven so the fat runs into a flask. If necessary the fat can be separated by the acid Werner-Schmid process but such extracted fat tends to give a slightly lower Reichert value. The dry fat is examined in the same way as butter fat, *i.e.*, the Reichert and Polenske values are determined. If only a small amount of sample is available the semi-micro Reichert method of Dyer *et al.* (p. 392) is suitable. In interpreting the results it should be borne in mind that slight changes occur in the fat during the ripening process, which may cause slight departures from the normal values obtained from butter fat.

ICE-CREAM

Ice-cream is made from fresh or processed milk, fat, sugar, a stabiliser to produce a smooth texture (*e.g.*, gelatine, alginate gums, agar) and an emulsifying agent (*e.g.*, G.M.S., egg, lecithin).

The Food Standards (Ice-Cream) Order, 1953, prescribed minima for fat (5 per cent.), sugar (10 per cent.) and milk solids other than fat or M.S.N.F. (7·5 per cent.). Following the publication of the "*Food Standards Committee Report on the Ice-Cream Standard*" (1957), *The Food Standards* (*Ice-Cream*) *Regulations*, 1959 (*S.I.* 1959, *No.* 472) came into operation in England and Wales. The main requirements of the latter order are:—

(1) **Ice-cream** must contain not less than 5 per cent. fat and 7·5 per cent. M.S.N.F.

(2) **Ice-cream containing fruit** must *either* conform to the above standard *or* the *fat* + *M.S.N.F.* must be not less than 12·5 per cent., which must *include* not less than 7·5 per cent. *fat* and 2 per cent. *M.S.N.F.*

(3) **Dairy ice-cream, dairy cream ice** or **cream ice** must conform to the standards in (1) and (2) above, whichever is appropriate but in addition the " fat " must consist entirely of " milk fat and no other fat (save as may be introduced by the use as an ingredient of any egg, any flavouring substance or any emulsifying or stabilising agent)."

(4) **Milk ice** (*including milk ice containing fruit*) must contain not less than 2·5 per cent. milk fat (other than introduced from egg, etc.—see (3) above) and not less than 7 per cent. M.S.N.F.

(5) **" Parev " (kosher) ice** must contain not less than 10 per cent. fat and no milk fat or other derivative of milk.

(6) No ice-cream shall contain any artificial sweetener.

In addition, the Labelling of Food (Amendment) Regulations, 1959 (S.I. 1959, No. 471) prohibit the labelling or advertising of ice-cream in a manner which is suggestive of butter, cream or milk or of anything connected with the dairy interest unless no fat other than milk fat is present (other than introduced from egg, etc.—see (3) above). In addition if fat other than milk fat is present, the ice-cream shall be labelled with the words " Contains non-milk fat " or " Contains vegetable fat " (the label may, however, bear a statement to the effect that the ice-cream contains skimmed milk solids). Neither of these two regulations applies to water ices (*e.g.*, ice lollies).

As a number of outbreaks of food poisoning have been attributed to ice-cream, *The Ice-Cream (Heat Treatment, etc.) Regulations*, 1947–1959 (*S.I.* 1959, *No.* 734; *see also M.O.H. Circulars* 8/51, 14/52 *and* 8/59) have been introduced in order to control the processing and storage. The main requirements of these regulations are:—

Hot mix ice-cream. The mixture shall not be kept for more than one hour above 45° F. before being heat-treated by *one* of the prescribed methods. Method 1 requires that " The mixture shall be raised to and kept at a temperature of not less than 150° Fahrenheit for at least thirty minutes." The Order also prescribes the following *alternative* minimum temperatures and times for the *pasteurisation:* Method 2, 160° F., ten minutes *or* Method 3, 175° F., fifteen seconds. As a *further alternative* a *sterilisation* process is prescribed, the minimum conditions being 300° F. and two seconds. The apparatus (thermometers, etc.) shall be installed, maintained and operated to the satisfaction of the appropriate authority.

After the heat-treatment the mixture shall (with certain provisos) be reduced to not more than 45° F. within one and a half hours and shall be kept at such a temperature until the freezing process is begun.

Water ices and ice lollies, whether containing milk solids or not, that are mixed before freezing and have a pH value of 4·5 or less are exempted from the heat-treatment provisions as it is unlikely that pathogenic organisms will multiply in such acid mixtures.

Cold mix ice-cream. The reconstituted mixture shall be converted into ice-cream within one hour of reconstitution.

Storage of ice-cream. In addition to the heat treatment provisions above, ice-cream must not be sold unless it has also been kept at 28° F. or less since freezing. If the temperature rises after freezing above 28° F., the mixture must be heat-treated and re-frozen before sale. For full details the reader should consult the regulations and the appropriate circulars (see references above).

Composition

A large proportion of the ice-cream sold in Britain comes within these analytical ranges:—

	Per cent. (w/w)
Total solids	34–40
Fat	10–14
Sugar (sucrose)	12–17
Milk-solids-not-fat	9–14
Miscellaneous ingredients . . .	0·3–1·0

The Overrun is usually between 80 and 110 per cent. (by volume)

Analysis of Ice-cream

The routine examination of ice-cream should include the determination of **total solids, fat, sugars, nitrogen, trace elements** and the assessment of the **milk solids other than fat** (M.S.N.F.) and the **bacteriological purity.** A more detailed analysis should include the determination of **ash** and the **acidity** and the examination for **preservatives, starch** (*e.g.*, custard ices), **flavours, colouring matter, " additives "** (*e.g.*, stabilisers), **added colour** and the **type of fat present.** Industrially the *overrun* is of special interest also. B.S.2472 : 1954 describes the determination of some of these constituents.

(a) *Preparation of sample for chemical analysis.* The sample should preferably be analysed immediately after receipt. Otherwise the sample should be placed in a screw-capped jar at $-$ 15° C. or below. Prior to the analysis, melt the sample in a water bath ($\leqslant$ 45° C.), shake and cool to room temperature (see B.S.2472 : 1954; 809 : 1949 and Amendment PD.3005). For the methylene blue test the sample should be taken from the outside of the block.

(b) *Total Solids.* Dry 5 g. on sand firstly on the water bath and then in the 100° oven ($\geqslant 1\frac{1}{2}$ hours).

(c) *Fat.* For most types of ice-cream the **Rose-Gottlieb method** is suitable. For this weigh out about 5 g., add 2 ml. of ammonia solution (s.g. 0·880), mix, add 8 ml. of warm water and re-mix. Maintain the mixture at 65° C. for fifteen minutes. Then add 15 ml. alcohol, mix again and cool. If each stage is carefully followed (see B.S.2472 : 1954) no lumps should be present and the fat can then be extracted with mixed ethers as in the method for milk. With a few ice-creams, *e.g.*, custard ices, the **Werner-Schmid method** may extract the fat more efficiently. For routine work the **Gerber method** is convenient.

The type of fats present can be assessed from the oil values of the extracted fats especially from the Reichert and Polenske values (see Butter and Margarine), which may have to be determined on the semi-micro scale (see also Bassette and Keeney, *J. Ass. Off. Agric. Chem.*, 1956, **39**, 469).

(d) *Sugars.* Determine the sugars before and after inversion after clearing with zinc ferrocyanide or lead acetate and sodium oxalate (see B.S.2472 : 1954) using Lane and Eynon's method. Calculate the sucrose content from the double titration. If the reducing sugars (calculated as lactose) approximately equal one-half of the M.S.N.F. (as determined by the methods below) it can be assumed for routine purposes that no other reducing sugars are present. Storage at room temperature may cause loss of sucrose due to the action of the diplococcus *Leuconostoc mesenteroides*, which converts it to dextran (*Analyst*, 1956, **81**, 204).

(e) *Nitrogen.* Determine on 5–8 g. by the macro-Kjeldahl method. As the protein present is mainly due to milk use N $\times$ 6·38 for the calculation of total protein.

(f) *Ash.* Incinerate 10 g. at 550° C. after drying on the water bath.

(g) *Milk solids other than fat* (*M.S.N.F.*). The approximate M.S.N.F. can be calculated from individual constituents such as the protein, calcium and lactose contents. The correct allowance for the constituent present from other sources (*e.g.*, calcium alginate, added lactose) would, however, be difficult to assess.

The method of Crowhurst (*Analyst*, 1956, **81**, 123) employing the *formol titration* is one of the simplest for assessing the M.S.N.F. content:—

To 10 g. of ice-cream in a porcelain dish add 1 ml. of phenolphthalein solution and titrate the acidity with N/10 sodium hydroxide solution. Calculate the acidity as lactic acid (% w/w). Add exactly 3 ml. of formalin to the neutralised mixture, stir thoroughly and titrate to neutrality again with N/10 sodium hydroxide (titration = x ml.). Titrate separately 3 ml. of the formalin solution (titration = y ml. N/10). Then the M.S.N.F. (%) = $(x - y) \times 5{\cdot}67$.

Hamence and Hart (*Analyst*, 1951, **76**, 429) proposed an absorptiometric method for the determination of the M.S.N.F. in which the tryptophan forms a blue colour with *p*-dimethylaminobenzaldehyde. This method is unsatisfactory in the presence of cocoa and soya flour.

For the **calcium content,** ash 5–10 g. of the ice-cream and precipitate as oxalate in the usual way. Then determine the calcium by igniting the precipitate and weighing or by dissolving in dilute sulphuric acid and titrating with N/50 permanganate (B.S.2472 : 1954). The average calcium content of the dry fat-free solids of milk is 1·88 per cent. (as CaO).

Lactose can be estimated by selective fermentation using yeast in a similar manner to that proposed for meat products (*J.A.O.A.C.*, 1940, **23**, 811) and bread (*Cereal Chem.*, 1950, **27**, 398; see p. 165). Such methods involve the determination of lactose after the other reducing sugars have been removed by the yeast (Lactose × 2 = M.S.N.F.).

(h) *Trace elements.* Lead and zinc are the most likely contaminating metals in ice-cream.

(i) " *Additives.*" Preservatives such as sulphur dioxide and benzoic acid are seldom met with in ice-cream. The detection or determination of other added materials is sometimes called for, *i.e.*, gelatine and alginates (*J.A.O.A.C.*, 1945, **28**, 97), gums (Jacobs' *Chemical Analysis of Foods and Food Products*, van Nostrand, N.Y.), saccharin (extract from a highly acid solution) and colouring matters.

(j) *Overrun.* When ice-cream is manufactured air is whipped in to improve the texture and in consequence the volume of the solidified product is greater than the volume of the original liquid mix (see " *Cronshaw's Dairy Information,*" *Dairy Ind., Lond.*, 1947). The percentage increase in volume is referred to as the *overrun*, which can be calculated from the formula:—

$$\text{Overrun (per cent.)} = \frac{\frac{1}{x} - \frac{1}{y}}{\frac{1}{x}} \times 100 = \frac{y - x}{y} \times 100$$

where x = s.g. of the solid ice-cream before melting and y = s.g. of the remelted ice-cream after removal of air.

Ice-cream is sold by volume and the overrun may be anything from 20–130 per cent. The Food Standards Committee therefore discussed the possibility of extending the present legal standards, which are calculated on a w/w basis, so that they included a minimum weight of total solids per gallon. In view of sampling and other difficulties they decided not to make a recommendation for its enforcement, however. For plant control it is convenient to prepare a chart which shows the weight of a " standard " vessel when filled with ice-cream at various percentages of overrun (*e.g.*, England, *Ice-Cream Rev.*, 1952, **35** (**11**), 40).

(k) *Bacteriological examination* (*Methylene blue test*). The Minister of Health has recommended that local authorities should use the *methylene blue test* for the provisional measurement of the bacteriological cleanliness of ice-cream (M.O.H. Circular 8/59). The method has been described and discussed in the Monthly Bulletins of the M.O.H. and P.H. Laboratory Service for March 1947, April 1948, July 1949 and October 1950. The main features of the method are:—

The *sampling* should be arranged so that the ice-cream reaches the laboratory in not more than six hours and the test should preferably be commenced at 5 p.m. on the same day. If necessary the sample should be packed in ice and if it has to be melted at room temperature this should not occupy more than one hour.

METHOD

N.B. The test must be carried out taking the usual aseptic precautions. The methylene blue test for milk is given in p. 357.

To a sterile reductase tube (marked at 10 ml.) add 7 ml. of ¼ strength Ringer solution, 1 ml. of methylene blue solution (as used for milk) and, using a pipette with a wide-bore tip, 2 ml. of the melted sample (preferably taken from the outside of the block of ice-cream). Insert a sterile rubber bung, invert the tube once and make up to the 10 ml. mark with more of the ice-cream sample (to make up for the volume of air lost). *At the same time*, prepare two similar control tubes:—

(a) *Ice-cream colour*. To 8 ml. of ¼ strength Ringer solution add 2+ ml. of the ice-cream so that the volume is 10 ml. after inverting.

(b) *Methylene blue colour*. Heat about 5 ml. of ice-cream in a boiling water bath for fifteen minutes and prepare another tube as in the test but using this sterile sample.

Then incubate the test and control tubes for seventeen hours *in a water bath* at 20° C. Note any decolorisation (at 0 hours) and incubate in a water bath at 37° C., reading and inverting once every half hour. Note the time taken for complete decolorisation of the methylene blue and assign to it one of the following provisional grades:—

Provisional grade	*Time taken to reduce methylene blue*
1	Fails to reduce in 4 hours
2	2½–4 hours
3	½–2 hours
4	0

The majority of samples from any one producer (preferably considered over a period) should fall into Grade 1 or 2. Although thermoduric organisms derived from ingredients may tend to put the ice-cream into a lower grade they are seldom responsible for repeated results at Grades 3 and 4 levels. Certain added flavours (*e.g.*, banana) however, appear to be capable of causing decolorisation of the methylene blue by chemical action. The test is only proposed for the *routine* grading of ice-cream and is to be used for indicating where further investigations into manufacturing practices are advisable.

Confirmatory evidence of a low grading can often be obtained from bacteriological data (count on agar at 37° C. for forty-eight hours should not exceed 50,000 per ml. and presumptive coliforms should be absent in 0·1 ml.). If too many samples from a manufacturer fall into Grade 2 or below, cultures should be made to identify the prevalent organism. Although the finding of Bact. coli (even of "fæcal" or Type I) in ice-cream is not necessarily a sign of direct fæcal contamination or of the liability to cause disease in the consumer, it is advisable to subculture any MacConkey tubes showing acid and gas into another MacConkey broth and incubate at 44° C. for twenty-four hours. Then if any of these 44° tubes suggests the presence of fæcal coli, the findings should be confirmed by the indole, methyl red and Voges-Proskauer tests, etc. (see "*The Bacteriological Examination of Water Supplies,*" Report on Public Health and Medical Subjects No. 71, H.M.S.O., 1956).

Chapter 14

OILS AND FATS

Lard—Suet—Olive Oil Group—Determination of the Fat-soluble Vitamins—Mineral Oil in Food

GENERAL METHODS FOR OILS AND FATS

The determinations which follow are those which are commonly used in the examination of oils and fats. In routine analysis determination of the iodine value, saponification value, unsaponifiable matter, acid value and peroxide value, coupled with qualitative tests for appropriate adulterants will usually be sufficient for confirming the identity and edibility of most oils and fats. Further tests are described by Williams (" Oils, Fats and Fatty Foods," 1950, Churchill, Lond.) and in the B.P. and B.S.684 : 1958.

Water. The moisture content of oils and fats should preferably be determined by Dean and Stark's distillation method using heptane or by the Karl Fischer method using n-propanol instead of methanol throughout (B.S.2511). The A.O.A.C. prescribes a drying method in which 5 g. is dried in a vacuum oven at a uniform temperature 20–25° C. above the b.pt. of water at the working pressure which should not exceed 100 mm. of mercury. This latter method determines "water + other volatile matter."

Specific gravity. Determine the s.g. by specific gravity bottle or pyknometer at 15·5° C. The B.P. prescribes the determination of density in terms of weight/ml. (often at 20° C.). Fats require higher temperatures.

Refractive index. For the determination of the refractive index, B.S.684 states the accepted standard temperatures for the expression of results: 20° C. for oils; 40° C. for solid fats and mixtures of fatty acids liquid at that temperature; 65° C. for solid fats, etc., not liquid at 40° C. (using sodium light of wavelength 589·3 mμ).

Melting point (slip point). Commercial fats are heterogeneous mixtures and do not have a sharp melting point. The methods of determination of the melting point are described in B.S.684 : 1958, p. 14 and B.P. 1958, p. 822.

Titre. The titre of an oil or fat is the solidifying point of the mixed fatty acids. The titre temperature is of value for characterising oils and fats and for assessing the hardness.

In the method the saponified oil is acidified and the melted fatty acids allowed to cool. After solid begins to separate, the temperature rises slightly due to the latent heat liberated and the highest temperature subsequently reached is taken as the titre. The following method uses glycerol-potash for the saponification, which is shorter than the alcoholic-alkali method of Dalican (*cf.* B.S.684 : 1958, p. 22):—

Method. Heat 19–20 g. of potassium hydroxide pellets with about 70 ml. glycerol in a 400 ml. beaker until the temperature of the solution reaches 150° C. (not more). During the heating the liquid can be gently stirred with a thermometer. Add 50 ml. of melted fat and maintain at 145–150° C. for fifteen minutes or until no separation occurs, stirring gently during the heating. Partially *cool,* pour into a litre beaker containing 450 ml. hot water and wash in the remaining liquid with about 50 ml. more of hot water. Then add 50 ml. dilute sulphuric acid (1 vol. conc. + 3 vols. water) and boil until the top layer of fatty acids melts and clears. Siphon off and reject the lower aqueous layer using two pieces of glass tubing joined by a piece of rubber tubing fitted with a spring clip. Add 200 ml. of hot water to the beaker and siphon off again. Repeat the washing twice more. Decant the fatty acids carefully from any remaining drops of water into a filter paper and filter into a flask in the 100° C. oven for at least thirty minutes. Pour the dried melted fatty acids into a large test tube (9 cm. × 3 cm.) fixed axially by means of a cork into a wide necked bottle. Suspend an accurately calibrated 0°–50° C. thermometer centrally in the tube by means of a piece of cotton. Allow the liquid to cool and when solid appears, stir with the thermometer, first a few times from left to right and then similarly from right to left. Then stir rapidly with a circular movement avoiding touching the tube. The temperature should fall at first, then remain stationary and then rise before falling away again. The titre is the highest temperature reached after the stationary period. Remelt the fatty acids and repeat the determination until two successive determinations agree to within 0·2° C.

Colour. The colour of oils is usually measured in the Lovibond Tintometer using a 1 in. or $5\frac{1}{4}$ in. cell. Fats should be melted and any oil or fat that is cloudy should be filtered at a temperature not greater than 60° C. During matching the sample should be at room temperature (for oils) or not more than 10° C. above the m.pt.

Iodine value. The iodine value of an oil or fat is defined as the weight of iodine absorbed by 100 parts by weight of the sample. The glycerides of the unsaturated fatty acids present (particularly of the oleic acid series) unite with a definite amount of halogen and the iodine value is therefore a measure of the degree of unsaturation. It is constant for a particular oil or fat, but the exact figure obtained depends on the particular technique employed. The ranges of figures for the iodine values of the various groups of oils and fats are as follows:—

Group	*Examples*	*Ranges of Iodine Values*
Waxes	—	Very small
Animal fats	Butter, dripping, lard	30–70
Non-drying oils	Olive oil, arachis oil, almond oil	80–110
Semi-drying oils	Cottonseed oil, sesame oil, soya oil	80–140
Drying oils	Linseed oil, sunflower oil	125–200

The iodine value is often the most useful figure for identifying an oil or at least placing it into a particular group. It should also be noted that the less unsaturated fats with low iodine values are solid at room temperature, or conversely, oils that are more highly unsaturated are liquids (showing there is a relationship between the melting points and the iodine values). A further point of interest is that, in general, the greater the degree of unsaturation (*i.e.*, the higher the iodine value), the greater is the liability of the oil or fat to go rancid by oxidation.

The iodine value is usually determined by Wijs' method:—

Wijs' solution. Dissolve 8 g. iodine trichloride in 200 ml. glacial acetic acid. Dissolve 9 g. iodine in 300 ml. carbon tetrachloride. Mix the two solutions and dilute to 1000 ml. with glacial acetic acid.

METHOD (B.P.)

Pour the oil into a small beaker, add a small rod and weigh out a suitable quantity of the sample by difference into a dry glass stoppered bottle of about 250 ml. capacity. The approximate weight in g. of the oil to be taken can be calculated by dividing 20 by the highest expected iodine value. Add 10 ml. of carbon tetrachloride to the oil or melted fat and dissolve. Add 20 ml. of Wijs' solution, insert the stopper (previously moistened with potassium iodide solution) and allow to stand in the dark for thirty minutes. Add 15 ml. of potassium iodide solution (10 per cent.) and 100 ml. water, mix and titrate with N/10 thiosulphate solution using starch as indicator just before the end-point (titration = a ml.). Carry out a blank at the same time commencing with 10 ml. of carbon tetrachloride (titration = b ml.).

$$\text{Iodine value} = \frac{(b - a) \times 1{\cdot}269}{\text{wt. (in g.) of sample}}$$

If $(b - a)$ is greater than $b/2$ the test must be repeated using a smaller amount of the sample.

The Hanus method for determining iodine values employing iodine monobromide gives similar results to the Wij method and is prescribed by the U.S.P. and is stated as an alternative procedure in B.S.684 : 1958. For the determination of the iodine value of the glycerides of halibut-liver oil, the B.P. prescribes the rapid pyridine bromide method of Rosenmund and Kuhnhenn (*Z. Nahr.-u. Genussm.*, 1923, **46**, 154). Toms (*Analyst*, 1928, **53**, 69) has devised a novel method in which the bromine absorbed by the oil on a microscope slide is weighed and the result can be correlated with those obtained by the Wijs' method by using an appropriate factor.

Saponification value. The saponification value of an oil or fat is defined as the number of mg. of potassium hydroxide required to neutralise the fatty acids resulting from the complete hydrolysis of 1 g. of the sample.

A soap is formed during the saponification, for example:—

$$\underset{\text{Stearin}}{C_3H_5(C_{17}H_{35}COO)_3} + 3KOH = \underset{\text{Glycerol}}{C_3H_5(OH)_3} + \underset{\text{Potassium stearate}}{3C_{17}H_{35}.COOK}$$

The esters of the fatty acids of low molecular weight require the most alkali for saponification, so that the saponification value is inversely proportional to the mean of the molecular weights of the fatty acids in the glycerides present. As many oils have somewhat similar values (*e.g.*, those in the olive oil series fall within the range 188–196), the saponification value is not, in general, as useful for identification purposes as the iodine value. The saponification value is of most use for detecting the presence of coconut-oil (S.V. 255) and butter fat (S.V. 225), which contain a high proportion of the lower fatty acids. Paraffin, which, of course, gives a negligible saponification value, can also be detected and estimated if present as an adulterant. The alcoholic potash reagent is made up as follows:—

Alcoholic solution of potassium hydroxide. Dissolve 35–40 g. potassium hydroxide in 20 ml. water and dilute to one litre with alcohol (95 per cent.). Allow to stand overnight and decant off the pure liquid.

METHOD (B.P.)

Weigh 2 g. of the oil or fat into a conical flask and add exactly 25 ml. of the alcoholic potassium hydroxide solution. Attach a reflux condenser and heat the flask in boiling water for one hour, shaking frequently. Add 1 ml. of phenolphthalein (1 per cent.) solution and titrate the excess alkali with N/2 hydrochloric acid (titration = a ml.). Carry out a blank at the same time (titration = b ml.).

$$\text{Saponification value} = \frac{(b - a) \times 28{\cdot}05}{\text{wt. (in g.) of sample}}$$

The highest accuracy in the titrations is most important.

Unsaponifiable Matter can be defined as the material present in oils and fats which after saponification of the oil or fat by caustic alkali and extraction by a suitable solvent remains non-volatile on drying at 80° C.

The unsaponifiable matter includes hydrocarbons, higher alcohols and sterols (*e.g.*, cholesterol, phytosterol). Most oils and fats of normal purity contain less than 2 per cent. of unsaponifiable matter.

Qualitative test. Holdé described a simple qualitative test for the detection of paraffin in fats which is also available for the detection of mineral oil in vegetable oils:—

To a test tube add 10 drops of the oil or melted fat and 5 ml. of N/2 alcoholic potash and heat in boiling water *with frequent agitation for some minutes* to ensure that the reaction is complete. Then to the hot soap solution add water, 0·5 ml. at a time, until 15 ml. have been added altogether, shaking and observing after each addition. The formation of a turbidity indicates the presence of mineral oils.

Determination of unsaponifiable matter. The full method for determining the unsaponifiable matter in oils in fats is described in B.S. 684 : 1958, p. 52 and in the 1958 B.P., p. 875. In routine work, however, the extraction of the unsaponifiable matter may conveniently be combined with the determination of the saponification value:—

After the titration of the saponification value, make the liquid alkaline again with at least 1 ml. of alcoholic potash, transfer to a separator and wash in with water (50 ml. less the volume of N/2 HCl used). Extract the solution while still just warm three times with 50 ml. quantities of ether (the first portion should be used to wash out the original flask). Pour each ether extract into another separator containing 20 ml. of water. After the third extract has been added, shake the combined ether extracts very gently with the first 20 ml. of wash water and then vigorously with two further 20 ml. quantities. Wash the ether extract twice with 20 ml. of aqueous N/5 potash and twice or more with 20 ml. of water until the wash water is no longer alkaline to phenolphthalein. Pour the ether extract into a weighed flask, evaporate off the solvent, dry the residue at not more than 80° C. and weigh to constant weight. Dissolve the unsaponifiable matter in neutral alcohol and titrate with N/10 alkali (not more than 0·1 ml. should be required to neutralise the free fatty acid present).

Acid value or Free Fatty Acids (F.F.A.). The acid value of an oil or fat is defined as the number of mg. of potassium hydroxide required to neutralise the free acid in 1 g. of the sample.

The acid value is a measure of the extent to which the glycerides in the oil have been decomposed by lipase action. The decomposition is accelerated by heat and light. As rancidity is usually accompanied by free fatty acid formation, the determination is often used as a general indication of the condition and edibility of oils:—

METHOD (B.P.)

Mix 25 ml. ether with 25 ml. alcohol (95 per cent.) and 1 ml. of phenolphthalein solution (1 per cent.) and neutralise with N/10 alkali (few drops). Dissolve 1–10 g. of the oil or *melted* fat in the mixed neutral solvent* and titrate with aqueous potassium (or sodium) hydroxide, shaking constantly until a pink colour which persists for fifteen seconds is obtained. The titration should preferably not exceed about 10 ml.

$$\text{Acid value} = \frac{\text{Titration (ml. N/10)} \times 5{\cdot}61}{\text{wt. of sample}}$$

The F.F.A. figure is usually calculated as oleic acid (1 ml. N/10 $\equiv$ 0·028 g. oleic acid).

B.S.684 : 1958, however, recommends that the F.F.A. in the following oils should be calculated as the acid which is more likely to be present, *viz.*:—

Palm oil as palmitic acid (1 ml. N/10 $\equiv$ 0·0256 g.).

Palm kernel, coconut and similar oils as lauric acid (1 ml. N/10 $\equiv$ 0·02 g.).

With most oils, rancidity is noticeable to the palate when the acidity as oleic acid is about 0·8–1·5 per cent.

Rancidity

Fats undergo changes during storage which result in the production of an unpleasant taste and odour, which is commonly referred to as rancidity. Rancidity is brought about by the action of air (*oxidative rancidity*) or by micro-organisms (*ketonic rancidity*). Oxidative rancidity is accelerated by exposure to heat and light, by moisture and by the presence of traces of certain metals (*e.g.*, copper, nickel, iron). It is now generally accepted that oxygen is taken up by the fat with the formation of compounds which react as peroxides. In general, the greater the degree of unsaturation (the higher the iodine value) the greater is the liability of the fat to oxidative rancidity. When the concentration of " peroxides " reaches a certain level, complex chemical changes occur and volatile products are formed which are mainly responsible for the rancid taste and odour.

With most oils and fats, the free acidity increases during storage, but with refined oils particularly, the F.F.A. figure is not necessarily related to the extent to which rancidity has progressed. On the other hand, although the " peroxides " are possibly not directly responsible for the taste and odour of rancid fats, the concentration of them as represented by the peroxide value is often useful for assessing the extent to which spoilage has advanced.

Reference: " Rancidity in Edible Fats " by Lea (Food Invest. Spec. Report No. 46, 1938, H.M.S.O.).

Peroxide value. The peroxide value is a measure of the peroxides contained in the oil (see above). During storage, peroxide

* Alternatively boiling neutral alcohol can be employed as solvent and the F.F.A. is then titrated whilst the liquid is still hot.

formation is slow at first during an induction period which may vary from a few weeks to several months according to the particular oil or fat, the temperature, etc., and this must be borne in mind when interpreting quantitative results.

The peroxide value is usually determined volumetrically by methods which have been largely developed by Lea. The following simple method is suitable for the routine testing of samples:—

METHOD I

Carry out the test in subdued daylight:—

Weigh out 1 g. (or less) of oil or fat into a clean dry boiling tube and while still liquid add 1 g. of powdered potassium iodide and 20 ml. of solvent mixture (2 vols. glacial acetic acid + 1 vol. chloroform). Place the tube in boiling water so that the liquid boils within thirty seconds, and allow to boil vigorously for not more than thirty seconds. Pour the contents quickly into a flask containing 20 ml. of potassium iodide solution (5 per cent.), wash out the tube twice with 25 ml. of water and titrate with N/500 sodium thiosulphate solution using starch near the end-point. A blank should be performed at the same time. The peroxide value is usually reported as the number of mls. of N/500 sodium thiosulphate per g. of sample. If the value so obtained is multiplied by 2, the figure then equals milliequivalents of peroxide oxygen per kg. of sample (ml. N/1 per kg.). Modifications of this method are described in B.S.684 : 1958 (see also *Analyst*, 1948, **70**, 403).

Fresh oils usually have peroxide values well below 5 ml. N/500 thiosulphate per g. A rancid taste is often just noticeable when the peroxide value is between 10 and 20. In interpreting such figures, however, it is necessary to take into account the particular oil or fat involved.

METHOD II

The method of Sully (*Analyst*, 1954, **79**, 86) for determining the peroxide value of oils, in which the oil is poured down a specially designed reflux condenser, appears to give consistent results:—

Apparatus

A 100 ml. r.b. flask with a ground glass joint is attached to a plain reflux tube, 75 cm. long and 9 mm. internal diameter, the upper 15 cm. of which can be cooled by a water jacket.

Procedure

Add 10 ml. of chloroform and 10 ml. of glacial acetic acid to the flask and boil the mixture to the top of the tube (where it is condensed by the water jacket). A micro gas flame close to the glass ensures steady boiling. When the mixture is refluxing steadily pour 1 g. of potassium iodide dissolved in 1·3 ml. water slowly down the condenser so that the refluxing is not interrupted. Re-dissolve any " precipitated " iodide by adding six drops of water. Then add 1 g. (or less) of the sample down the condenser, again without interrupting the refluxing, and turn off the condenser water so that all the sample is washed into the flask. Boil the mixture for four minutes with the condenser water running. Then remove the flask, cool rapidly, add 50 ml. of water and titrate the liberated iodine with thiosulphate using starch as before.

A colorimetric method for determining the peroxide values of fats has been described by Swoboda and Lea (*Chem. & Ind.*, 1958, No. 33, 1090) in which the iodine liberated by the peroxides in an acidic aqueous phase is converted to a relatively stable cadmium-iodide complex. The amount of iodine can then be measured absorptiometrically as the blue starch-iodide complex (*cf.* Lea, *J. Sci. Food Agric.*, 1952, **3**, 586).

Kreis Test. The Kreis test involves the production of a red colour due to the extremely sensitive reaction between phloroglucinol and the epihydrin aldehyde.

Qualitative Test. Shake 10 ml. of the oil or melted fat vigorously with 10 ml. of 0·1 per cent. phloroglucinol solution in ether and 10 ml. of concentrated hydrochloric acid for twenty seconds. A pink colour indicates incipient rancidity. If the oil is diluted 1 in 20 with kerosene and the test is still positive the rancidity will probably be evident to the taste and smell (*J. Ind. Eng. Chem.*, 1918, **10**, 471).

Quantitative Test. (Walters, Muers and Anderson, *J.S.C.I.*, 1938, **57**, 53). *N.B.* All volumes must be measured exactly by pipette. Weigh out 3 ml. of oil or melted fat into a test tube, add 1 ml. of phloroglucinol solution (0·5 per cent. w/v in amyl acetate) and stir vigorously for a minute. Add 2 ml. of a solution containing 10 g. of trichloracetic acid dissolved in 3·28 ml. of amyl acetate and immerse the tube in a water-bath at exactly 45° C. for fifteen minutes, keeping the mixture stirred continuously. Remove the tube and dilute immediately with 10 ml. of an ice-cold solution made by diluting 1 volume of the above trichloracetic acid solution with 2 volumes of amyl acetate. Match the colour in the Lovibond tintometer immediately and record the number of red units (R). Carry out a blank at the same time, using amyl acetate in place of the phloroglucinol solutions (reading R′ red units).

$$\text{Kreis value} = T = \frac{R - R'}{lc}$$

where l = cell length in cm.
and c = concentration of oil in g. per ml. of final solution.

In view of the highly sensitive nature of the Kreis reaction, this test gives an indication of incipient rancidity. Walters and his co-workers found a value for T of about 20 for fresh butter fat (*cf.* Pyke, *Analyst*, 1935, **60**, 515).

Ketone or Perfume Rancidity. A form of rancidity due to ketones, caused by moulds and encouraged by moisture, sometimes occurs in butter fat and in coconut and palm kernel oils. A characteristic taste and odour is produced. Taufel and Thaler (*Chem. Z.*, 1932, **56**, 265) have described a method for the detection of ketone (or perfume) rancidity which is based on the colour produced when salicylaldehyde reacts with the ketones in the presence of sulphuric acid.

Tests for the susceptibility to oxidation of oils and fats. Industrially, it is sometimes important to get an idea of whether a particular oil or fat is likely to be able to withstand storage over a period. The probable course of the rancidity can be predicted by methods in which the oxidation is artificially accelerated by passing air through the sample or by raising the temperature (or both). Samples can be taken off periodically and examined for F.F.A., peroxide value, etc., at regular intervals. In the Schaal Oven Test (see *Oil and Soap*, 1938, **15**, 184) the sample is incubated at 65° C. and examined daily. In the rather more elaborate Swift Stability Test of King *et al.* (*ibid.*, 1933, **10**, 1) a sample of lard is aerated at 98° C. and the rancidity is assessed at hourly intervals. The authors of the paper record the number of hours for the peroxide value of the lard to reach 20 milli-equivalents of oxygen per kg. Mehlenbacher (*ibid.*, 1942, **19**, 137) reduces the induction period still further by carrying out the aeration at 110° C. The oxidation can also be accelerated by the addition of certain metals, *e.g.*, copper.

Lard

The definition and composition of lard have changed somewhat in recent years by reason of the inclusion of the fat of all parts of the pig, instead of only that from the omentum, as formerly. The U.S. Department of Agriculture gives the following definitions: Lard is the rendered fresh fat from hogs in good health at the time of slaughter; it must be clear, free from rancidity, and contain not more than 1 per cent. of substances other than fatty acids and fat. Leaf lard is similarly defined as lard rendered at a moderately high temperature from the internal fat of the abdomen of the hog, excluding that adherent to the intestines, and having an iodine number not greater than sixty. Neutral lard is lard rendered at low temperatures.

Lard consists of glycerides of the following fatty acids: oleic, linolenic, lauric, myristic, palmitic and stearic. There is also usually present a minute amount of tissue substance, a trace of moisture and unsaponifiable matter. In genuine lard the melting point affords an indication of the origin of the fat. Leaf lard usually has a melting point of 44°–45° C., lard from the kidneys 39°–42° C., that from the back and other parts 34°–39° C.; but these figures are, of course, somewhat variable. The definition of "Lard" in the Oils and Fats Order, 1949, stated that the melting point should be not less than 25° C. when determined by a specified method.

CONSTANTS OF LARD

	Usual range	*B.P.C.* 1959 *standards*
Specific gravity at 40° C.	0·902–0·907	—
Melting point (° C.)	34–45	33–41 (after treatment)
Refractive index at 40° C.	1·4583–1·4610	—
= Zeiss butyro reading	48·5–52	—
Iodine value	51–70	50–66
Saponification value	192–200	192–198
Unsaponifiable matter (%)	0·2–0·7	0·5 (max.)
Acid value	0·2–1·5	1·2 (max.)

The B.P.C. also requires the tests for cottonseed oil and sesame oil to be negative.

Analysis of Lard

In the examination of lard the melting point and specific gravity figures are of relatively little value as most of the likely adulterants give similar values. The refractive index and the iodine value are perhaps the most important constants to determine and when supplemented by qualitative tests will usually suffice for the identification of pure or adulterated samples.

Water. If lard appears cloudy on melting the presence of water is suggested.

Iodine value. It has been shown that the iodine value varies according to the part of the animal from which the fat was obtained. Also Ellis *et al.* (*J. Biol. Chem.*, 1925, **66**, 101; 1926, **69**, 219) have shown that (a) the iodine value of lard from young hogs is higher than that of mature animals and (b) oils assimilated by the pig are transferred to its fat with little change in the fatty acids. Ellis quotes a figure of 93·2 for the iodine value of lard obtained from pigs fed on soya beans (iodine value of soya oil = 128). Sutton *et al.* (*Analyst,* 1940, **65**, 623) have presented a useful survey of the constants of lard. All but 5 of the 113 samples which they examined had iodine values within the range 57·0–72·9. In view of the above findings, samples of lard with iodine values in excess of the B.P.C. maximum of 66 cannot be regarded as necessarily adulterated with vegetable oil.

Refractive index. The refractive index at 40° C. is sometimes of value in detecting adulteration.

Unsaponifiable matter. For routine purposes Holdé's qualitative test (p. 420) should be adequate.

Vegetable oils. The most reliable test for vegetable fats in lard is the separation of the sterol acetates from the unsaponifiable matter as described on p. 397. The finding of any phytosterol would be conclusive evidence of the presence of vegetable fat. Modern lard substitutes and cooking fats consist usually of hydrogenated oils. These may be of animal or vegetable origin, and their detection in a lard mixture is sometimes difficult. The finding of any significant proportion of *iso*-oleic acid or other solid unsaturated acid is, however, a definite indication of the presence of such hardened oil.

Hydrogenated Oils. Oils are "hardened" by causing a proportion of the unsaturated acids to combine with hydrogen in the presence of a nickel catalyst so that saturated acids are produced. Many of the analytical figures change during the process, *e.g.*, the iodine value and refractive index get less and the melting point and solidifying point are raised.

Apart from other changes which occur during hydrogenation, an isomer of oleic acid is produced which can be detected by its property of forming a lead salt which is insoluble in solvents (those of other unsaturated acids are soluble). In the following method of Cocks, Christian and Harding (*Analyst*, 1931, **56**, 368) the insoluble lead salts are weighed and a portion of them is taken for the determination of the iodine value:—

About 5 g. of the fat are saponified with alcoholic potash, and the fatty acid liberated by the addition of dilute sulphuric acid. The fatty acids are washed by boiling with water and dried by passing through dry filter paper. Then 3·5 g. of dry, freshly prepared fatty acids are dissolved in 50 ml. of alcohol and heated to boiling point; to the solutions are added 50 ml. of boiling alcoholic solution of lead acetate (3·45 g. lead acetate for lard or other fat containing upwards of 25 per cent. of total solid acids, 1 g. for oils containing less than 25 per cent.). The mixture is brought to the boil and allowed to stand overnight at a temperature of 15°–20° C. Next morning the lead salts are collected on a Buchner funnel and, after changing the receiver, washed with 100 ml. of petroleum spirit. The petroleum spirit is removed by distillation and the residue dissolved in 20 ml. of boiling alcohol and 1 drop of glacial acetic acid added from a pipette. The solution is allowed to crystallise at 15°–20° C. for three hours. The lead salts so precipitated are filtered off and washed with 20 ml. of cold alcohol and added to the main bulk of lead salts. These are suspended in a little water, and decomposed by addition of a small excess of dilute nitric acid in the presence of ether. The ethereal solution of the acids is washed with water to remove mineral acid, then evaporated in a tared flask and the total solid fatty acids weighed. A portion of this is taken for determination of the iodine value. From the iodine value the proportion of unsaturated acids is calculated on the basis of an iodine value of 90 (unless rape oil is known to be present, the erucic acid of which has iodine value 75·5). Natural oils other than rape oil will contain less than 3 per cent. of solid unsaturated acids, except tallow, which may show 5 or 6 per cent. Rape oil contains about 50 per cent. Hardened fats usually show from 20 to 50 per cent. of solid unsaturated acids (*cf. Analyst*, 1924, **49**, 460).

Beef fat, Mutton fat. The detection of beef or mutton fat in lard is one of the difficult problems which may arise in the examination of fats. Formerly much stress was laid upon the form of the crystals deposited from ether, but experience shows that the observations must be interpreted with caution, and a positive result ought to be supported by other evidence, such as an abnormal iodine value.

Two ml. of the melted fat are dissolved in 10 ml. of ether in a cylinder closed with a plug of cotton-wool and the solution set aside in a cool place until a deposit of white crystals appears. The ether is poured off and a few of the crystals are rapidly transferred to a microscope slide with a drop of neutral oil and examined, first under a low power and then under a high power, 200–250 diameters. The characteristic appearance of lard crystals is in large tufts, each crystal having chisel-shaped ends, or as oblong plates with oblique ends, whereas beef or mutton fat gives fan-like groups with pointed ends. It has been shown by Hehner and Mitchell that the form of the crystals depends on the proportion of stearic acid and that some genuine lards give crystals resembling those of beef fat. Repeated crystallisation of genuine lard from ether causes the crystals to assume the beef-like form; after five or six crystallisations quite misleading results will be obtained. The important matter is to note the ends. Lard often forms beef-like tufts, but when examined under the high power the ends of the crystals are always oblique, not pointed. Sutton *et al.* (*loc. cit.*) have published photomicrographs of the crystals deposited from ether by mixtures of lard and beef fat. Hydrogenated fats often yield tufts of crystals closely resembling those of beef or mutton.

Bömer's test. Bömer has described a useful confirmatory test for beef fat, depending on the melting point of the distearin, which in lard is 68·5° C. and in beef is 63·3° C., although these values are not obtained on the mixed glycerides:—

Fifty ml. of the fat are dissolved in an equal volume of acetone, and the solution is cooled to 15° C. and allowed to stand for an hour, then the crystals are rapidly filtered off and dried in a desiccator over sulphuric acid. They are now re-dissolved and crystallised from ether. The melting point of the crystals is taken, then they are saponified, and the melting point of the liberated fatty acids is observed. The melting point of the glycerides should be more than 62° C. and of the fatty acids not less than 56° C., whereas the corresponding figures for beef are much lower. The presence of hydrogenated fats may vitiate this test, but otherwise it is reliable when the amount of beef present is not less than 10 per cent., and is favourably reported on by Prescher (*Z. Unters. Nahr. Genussm.*, 1915, **29**, 433), who investigated many samples by this and other methods.

Constants obtained from Lard and Some Possible Adulterants of Lard

	*Lard	Beef fat	Cotton stearine	Coconut oil	Palm-kernel oil
Saponification value	192–200	194–200	194–199	245–265	245–255
Iodine value	51–70	35–44	88–92	8–10	10–15
Refractive index 40°	1·458–1·461	1·455–1·459	1·475	1·440	1·440
Melting point of fatty acids	37°–47°	40°–44°	28°–37°	18°–25°	22°–28°
Reichert value	Less than 1·0	Less than 1·0	Less than 1·0	6·7–8·4	5–6·5
Polenske value	Trace	Trace	—	16	10

* See also p. 425

The interpretation of Bömer values has been discussed by Sutton *et al.* (*loc. cit.*).

Cottonseed oil. Cottonseed oil in lard can be detected by means of Halphen's test (p. 436). In interpreting the results it must be borne in mind that lard from pigs which have been fed on cottonseed cake sometimes produces a slight positive reaction.

Rancidity. The edibility of lard can be assessed from the determination of the F.F.A. (p. 420), the peroxide value (p. 421) and the Kreis test (p. 423).

Lard Substitutes, Cooking Fats

"Cooking Fats" were defined in The Oils and Fats Order, 1949, as "any fat or mixtures of oils and fats fit for human food or for use in manufacture, preparation, or treatment of human food, having a melting-point which when determined by the method specified in the Fourth Schedule to this Order is not less than 30 degrees Centigrade; but does not include dripping, premier jus, lard, margarine, butter, vegetarian butter, cocoa butter, horse fat, technical tallow, or any unrendered animal fat." These mixtures may contain hydrogenated vegetable or whale oils together with possibly coconut oil, stearin or tallow. They may be examined by the methods applicable to lard, with the addition of the Reichert-Polenske values. Lard substitutes should be free from water and the F.F.A. should be low.

Dripping

According to The Oils and Fats Order, 1949, "Dripping" means "unbleached, unadulterated fat: (a) produced from, or by the rendering or processing of, the fat or bones of sheep or oxen; (b) of a sweet taste; (c) of a sweet smell; (d) untreated by any chemical process; and (e) containing not less than 99 per cent. of saponifiable matter and not more than 1·5 per cent. of free fatty acids; but does not include imported premier jus."

Most samples of dripping have iodine values between 35 and 44.

Suet

Legislation

Standard for Block Suet and Shredded Suet

The Food Standards (Suet) Order, 1952 (S.I. 1952 No. 2203)

"1. Block suet shall consist of rendered beef suet, shall be free from fibrous tissue and shall contain not less than 99 per centum by weight of beef fat.

2. Shredded suet shall consist of rendered beef suet with farinaceous material. It shall be free from fibrous tissue, shall be shredded, flaked or otherwise comminuted and shall contain not less than 83 per centum by weight of beef fat."

The Code of Practice for shredded suet (C.P. 12) infers that with certain reservations there is no justification for claims (a) that shredded suet is more economical than ordinary suet, (b) that it is " richer in fat," (c) that it is more easily digestible or (d) that there are special nutritional or dietary properties.

Suet may be defined as the fatty tissue situated about the loins or kidneys of oxen or sheep. It consists of an accumulation of microscopic particles of fat, each contained in a membranous bag. There is present about 1 per cent. of cellular tissue with some 4 per cent. of water and 95 per cent. of fat. The fat, when run out by heating, forms tallow. It is difficult or impossible to differentiate chemically between beef and mutton fats, though the superior taste and smell of beef may be recognised. The usual constants are given in the table, but these fats are not often adulterated.

Specific gravity at 15·5° C.	0·940–0·950
Melting point (°C.)	47°–54°
Saponification value	194–200
Iodine value	35–44
Refractive index 40° C.	1·455–1·459
Reichert value	0·5–1·0

See also p. 427.

The microscopical appearance of the crystals obtained from an ether solution of these fats has been mentioned on p. 427.

Shredded suet. Suet can be shredded because of the membranous substance present, while tallow cannot be so divided without the addition of starchy matter. This has led to the introduction of prepared suets. The usual method of preparation is to force the fat through fine holes so that it forms thin fragments. The fragments are coated with rice (or wheat) flour, which prevents the pieces from caking together again and is also said to assist in retarding rancidity. The analytical figures for most samples of shredded suet fall within the following percentage ranges:—

moisture 1·8–2·7, dry rice flour 12–15 and beef fat 83–86.

Analysis of Shredded Suet

Sampling. As some of the cereal tends to separate it is especially important to mix the whole sample to ensure uniformity.

Water. Dry 5 g. at 100° C. for two to three hours.

Fat. Filter a weighed quantity on a tared filter paper in the 100° C. oven until most of the fat has run through into a Soxhlet flask. Then complete the extraction in a Soxhlet apparatus using light petroleum (legal minimum for fat is 83 per cent.). Then determine the iodine value, etc. to confirm the presence of beef fat and also the F.F.A., etc., to assess the edibility.

Starch. The dry cereal filler can be obtained by difference. Confirm the figure by weighing the residue from the fat extraction. Identify the starch microscopically.

SALAD OILS

The term " salad oil " is not necessarily synonymous with olive oil. The better qualities of salad oil consist of pure olive oil, but the lower quality products may contain other oils such as arachis or teaseed oil.

The examination of salad oil should include the determination of the usual constants. As the constants for several of the possible oils are somewhat similar it is necessary to apply special tests for the presence of arachis oil (p. 435), cottonseed oil (p. 436), sesame oil (p. 436) and teaseed oil (p. 437). It should be borne in mind that the iodine value of olive oil is lower than that of the other oils.

Olive Oil

Olive oil is defined in the B.P. as the fixed oil expressed from the ripe fruits of *Olea europœa.* Its colour is described as pale yellow or greenish-yellow. The B.P. requirements include:—

Weight per ml. at 20° C.	0·910–0·913
Refractive index at 20° C.	1·468–1·471
Iodine value	79–88
Saponification value	190–195
Acid value	2·0 (max.)

The B.P. also requires arachis oil, cottonseed oil and sesame oil (see pp. 435–437) to be absent. The 1948 B.P. required olive oil to show a negative reaction to Fitelson's test for teaseed oil (p. 437). For routine purposes it is advisable to apply in addition to these tests the rapid Holdé test (p. 420) for the detection of mineral oil, *e.g.*, liquid paraffin.

Complaints occasionally arise in cold weather in relation to a deposit which forms at the bottom of the bottle. This is quite normal and is merely due to the deposition of stearine which occurs below 10° C.

Arachis Oil

Arachis oil (Ground-nut oil, Nut oil, Peanut oil) is defined in the B.P. as the fixed oil obtained from the seeds of *Arachis Hypogœa.* The B.P. requirements include:—

Weight per ml. at 20° C.	0·911–0·915
Refractive index at 20° C.	1·468–1·472
Iodine value	85–100
Saponification value	188–196
Acid value	2·0 (max.)

The B.P. also requires the tests for cottonseed and sesame oil (see p. 436) to be negative. The B.P. test for "other vegetable oils" is essentially Evers' semi-quantitative test for arachis oil (p. 435), but the acid used is 1·5 ml. of acetic acid (33·5 per cent.) instead of hydrochloric acid. The clouding point should be not lower than 37° C. It is also advisable to apply Fitelson's test for teaseed oil (p. 437) and Holdé's test (p. 420).

Cottonseed Oil

Cottonseed oil is defined in the B.P. as the fixed oil obtained from the seeds of various cultivated species of *Gossypium* and refined. Its colour is described as pale yellow or yellow and the odour as "odourless, or nearly odourless." The crude oil is dark in colour and has a characteristic odour. After refining both are lessened but the taste and smell are always recognisable. The B.P. requirements include:—

Weight per ml. at 20° C.	0·915–0·920
Refractive index at 20° C.	1·472–1·474
Iodine value	103–107
Saponification value	190–198
Acid value	0·5 (max.)

Freezing-point. At a temperature below 12° C. particles of solid fat begin to separate and the oil congeals at a temperature between 0° and − 5° C.

The B.P. also requires the test for sesame oil (p. 436) to be negative. Cottonseed oil is readily identified by Halphen's test—a pink or red colour should develop within thirty minutes.

The titre (p. 417) of the fatty acids of cottonseed oil is unusually high, generally above 32° C. (olive oil gives a titre of about 23° C.). Note also the higher iodine value (see Table, p. 432).

Rape Oil, Colza Oil

Rape oil (Colza oil) is expressed from the seeds of various species of *Brassica.* Reference to the table of constants (p. 432) shows that the saponification value is lower than that of other vegetable oils. Rape oil also has an unusually high viscosity and may be used for

TABLE SHOWING COMPARISON OF CONSTANTS FOR SALAD OILS AND ADULTERANTS

Oil	Specific gravity at 15·5° C.	Refractive index 40° C.	Iodine value	Saponification value	F.F.A. as oleic acid (%)	Unsaponifiable matter (%)	Titre ° C.	Special tests
Almond	0·915–0·920	1·4624–1·4650	95–100	188–196	0·4–1·0	0·4–1·0	—	—
Apricot-kernel	0·918–0·921	1·4640–1·4650	98–110	188–192	0·4–1·0	0·5–1·0	—	Bieber's test (p. 437)
Arachis	0·917–0·919	1·4625–1·4645	85–100	100–196	0·4–1·0	0·4–1·0	30	Bellier's test (p. 435)
Cottonseed	0·921–0·925	1·4640–1·4655	103–107	190–198	0·5–1·5	0·4–1·0	33	Halphen's test (p. 436)
Olive	0·915–0·919	1·4605–1·4635	79–88	190–195	0·3–1·0	0·5–1·0	23	—
Peach-kernel	0·918–0·920	1·4640–1·4650	93–105	189–192	0·4–1·0	0·5–1·0	—	Bieber's test (p. 437)
Poppyseed	0·924–0·926	1·4680–1·4690	130–138	192–196	0·5–2·0	0·5–1·5	16	—
Rape	0·914–0·916	1·4653–1·4660	95–105	170–176	0·3–2·0	0·4–1·0	—	Tortelli and Fortini's test (p. 437)
Sesame	0·921–0·924	1·4650–1·4665	103–116	188–195	0·4–2·0	0·7–1·2	23	Baudouin's test (p. 436)
Sunflower	0·924–0·926	1·4660–1·4680	125–140	190–194	0·5–2·0	0·5–1·0	—	—
Teaseed	0·915–0·920	1·4605–1·4640	80–90	188–196	0·3–6·0	0·4–1·0	14	Fitelson's test (p. 437)

N.B. The principal B.P. requirements for the official oils are stated on other pages.

viscometer standardisation. The oil can be detected by Tortelli and Fortini's test (p. 437).

Fish oils can be detected by the hexabromide test (p. 438).

Sesame Oil

Sesame oil is defined in the B.P. as the fixed oil expressed from the seeds of *Sesamum indicum* L. The B.P. describes sesame oil as a pale yellow oil which does not solidify when cooled to 0° C. The B.P. requirements include:—

Weight per ml. at 20° C.	0·916–0·919
Refractive index at 20° C.	1·472–1·476
Iodine value	103–116
Saponification value	188–195
Acid value	2·0 (max.)

The B.P. also requires Halphen's test for cottonseed oil (p. 436) to be negative. The red colour formed in Baudouin's identity test (p. 436) is due to the presence of a phenolic substance sesamol in sesame oil. This substance is, however, destroyed by heat and no reaction takes place with oils which have been heated for half an hour at 200° C. The iodine value of sesame oil is somewhat higher than that of olive oil.

Teaseed Oil

This oil is expressed from the seed of *Thea sasangua,* which is cultivated in China. The constants of teaseed oil are rather similar to those of olive oil (see Table, p. 432). Teaseed oil can be detected by Fitelson's test (p. 437) and by the low titre of about 14° C. (*cf.* olive oil, 23° C.). Teaseed oil may be used as a salad oil and is particularly valuable as a frying oil.

Almond Oil

Almond oil is defined in the B.P. as the fixed oil obtained by expression (without the application of heat) from the seeds of *Prunus amygdalus.* var. *dulcis* or of *Prunus amgydalus.* var. *amara.* The colour is described as pale yellow and the taste as bland and nutty. The B.P. requirements include:—

Weight per ml. at 20° C.	0·909–0·914
Refractive index at 20° C.	1·470–1·473
Iodine value	95–100
Saponification value	188–196
Acid value	2·0 (max.)

Freezing-point. Remains clear after keeping at − 10° C. for three hours; does not congeal until the temperature has been reduced to about − 18° C.

The B.P. also requires the test for the following oils to be negative —cottonseed oil (p. 436), sesame oil (p. 436), apricot-kernel and peach-kernel oil (Bieber's test, p. 437) and arachis oil (clouding point not above 4° C., p. 435).

Ground almonds are prepared from the sweet almond, often with some small proportion of the bitter almond. The product normally contains:—

Fixed oil	55–60 per cent.
Ash	3–4 ,,
Moisture	4–6 ,,

If any bitter almond is present there will also be a small proportion of hydrocyanic acid. Processes exist for the debittering of bitter almonds, and it is possible that some ground almonds contain such a product. The principal analytical problem presented by ground almonds is the detection of foreign nuts. Two methods are available:— first, a microscopic examination of the de-fatted material, and second, tests on the extracted oil. Almonds are said to contain no starch. Nevertheless, on microscopical examination one or two small starch grains may be found in most fields. Barcelona nuts, cob nuts and peanuts contain larger amounts of starch, so that if they are present in any substantial quantity the starch is easily detected. Chestnuts contain much starch. Examination of the ash is of interest, for although the total amount of the ash of the different nuts is not of much diagnostic value, cob nuts and others contain traces of manganese, and Brazil nuts contain barium, whereas almonds contain no barium, and only a minute quantity of manganese. For the examination of the oil, it is better to separate some by pressure rather than by solvent. If a solvent is used, particular care should be taken to remove all of it before applying the Bieber and Evers' tests.

Peach-kernel oil and **Apricot-kernel oil** have been sold as adulterants of or substitutes for almond oil. These oils can be detected by Bieber's test (p. 437). Also on shaking with an equal volume of nitric acid, peach-kernel oil gives a dark orange colour, apricot-kernel oil an orange-yellow and almond oil a pale yellowish-brown. The iodine value of apricot-kernel oil is slightly higher than that of peach-kernel or almond oil (see also Table of constants— p. 432).

Special Tests for Individual Oils

Detection and Determination of Arachis Oil in Other Oils

The oil values of arachis oil are very similar to those obtained with a number of other oils. Tests can be employed, however, for its detection which are based on the separation of arachidic acid from the glyceride present in arachis oil. Bellier (*Ann. Chim. Anal.*, 1899, **4**, 4) reported on the relative insolubility of arachidic acid in 70 per cent. alcohol when compared with stearic and palmitic acids. Evers and co-workers (*Analyst*, 1912, **37**, 487; 1937, **62**, 96) have described tests for the detection and estimation of arachis oil in olive and other oils which are based on this test (see also B.P. 1958, p. 871; B.S. 684 : 1958, p. 97):—

Semi-Quantitative Test

Boil 1 ml. of the oil with 5 ml. of 1·5 N alcoholic potash (8·416 per cent. KOH in 95 per cent. alcohol) in a small flask fitted with a long air condenser for ten minutes. Add 50 ml. of alcohol (70 per cent. v/v) and 0·8 ml. of hydrochloric acid (s.g. 1·16) and warm to dissolve any precipitate which may have formed. Cool, stirring continuously with a thermometer in the liquid, so that the temperature falls at the rate of about 1° C. per minute. Arachis oil produces a turbidity at 39–40° C., whereas with olive oil and almond oil the liquid remains clear at 9° C. and 4° C. respectively. Evers' 1937 paper gives a graph which relates the clouding points of mixtures of olive and arachis oils which can be employed for quantitative estimations provided only these two oils are present. The clouding points of various oils are as follows:—arachis oil 39–40°; rape seed oil 22·5°; sesame oil 15°; cottonseed oil 13°; olive-kernel oil 10°; maize oil 7·5°; olive oil 6–9°; almond oil − 1 to + 1°; apricot-kernel oil − 4 to − 8°.

Quantitative Method

Five g. of the oil are saponified under a reflux condenser with 25 ml. alcoholic potash (80 g. potassium hydroxide dissolved in 80 ml. of water and diluted to 1 litre with 90 cent. alcohol). To the hot solution is added 7·5 ml. of acetic acid (1 volume of glacial acetic acid to 2 volumes of water) and 100 ml. of 70 per cent. alcohol, containing 1 per cent. of hydrochloric acid by volume. The solution is cooled to 12°–14° C. for an hour, then filtered and washed with acidified 70 per cent. alcohol, the washing being continued until the filtrate gives no turbidity with water. The total volume of wash liquid should be measured. The precipitate is dissolved according to its bulk in 25–70 ml. of hot 90 per cent. alcohol and cooled to 15°. The solution is allowed to stand at this temperature for three hours, filtered and the crystals washed with a measured volume of 90 per cent. alcohol and finally with 50 ml. of 70 per cent. alcohol. Then the crystals are washed into a tared flask with ether, dried and weighed and their m.p. noted. If the melting point is lower than 71° the product should be re-crystallised from 90 per cent. alcohol. To the weight of arachidic and lignoceric acid so obtained is added a correction for the solubility in 90 per cent. alcohol plus a correction for the total volume of 70 per cent. alcohol used. The corrected weight of acids is multiplied by the factor 17, 20 or 22 according as the melting point is 71°, 72° or 73° to convert it to percentage of arachis oil. The corrections to be added for each 100 ml. are (1) for solubility in 90 per cent. alcohol:—

Weight of Mixed Acids	Temperature of Washing Liquor		
	15°	17·5°	20°
0·05 gm.	+0·031	0·040	0·046
0·10 ,,	0·036	0·045	0·052
0·20 ,,	0·048	0·056	0·062
0·40 ,,	0·061	0·071	0·078
0·60 ,,	0·066	0·080	0·088
0·80 ,,	0·069	0·083	0·092
1·00 ,,	0·071	0·084	0·091

and (2) for the 70 per cent. alcohol:—

Weight of Acids corrected for 90 per cent. Alcohol	Melting Point of Crystals		
	71°	72°	73°
Less than 0·02 gm.	0·006	0·005	0·004
0·02–0·05	0·007	0·006	0·005
0·05–0·08	0·009	0·007	0·005
0·08–0·10	0·011	0·007	0·005
Over 0·10	0·013	0·008	0·006

Halphen's test for cottonseed oil. Prepare Halphen's reagent by mixing equal volumes of amyl alcohol and a 1 per cent. (w/v) solution of precipitated sulphur in carbon disulphide. For the test mix 2·5 ml. of the oil or melted fat with 2·5 ml. of the reagent in a tall one ounce screw capped media bottle. Screw in the cap and heat the tube in boiling water for 30 minutes. In the presence of from 2 per cent. and upwards of cottonseed oil a rose colour is produced, the intensity of which is approximately proportional to the amount present. As the chromogenic constituent varies in different specimens of the oil, it is necessary to use the same oil always in making colour comparisons and to interpret them quantitatively with caution. The colour is less with oil which has been strongly heated, and is no longer produced by oil which has been exposed to a temperature of 250°. Hydrogenation wholly or partly destroys the chromogenic bodies according to the stage to which the process has been carried. The indications of Halphen's test must be interpreted with special care in the case of butter or lard, since the feeding of the cows or pigs on cotton cake may cause a slight indication. Certain uncommon oils (kapok and baobab) are said to give the rose colour with Halphen's reagent.

Baudouin's test for sesame oil. Shake 2 ml. of the oil or melted fat with 1 ml. of concentrated hydrochloric acid containing 1 per cent.

w/v of sucrose and set aside for five minutes. In the presence of 1 per cent. or more of sesame oil, a reddish colour develops in the lower layer. Many oils give a dirty brown tint with this test, so that in cases of doubt a comparison should be made with a known oil and the same reagents. Butter or lard produced from animals fed on sesame cake may give a faint pink. There is always the possibility that dyes may be present, in butter for instance, which give a pink colour with the acid, but this can be checked by carrying out the test without the sucrose. Certain olive oils from North Africa give positive reactions to the Baudouin reaction and such oils should be examined by the following modified test: Shake 5 ml. of the oil in a flask with 5 ml. of a mixture of 9 volumes of ethanol (90 per cent. v/v) and 1 volume of 0·88 ammonia and heat the mixture on the water bath until free from alcohol and ammonia. Then if the Baudouin test is carried out on the product genuine oils give no reaction.

Fitelson's Test for Teaseed Oil (*J.A.O.A.C.*, 1936, **19**, 493; 1937, **20**, 418)

Mix together 0·8 ml. acetic anhydride, 1·5 ml. chloroform and 0·2 ml. concentrated sulphuric acid in a test tube. Cool in ice water, add about 0·22 g. (7 drops) of the oil, cool again and mix. If the liquid appears cloudy add acetic anhydride drop by drop, shaking after each addition until clear. Place the tube in ice water for five minutes. Teaseed oil produces a deep green colour by reflected light and a brown by transmitted light. Olive oil produces a green colour in both. Then add 10 ml. of dry ether and mix by one inversion of the tube. Teaseed oil gives a brown colour which changes to red within a minute and later fades. Olive oil gives an initial green colour, which slowly fades to a brownish-grey. Some Tunisian olive oils, however, give a transient faint pink colour (*cf. Quart. J. Pharm.*, 1937, **10**, 242).

Bieber's Test for the Detection of Kernel Oils in Almond Oil

Mix together (cautiously and keeping cool) equal parts by weight of water, concentrated sulphuric acid and fuming nitric acid (s.g. approx. 1·45). Then shake 5 ml. of the sample of almond oil vigorously with 1 ml. of the acid mixture and allow to stand for fifteen minutes. In the presence of about 30 per cent. of apricot-kernel oil or peach-kernel oil the whitish mixture produced develops a pink colour. Genuine almond oil only gives a brownish tinge.

Detection of Rape Oil

Tortelli and Fortini (*Chem. Zeit.*, 1910, **24**, 689) described a method for the detection of rape oil which depends on the presence of about 50 per cent. of erucic acid in the oil. The lead salt of this solid unsaturated acid is only partially soluble in ether, but lead oleate is soluble: Twenty g. of the oil are saponified with 6 g. of potassium hydroxide in 50 ml. of alcohol and the solution neutralised to phenolphthalein by titration with 10 per cent. acetic acid, then slowly poured into 300 ml. of boiling water containing 20 g. of lead acetate. The solution is cooled while being rotated so as to cause the soap to stick to the sides of the beaker, then the water is poured off and the soap washed three times with 200 ml. of warm water (60°–70°) and dried with filter paper. The dried soap is heated and shaken with 80 ml. of ether under a reflux condenser for half an hour, then the flask is cooled, corked and kept at exactly 15° C. for an hour. The solution is now filtered into a separator, the funnel being kept covered as closely as possible to avoid loss by evaporation. The filter and residue are re-treated with 40 ml. of ether, cooled again for an hour and filtered into the separator. The mixed ether solution of the lead soap is decomposed by 150 ml. of 10 per cent. hydrochloric acid and then the ether is washed with two 100 ml. quantities of water and evaporated at a low temperature (spontaneous evaporation is best) in a dry flask. The liquid acids are dissolved in 40 ml. of strong alcohol and a slight excess of saturated aqueous solution of sodium carbonate is added, *i.e.*, until sodium carbonate just begins to separate. The alcohol is now distilled off and the residual sodium soap is dried in a vacuum desiccator for forty-eight hours, then heated with 50 ml. of absolute alcohol and quickly filtered. The residue is re-treated with absolute alcohol until all the soap is dissolved. The alcoholic soap solution is evaporated and the residue dried over sulphuric acid. A quantity of 0·5 g. of the dry soap is dissolved in a large test tube by heating with 20 ml. of absolute alcohol. Then the solution is cooled slowly while it is stirred with a thermometer, and the temperature at which it becomes turbid is noted. The turbidity temperatures are as under:—

Oil	Turbidity temperature
Olive oil	24°–20°
Rape oil	50°–45°
Olive + 50 per cent. rape oil	40°–35°
+ 30 ,, ,,	35°–30°
+ 20 ,, ,,	35°–30°
+ 10 ,, ,,	34°–30°
Sesame oil	20°–18°
Arachis oil	22°–18°
Cotton-seed oil	16°–14°

The form of the deposit also may afford some information; it is flocculent and gelatinous in the case of olive oil, whilst in the case of rape oil it consists of minute crystalline needles which float in the liquid.

Hexabromide Test for the Detection of Fish Oils

Dissolve 0·5 ml. of oil in 10 ml. of dry ether, add 10 ml. of a mixture containing 28 vols. glacial acetic acid, 1 vol. bromine and 4 vols. nitrobenzene and shake vigorously.

If only rape oil is present, the mixture remains clear, but in the presence of fish oil or a drying oil such as linseed, a precipitate will soon appear. If this test be applied to the oil directly without dilution with ether, drying oils or fish oils give an immediate precipitate, rape oil and soya-bean oil give a turbidity, whereas other non-drying vegetable oils give no precipitate or only the slightest turbidity. Animal fats may give a turbidity when tested in this way.

Fat-soluble Vitamins

VITAMIN A

Vitamin A is usually determined by one of the following processes: (1) the rapid Carr-Price method, which is useful as a preliminary test and (2) the more accurate spectrophotometric method.

When the expected concentration of vitamin A in the oil is high, the blue value or the absorption spectrum can be determined directly on a solution of the sample, but with most samples, the determination should be made on a solution of the unsaponifiable matter.

Carr-Price Blue Value (*Biochem. J.*, 1926, **20**, 49T)

Antimony Trichloride Reagent (*B.P.* 1932). Wash chloroform two or three times with its own volume of water, and dry it over anhydrous potassium carbonate; pour off and distil, rejecting the first 10 per cent. of distillate. During drying and distillation, protect the chloroform from light. Wash antimony trichloride with the pure dry chloroform, until the washings are clear. Prepare a solution, saturated at 20°, of the washed, antimony trichloride with the pure dry chloroform. The solution contains 21–23 per cent. w/v. of $SbCl_3$ and should be kept in an amber-coloured well-stoppered bottle. To assay, mix 1 ml. with a solution of 2 g. of sodium potassium tartrate in 20 ml. of water, rotate the mixture, add 2 g. of sodium bicarbonate and titrate with N/10 iodine (1 ml. N/10 Iodine = 0·01141 g. $SbCl_3$).

METHOD

Pipette 0·2 ml. of the prepared solution (either 20 per cent. w/v of the oil in chloroform, or the solution of the unsaponifiable matter, prepared as described on p. 428, but using chloroform as solvent instead of cyclohexane) into a 1 cm. cell placed on the platform of a Lovibond tintometer. Add 2 ml. of the antimony trichloride reagent rapidly and observe the maximum intensity produced against the blue glass. For maximum accuracy the tintometer reading should lie between 4·0 and 6·0 blue units. Calculate the value to 1 g. of sample and multiply the figure obtained by 1·28 to give International Units per g. (*cf. Analyst*, 1940, **65**, 484). In the A.O.A.C. method, the colour is measured absorptiometrically at 620 mμ and the reading is compared against a standard curve representing the U.S.P. Vitamin A Reference Standard.

The Spectrophotometric Method

Vitamin A can be accurately determined by measuring the extinction in cyclohexane or alcohol at certain wavelengths.

THE INTERNATIONAL UNIT OF VITAMIN A

The Unit = 0·344 μg. vitamin A Acetate = 0·3 μg. vitamin A alcohol.

I.U. per g. = 1900 $E^{1\%}_{1\,cm.}$ at 326·5 mμ (for vitamin A alcohol).

METHOD (*B.P.* 1958)

Dissolve a suitable quantity of the sample in cyclohexane and determine the wavelength of maximum absorption by measuring extinctions between 323 and 331 mμ. Then measure the extinctions at the wavelengths given in the following table and calculate the relative extinctions.

RELATIVE EXTINCTIONS IN CYCLOHEXANE

Wavelength (mμ)	Vitamin A Acetate	Wavelength (mμ)	Vitamin A Alcchol
300	0·555	300	0·553
312·5	0·857	312·5	0·857
327·5 (max.)	1	326·5 (max.)	1
337·5	0·857	336·5	0·857
345	0·695	345	0·667
360	0·299	360	0·260

The wavelength of maximum absorption should be within $\pm$ 1·5 mμ of the stated value and the observed relative extinctions at the other five wavelengths within 0·02 of those in the table. When these conditions are not fulfilled, irrelevant absorption is present.

In some instances, the interfering substances may be removed by fractionating on a chromatographic column (Gridgeman *et al.*, *Analyst*, 1948, **73**, 662). If, however, the sample contains preponderantly pure or esterified vitamin A and its absorption curve is not grossly distorted relative to that of the pure vitamin it is possible to apply correction equations. This simple geometrical procedure involves the measurement of the absorption at three wavelengths between 310 and 340 mμ, the assumption being made that the irrelevant absorption is linear over this comparatively short range (see Morton and Stubbs, *Analyst*, 1946, **71**, 348). This method can be applied to the liver oils of halibut and cod:—

HALIBUT-LIVER OIL

$E_{327 \cdot 5\ m\mu}$ (corrected) = 7 ($E_{327 \cdot 5\ m\mu}$ − 0·405 $E_{312 \cdot 5\ m\mu}$ − 0·595 $E_{337 \cdot 5\ m\mu}$).

Then calculate the potency of vitamin A in I.U. per g. by multiplying the corrected value for $E^{1\%}_{1\,cm.}$ at 327·5 mμ by 1900.

COD-LIVER OIL generally shows irrelevant absorption near 308 and 317 mμ due to conjugated tetra-ene acids and it is preferable to carry out the determination on the unsaponifiable matter:—

Boil 1 g. of oil with 10 ml. of freshly prepared N/2 alcoholic potassium hydroxide for ten minutes. Add 20 ml. of water, cool and extract with four successive portions of 25 ml. ether. Wash the bulked ether extracts by gentle rotation (avoid violent shaking), successively with 10 ml. of water, with 10 ml. of N/2 aqueous potassium hydroxide, and with successive quantities of 10 ml. of water until the washings no longer give a pink colour with phenolphthalein. Filter the ethereal solution into a flask and wash the filter with ether. Evaporate off the ether by warming on a water-bath in a current of oxygen-free nitrogen. Dissolve the residue in cyclohexane and measure the extinction at 312·5, 326·5 and 336·5 mμ (*cf.* B.P. Addendum 1960, pp. 74–76). $E_{326 \cdot 5\ m\mu}$ (corrected) = 7 ($E_{326 \cdot 5\ m\mu}$ − 0·422 $E_{312 \cdot 5\ m\mu}$ − 0·578 $E_{336 \cdot 5\ m\mu}$) vitamin A (I.U./g.) = 1900 × (corrected value of $E^{1\%}_{1\,cm.}$ at 326·5 mμ).

MARGARINE. See p. 400.

MISCELLANEOUS REFERENCES (VITAMIN A)

Gridgeman (*Chem. & Ind.* 1947, p. 555, 574; *Analyst*, 1951, **76**, 449); Prokhovnik (*ibid.*, 1952, **77**, 185); Bagnall and Stock (*ibid.*, 1952, **77**, 356, 396); Dowler and Laughland (*Analyt. Chem.*, 1952, **24**, 1047).

VITAMIN D

The World Health Organisation (1949/50) has defined the International Unit of vitamin D as the activity of 0·025 μg. of vitamin D_3 (activated 7—dehydrocholesterol). $E^{1\%}_{1\,cm.}$ at 265 mμ for recrystallised vitamin D_3 is given as 490 (ethanol), which corresponds to a molecular extinction coefficient of 18,800.

The estimation of vitamin D in foods presents a number of difficulties. Nield, Russell and Zimmerli (*J. Biol. Chem.*, 1940, **146**, 73;

1943, **148**, 245) however, found that if acetyl chloride is mixed with the Carr-Price reagent, the blue colour quickly fades and an orange colour is developed which is suitable for quantitative measurements. The reagent, which must be fresh and free from alcohol, contains 20 per cent. w/v of antimony trichloride in chloroform (see vitamin A reagent above) with 4 per cent. pure acetyl chloride added. For the method, 9 volumes of this reagent are added to 1 volume of the unsaponifiable matter in chloroform and the value for $E_{1\,cm.}^{1\%}$ at 500 mμ is measured. This extinction × 1800 represents an approximate measure of vitamin D per g. Cox (*Analyst*, 1950, **75**, 521) suggests that this method is suitable for control purposes provided that the amount of vitamin A is not more than five times that of the vitamin D. If possible a suitable correction should be made for absorption due to other sterols.

Rogers (*J. Pharm. Pharmacol.*, 1954, **6**, 780) reduces the absorption due to other substances present by using as reagent a solution of 20 per cent. antimony trichloride in ethylene dichloride with 2 per cent. acetyl chloride added. The extinction at 500 mμ is recorded every half minute for six minutes and the resultant graph is extrapolated to zero time.

Stross and Brealey (*J. Pharm. Pharmacol.*, 1955, **7**, 739) have compared the results of vitamin D determinations obtained by various methods (including biological assays).

VITAMIN E

Vitamin E (tocopherols) occurs in the fatty portion of some foods such as cereals. For the determination of vitamin E the method of Emmerie and Engel (*Rec. Trav. Chim. Pays-Bas.*, 1939, **58**, 283) is useful for routine purposes. This depends upon the oxidation of the tocopherol in alcoholic solution by ferric chloride and the subsequent measurement (absorptiometrically) of the red colour produced when the resultant ferrous iron reacts with αα′-dipyridyl:—

Extract the unsaponifiable matter from the fat as described under vitamin A, dissolve in pure absolute alcohol and dilute to such a volume that 1 ml. contains about 0·1 mg. tocopherol (the unsaponifiable matter can be more readily dissolved by adding 2 ml. benzene and diluting with alcohol). To 1 ml. or more of the solution add 1 ml. of 0·2 per cent. ferric chloride in pure alcohol and 1 ml. of alcoholic 0·5 per cent. αα′-dipyridyl and dilute to 25 ml. Measure the red colour absorptiometrically after ten minutes at 520 mμ and calculate the amount of tocopherol present from the standard graph which is prepared in the following manner: Hydrolyse 0·1 g. of α-tocopherol acetate (= 0·091 g. α-tocopherol) with alcoholic potassium hydroxide, neutralise and dilute with alcohol to 100 ml. To 0, 0·1, 0·2. . . . 1·0 ml. portions of this solution add the ferric chloride and dipyridyl reagents, dilute to 25 ml. with alcohol and measure the colour at 520 mμ as described above (see also *Analyst*, 1939, **64**, pp. 216 & 446 and 1940, **65**, 370).

Materials such as cholesterol and carotenoids interfere in the above reaction, but various workers have put forward methods of purification. The Vitamin E Panel of the Analytical Methods Committee of the S.A.C. (*Analyst*, 1959, **84**, 356) have reviewed the methods which have been proposed for the separation and determination of tocopherols and have described a method which can be applied to oils, foods and feeding stuffs. The complete method involves (a) solvent extraction of the sample; (b) saponification and extraction of the unsaponifiable matter; (c) separation of most of the steroids by freezing from solution in methanol; (d) removal of carotenoids and residual steroids on a column of floridin earth; (e) separation of the individual tocopherols by two-dimensional paper chromatography followed by elution and (f) determination of the tocopherol concentration in the eluates by a rapid modification of the Emmerie-Engel method. The factors to be employed for several different tocopherols are given (*cf. Anal. Chem.*, 1947, **19**, 902; *Analyst*, 1959, **84**, pp. 297, 567 and 796; *Lab. Practice.*, 1960, **9**, 427).

Mineral Oil in Food

Maximum limits for mineral oil in foods are prescribed in The Mineral Oil in Food Order, 1949 (S.I. 1949 No. 614 as amended by S.I. 1950 No. 1239, S.I. 1954 No. 1044 and S.I. 1955 No. 1901). For the purposes of the Order " Mineral oil " means any hydrocarbon product, whether liquid, semi-liquid or solid, derived from any substance of mineral origin and includes liquid paraffins, white oils, petroleum jellies and hard paraffins.

Essentially the Order states that, with certain exceptions, mineral oil shall not be used in the composition or preparation of food. The main exceptions are as follows:—

	Mineral Oil (Maximum % w/w)
Dried fruit	0·5
Citrus fruit	0·1
General limit (when used as greasing agent)	0·2 *
Chewing compounds (*e.g.*, chewing gum)	12·5 (of microcrystalline wax†)

* " by reason of........its use as a lubricant or greasing agent on some surface with which the article of food has necessarily to come into contact in the course of its preparation.........."

† " Microcrystalline wax " means a mixture of solid hydrocarbons obtained from any substance of mineral origin: colour—white to pale amber; almost tasteless and odourless; m.pt. not below 160° F.; iodine number not exceeding 4·0 and conforms to the B.P. test for sulphur compounds in liquid paraffin.

There is evidence which suggests that ingestion of mineral oil may lead to liver and other damage (*J. Physiol.*, 1944, **103**, 306). Also in view of possible potential carcinogenic effects of impurities present, mineral oils of low quality should not be present in foods.

Determination of mineral oil in foods. The determination of mineral oil is perhaps most commonly required in confectionery and baked goods for the manufacture of which a greasing agent is often necessary. In general, the mineral oil present is usually estimated by extracting the fatty matter from the food with an organic solvent and after removal of the latter the unsaponifiable matter is determined in the extracted material. Complete extraction of the mineral oil from baked goods is difficult to achieve and for such products, Cookson and Coppock (*J. Sci. Food Agric.*, 1951, **2**, 434) recommend the use of cold carbon tetrachloride for the preliminary extraction. The total unsaponifiable matter will include that which is natural to any edible oils present in the sample, so for more accurate work, it is preferable to separate sterols, etc. on a column of alumina and wash through the mineral oil using light petroleum (see Williams, "*Oils, Fats and Fatty Foods*", 3rd Edn., pp. 105–7, 1950, Churchill, Lond.; *Añalyst*, 1938, **63**, 84; *J. Soc. Chem. Ind.*, 1949, **68**, 274).

It should be borne in mind that many foods are capable of absorbing mineral hydrocarbons from waxed wrapping papers.

Chapter 15

MISCELLANEOUS

Salt—Iodised Salt—Canned Soups—Dried Soups—Butter Confectionery—Saccharin Tablets

SALT

Code of Practice C.P.11 states that **non-compounded salt** should normally be described as Household Salt or Cooking Salt and possibly as " Table Salt " (provided no reference is made to its free running qualities). **Table salt** is the common name for salt containing magnesium carbonate and the words "prepared" or "free-running" may be used. A prominent indication of the free-flowing properties of a "Table Salt" should be given in the following form: " To improve the free-running qualities of this salt a very small proportion of magnesium carbonate has been added."

Iodised salt should also be described as " prepared " or " free-running " and the disclosure on the label should be in the following form: " This salt contains magnesium carbonate (to improve the free-running qualities) and potassium iodide. The iodine content is from 433 to 725 micrograms per ounce." Alternatively sodium iodide may be used.

Analysis of Salt

For routine purposes the *methods* prescribed in the B.P. monograph for sodium chloride are useful for assessing the general qualities of samples of salt: *viz.*, assay (titrate 0·25 g. with N/10 silver nitrate using potassium chromate; minimum 99·5% as NaCl); loss on drying at 130° C. (maximum 1·0 per cent.); limit tests for acidity and alkalinity, arsenic, barium, calcium and magnesium iron, lead, bromide and iodide and sulphate. It should be borne in mind, however, that table salt should not necessarily be expected to comply with the B.P. standards, *e.g.*, table salt containing added magnesium carbonate (see above) is unlikely to comply with the limit test for magnesium.

DETERMINATION OF IODINE IN IODISED SALT (Andrew and Mandeno, *Analyst,* 1935, **60**, 801)

Dissolve 50 g. iodised salt in water in a 250 ml. graduated flask, make up to the mark and filter. Transfer 200 ml. of filtrate to a conical flask, make faintly acid to methyl orange and add 1 ml. of saturated bromine water. Add a few glass beads, boil until salt begins to separate and then add sufficient water to just dissolve the solid (after cooling). Add 2 ml. of N/1 hydrochloric acid and 0·2 g. potassium iodide and titrate the liberated iodine (= 40 g. of sample) with N/200 thiosulphate using starch. Calculate the result as micrograms of iodine per ounce of sample (see Code of Practice).

1 ml. N/200 thiosulphate = 0·000105 g. iodine.

N.B.—P.p.m. × 28·35 = μg./oz.

A further method for determining iodine in iodised salt using permanganate as oxidising agent is described by Monier-Williams (" *Trace Elements in Food,*" 1949, p. 332, Chapman and Hall, Lond.).

The amount of iodine in iodised salt falls during storage (especially in cardboard containers).

CANNED SOUPS

Code of Practice

Minimum quantities for various ingredients are stated in the Code of Practice (C.P.4) for all the main types of canned soups. The code was drawn up when the raw material supply position was difficult and "without prejudice to amendment as and when ingredients become more freely available." The following represent some of the main requirements (all percentages w/w):—

Note. If the label bears a reference to dilution, the relevant minima must be increased proportionately.

Meat soups. General minimum of 6 per cent. of raw fat-free meat (excluding bone and fat). *In addition,* " oxtail soup " must contain at least 2 per cent. of whole oxtail (including bone) and for " kidney soup " two-thirds of the meat shall be kidney.

Cream soups. The code, as amended by Circular MF 3/54, states the following alternatives: *either* 2·5 per cent. butter-fat (min.) *or* 1·5–2·5 per cent. butter-fat and 3 per cent. total fat (min.) *or* 0–1·5 per cent. butter-fat and 3·5 per cent. total fat (min.).

Vegetable soups. Mixed vegetable soups to contain at least four different vegetables with no one unduly predominant. With single-named vegetable soups (*e.g.,* " celery soup ") the named vegetable to exceed the sum of the bulk weight of other vegetables. " Green Pea Soup " should not be made with processed peas.

Clear soups. Unless the word "Vegetable" appears in the title, meat should be the predominant single ingredient in the preparation of "Consommé" or "Julienne." "Julienne" may contain 3–6 per cent. strip or diced vegetables.

Broths should contain extractives from meat or protein (N × 6·25). The named meat should be the only meat used, except in the case of "Chicken Broth," in which a quarter of the meat may be derived from other meats. Tomato broth should contain at least 20 per cent. tomatoes.

Miscellaneous. "Brown," "Windsor" and "Eton" soups should be thick brown soups (the basis to be a roux to which broth has been added). "Minestrone" must contain macaroni, spaghetti or vermicelli, and should contain strip or diced assorted vegetables, but meat is optional. "Scotch Broth" should contain at least 6 per cent. of meat (including bone and fat) or 3 per cent. boned meat. The predominating meat should be mutton and barley and an assortment of vegetables should be present. Mushrooms should be the predominant single ingredient in "Mushroom Soup."

Analysis of Canned Soups

In general, canned soups should normally be examined for trace metals (*e.g.*, Pb, Sn, Cu). With cream soups the fat content should be determined (see below). From the point of view of assessing the general quality the following determinations are also useful: total solids, protein, ash, salt and the pH value. Apart from the Code of Practice, standards for canned soups were stated in The Meat Products, Canned Soup and Canned Meat (Control and Maximum Prices) Order, 1946. Although now revoked the requirements of this Order are still useful as a guide to general quality (see also Huntley, *Chem. & Ind.*, 1951, p. 227; and McLachlan, *Food Proc. & Packaging*, 1961, **30**, 163):—

Variety of Canned Soup	Minimum Standard of Quality		
	Solids (excluding bone)	Protein (N × 6·25)	Fat
	Per cent. (w/w)	Per cent. (w/w)	Per cent. (w/w)
Meat	12	2·5	—
Mulligatawny	13	2	2
Cream	10	—	3·5
Vegetable	10	1	—
Clear	7	6	—

(S.R. & O. 1946 No. 1355, as amended by S.R. & O. 1946 No. 1727.)

Determination of Fat in Canned Cream Soups (*Analyst*, 1954, **79**, 509)

Weigh out 5–10 g. sample into an extraction tube. Add 2 ml. alcohol, mix, add 10 ml. hydrochloric acid (25 ml. conc. + 11 ml. H_2O) and mix well again. Place in a water bath maintained at 75° C. for thirty to forty minutes or until the liquid is virtually clear. Cool, add 10 ml. alcohol and shake with 25 ml. ether. Then add 25 ml. light petroleum, shake again, blow the fat solution into a weighed flask, etc., and continue as usual. If the total fat falls below 3·5 per cent., the proportion of butter-fat should be assessed by applying the semi-micro Reichert process (p. 392) to the fat extracted from a larger portion of the sample.

SOUP POWDERS, DEHYDRATED SOUPS

Dried soups are either prepared by mixing the dry ingredients together, or by drying a liquid soup. Besides powders and powders compressed into blocks, dehydrated soups containing larger pieces such as strip, diced and kibbled vegetables have been marketed for some time. Such samples should normally be examined for **arsenic, lead, sulphur dioxide** and **added colour**. A guide to the types of analysis which are carried out in industrial laboratories has been presented by Devey (*Food Tr. Rev.*, 1958, **28** (7), 6; (8), 18), who suggests a maximum moisture content of 5 per cent. (see also McLachlan, *loc. cit.*).

Dehydrated mushroom soups, which contained *Boletus edulis* (an edible fungus which is not a "true" mushroom, but gives a somewhat similar flavour when reconstituted) were the subject of court proceedings in 1959. The cases were dismissed and it was subsequently ruled that *Boletus* could be described on the label as "dehydrated mushrooms," but illustrations of the common mushroom should not appear on the packet (*Brit. Food J.*, 1959, **61**, 69 and 99).

"BUTTER" CONFECTIONERY

Code of Practice (*Circular MF* 21/51)

Where the word "butter" is used without qualification for sugar confectionery, *e.g.*, "butterscotch," "butta mint," the butter-fat content should be not less than 4 per cent. Where the product consists of two parts this minimum applies to the part with which the butter-fat is mixed. Where the butter-fat is less than 4 per cent. the word "butter" if used shall be qualified by the term "flavoured."

Analysis

Total fat. The total fat can usually be determined by the Rose-Gottlieb method (p. 341) commencing by *boiling* with alcohol and ammonia.

Butter fat. Dissolve the sample in water and extract with ether/light petroleum. With some samples a preliminary treatment with hydrochloric acid may assist with the liberation of the fat but tends to give a darker extract. A purer fat can be obtained by re-extracting the dried fat with ether/light petroleum mixture, filtering and then drying again. By carrying out the semi-micro Reichert process (p. 392) on the purified fat it is possible to calculate the proportion of butter-fat in the extracted fat (p. 394). The percentage of butter-fat in the whole sweet can then be readily calculated from the total fat figure.

SACCHARIN TABLETS

Saccharin (o-benzoicsulphimide), $C_7H_5O_3NS$.

Saccharin is very sparingly soluble in water (1 in 290) and its melting point is 226–230° C. (B.P.). Saccharin tablets are prepared by mixing saccharin with sufficient sodium bicarbonate or by using the sodium salt, $C_7H_4O_3NSNa, 2H_2O$ (solubility 1 in 1·5). One gram of saccharin has approximately the same sweetening power as 550 grams of sucrose.

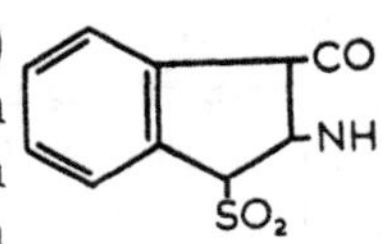

Saccharin

Legislation. The standards for saccharin tablets are prescribed in the Food Standards (Saccharin Tablets) Order, 1953 (S.I. 1953 No. 1310):—

"3. A saccharin tablet or other sweetening tablet containing saccharin:—

(i) shall contain not less than 0·18 grain and not more than 0·22 grain of saccharin or the equivalent weight of soluble saccharin;

(ii) may contain as excipient sodium bicarbonate with or without other suitable substances, the total amount of excipient not to exceed four times the maximum quantity of saccharin;

(iii) shall not contain more than 5 per cent. water-insoluble matter nor less bicarbonate than that required to render the saccharin completely soluble."

Analysis of Saccharin Tablets

Tablet weight. Determine the average weight by weighing 20 tablets on an analytical balance.

Qualitative test for saccharin. Mix about half of a powdered tablet with 50 mg. resorcinol. Add 10 drops of concentrated sulphuric acid and heat gently until it assumes a dark green colour. Cool carefully, add 10 ml. water and excess of 5N sodium hydroxide solution. Saccharin produces a fluorescent green colour.

DETERMINATION OF SACCHARIN—METHOD I

Place 0·4–0·5 g. of the powdered tablets in a Kjeldahl digestion flask and add 10 ml. of sodium hydroxide solution (30 per cent.). Boil gently for two minutes, *cool*, add 15 ml. of concentrated hydrochloric acid and boil for fifty minutes under reflux. Cool, rinse the condenser with 50 ml. of water and pass a current of air through the flask to remove acid vapour. Connect with an ammonia distillation apparatus, add 20 ml. of the sodium hydroxide solution (30 per cent.) and distil into 50 ml. N/10 sulphuric acid and back titrate the excess acid with N/10 alkali to methyl red (alternatively distil into boric acid and titrate with N/10 acid—see p. 34). Express the result as grains of saccharin per tablet.

1 gramme = 15·432 grains.

1 ml. N/10 = 0·01832 g. $C_7H_5O_3NS$ = 0·02052 g. $C_7H_4O_3NSNa$

(*cf. J. Soc. Chem. Ind. Lond.*, 1918, **52**, 377; B.P; B.P.C; U.S.P.).

DETERMINATION OF SACCHARIN—METHOD II

Parikh and Mukherji (*Analyst*, 1960, **85**, 25) precipitate the saccharin as the silver salt at a pH below 6 and the unused silver nitrate is titrated using thiocyanate:—

Weigh out a portion of the powdered tablets equivalent to about 0·6 g. of saccharin and shake with water in a 50 ml. graduated flask for five minutes, make up to the mark and filter, rejecting the first runnings. To 25 ml. of filtrate add 2 ml. of dilute acetic acid (10 per cent. w/w) and 25 ml. of N/10 silver nitrate. Shake, heat in boiling water for two minutes and cool in ice. Filter off the precipitate of the silver salt of saccharin on a No. 4G ground-glass filter and wash with three 5 ml. portions of ice-cold water. Acidify the filtrate with 10 ml. of nitric acid, add 2 ml. of ferric alum solution and titrate the unused silver nitrate with N/10 thiocyanate. Carry out a blank at the same time. The difference between the two titrations represents the amount of silver nitrate which has reacted with the saccharin. Express the result as grains per tablet.

1 ml. N/10 $AgNO_3$ = 0·01832 g. saccharin.

Alkalinity. Dissolve some of the powdered tablets by boiling in excess of N/10 acid, cool and back titrate to phenolphthalein. Also determine the alkalinity of the ash. Express the alkalinity as sodium bicarbonate and also calculate the equivalence to saccharin.

Water-insoluble matter. Stir a weighed quantity of the powdered tablets in hot water (80° C.), filter on a tared filter paper, dry and weigh the residue of insoluble matter.

Metallic contamination. Determine arsenic and lead.

APPENDIX I

The Public Health (Preservatives, etc., in Food) Regulations, 1925-1958

Part I

2.—(1) " Cream " means that portion of milk rich in milk-fat which has been separated by skimming or otherwise and is intended for human consumption;

" Ortho-phenylphenol " includes ortho-phenylphenate;

" Preservative " means any substance which is capable of inhibiting, retarding or arresting the process of fermentation, acidification, or other decomposition of food or of masking any of the evidences of putrefaction; but does not include common salt (sodium chloride), saltpetre (sodium or potassium nitrate), sugars, lactic acid, acetic acid or vinegar, glycerine, alcohol or potable spirits, herbs, hop extract, spices and essential oils used for flavouring purposes or any substance added to food by the process of curing known as smoking;

" Thickening substance " means sucrate of lime, gelatine, starch paste or any other substance, which when added to cream is capable of increasing its viscosity but does not include cane or beet sugar;

" Sulphur dioxide " includes sulphites, and " benzoic acid " includes benzoates.

(2) Percentages shall be calculated by weight.

(3) Sulphites shall be calculated as sulphur dioxide (SO_2) and benzoates as benzoic acid (C_6H_5COOH).

(3A) Ortho-phenylphenate shall be calculated as orthophenylphenol.

Part II

Sale of Articles of Food and Preservatives

3. The Local Authority and every officer of a Local Authority authorised in writing by the Authority shall enforce and execute this Part of these Regulations, and for this purpose shall make such enquiries and take such other steps as may seem to them or him to be necessary for securing the due observance of this Part of the Regulations in the district of the local authority.

4.—(1) No person shall manufacture for sale or sell any article of food which contains any added preservative:

Provided that—

(i) any article of food specified in Part I of the said Schedule may contain preservative of the nature and in the proportion therein specified;

(ii) where an article of food specified in Part I of the said Schedule otherwise than in item 4 thereof is used in the preparation of any other article of food, the latter article may contain any preservative necessarily introduced by the use of the former article, but the total proportion of any one preservative contained in any article of food specified in that Part of the Schedule shall not exceed the proportion therein specified.

(iii) The provisions of this Article shall not apply so as to prohibit the presence of sulphur dioxide in any article of food other than meat if it is shown either—

(a) that the article not being an article specified in Part I of the said Schedule is intended to be used in the preparation of an article which is so specified, or

(b) that the article being itself an article so specified, other than fruit or fruit pulp, is intended to be so treated before it is sold or exposed for sale by retail as to comply with the provisions of the Schedule as regards the proportion of sulphur dioxide contained.

(iv) The provisions of this Article shall not apply

(a) so as to prohibit the presence in bacon, ham and cooked pickled meat of any added sodium or potassium nitrite.

(b) so as to prohibit the presence in any article of food of sodium or potassium nitrite introduced in the preparation of such article by the use of any bacon, ham or cooked pickled meat containing sodium or potassium nitrite.

(c) so as to prohibit citrus fuit from containing diphenyl in proportions not exceeding one hundred parts per million or orthophenylphenol in proportions not exceeding seventy parts per million or an admixture of diphenyl and ortho-phenylphenol if the sum of the quantity of each of those substances, when expressed as a percentage of one hundred parts per million and seventy parts per million respectively, does not exceed one hundred.

(d) so as to prohibit any article of food from containing diphenyl or ortho-phenylphenol or an admixture thereof introduced in the preparation of such article by the use of any citrus fruit containing diphenyl or ortho-phenylphenol or an admixture thereof in proportions not exceeding those specified in sub-paragraph (c) of this paragraph.

(e) so as to prohibit any fruit specified in the first column of the following table from containing ortho-phenylphenol in proportions not exceeding those specified in relation thereto in the second column of the table, namely—

Apples, pears or pineapples . .	10 parts per million
Peaches	20 parts per million
Melons	125 parts per million.

(f) so as to prohibit any article of food from containing ortho-phenylphenol introduced in the preparation of such article by the use of any fruit specified in the first column of the following table in proportions not exceeding those specified in relation thereto in the second column of the table, namely—

Apples, melons, pears or pineapples	10 parts per million
Peaches	20 parts per million.

(4) No person shall manufacture for sale or sell any cooked pickled meat intended for human consumption, other than bacon and ham, which contains sodium or potassium nitrite in proportions exceeding two hundred parts per million calculated as sodium nitrite.

THE FIRST SCHEDULE

PART I.—ARTICLES OF FOOD WHICH MAY CONTAIN PRESERVATIVE AND NATURE AND PROPORTION OF PRESERVATIVE IN EACH CASE:—

The articles of food specified in the first column of the following table may contain the preservative specified in the second column in proportions not exceeding the number of parts (estimated by weight) per million specified in the third column:—

Food.	Preservative.	Parts per Million.
1. Sausages and sausage meat containing raw meat, cereals and condiments.	Sulphur dioxide	450
2. Fruit and fruit pulp (not dried) for conversion into jam or crystallised glacé or cured fruit:		
(a) Cherries	Do.	3,000
(b) Strawberries and raspberries	Do.	2,000
(c) Other fruit	Do.	1,500
3. Dried fruit:		
(a) Apricots, peaches, nectarines, apples and pears.	Sulphur dioxide	2,000
(b) Raisins and sultanas	Do.	750
4. Unfermented grape juice and non-alcoholic wine made from such grape juice if labelled in accordance with the rules contained in the Second Schedule to these Regulations.	Benzoic acid	2,000
5. Other non-alcoholic wines, cordials and fruit juices, sweetened or unsweetened.	Either sulphur dioxide	350
	or benzoic acid	600
6. Jam (including marmalade and fruit jelly prepared in the way in which jam is prepared, but not including marmalade made from citrous fruits).	Sulphur dioxide	100
7. Crystallised glacé or cured fruit, including candied peel.	Do.	100
(a) Fruit and fruit pulp not otherwise specified in this Schedule.	Do.	350
8. Sugar (including solid glucose and cane syrups).	Do.	70
(a) Cornflour (maize starch) and other prepared starches.	Do.	100
9. Corn syrup (liquid glucose)	Do.	450
10. Gelatine	Do.	1,000
11. Beer	Do.	70
12. Cider	Do.	200
13. Alcoholic wines	Do.	450
14. Sweetened mineral waters	Either sulphur dioxide	70
	or benzoic acid	120
15. Brewed ginger beer	Benzoic acid	120
16. Coffee extract	Do.	450
17. Pickles and sauces made from fruit or vegetables.	Do.	250
18. Dehydrated Vegetables:		
(a) Potatoes	Sulphur dioxide	550
(b) Runner beans, turnips, spinach, swedes, peas, parsnips, carrots, mixed vegetables.	Do.	2,000
(c) Cabbage	Do.	3,000

THE SECOND SCHEDULE

LABELLING OF ARTICLES OF FOOD CONTAINING PRESERVATIVE AND OF PRESERVATIVES

1. The articles of food containing preservative to which the Rules as to labelling set out in this Schedule apply are sausages, sausage-meat, coffee extract, pickles and sauces, and (where the proportion of benzoic acid exceeds 600 parts per million) grape juice and wine.

2.—(1) Where any of the said articles of food contains preservative it shall bear a label on which is printed the following declaration or such other declaration substantially to the like effect as may be allowed by the Minister:

(a) CONTAIN(S) PRESERVATIVE

(2) The declaration shall be completed by inserting at (a) the word " This " or " These," followed by the name of the food as used in paragraph 1 of this Schedule.

(3) In the case of grape juice or wine to which these Rules apply there shall be added to the declaration the words " and is not intended for use as a beverage."

3.—(1) An article sold as a preservative shall bear a label on which is printed the following declaration or such other declaration substantially to the like effect as may be allowed by the Minister:—

THIS PRESERVATIVE CONTAINS (a) PER CENT. OF SULPHUR DIOXIDE

(2) Where the article contains benzoic acid the words " Benzoic Acid " shall be substituted for the words " Sulphur Dioxide."

(3) The declaration shall be completed by inserting at (a) in words and figures, excluding fractions (*e.g.*, " seventy (70) ") the true percentage of the sulphur dioxide or benzoic acid present in the article.

Food Standards Committee Report on Preservatives in Food, 1959

PART V: SUMMARY OF RECOMMENDATIONS

A. Recommendations for Amendment of the Regulations

(i) The definition of " preservative " for the purpose of the Regulations should be amended to clarify the intended scope of the Regulations and to amend the list of substances excluded from the definition.

(ii) The definitions of " food " and " human consumption " should be made to correspond with those in the Food and Drugs Act, 1955.

(iii) Regulation 4(I) should be replaced by a provision that no person shall sell, offer or expose for sale or have in possession for sale by retail any food containing a non-permitted preservative or a permitted preservative in excess of the amount specifically allowed.

(iv) A similar provision should apply to sales etc., other than by retail, and to importation, unless the food (being a Scheduled food or intended for use in the preparation of a Scheduled food, but not being fruit or fruit pulp for manufacturing purposes containing sulphur dioxide or a food which is pre-packed for retail sale) is accompanied by a statement in prescribed form of the actual content of permitted preservative.

(v) Provision should be made that food containing a non-permitted preservative may be treated as unfit for human consumption for purposes of Section 9 of the Food and Drugs Act.

(vi) Regulation 4(I)(ii) should be replaced by a provision that any compound food (not being a Scheduled food) which contains one or more Scheduled foods (other than fruit and fruit pulp for manufacturing purposes and unfermented grape juice products for sacramental purposes) as ingredients may contain preservative of the nature and to the extent permitted in the Scheduled foods and in proportion to the amount of each Scheduled food which it contains.

(vii) A tolerance of 50 p.p.m. of sulphur dioxide should be allowed in any compounded food (not being a Scheduled food) permitted to contain sulphur dioxide by virtue of the previous provision.

(viii) The First Schedule should be amended to give effect to the recommendations made in Part IV of the report and summarised in the following table:—

Scheduled Food	Preservative permitted	Maximum amount permitted (parts per million)
Wine	sulphur dioxide	450
Beer	sulphur dioxide	70
Cider and perry	sulphur dioxide	200
Soft drinks for consumption after dilution and soft drinks having a comparable carbohydrate content for consumption after or without dilution	sulphur dioxide or benzoic acid	350 800
Other soft drinks for consumption without dilution	sulphur dioxide or benzoic acid	70 160
Fruit juices, sweetened or unsweetened, whether concentrated or not	sulphur dioxide or benzoic acid	350 800
Unfermented grape juice products for sacramental purposes	benzoic acid or methyl- or propyl-p-hydroxybenzoate	2,000 2,000
Vinegar	sulphur dioxide	70
Bread	propionic acid	0·3% of flour weight
Flour confectionery	sorbic acid	1,000
Cheese	sorbic acid or nisin	1,000 (a)
Citrus fruit	diphenyl or ortho-phenylphenol	100 70
Fruit and fruit pulp (other than tomato pulp) for manufacturing purposes	sulphur dioxide	3,000
Tomato pulp	methyl- or propyl-p-hydroxybenzoate	900
Dried fruit, including raisins and sultanas	sulphur dioxide	2,000

Scheduled Food	Preservative permitted	Maximum amount permitted (parts per million)
Crystallised and glacé fruit	sulphur dioxide	100
Candied peel and cut and drained (syruped) peel	sulphur dioxide	100
Syruped fruit	sulphur dioxide	100
Fruit and fruit pulp not elsewhere specified	sulphur dioxide	350
	or benzoic acid	800
Raw peeled potatoes	sulphur dioxide	50
Dehydrated potato	sulphur dioxide	550
Dehydrated cabbage	sulphur dioxide	2,500
Other dehydrated vegetables	sulphur dioxide	2,000
Bacon and ham	sodium nitrate	(a)
	and sodium nitrite	(b)
Cooked pickled meat	sodium nitrate	(a)
	and sodium nitrite	200
Uncooked pickled meat	sodium nitrate	(a)
	and sodium nitrite	500
Sausages and sausage meat	sulphur dioxide	450
Butchers' minced meat (c)	sulphur dioxide	450
Sugar and sugar syrups	sulphur dioxide	70
Solid hydrolysed starch	sulphur dioxide	70
Hydrolysed starch syrups	sulphur dioxide	450
Prepared starches	sulphur dioxide	100
Jam, including preserves sold for special dietetic purposes	sulphur dioxide	100
Marzipan	sorbic acid	1,000
Gelatin	sulphur dioxide	1,000
Liquid pectin	sulphur dioxide	250
Liquid rennet	benzoic acid or	2,000
	methyl- or propyl-p-hydroxybenzoate	2,000
Flavouring syrups	benzoic acid or	800
	methyl- or propyl-p-hydroxybenzoate	800
Liquid coffee (or coffee and chicory) extract	benzoic acid or	450
	methyl- or propyl-p-hydroxybenzoate	450
Drinking chocolate concentrate	benzoic acid or	700
	methyl- or propyl-p-hydroxybenzoate	700
Pickles and sauces	sulphur dioxide	100
Dry root ginger	sulphur dioxide	150
Solutions of permitted food colours	benzoic acid or	2,000
	sorbic acid	1,000
Silicone anti-foam emulsion	benzoic acid or	2,000
	sorbic acid	1,000
Canned foods as defined	nisin	(a)

Notes: (a) no statutory limit recommended

(b) no statutory limit recommended pending review—see recommendation (xiv)

(c) applicable to Scotland only and limited to the months of June, July, August and September.

(ix) Provision should be made that where, as indicated in the above table, an alternative preservative is allowed and a mixture of preservatives is used, the amount of each preservative present shall be such that, when expressed as a percentage of the amount respectively permitted, the sum of the two does not exceed 100.

(x) Provision should be made for the use of alternative forms of permitted preservatives as indicated in Part III of the report and summarised in the following table:—

Form of preservative to which the limit applies	Alternative forms in which the preservative may be used in appropriate amounts
sulphur dioxide	sulphurous acid and sodium, potassium and calcium sulphites
benzoic acid	the sodium and potassium salts
sodium nitrate	potassium nitrate
sodium nitrite	potassium nitrite
ortho-phenylphenol . . .	the sodium salt
methyl-p-hydroxybenzoate . .	the sodium salt
propyl-p-hydroxybenzoate . .	the sodium salt
sorbic acid	the sodium salt
propionic acid	the sodium and calcium salts

(xi) Specifications of purity should be prescribed for permitted preservatives.

B. Other Recommendations

(xii) Arrangements should be made for keeping under review the use of preservatives in food and for advising on such further changes in the Regulations as may be necessary from time to time.

(xiii) The provision of the Regulations dealing with the labelling of preservatives and of foods containing preservative should be considered during a review of the Labelling of Food Order.

(xiv) Further consideration should be given to regulating the nitrite content of bacon and ham as soon as the necessary technical evidence is available.

APPENDIX II

Emulsifying and Stabilising Agents

In 1956, the F.S.C. adopted a Report on Emulsifying and Stabilising Agents which they had received from their Preservatives Sub-Committee. Emulsifying and Stabilising Agents are defined in paragraph 2 of the Report as " Substances capable of facilitating the uniform dispersion of oils and fats in aqueous media, or vice versa, and of stabilising such emulsions." Most of the substances covered in the Report do act as emulsifiers and stabilisers, but some of them are used for other properties they possess, *e.g.*, as crumb-softeners or anti-staling agents, or to produce or break foams. The Report adds that when emulsifying agents were first used they were few in number and those used were either natural substances, particularly gums or synthetic substances of simple constitution. Progress in chemical synthesis, however, has now made available a large number of new products with characteristics suitable for virtually every requirement.

Under the heading " Classification of Emulsifying Agents," the F.S.C. Report states:—

" 22. We recommend that any statutory definition of ' emulsifying agents ' for the purpose of regulating the use of such substances should specifically exclude the following:—

agar, alginic acid, calcium and sodium alginates, carrageen, edible gums, dextrin, sorbitol, pectin, sodium and calcium pectate, ' emulsifying salts ' (sodium citrate, phosphates and tartrate), calcium lactate, lecithin, albumen, gelatin, quillaia, saponin, hydrolysed proteins, modified starches.

23. The remainder, which are listed in Appendix C, have been examined by the Pharmacology Panel of the Sub-Committee and classified as follows:—

Class A. Substances for which the available biological data indicate that their use in foods in the amounts customarily employed does not appear to present a health hazard. Examples of Class A substances quoted in Appendix C to the Report are:—

*stearyl tartrate, *super-glycerinated fats, *diacetyl tartaric acid esters of super-glycerinated fats, *monostearin sodium sulphoacetate, *synthetic lecithin, *methyl cellulose, *methyl ethyl cellulose, *sodium carboxymethyl cellulose.

Class B(i). Substances which, judged on their chemical formulation, do not seem likely to present a health hazard if used in food but in respect of which the biological data, though consistent, are not sufficiently comprehensive to justify their classification as ' A.' Examples of Class B(i) substances quoted in Appendix C to the Report are:—

*propylene glycol alginate, *propylene glycol stearate, *sorbitan esters

* See next page.

of fatty acids, brominated edible vegetable oils, methylpolysiloxane, methylphenylpolysiloxane.

Class B(ii). Substances which must at present be regarded with some suspicion either because the biological data are inadequate or conflicting or because the precise chemical composition is not known with certainty. Examples of Class B(ii) substances quoted in Appendix C to the Report are:—

lauryl sulphate, sodium dioctylsulphosuccinate, pentaerythritol mono- and di-stearate, polyoxyethylene stearate, polyethylene sorbitan monostearate, polyglycerol esters of fatty acids, polyglycerol esters of polymerised fatty acids, polymerised oils, sodium carboxymethyl galactomannose.

Class C. Substances the biological data relating to which, or the chemical formulation of which, indicate that their use in food is likely to present a health hazard.

24. On the evidence available, no substance was put in Class C. . . ."

The F.S.C.'s " Summary of Recommendations" are stated in paragraph 41:

" (i) that the addition to foods of emulsifying and stabilising agents as defined in paragraph 2 should be subject to statutory regulations;

(ii) that, except for the common substances enumerated in paragraph 22, official approval to the use of such agents should be restricted for the present to the substances specified in paragraphs 30 and 31 " **(i.e., Class A and B(i) substances marked with an asterisk in the lists above)**;

" (iii) that, in the case of bread, approval should be restricted to permit only the addition of super-glycerinated fats and stearyl tartrate;

(iv) that the addition of emulsifying agents to milk and of thickening agents to cream should be prohibited;

(v) that specifications of composition and purity should be prescribed for each of the permitted substances;

(vi) that emulsifying and stabilising agents sold for use in food should bear a statement of their chemical nature in addition to any trade name and indicate that they are suitable for use in food."

Methods for the determination of polyoxyethylene esters in bread and rolls have been described by Garrison *et al.* (*J. Ass. Off. Agric. Chem.*, 1960, **43**, 531).

APPENDIX III

Claims Regarding Vitamin and Mineral Contents

Where a food advertisement or label claims or suggests that the food concerned contains vitamins or minerals, it must, under the Labelling of Food Order disclose the quantity of each present (in each ounce or fluid ounce) in the following terms:—

VITAMINS

Substance	To be calculated as
Vitamin A	International units of vitamin A
Carotene	International units of vitamin A, on the basis that 0·6 micrograms of beta-carotene is equivalent to one international unit of vitamin A
Vitamin B_1 . . . . . . Aneurin . . . . . . Aneurin hydrochloride . . . Thiamin . . . . . . Thiamin hydrochloride . . .	Milligrams of aneurin hydrochloride
Vitamin B_2 . . . . . . Riboflavin	Milligrams of riboflavin
Nicotinic acid, nicotinic acid amide and their active derivatives. . Niacin . . . . . . Niacinamide . . . . . Nicotinamide	Milligrams of nicotinic acid or the chemically equivalent quantity of nicotinic acid in milligrams
Vitamin C . . . . . . Ascorbic acid	Milligrams of ascorbic acid
Vitamin D	International units of vitamin D
Vitamin D_2 . . . . . . Calciferol	International units of vitamin D
Vitamin D_3	International units of vitamin D

International Units

1 I.U. of vitamin A $\equiv$ 0·344 μg. *all-trans* vitamin A acetate.

$\equiv$ 0·3 μg. *all-trans* vitamin A alcohol.

1 I.U. of vitamin D $\equiv$ 0·025 μg. crystallise vitamin D_3.

MINERALS

Substance	To be calculated and specified as
Calcium	Milligrams of calcium.
Iodine	Micrograms of iodine.
Iron	Milligrams of iron.
Phosphorus	Milligrams of phosphorus.

Suggested Code of Practice in Framing Labels and Advertisements

(*"The Advertising Labelling and Composition of Food," M.O.F.*, 1949, *H.M.S.O.*)

1. Unless the amount of the food that would ordinarily be consumed in one day contains at least *one-sixth of the daily requirement* of the vitamin or mineral—

(a) no claim based on its presence should be made, and

(b) no reference to its presence is justified in any advertisement for the general public, or on any label.

(The inclusion of a mineral salt in the statement of ingredients required by the Labelling of Food Order will not of itself be regarded as implying the presence of a scheduled mineral).

2. Unless the amount of the food that would ordinarily be consumed in one day contains at least *one-half of the daily requirement* of the vitamin or mineral—

no claims are justified which imply that the food is a "rich" or "excellent" source of the vitamin or mineral.

3. Unless the amount of the food that would ordinarily be consumed in one day contains at least the *full daily requirement* of the vitamin or mineral—

no reference is justified to the value of the food for the prevention or cure of disease due to the lack of the vitamin or mineral present in the food.

4. Claims based on the presence of more than one vitamin or mineral should not be made unless each such vitamin or mineral is present in the proportion necessary to justify the claim.

5. There is no evidence of a deficiency of phosphorus in the ordinary mixed diet of this country. Statements in advertisements or on labels suggesting the contrary or that it is desirable to supplement the dietary phosphorus intake should not be made.

6. The following table, which includes the figures laid down by the Third Session of the Technical Commission on Nutrition, 1937, of the League of Nations Health Organisation,* shows the normal daily requirement of the vitamins and minerals scheduled in the Labelling of Food Order:—

Vitamin A . .	3,000 I.U.	Calcium . .	0·75 gram
,, B_1 . .	300 I.U.	Iodine . . .	0·1 mg.
,, B_2 . .	1·8 mg.	Iron . . .	10·0 mg.
Nicotinic Acid .	12·0 mg.	Phosphorus . .	0·75 gram
Vitamin C . .	30·0 mg.		
,, D . .	500 I.U.		

The statement refers to the foods commonly consumed by normal healthy adults and may not in all cases be applicable to foods used for special purposes.

* *cf.* B.M.A. "Report of the Committee on Nutrition" (1950).

APPENDIX IV

Ilford Spectrum Filters for Absorptiometry

Number	Colour of Filter	Peak Wavelength (mμ)	Transmission Region (mμ)
600	Spectrum Deep Violet	405	380–450
601	Spectrum Violet	425	380–470
602	Spectrum Blue	470	440–490
603	Spectrum Blue-Green	490	470–520
604	Spectrum Green	520	500–540
605	Spectrum Yellow-Green	550	530–570
606	Spectrum Yellow	580	560–610
607	Spectrum Orange	600	575 onwards, with absorption increasing from 600
608	Spectrum Red	660	620 into infra-red
609	Spectrum Deep Red	690	650 into infra-red
621	Bright Spectrum Violet	445	340–515
622	Bright Spectrum Blue	470	375–530
623	Bright Spectrum Blue-Green	490	460–545
624	Bright Spectrum Green	520	490–575
625	Brighter Spectrum Yellow-Green	540	510–590
626	Bright Spectrum Yellow	575	545–620

1 *millimicron* ($m\mu$) = 10^{-6}*mm.* = 10 Å

APPENDIX V

Factors for Volumetric Analysis

ACID SUBSTANCES	*FACTOR EQUIVALENT*
Acetic acid	1 ml. N/10 ≡ 0·006005 g. $CH_3.COOH$
Benzoic acid	1 ml. N/10 ≡ 0·001221 g. $C_6H_5.COOH$
Boric acid	1 ml. N/10 ≡ 0·006184 g. H_3BO_3
Citric acid	1 ml. N/10 ≡ 0·007003 g. $C_6H_8O_7,H_2O$
Hydrochloric acid	1 ml. N/10 ≡ 0·003647 g. HCl
Lactic acid	1 ml. N/10 ≡ 0·009008 g. $C_3H_6O_3$
Malic acid	1 ml. N/10 ≡ 0·006706 g. $C_4H_6O_5$
Oleic acid	1 ml. N/10 ≡ 0·028245 g. $C_{18}H_{34}O_2$
Phosphoric acid (ortho)	1 ml. N/10 ≡ 0·004900 g. H_3PO_4
Potassium hydrogen tartrate (Cream of tartar)	1 ml. N/10 ≡ 0·01882 g. $C_4H_5O_6K$
Sulphuric acid	1 ml. N/10 ≡ 0·004904 g. H_2SO_4
Tartaric acid	1 ml. N/10 ≡ 0·007505 g. $C_4H_6O_6$

ALKALIS

Ammonia	1 ml. N/10 ≡ 0·001703 g. NH_3
Barium hydroxide	1 ml. N/10 ≡ 0·008569 g. $Ba(OH)_2$
Calcium carbonate	1 ml. N/10 ≡ 0·005004 g. $CaCO_3$
Calcium oxide	1 ml. N/10 ≡ 0·002804 g. CaO
Potassium carbonate	1 ml. N/10 ≡ 0·006910 g. K_2CO_3
Potassium hydroxide	1 ml. N/10 ≡ 0·005610 g. KOH
Potassium oxide	1 ml. N/10 ≡ 0·004710 g. K_2O
Sodium bicarbonate	1 ml. N/10 ≡ 0·008402 g. $NaHCO_3$
Sodium carbonate	1 ml. N/10 ≡ 0·005300 g. Na_2CO_3
Sodium hydroxide	1 ml. N/10 ≡ 0·004001 g. NaOH
Sodium oxide	1 ml. N/10 ≡ 0·003100 g. Na_2O

MISCELLANEOUS

Calcium	1 ml. N/10 permanganate	≡ 0·002004 g. Ca
Calcium carbonate	1 ml. N/10 permanganate	≡ 0·005004 g. $CaCO_3$
Calcium oxide	1 ml. N/10 permanganate	≡ 0·002804 g. CaO
Chloride	1 ml. N/10 silver nitrate	≡ 0·003546 g. Cl
Iodide	1 ml. N/10 silver nitrate	≡ 0·01269 g. I
Ferrous iron	1 ml. N/10 permanganate	≡ 0·005585 g. Fe
Oxalic acid	1 ml. N/10 permanganate	≡ 0·006303 g. $C_2H_2O_4,2H_2O$
Potassium chloride	1 ml. N/10 silver nitrate	≡ 0·007456 g. KCl
Potassium iodide	1 ml. N/10 silver nitrate	≡ 0·01660 g. KI
Sodium chloride	1 ml. N/10 silver nitrate	≡ 0·005845 g. NaCl
Sodium nitrite	1 ml. N/10 permanganate	≡ 0·003451 g. $NaNO_2$
Sodium sulphite	1 ml. N/10 iodine	≡ 0·01261 g. $Na_2SO_3,7H_2O$
Sulphur dioxide	1 ml. N/10 iodine	≡ 0·003203 g. SO_2
Sulphurous acid	1 ml. N/10 iodine	≡ 0·004054 g. H_2SO_3

N/10 Iodine	12·69 g. I (+18 g. KI) per litre
N/10 Potassium permanganate	3·161 g. $KMnO_4$ per litre
N/10 Sodium thiosulphate	24·82 g. $Na_2S_2O_3,5H_2O$ per litre
N/10 Silver nitrate	16·99 g. $AgNO_3$ per litre

ATOMIC WEIGHTS

Al	26·98	Cr	52·01	N	14·008
Sb	121·76	Co	58·94	O	16·0000
As	74·91	Cu	63·54	P	30·975
Ba	137·36	F	19·00	Pt	195·23
Bi	209·00	H	1·0080	K	39·10
B	10·82	I	126·91	Si	28·09
Br	79·916	Fe	55·85	Ag	107·880
Cd	112·41	Pb	207·21	Na	22·997
Ca	40·08	Mg	24·32	S	32·066
C	12·010	Mn	54·93	Sn	118·70
Cl	35·457	Hg	200·61	Zn	65·38

APPENDIX VI

Weights and Measures

Relations of Metric and Imperial Measures

MASS

1 gramme	=	15·43 grains
1 pound	=	453·59 grammes
1 ounce	=	28·35 grammes
1 grain	=	0·0648 gramme

CAPACITY

1 litre	=	1·76 pints
1 pint	=	568·25 millilitres
1 fluid ounce	=	28·41 millilitres
1 fluid drachm	=	3·5515 millilitres

LENGTH

1 metre	=	39·37 inches
1 inch	=	25·40 millimetres

Imperial System

MASS

1 ounce	=	437·5 grains
1 pound	=	7000 grains

VOLUMES

1 pint	=	20 fluid ounces
1 fluid ounce	=	8 fluid drachms

INDEX

LOGARITHMS

LOGARITHMS

	0	1	2	3	4	5	6	7	8	9	Proportional parts. 1	2	3	4	5	6	7	8	9
10	0000	0043	0086	0128	0170	0212	0253	0294	0334	0374	4	8	12	17	21	25	29	33	37
11	0414	0453	0492	0531	0569	0607	0645	0682	0719	0755	4	8	11	15	19	23	26	30	34
12	0792	0828	0864	0899	0934	0969	1004	1038	1072	1106	3	7	10	14	17	21	24	28	31
13	1139	1173	1206	1239	1271	1303	1335	1367	1399	1430	3	6	10	13	16	19	23	26	29
14	1461	1492	1523	1553	1584	1614	1644	1673	1703	1732	3	6	9	12	15	18	21	24	27
15	1761	1790	1818	1847	1875	1903	1931	1959	1987	2014	3	6	8	11	14	17	20	22	25
16	2041	2068	2095	2122	2148	2175	2201	2227	2253	2279	3	5	8	11	13	16	18	21	24
17	2304	2330	2355	2380	2405	2430	2455	2480	2504	2529	2	5	7	10	12	15	17	20	22
18	2553	2577	2601	2625	2648	2672	2695	2718	2742	2765	2	5	7	9	12	14	16	19	21
19	2788	2810	2833	2856	2878	2900	2923	2945	2967	2989	2	4	7	9	11	13	16	18	20
20	3010	3032	3054	3075	3096	3118	3139	3160	3181	3201	2	4	6	8	11	13	15	17	19
21	3222	3243	3263	3284	3304	3324	3345	3365	3385	3404	2	4	6	8	10	12	14	16	18
22	3424	3444	3464	3483	3502	3522	3541	3560	3579	3598	2	4	6	8	10	12	14	15	17
23	3617	3636	3655	3674	3692	3711	3729	3747	3766	3784	2	4	6	7	9	11	13	15	17
24	3802	3820	3838	3856	3874	3892	3909	3927	3945	3962	2	4	5	7	9	11	12	14	16
25	3979	3997	4014	4031	4048	4065	4082	4099	4116	4133	2	3	5	7	9	10	12	14	15
26	4150	4166	4183	4200	4216	4232	4249	4265	4281	4298	2	3	5	7	8	10	11	13	15
27	4314	4330	4346	4362	4378	4393	4409	4425	4440	4456	2	3	5	6	8	9	11	13	14
28	4472	4487	4502	4518	4533	4548	4564	4579	4594	4609	2	3	5	6	8	9	11	12	14
29	4624	4639	4654	4669	4683	4698	4713	4728	4742	4757	1	3	4	6	7	9	10	12	13
30	4771	4786	4800	4814	4829	4843	4857	4871	4886	4900	1	3	4	6	7	9	10	11	13
31	4914	4928	4942	4955	4969	4983	4997	5011	5024	5038	1	3	4	6	7	8	10	11	12
32	5051	5065	5079	5092	5105	5119	5132	5145	5159	5172	1	3	4	5	7	8	9	11	12
33	5185	5198	5211	5224	5237	5250	5263	5276	5289	5302	1	3	4	5	6	8	9	10	12
34	5315	5328	5340	5353	5366	5378	5391	5403	5416	5428	1	3	4	5	6	8	9	10	11
35	5441	5453	5465	5478	5490	5502	5514	5527	5539	5551	1	2	4	5	6	7	9	10	11
36	5563	5575	5587	5599	5611	5623	5635	5647	5658	5670	1	2	4	5	6	7	8	10	11
37	5682	5694	5705	5717	5729	5740	5752	5763	5775	5786	1	2	3	5	6	7	8	9	10
38	5798	5809	5821	5832	5843	5855	5866	5877	5888	5899	1	2	3	5	6	7	8	9	10
39	5911	5922	5933	5944	5955	5966	5977	5988	5999	6010	1	2	3	4	5	7	8	9	10
40	6021	6031	6042	6053	6064	6075	6085	6096	6107	6117	1	2	3	4	5	6	8	9	10
41	6128	6138	6149	6160	6170	6180	6191	6201	6212	6222	1	2	3	4	5	6	7	8	9
42	6232	6243	6253	6263	6274	6284	6294	6304	6314	6325	1	2	3	4	5	6	7	8	9
43	6335	6345	6355	6365	6375	6385	6395	6405	6415	6425	1	2	3	4	5	6	7	8	9
44	6435	6444	6454	6464	6474	6484	6493	6503	6513	6522	1	2	3	4	5	6	7	8	9
45	6532	6542	6551	6561	6571	6580	6590	6599	6609	6618	1	2	3	4	5	6	7	8	9
46	6628	6637	6646	6656	6665	6675	6684	6693	6702	6712	1	2	3	4	5	6	7	7	8
47	6721	6730	6739	6749	6758	6767	6776	6785	6794	6803	1	2	3	4	5	5	6	7	8
48	6812	6821	6830	6839	6848	6857	6866	6875	6884	6893	1	2	3	4	4	5	6	7	8
49	6902	6911	6920	6928	6937	6946	6955	6964	6972	6981	1	2	3	4	4	5	6	7	8
50	6990	6998	7007	7016	7024	7033	7042	7050	7059	7067	1	2	3	3	4	5	6	7	8
51	7076	7084	7093	7101	7110	7118	7126	7135	7143	7152	1	2	3	3	4	5	6	7	8
52	7160	7168	7177	7185	7193	7202	7210	7218	7226	7235	1	2	2	3	4	5	6	7	7
53	7243	7251	7259	7267	7275	7284	7292	7300	7308	7316	1	2	2	3	4	5	6	6	7
54	7324	7332	7340	7348	7356	7364	7372	7380	7388	7396	1	2	2	3	4	5	6	6	7
	0	1	2	3	4	5	6	7	8	9	1	2	3	4	5	6	7	8	9

LOGARITHMS

	0	1	2	3	4	5	6	7	8	9	Proportional parts. 1	2	3	4	5	6	7	8	9
55	7404	7412	7419	7427	7435	7443	7451	7459	7466	7474	1	2	2	3	4	5	5	6	7
56	7482	7490	7497	7505	7513	7520	7528	7536	7543	7551	1	2	2	3	4	5	5	6	7
57	7559	7566	7574	7582	7589	7597	7604	7612	7619	7627	1	2	2	3	4	5	5	6	7
58	7634	7642	7649	7657	7664	7672	7679	7686	7694	7701	1	1	2	3	4	4	5	6	7
59	7709	7716	7723	7731	7738	7745	7752	7760	7767	7774	1	1	2	3	4	4	5	6	7
60	7782	7789	7796	7803	7810	7818	7825	7832	7839	7846	1	1	2	3	4	4	5	6	6
61	7853	7860	7868	7875	7882	7889	7896	7903	7910	7917	1	1	2	3	4	4	5	6	6
62	7924	7931	7938	7945	7952	7959	7966	7973	7980	7987	1	1	2	3	3	4	5	6	6
63	7993	8000	8007	8014	8021	8028	8035	8041	8048	8055	1	1	2	3	3	4	5	5	6
64	8062	8069	8075	8082	8089	8096	8102	8109	8116	8122	1	1	2	3	3	4	5	5	6
65	8129	8136	8142	8149	8156	8162	8169	8176	8182	8189	1	1	2	3	3	4	5	5	6
66	8195	8202	8209	8215	8222	8228	8235	8241	8248	8254	1	1	2	3	3	4	5	5	6
67	8261	8267	8274	8280	8287	8293	8299	8306	8312	8319	1	1	2	3	3	4	5	5	6
68	8325	8331	8338	8344	8351	8357	8363	8370	8376	8382	1	1	2	3	3	4	4	5	6
69	8388	8395	8401	8407	8414	8420	8426	8432	8439	8445	1	1	2	2	3	4	4	5	6
70	8451	8457	8463	8470	8476	8482	8488	8494	8500	8506	1	1	2	2	3	4	4	5	6
71	8513	8519	8525	8531	8537	8543	8549	8555	8561	8567	1	1	2	2	3	4	4	5	5
72	8573	8579	8585	8591	8597	8603	8609	8615	8621	8627	1	1	2	2	3	4	4	5	5
73	8633	8639	8645	8651	8657	8663	8669	8675	8681	8686	1	1	2	2	3	4	4	5	5
74	8692	8698	8704	8710	8716	8722	8727	8733	8739	8745	1	1	2	2	3	4	4	5	5
75	8751	8756	8762	8768	8774	8779	8785	8791	8797	8802	1	1	2	2	3	3	4	5	5
76	8808	8814	8820	8825	8831	8837	8842	8848	8854	8859	1	1	2	2	3	3	4	5	5
77	8865	8871	8876	8882	8887	8893	8899	8904	8910	8915	1	1	2	2	3	3	4	4	5
78	8921	8927	8932	8938	8943	8949	8954	8960	8965	8971	1	1	2	2	3	3	4	4	5
79	8976	8982	8987	8993	8998	9004	9009	9015	9020	9025	1	1	2	2	3	3	4	4	5
80	9031	9036	9042	9047	9053	9058	9063	9069	9074	9079	1	1	2	2	3	3	4	4	5
81	9085	9090	9096	9101	9106	9112	9117	9122	9128	9133	1	1	2	2	3	3	4	4	5
82	9138	9143	9149	9154	9159	9165	9170	9175	9180	9186	1	1	2	2	3	3	4	4	5
83	9191	9196	9201	9206	9212	9217	9222	9227	9232	9238	1	1	2	2	3	3	4	4	5
84	9243	9248	9253	9258	9263	9269	9274	9279	9284	9289	1	1	2	2	3	3	4	4	5
85	9294	9299	9304	9309	9315	9320	9325	9330	9335	9340	1	1	2	2	3	3	4	4	5
86	9345	9350	9355	9360	9365	9370	9375	9380	9385	9390	1	1	2	2	3	3	4	4	5
87	9395	9400	9405	9410	9415	9420	9425	9430	9435	9440	0	1	1	2	2	3	3	4	4
88	9445	9450	9455	9460	9465	9469	9474	9479	9484	9489	0	1	1	2	2	3	3	4	4
89	9494	9499	9504	9509	9513	9518	9523	9528	9533	9538	0	1	1	2	2	3	3	4	4
90	9542	9547	9552	9557	9562	9566	9571	9576	9581	9586	0	1	1	2	2	3	3	4	4
91	9590	9595	9600	9605	9609	9614	9619	9624	9628	9633	0	1	1	2	2	3	3	4	4
92	9638	9643	9647	9652	9657	9661	9666	9671	9675	9680	0	1	1	2	2	3	3	4	4
93	9685	9689	9694	9699	9703	9708	9713	9717	9722	9727	0	1	1	2	2	3	3	4	4
94	9731	9736	9741	9745	9750	9754	9759	9763	9768	9773	0	1	1	2	2	3	3	4	4
95	9777	9782	9786	9791	9795	9800	9805	9809	9814	9818	0	1	1	2	2	3	3	4	4
96	9823	9827	9832	9836	9841	9845	9850	9854	9859	9863	0	1	1	2	2	3	3	4	4
97	9868	9872	9877	9881	9886	9890	9894	9899	9903	9908	0	1	1	2	2	3	3	4	4
98	9912	9917	9921	9926	9930	9934	9939	9943	9948	9952	0	1	1	2	2	3	3	4	4
99	9956	9961	9965	9969	9974	9978	9983	9987	9991	9996	0	1	1	2	2	3	3	3	4
	0	1	2	3	4	5	6	7	8	9	1	2	3	4	5	6	7	8	9

www.ingramcontent.com/pod-product-compliance
Lightning Source LLC
LaVergne TN
LVHW010621300926
849162LV00001B/19
* 9 7 8 0 8 2 0 6 0 1 4 1 0 *